The Handbook of Environmental Chemistry

Founded by Otto Hutzinger

Editors-in-Chief: Damià Barceló • Andrey G. Kostianoy

Volume 13

The Handbook of Environmental Chemistry
Recently Published and Forthcoming Volumes

Waste Water Treatment and Reuse in the Mediterranean Region
Volume Editors: D. Barceló and M. Petrovic
Vol. 14, 2011

The Ebro River Basin
Volume Editors: D. Barceló and M. Petrovic
Vol. 13, 2011

Polymers – Opportunities and Risks II: Sustainability, Product Design and Processing
Volume Editors: P. Eyerer, M. Weller, and C. Hübner
Vol. 12, 2010

Polymers – Opportunities and Risks I: General and Environmental Aspects
Volume Editor: P. Eyerer
Vol. 11, 2010

Chlorinated Paraffins
Volume Editor: J. de Boer
Vol. 10, 2010

Biodegradation of Azo Dyes
Volume Editor: H. Atacag Erkurt
Vol. 9, 2010

Water Scarcity in the Mediterranean: Perspectives Under Global Change
Volume Editors: S. Sabater and D. Barceló
Vol. 8, 2010

The Aral Sea Environment
Volume Editors: A.G. Kostianoy and A.N. Kosarev
Vol. 7, 2010

Alpine Waters
Volume Editor: U. Bundi
Vol. 6, 2010

Transformation Products of Synthetic Chemicals in the Environment
Volume Editor: A.B.A. Boxall
Vol. 2/P, 2009

Contaminated Sediments
Volume Editors: T.A. Kassim and D. Barceló
Vol. 5/T, 2009

Biosensors for the Environmental Monitoring of Aquatic Systems
Bioanalytical and Chemical Methods for Endocrine Disruptors
Volume Editors: D. Barceló and P.-D. Hansen
Vol. 5/J, 2009

Environmental Consequences of War and Aftermath
Volume Editors: T.A. Kassim and D. Barceló
Vol. 3/U, 2009

The Black Sea Environment
Volume Editors: A. Kostianoy and A. Kosarev
Vol. 5/Q, 2008

Emerging Contaminants from Industrial and Municipal Waste
Removal Technologies
Volume Editors: D. Barceló and M. Petrovic
Vol. 5/S/2, 2008

Fuel Oxygenates
Volume Editor: D. Barceló
Vol. 5/R, 2007

The Rhine
Volume Editor: T. P. Knepper
Vol. 5/L, 2006

Persistent Organic Pollutants in the Great Lakes
Volume Editor: R. A. Hites
Vol. 5/N, 2006

The Ebro River Basin

Volume Editors: Damià Barceló · Mira Petrovic

With contributions by

J. Armengol · J. Artigas · C. Barata · D. Barceló · J.A.C. Barth · R.J. Batalla · J.M. Bayona · A. Bellin · S. Blenkinsop · C.I. Bovolo · A. Burton · E. Capri · L. Carrasco · M. Casado · J. Damasio · R.M. Darbra · M.S. Díaz-Cruz · S. Díez · J. Dolz · C. Duran · E. Eljarrat · M. Faria · M. Farré · H.J. Fowler · L. Galbiati · A. Ginebreda · P. Grathwohl · M. Gros · P.D. Hansen · M. Köck-Schulmeyer · M. López de Alda · B. Majone · R. Marcé · E. Martínez · I. Muñoz · A. Navarro · A. Navarro-Ortega · P. Négrel · A. Olivares · S. Pelayo · M. Pérez, · E. Petelet-Giraud · M. Petrovic · B. Piña · C. Postigo · J. Prats · L. Quirós · D. Raldúa · A.M. Romaní · S. Sabater · M. Sánchez-Juny · R. Tauler · M. Terrado · C. Vale · D. Vericat · M. Zambrano-Bigiarini

Springer

Editors
Prof. Dr. Damià Barceló
Department of Environmental Chemistry
IDAEA-CSIC
c/Jordi Girona 18–26
08034 Barcelona, Spain
and
Catalan Institute for Water Research (ICRA)
H20 Building
Scientific and Technological Park of the
University of Girona
Emili Grahit, 101
17003 Girona, Spain
dbcqam@cid.csic.es

Prof. Dr. Mira Petrovic
Department of Environmental Chemistry
IDAEA-CSIC
c/Jordi Girona, 18–26
08034 Barcelona, Spain
and
Catalan Institution for Research and
Advanced Studies (ICREA)
Passeig Lluís Companys, 23
08010 Barcelona, Spain
mira.petrovic@idaea.csic.es

The Handbook of Environmental Chemistry
ISSN 1867-979X e-ISSN 1616-864X
ISBN 978-3-642-18031-6 e-ISBN 978-3-642-18032-3
DOI 10.1007/978-3-642-18032-3
Springer Heidelberg Dordrecht London New York

Cover design: SPi Publisher Services

Printed on acid-free paper

Springer is part of Springer Science+Business Media (www.springer.com)

The Handbook of Environmental Chemistry Also Available Electronically

The Handbook of Environmental Chemistry is included in Springer's eBook package *Earth and Environmental Science*. If a library does not opt for the whole package, the book series may be bought on a subscription basis.

For all customers who have a standing order to the print version of *The Handbook of Environmental Chemistry*, we offer free access to the electronic volumes of the Series published in the current year via SpringerLink. If you do not have access, you can still view the table of contents of each volume and the abstract of each article on SpringerLink (www.springerlink.com/content/110354/).

You will find information about the

- Editorial Board
- Aims and Scope
- Instructions for Authors
- Sample Contribution

at springer.com (www.springer.com/series/698).

All figures submitted in color are published in full color in the electronic version on SpringerLink.

Aims and Scope

Since 1980, *The Handbook of Environmental Chemistry* has provided sound and solid knowledge about environmental topics from a chemical perspective. Presenting a wide spectrum of viewpoints and approaches, the series now covers topics such as local and global changes of natural environment and climate; anthropogenic impact on the environment; water, air and soil pollution; remediation and waste characterization; environmental contaminants; biogeochemistry; geoecology; chemical reactions and processes; chemical and biological transformations as well as physical transport of chemicals in the environment; or environmental modeling. A particular focus of the series lies on methodological advances in environmental analytical chemistry.

Series Preface

With remarkable vision, Prof. Otto Hutzinger initiated *The Handbook of Environmental Chemistry* in 1980 and became the founding Editor-in-Chief. At that time, environmental chemistry was an emerging field, aiming at a complete description of the Earth's environment, encompassing the physical, chemical, biological, and geological transformations of chemical substances occurring on a local as well as a global scale. Environmental chemistry was intended to provide an account of the impact of man's activities on the natural environment by describing observed changes.

While a considerable amount of knowledge has been accumulated over the last three decades, as reflected in the more than 70 volumes of *The Handbook of Environmental Chemistry,* there are still many scientific and policy challenges ahead due to the complexity and interdisciplinary nature of the field. The series will therefore continue to provide compilations of current knowledge. Contributions are written by leading experts with practical experience in their fields. *The Handbook of Environmental Chemistry* grows with the increases in our scientific understanding, and provides a valuable source not only for scientists but also for environmental managers and decision-makers. Today, the series covers a broad range of environmental topics from a chemical perspective, including methodological advances in environmental analytical chemistry.

In recent years, there has been a growing tendency to include subject matter of societal relevance in the broad view of environmental chemistry. Topics include life cycle analysis, environmental management, sustainable development, and socio-economic, legal and even political problems, among others. While these topics are of great importance for the development and acceptance of *The Handbook of Environmental Chemistry,* the publisher and Editors-in-Chief have decided to keep the handbook essentially a source of information on "hard sciences" with a particular emphasis on chemistry, but also covering biology, geology, hydrology and engineering as applied to environmental sciences.

The volumes of the series are written at an advanced level, addressing the needs of both researchers and graduate students, as well as of people outside the field of "pure" chemistry, including those in industry, business, government, research establishments, and public interest groups. It would be very satisfying to see these volumes used as a basis for graduate courses in environmental chemistry. With its high standards of scientific quality and clarity, *The Handbook of*

Environmental Chemistry provides a solid basis from which scientists can share their knowledge on the different aspects of environmental problems, presenting a wide spectrum of viewpoints and approaches.

The Handbook of Environmental Chemistry is available both in print and online via www.springerlink.com/content/110354/. Articles are published online as soon as they have been approved for publication. Authors, Volume Editors and Editors-in-Chief are rewarded by the broad acceptance of *The Handbook of Environmental Chemistry* by the scientific community, from whom suggestions for new topics to the Editors-in-Chief are always very welcome.

Damià Barceló
Andrey G. Kostianoy
Editors-in-Chief

Volume Preface

The book on *The Ebro River Basin* is based on the scientific developments and results achieved within the European Union (EU) FP6 funded project AquaTerra – *Integrated modeling of the river-sediment-soil-groundwater system; advanced tools for the management of catchment areas and river basins in the context of global change*. Integrated Project AquaTerra was established with the primary objective of laying foundations for a better understanding of the behavior of environmental pollutants and their fluxes with respect to climate and land use changes. Environmental topics covered a wide range of disciplines from about 250 researchers across Europe and five main study areas (catchments of the Ebro, Meuse, Elbe and Danube Rivers and the Bréville Spring).

The results presented in this book were generated within the sub-project ***Basin*** that had as main objectives the following topics:

- Soil–sediment–groundwater-related issues, as experienced in River Basin practise, including those related to implementation of the Water Framework Directive (WFD) and Groundwater Daughter Directive (GWDD)
- Missing knowledge and lacking understanding of soil–groundwater–river functioning hindering the setup of appropriate measures to mitigate (effects of) perturbations, such as:
  - Climate change
  - Land-use practises and associated soil and groundwater degradation
  - Contamination (of soil, sediment and groundwater, in interaction with river systems)

Finally, the book is in the line with the aim of the ***Basin*** sub-project of connecting the scientific research to practical research cases and demands of River Basin Managers and associated stakeholders.

The book covers a wide range of topics related to the functioning of the Ebro river basin:

- Hydrology and sediment transport and their alterations caused by climate change
- Aquatic and riparian biodiversity in the Ebro watershed
- Occurrence and distribution of a wide range of priority and emerging contaminants (pharmaceuticals, drugs of abuse, polar pesticides, etc.)

- Effects of chemical pollution on biota
- Integration of climate change scenarios with several aspects of the Ebro's hydrology and potential impacts of climate change on pollution

We hope the book will be of interest to a broad audience of analytical chemists, environmental chemists, water managers, operators and technologists working in the field.

We would like to specially thank Dr. Jürgen Büsing and Dr. Cathy Eccles, scientific officers at EC responsible for AquaTerra, whose suggestions and directions were of great help in successful realization of the project.

Finally we would like to thank all the contributing authors of this book for their time and effort in preparing this comprehensive compilation of research papers.

Barcelona, September 2010

Damià Barceló
Mira Petrovic

Contents

The Physical Framework and Historic Human Influences in the Ebro River

A.M. Romaní, S. Sabater, and I. Muñoz

Abstract The river Ebro watershed is highly diverse, including high mountain sub-watersheds (Pyrenees) flowing on siliceous material to slow flow meandering areas in the middle reach and canyon type channel at the lower part of the main Ebro River. The large depression at the middle part of the watershed, draining a calcareous gypsum soil, determines high conductivity values of the river water. The geography of the river also determines a large range of climatic conditions from the Atlantic climate type to the semi-arid Mediterranean climate. At the same time, vegetation is also highly diverse including boreal species and Mediterranean species. However, the biogeochemical characteristics of the river water are highly influenced by anthropogenic activities. The main effects are those due to discharge regulation (i.e., the construction of the large reservoirs) and agriculture (determining increases in nitrate concentration).

Keywords Biogeochemistry, Ebro River watershed, Human settlements, Land use, Nutrient content, Physiography

A.M. Romaní (✉)
Institute of Aquatic Ecology and Department of Environmental Sciences, University of Girona, Campus de Montilivi, 17071 Girona, Spain
e-mail: anna.romani@udg.edu

S. Sabater
Catalan Institute of Water Research, University of Girona, Girona, Spain

I. Muñoz
University of Barcelona, Barcelona, Spain

D. Barceló and M. Petrovic (eds.), *The Ebro River Basin*,
Hdb Env Chem (2011) 13: 1–20, DOI 10.1007/698_2010_66,

Contents

1 Introduction

Rivers of the Iberian Peninsula can be separated by those flowing to the Atlantic and those flowing to the Mediterranean Seas. The separation between these two large basins is asymmetrical with the Mediterranean basin encompassing 182,661 km^2 (31% of the total surface area) and the Atlantic 400,839 km^2 (69% of the surface area). The largest rivers flow to the Atlantic, which include the Duero, Tagus, Guadiana, and Guadalquivir. The Ebro is the only large river in the Iberian Peninsula that flows into the Mediterranean.

The Ebro River basin is located in the NE of the Iberian Peninsula, occupying a total surface of 85,362 Km^2. Most of the watershed surface area is in Spanish territory, but small parts drain in Andorra and in France (445 km^2 and 502 km^2, respectively). The Ebro River is the largest hydrographic basin in Spain, accounting for 17.3% of its total surface area.

The Cantabrian Mountains and the Pyrenees in the North, the Iberian System in the South-East, and the Coastal Catalan mountains in the East are the natural limits of the Ebro River basin. Traditionally, the river source was supposed to be at Fontibre (name derived from *Fontes Iberis* in latin, "Springs of Iberia") at 880 m.a.s.l., near Reinosa in Cantabria. Nowadays, the river source is placed at 1,980 m.a.s.l., the water coming from a source in Peñalara (27 km upstream from Reinosa). The main river channel is 910 km in length, flows NW-SE, from the Cantabrian Mountains to the Mediterranean Sea, where it forms a delta.

The Ebro collects water from the Pyrennes and Cantabrian mountains in the left margin, where relevant tributaries such as the rivers Aragón, Gállego, and Cinca-Segre enter the main channel. In the right margin the river receives tributaries of lower discharge coming from the Iberian System, such as the rivers Oja, Iregua, Jalón, Huerva, Guadalope, and Matarranya. In total, the drainage network accounts for 12,000 km in length. The main channel flows closer to the Iberian mountains than to the Pyrenees, shaping an asymmetrical watershed of about 50,000 km^2 at its left margin and about 30,000 km^2 at its right margin. The Ebro River basin occupies one of the largest depressions in the Iberian Peninsula external to the central Meseta. The Ebro Delta occupies about 330 km^2, 20% of them being naturally protected areas, the rest being urban and agricultural areas, rice crops the most important.

Spotted in the river basin there are several small lakes, mainly in the mountainous zones of the Pyrenees, including the karstic lake of Montcortès (Lleida). In the mid to lower Ebro basin, several endorreic lagoons occur, such as Sariñena lagoon, and the salty lagoons of Chiprana and Gallocanta at the Monegros (Zaragoza). The Ebro River channel is submitted to the regulation effect of many reservoirs scattered all over the river network. The most important ones are those located in the mid to lower part of the basin (i.e., Mequinenza and Ribaroja) which have produced long-lasting decreases in sediment transport to the Delta.

This chapter describes the main physical characteristics of the Ebro River, including the watershed orography, the biogeography and vegetation, the climatic and hydrological characteristics, and the soil type and biogeochemistry of river Ebro waters. The Ebro watershed has historically served as nucleus and connection for humans; human settlements are known since pre-historic years and nowadays the river water chemistry cannot be understood without the anthropogenic effects. Therefore, the potential effects of human activities at the Ebro watershed are analyzed.

2 Watershed Relief and Drainage Network

The river Ebro basin has a triangular shape, where the larger sides are the Iberian range and the Pyrenees, the two converging in the north-east (Fig. 1). In between, a depression increases in width from the west to the east. Just before the river mouth the Ebro crosses the Catalan Mountain Range.

Along the first 240 Km the main Ebro channel flows down from the Cantabrian Mountains. In that section, the rivers form meanders and rocky canyons of rapid current velocity. In the middle reach, for about 510 km, between Conchas de Haro and Mequinenza, the river flows over the main plain with many meanders. After the city of Tudela, water is diverted from the main channel to two irrigation channels: the Tauste Canal, at the left margin, and the Imperial Canal (Canal Imperial de Aragón) at the right margin. Waters from the Imperial Canal flow again into the Ebro downstream the city of Zaragoza.

At the middle Ebro, the main tributaries coming from the Pyrenees are larger and with higher discharge than the tributaries from the right margin. The right margin tributaries are perpendicular to the main channel, composing a parallel net. The tributaries from the left margin present different typologies; while in the upper part they are also perpendicular; in the lower part have a dendritic network. The main tributary from the right margin is the river Jalón, while those from the left margin are the Aragón, Gállego, and Cinca-Segre. The Aragón River has its headwaters in the Canfranc Valley at the Pyrenees (1,758 m.a.s.l.), and flows to the southwest collecting waters from tributaries such as the Salazar, Urrubi, and Arga. The Gállego headwaters collect waters from the Panticosa and Sallent valleys, and

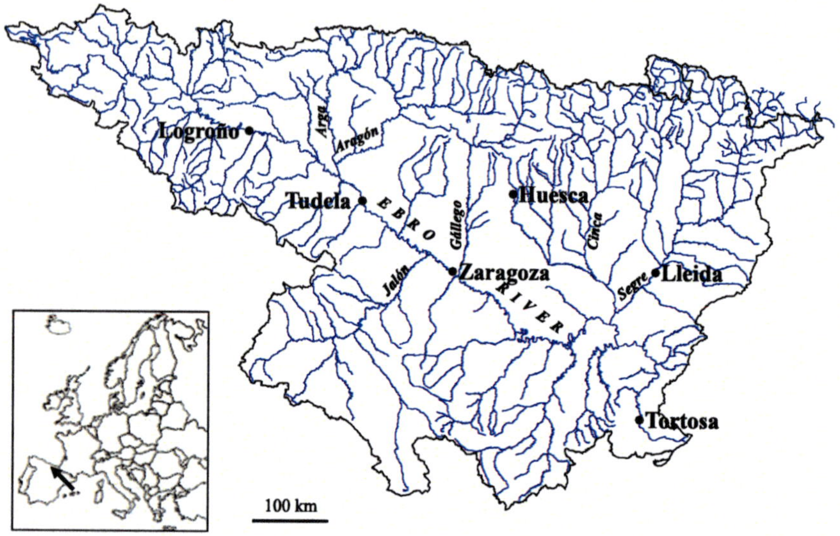

Fig. 1 Map of the Ebro River watershed showing the main cities and tributaries. The position of the Ebro with respect to the Iberian Peninsula is also indicated

after the sub-pyrenaic line it goes to the west, flowing into the main Ebro channel at the city of Zaragoza. The river Segre is the longest tributary of the Ebro, and collects waters from its Pyrenean tributaries. The Valira (crossing Andorra's Principate), the Noguera Ribagorçana, and Noguera Pallaresa are the main tributaries. The river Cinca enters the Segre before the joint confluence in the Ebro. The Cinca collects waters from the Ara at the Ordesa National Park, and later on from the Essera. Close to the reservoir of Mequinenza other tributaries apart from the Segre-Cinca enter the Ebro. The most relevant are the Valcuerna (coming from the esteparic gypsum area of the Monegros), and the Guadalope and Matarranya (from the right margin). In this area the three large reservoirs, Mequinenza, Ribaroja, and Flix, are built in the main channel and exert a heavy impact on the hydrology and biogeochemistry of the Ebro River.

The river downstream the reservoirs is about 120 km length, forms canyon meanders, and is very deep. The river widens up again at Móra d'Ebre, and after crossing the littoral Catalan mountains reaches the city of Tortosa and flows to the Mediterranean Sea in the Ebro Delta (30 km).

3 Palaeogeography and Soil Types of the River Ebro Watershed

Many of the physiographic characteristics of the river Ebro can be understood from the origin and evolution of its watershed [1]. The Ebro Basin underwent a long period of closed intramountain drainage as a result of tectonic topography generation at the Pyrenees, the Iberian Range, and the Catalan Coastal Range. In the late Oligocene, the Catalan Coastal Range underwent extension but the Ebro Basin remained closed. Dry climatic conditions probably lowered the lake level and contributed to extend this endorreic basin stage [2]. Replenishment of the Ebro River basin was mainly due to alluvial deposits from the Pyrenees [3] and from the center of the Iberian Peninsula, including an important contribution of marls and evaporitic materials and large proportions of gypsum and halite [4]. The Tertiary Ebro basin gradually opened to the Mediterranean at the later Miocene (between 13 and 8.5 Ma), as a result of lake capture by escarpment erosion and lake level rise associated with sediment accumulation and wetter climatic conditions. Sea level changes in the Mediterranean did not exert major impacts on the large-scale drainage evolution of the Ebro Basin [2].

The groundwater flow was relevant for the formation and transformation of evaporitic lacustrine facies in the Iberian Range and Ebro basin. The Triassic gypsum and marl formed the impermeable substratum of the overlying Jurassic and Cretaceous carbonated aquifers. The water discharged from the aquifer to springs or wetlands (saline lakes) had a high mineralization with a dry residue of over 1,000 mg/L (dominated by calcium sulfates). On the left margin of the Ebro River more than 60 depressions occur, where mineralised lakes form (locally referred to as "saladas," i.e., Gallocanta Lake) [5, 6]. During the Miocene, the hydrogeological functioning was similar to the present, allowing the groundwater and the dissolved salts to accumulate in large areas of diffuse discharge, creating lakes where the evaporites would precipitate [7].

Subsidence at the Ebro River basin occurred along the Tertiary and Quaternary due to the solution of underlying evaporitic formations (halite and gypsum) [8, 9], as observed for the Gállego River [10]. Sediment analysis from the central Ebro valley (geochemistry and pollen analysis from lake sediment records) indicate that, at least for some intervals during full glacial times, some lakes experienced more positive water balance than the one they show nowadays. These data are coherent with the hypothesis that, at least for some periods, the ice-age climate in the western Mediterranean region was characterized by cold winters, with relatively higher humidity. Increased flow from the Pyrenean Rivers during the early deglaciation could also have played a significant role in the paleohydrological cycle in the central Ebro valley [11].

The Ebro headwaters flow on calcareous substratum, specifically sandstone and calcium marls, from the Triassic, Cretacic, and Jurassic. During the Quaternary, at the plain of La Virga (Reinosa), a shallow lake accumulated the deposits of siliceous sandrocks. This old highland lake is now the Embalse del Ebro reservoir. From that point downstream to Conchas de Haro the main channel flows on calcareous rocks from the Cretacic, highly resistant to the erosion.

At the medium reach, the river flows into the Iberian Depression, with marl and gypsum Miocene deposits in some areas. The dissolution of evaporitic sediments (gypsum, halite, and sodium-sulfates) gives rise to numerous sinkholes. However, subsidence is also being masked by morpho-sedimentary dynamic processes such as aggradation and erosion [12].

4 Watershed Vegetation and Biogeography

Among the natural areas in the Ebro River basin, a broad spectrum of landscapes is scattered, ranging from boreal-alpine coniferous forests, mixed deciduous forests, Mediterranean evergreen and mixed forest and shrubs, and semi-arid treeless formations.

In the Tertiary Ebro depression the aridity of the climate causes the vegetation cover to be low, with lower richness and diversity and lower productivity than the riparian vegetation, although some important biomass can be reached by several species such as *Juniperus thurifera*, *Quercus ilex* subsp. *Ballota*, and *Pinus halepensis*. Paleartic and cosmopolitan species are characteristic along the river valley at the riparian zone. Some species usually found at the Ebro headwaters, such as *Cornus sanguinea* and *Brachypodium sylvaticum*, are dispersed through the corridor of the river channel because of the humid conditions that may be found at the riparian zone. In the plain, instead, species are mainly Mediterranean as well as iranoturanian, iberonorth African, and endemic species typical from arid gypsum substratum. The river water availability is responsible for the close proximity of biogeographically very distinct species at the depression from those found at the riparian zone. Although vegetation in the Ebro River valley is less altered than in other Iberian river basins, the actual forests represent only the 3.1% of the potential forested surface area [13].

Phyletic links of apparent endemic species of the central Ebro valley with easternmost species were revealed after studying the insect communities at the Monegros region. These have a pre-Pleistocene origin of their relict distributions, associated with the persistence of steppe habitats over gypsiferous soils in the area since the Late Tertiary. Distributions of phytophages and their parasitoids on plants such as *Krascheninnikovia ceratoides* or *Juniperus thurifera* supported the continuity of their presence in the central Ebro valley through the Quaternary [14].

5 Climate and Hydrology

The topography of the Ebro River basin determines a Mediterranean climate with continental characteristics in most of the river basin as well as a semi-arid climate in the center of the depression. At the western extreme of the basin (Pyrenees and Iberian mountains), there is an oceanic climate. The central part of the basin is isolated from the oceanic influence because of the surrounding mountains. This results in the increase of the continentality of the climate and the drastic decrease of the rainfall. In the central part of the river basin (Zaragoza, Alcañiz, and Lleida urban areas), aridity is the main climatic characteristic.

The mean annual precipitation in the Ebro River basin is of 622 mm (mean from years 1920–2000), but with a high monthly and annual variability. Long periods of low precipitation are usual in winter and the end of autumn, especially in the plain. Higher rainfall occurs in spring and autumn. The rainfall is also irregularly distributed in the river basin, where it varies from about 900 mm in the Atlantic headwaters, 950 mm in the west-central-Pyrenees, 800 mm at the eastern Pyrenees, and 500 mm at the southern Mediterranean zone. Record rainfall values in the watershed range from about 3,000 mm/year at the Pyrenees and less than 100 mm/year at the central plain. The analysis of the rainfall records along the period 1916–2000 did not show clear evidence of rainfall decrease in the Ebro River basin excepting a slight decrease at the south of Zaragoza [15, 16].

The temperature patterns follow a transition from the more oceanic western area (milder temperatures) to the central depression (high temperatures in summer and intense cold and fog in winter). The northwest-southeast cold and dry wind ("cierzo") is characteristic of the central depression of the Ebro basin, especially in spring. The "cierzo" at the middle Ebro valley (around Zaragoza) can lead to soil erosion and salt transport [17]. The wind intensity at this area is highly correlated with evapotranspiration [18]. A mild warm wind is sometimes occurring (especially in summer) following the opposite direction (southeast-northwest).

The Ebro River water temperature ranges from an average of 13°C at the headwaters to 17°C at the lower part, showing a clear seasonal pattern. In summer, after the large reservoirs of the middle Ebro River basin (Mequinenza, Ribaroja, and Flix), a slight decrease of river water temperature is registered due to the thermic inertia of the water in the reservoirs [19, 65]. In the autumn-winter months the inverse effect is detected. The more drastic intra-annual changes of river water temperature are those recorded in the Matarranya tributary, ranging from a minimum of 2.5°C to maximum values of 31°C. Analyses of stromatolithic microbial mats at the Ebro delta revealed long-term effects of El Niño Southern Oscillation event in this area [20]. Similar results were obtained after the analysis of sediment records in an endorreic saline lake at the Ebro Depression [21].

The different climate determines differences in the discharge regime. The tributaries from the Cantabric Mountains and from the western Pyrenees show a pluvial oceanic regime. The strong effect of snow retention at the central and eastern Pyrenees defines a nivopluvial regime in the corresponding tributaries. The tributaries of the Pyrenees have a nival regime with a maximum in spring and a relatively constant flow in summer. In an eastward direction, the hydrological regime is more continental. Rainfall-fed Atlantic rivers have much higher inputs with a slight decrease in summer flow. Moving to the southeast, the Atlantic influence disappears as the Mediterranean and continental character being stronger. In these tributaries, there is no snow retention. The Mediterranean tributaries have a rainfall-based flow regime with maxima in spring and autumn and a minimum in summer. The Mediterranean pluvial regime is extreme in the Guadalope and Matarranya watersheds. Because of the length and complexity of the landscape, flow regimes of some rivers also can vary. For example, headwaters of the Ebro are in the karstic area of Fontibre with an Atlantic influence. Downstream in the Ebro Depression, the flow regime progressively shifts to a Mediterranean type. The lower Ebro has a pluvio-nival flow regime after rivers from the Pyrenees enter the system.

Mean discharge at the Ebro River is 400 m^3/s, the tributaries from the left margin showing a higher discharge than those from the right margin (Table 1). Higher

Table 1 Most significant sub-watersheds of the Ebro River basin. Mean annual discharge considering the river at "natural" regime avoiding detraction, inputs for translocations, regulation or evaporation in the reservoirs. Data from CHE

Sub-watershed	Hm^3	%
Segre	6,356	34.9
Cinca	2,915	16.0
Aragón	4,521	24.8
Arga	1,697	9.3
Gállego	1,087	6.0
Jalón	551	3.0
Rest of watersheds	5,712	31.3
Ebro total	18,217	100.0

discharge is recorded from October to March on the average, although it can be delayed downstream until May. The maximum peak flow recorded on the Ebro was 12,000 m^3/s in 1907 in Tortosa [22]. The higher discharge in this period is due to the oceanic climate, while high spring discharge is related to the snowmelt from the Pyrenees. The lower discharge is recorded from July to October (see [66, 67] for more information).

The joint effect of the different hydrological regimes is diluted in the regime of the main Ebro channel. The Ebro is one of the rivers with less interannual variation among all the Iberian rivers. The relevance of groundwater inputs also smoothes the discharge regime. Groundwater influence is especially relevant from the Jalón to the Matarranya at the right margin, and at the Ega, Arga, Irati, and Alcanadre at the left margin. Three large aquifer zones are defined in the Ebro River basin: the Pyrenean (deep karstic aquifer), the alluvial (detritic-like aquifer with sand and cobbles), and Iberian (up to the river basins of Jalón, Guadalope, and Matarranya, calcium karstic sediments with low permeable marls).

In spite of this low variability, changes in land use over the last century (from 1945 to nowadays) are related to a decrease in discharge. The historical flow record of the Ebro River at the mouth (at Tortosa, mean annual runoff of 13,408 hm^3) shows a gross decrease of nearly 40% of the mean annual flow in the last 50 years. This decrease has been attributed to changes in land use and changes in climatic variables related to runoff generation [23]. Farm abandonment and increase in forest cover in the river headwaters and its associated increase of evapotranspiration and the increase in water consumption for irrigation for intensive agriculture may be related with this flow decrease [24]. Gallart and Llorens [25] concluded that two-thirds of the decrease may be attributed to irrigation and climate change but one-third must be attributed to the hydrological role of extensive increase of forest cover in headwaters.

The regulation of the Ebro River in the 1960s completed an irreversible change of the discharge pattern. The dams substantially altered flood timing, particularly of the flood peaks [26, 27]. Batalla et al. [28] analyzed flow records from 22 rivers to determine the effects of reservoirs on flow regime (flood frequency, flow duration of mean daily flows, monthly regime, and annual runoff) before and after dam construction. This research shows that variability of the mean daily flows was reduced in most cases due to the storage of winter floods and increased baseflows in summer (related to irrigation). Monthly flow patterns ranged from the absence of change to the complete inversion of the seasonal pattern (due to releases for irrigation in summer, formerly the season of lowest flows), in post-dam situations. The actual effect of a reservoir upon flow regime will depend not only upon reservoir capacity but also on the reservoir operation. The ratio between the reservoir capacity and the mean annual runoff is proposed as an indicator of the degree to which reservoirs could change flow patterns [28]. This index can serve as an indicator of the likelihood of dam-induced hydrologic changes, which in turn could affect channel morphology, sediment transport, and river ecology (see [66, 67]).

6 Human Influences at the River Ebro Watershed

6.1 Historic Human Settlements

The Ebro River basin crosses several cities of the provinces of Cantabria, Castilla and León, La Rioja, Navarra, Aragón, and Catalonia. Among those, the most relevant are Miranda de Ebro, Haro, Logroño, Tudela, Alagón, Zaragoza, Caspe, and Tortosa. Most of these cities are historically relevant. The Ebro River basin has been historically inhabited from the Paleolithic [29]. Pre-historic records are found in the Pyrenees and pre-Pyrenees (with remains such as dolmens and megalithic graves), as well as in the Iberian Mountains in La Rioja province and at the Guadalope basin. Ibers and Celts were occupying the Ebro River basin from the fifteenth to the third century BC. About 200 years BC the Romans entered to the basin from the south, at Matarranya and Guadalope basins and settled in cities such as Zaragoza, Huesca, and Teruel. When the arabs came to the Peninsula (year 711), they also settled in the Ebro River basin as their northeast site in the Iberian Peninsula. Zaragoza and Tortosa were the most important cities, easily connected by the Ebro River. Irrigation ditches and iron and copper industries developed during this period. Arabs also settled at Tudela, Calatayud, Huesca, and Barbastro. The "Reino de Aragón" (Aragón Kingdom) occupied most of the Ebro River basin. It was initiated around 1000 AD and was responsible for both the cultural and technical development of the population.

During the history of the human settlement in the Ebro basin, the river channel has played an important role as a frontier line but also as a communication line. The Ebro River basin was a crucial scenario for hard and bloody battles, such as the Ebro battle which occurred at the lower part of the Ebro basin During the Spanish Civil War (1936–1939). Hydrology played a military role during this battle, with sudden openings of upstream dams in order to interrupt the crossing of the infantry.

Agricultural development and navigation triggered the construction of several infrastructures in early periods. The Canal Imperial was first projected in 1446, and began the construction with "Fernando el Católico" (1510). With the Conde de Arana, Minister of Charles III of Spain, the Pignatelli dam was finished (1789). The first idea was the connection of the Cantabrian with the Mediterranean seas, the channel serving as navigation and irrigation, but finally it is flowing only in 108 km parallel to the main channel at its right margin, from El Bocal (Navarra) to Fuentes de Ebro (Zaragoza).

6.2 Land-Use

The population in the basin is about 2.8 million people, with a density of 33 inhabitants per km^2. This is a much lower population density than the Spanish mean (78 inhabitants per km^2). The population is heterogeneously distributed, nearly half of the population is concentrated in the cities of Zaragoza, Vitoria, Logroño, Pamplona, Huesca, and Lleida, in the center of the Ebro valley. However, in the Pyrenees and in the Iberian System the population density is very low (most spots having less than 2,000 inhabitants). Altogether, nearly 40% of the Ebro territory is uninhabited (less than 5 inhabitants per km^2). Poorly populated zones are in the Pyrenees, in several areas of the left margin, and in the desert landscapes at the center of the valley. However, since the mid twentieth century land use changes occurred in the Pyrenees due to the progressive abandonment of rural activities and the improvement of life standards. The consequences have been the spontaneous recuperation of woodland after decrease of human pressure and the afforestation works for avoiding erosion leading to a relevant increase in forest cover [25].

Land use in the Ebro River basin has been traditionally based on agricultural crops, such as vineyards, orchards, and maize. Up to 783,948 Ha are dedicated to agriculture, and mainly in the mid and lower Ebro sections, are irrigated. Nowadays, industry is a relevant activity at the most important cities (e.g., Zaragoza, Pamplona). Hydroelectric energy production uses about 8,297 m^3/s in 340 hydroelectric plants at the Ebro River basin. Water of the Ebro River is also used for cooling nuclear and thermic plants. Urban water demand is 5% of that used for agriculture.

6.3 Economic Activity and Management

The main economic use of the Ebro River has been hydropower and irrigation. The Ebro River has 187 reservoirs impounding 57% of the mean annual runoff. Such a large number of reservoirs deeply alter the fluvial regimes. None of the major dams in the basin was built for flood control, but the sheer volume of the impoundments affects the flood magnitude. Diverted water is used mainly for hydropower production and for irrigation. All the dams were constructed during the twentieth century, with the 67% of the reservoir capacity built in the period 1950–1975, when 5,200 Mm^3 of water were impounded. The 24 reservoirs with capacity between 50 and 500 Mm^3 have a total storage capacity of 4,200 Mm^3, over half the total basin storage and equivalent to 30% of the total annual runoff. Three reservoirs have more than 500 Mm^3 of the capacity: Ebro and Mequinenza in the main channel and Canyelles in the Noguera Ribagorçana River. Only the Ebro River headwaters tributaries still have a natural flow regime [30].

Other uses include the cooling of nuclear (at Ascó and Santa María de Garoña) and thermic centrals (Andorra, Escucha and Escatrón), as well as some aquiculture activity is developed in the Ebro River basin (about 80 fisheries, which mainly commercially produce trouts).

The Hydrographical Ebro Confederation (CHE, "Confederación Hidrográfica del Ebro") was the first organization for the managing of river water in Spain (created at 1926), at first with the objective of organizing irrigation for agriculture activities. Nowadays this organism is responsible for the control of many catchment master plans through the development of the Water Framework Directive. Catchment master plans include the ICA Network (aimed to the control of the Water quality) including the control of surface waters, groundwaters, network for quality alert as well as register of protected areas, analyses of impacts on surface waters, and monitoring of quality in reservoirs. The CHE have been historically collected and analyzed water quality parameters (including nutrient content, temperature, pH, conductivity, and several contaminants) from the Ebro watershed, providing monthly data for many stations and most parameters. These data show time evolution of the biogeochemistry of the Ebro River water and permit an analysis of relevant key factors explaining their variability. Specifically, Lasaletta et al. [31] analyzed the trend pattern of nitrate throughout a 25-year period (1981–2004), showing a significant positive correlation between agricultural cover and nitrate concentration in the water. Furthermore, for 46% of the sites explored, a trend of increase in nitrate concentration is found. This positive trend has been mainly related to the agricultural practices in the watershed, such as the application of intensive agriculture, while no significant changes in agricultural land use cover have been observed. Although improvements in agricultural techniques have been made, irrigation and fertilitzation management have still to improve in order to control water nitrate content [32]. Instead, a possible contribution of atmospheric N deposition in water nitrate content is also suggested [33].

7 River Ebro Water Biogeochemistry and Occurrence of Pollutants

The river Ebro is characterized by high water conductivity mainly because of its geology. The abundance of gypsum is mainly responsible for the salinity increase of the river water mostly due to the inputs of chlorides and sulfates in the Ebro

Depression [34]. Conductivity increases from the headwaters (about 200 μS/cm) to the meandering zone down Zaragoza (1,200 μS/cm in average, Table 2), reaching about 2,500 μS/cm in some periods [35]. The entrance of the Gállego River at the city of Zaragoza (showing conductivities of 1,600 μS/cm in average, Fig. 2) could be in part responsible for the conductivity increase at the middle Ebro. The Gállego drains a large extension of agricultural irrigated areas, and receives the sewage of two large paper mills, while in its lower part drains a gypsum area. The great organic pollution at the lower Gállego River (close to Zaragoza) might be responsible for the lowest oxygen values (mean of 6.66 mg/L for the period 1980–2005), with anoxia episodes (Fig. 2). However, a slight recovery (increase) in oxygen content was recorded since 2003 (Fig. 2).

The river flows quietly at the meandering zone. In this area, the largest values of suspended solids are measured (Table 2). The large development of phytoplankton in this area (near Sástago) [35] might be responsible for the higher dissolved oxygen measured in the flowing water (Fig. 2). The river water conductivity decreases below the large reservoirs up to 900 μS/cm (Table 2). This decrease can be related to the input of the Segre waters as well as to the decantation effect of the large reservoirs [36]. In contrast to the Gállego River, a much lower conductivity is received from the waters of Aragón and Segre tributaries (450 and 550 μS/cm respectively) that collect water from the Pyrenees (Fig. 2).

The larger content of nitrate (from 1.8–3.4 mg/L N-NO_3) than phosphate (10–40 μg/L P-PO_4) in the river Ebro (Fig. 3, Table 2) has been mainly related to the agricultural activities in the watershed. Nutrient excess is a relevant feature of water biogeochemistry in the middle and lower part of the river Ebro. The contribution of non-point sources, such as agricultural fertilizers and pig farming manure, is extremely relevant. The analysis of nutrient loads (nitrate, phosphate, and dissolved organic carbon-DOC) in the alluvial aquifer of the central part of the river basin (from Tudela to Zaragoza) concludes that non-point sources contribute annual nitrate loads of 25 Tm NO_3/day [34]. These values are maximal during summer, coinciding with the most intense irrigation period. Industrial activities at the Ebro River basin are concentrated in Zaragoza, Vitoria, Pamplona, Logroño, Huesca, Monzón, Lleida, and Tortosa. Zaragoza and surroundings have been qualified as the most polluted areas in the Ebro. Agricultural non-point sources

Table 2 Physical and chemical parameters at three sites of the main Ebro channel, Logroño (kilometric point 609), Sástago (kilometric point 285.5), and Tortosa (kilometric point 43). Values are means and SD ($n = 4$) for data collected in 2008 and 2009 (spring and autumn)

	Conductivity (μS/cm)	Temperature (°C)	Oxygen (mg/L)	Suspended solids (mg/L)	N-NO3 (mg/L)	P-PO4 (μg/L)
Logroño	615	12.75	10.16	15.97	2.85	33.6
	(161)	*(5.75)*	*(2.20)*	*(20.78)*	*(0.34)*	*(34.1)*
Sástago	1,231	14.87	8.78	38.42	3.97	19.9
	(551)	*(7.49)*	*(2.62)*	*(21.28)*	*(1.68)*	*(15.9)*
Tortosa	949	17.65	9.49	8.38	3.13	42.4
	(264)	*(7.03)*	*(1.48)*	*(5.10)*	*(1.91)*	*(34.8)*

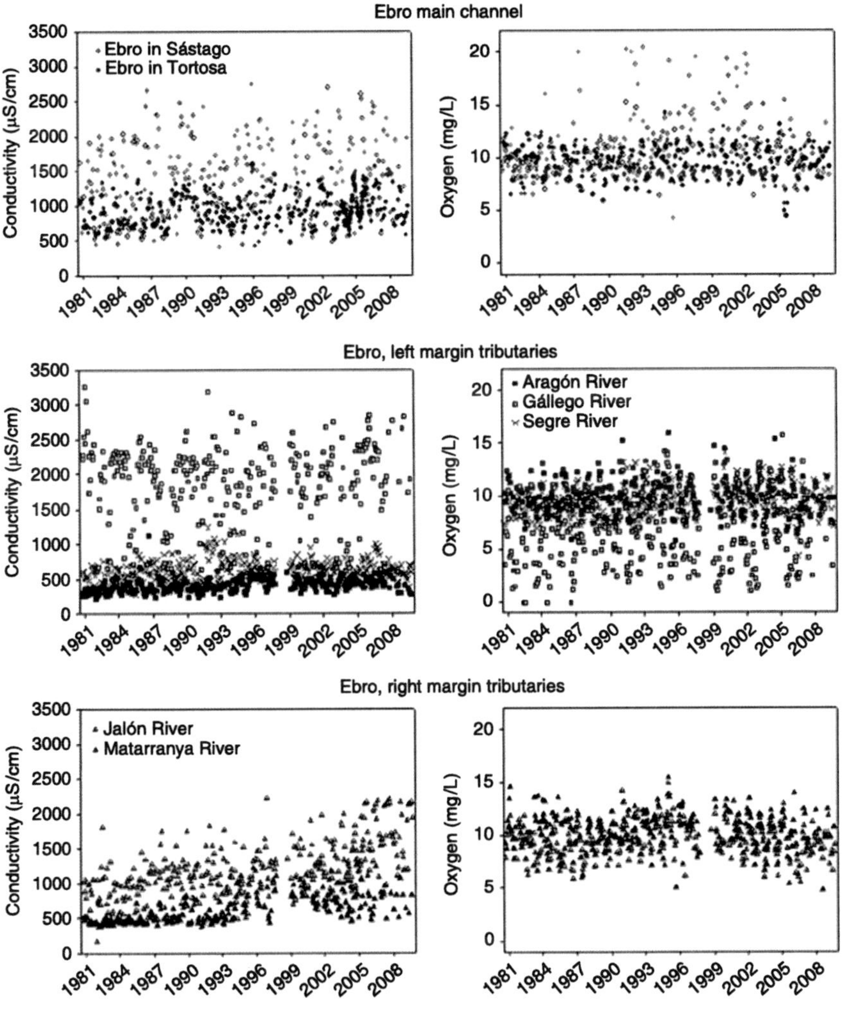

Fig. 2 Conductivity and oxygen content for the period 1981–2009 at the main Ebro channel (Sástago and Tortosa sites), at Aragón Gállego and Segre left margin tributaries, and at the Jalón and Matarranya right margin tributaries. Data are monthly values

account for 64% of nitrate loads generated in this central area of the Ebro River basin, while urban and industrial point sources are responsible for the 88% phosphate and 71% DOC loads [34].

Nutrient concentration has a marked seasonal variation in the middle Ebro. Nutrient loads transported by the river are relevant during the high flow season (December–April), while nitrate inputs from agriculture can also be relevant. In this period, crop fertilization is carried out, precipitation is higher, and plant nitrogen uptake is lower [37]. Although nutrient loads are high, dilution causes NO_3 and

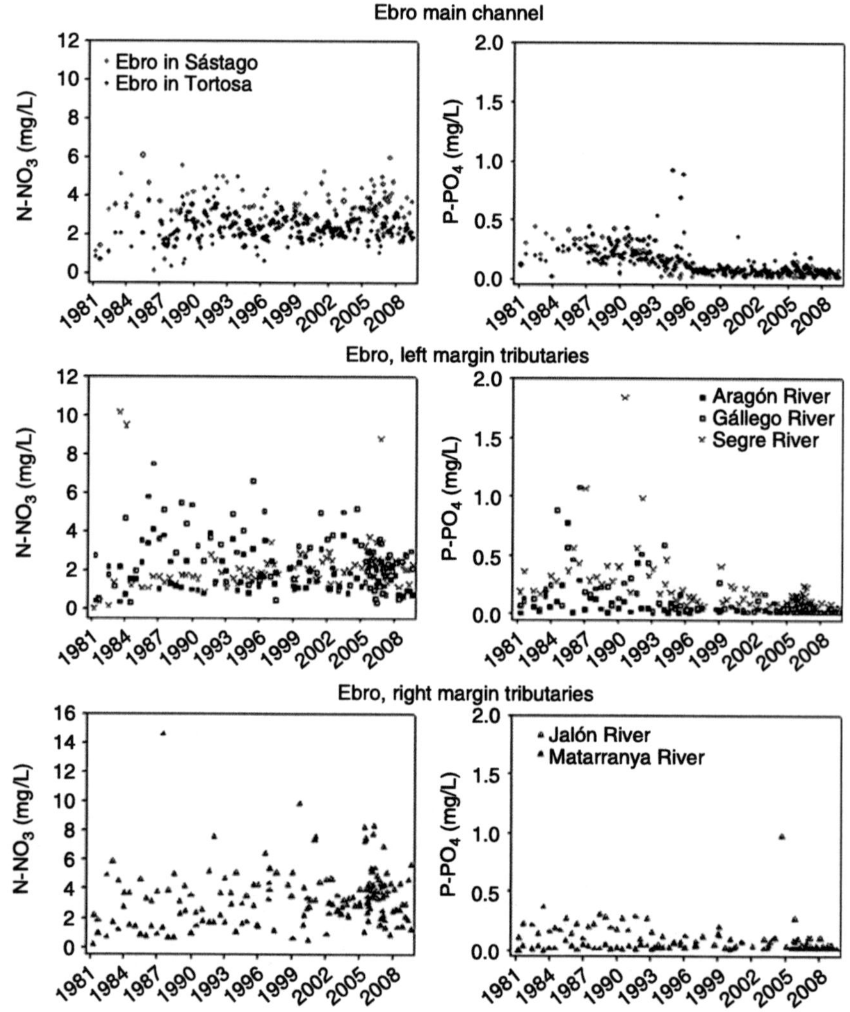

Fig. 3 Nitrate and phosphate river water concentrations for the period 1981–2009 at the main Ebro channel (Sástago and Tortosa sites), at Aragón Gállego and Segre left margin tributaries, and at the Jalón and Matarranya right margin tributaries. Data are monthly values

DOC concentration to be relatively low, and oxygen content relatively high. During this period, phosphates are high, probably because of the combined leaching from the basin, mobilization from the sediment, and lower biological consumption at lower temperatures.

In contrast, during the low flow season (June–October) important nutrient loads from both point and non-point sources are relevant. Summer irrigation drives nitrate inputs to stream waters [38]. The lower dilution capacity of the river causes higher concentrations of nitrate and DOC, as well as an increase in phosphate content with respect to spring months. In this period, oxygen concentration decreases (often below 4 mg/L), and the river presents eutrophication symptoms such as very high turbidity, cyanobacterial foams, or fish migration from deep to riffle zones [34].

Bouza-Deaño et al. [39] and Ibáñez et al. [40] described a decreased trend of phosphate concentration and a pH increase in the last 25 years along the river. Mean phosphate concentrations at Tortosa ranged from 0.08 to 0.27 mg/L P-PO_4, in the period 1980–2005 and decreased down to 0.02–0.06 mg/L in 2008–2009 (Fig. 3). A decrease in phosphate concentration has also been observed at the river Ebro tributaries (Fig. 3). The construction of water treatment plants (at Tudela and Zaragoza) in the early 1990s can be the main cause for decrease in phosphate [34, 41], the decrease in DOC, and the increase in oxygen content.

Other pollutant sources are mining activities in the northern part of the river, historical mercury pollution coming from chloroalkali industry, production and utilization of solvents and chlorinated pesticides, and usage of flame retardants in car and electrical plants in the middle-lower course. Analysis of priority pollutants along the whole river basin showed the detection of organic compounds on sediments such as PAHs (polycyclic aromatic hydrocarbons), APs (alkylphenols), and PBDEs (polybrominated diphenyl ethers), while organotin and organochlorine compounds (DDTs and chlorobenzenes) have also been detected in several sampling points (concentration range of target compounds was between 0.01 and 2,332 μg/kg dry weight). In contrast, fish mainly accumulated organochlorine compounds and PBDEs [42]. High levels of organochlorine pollutants have been detected in sediments of the middle and lower Ebro [42, 43]. Bioaccumulation of polybrominated diphenyl ethers (PBDEs) and hexabromocyclododecane (HBCD) in fish (*Barbus graellsii* and *Alburnus alburnus*) and concentrations in sediment were also detected after the heavily industrialized city of Monzón [44, 45]. At the lower part of the Ebro basin, high concentrations (20–225 ng/L) of atrazine have been recorded, other pesticides reaching maximum values of 150 ng/L. Annual loads of herbicides were directly correlated with field application and stream discharge [46].

Endocrine-disrupting effects on carps (*Cyprinus carpio*) have been detected at some hot spots in the medium and low course of the Ebro (downstream Zaragoza, at Flix reservoir, and at the Imperial channel) [47]. Carps from industrialized areas also showed contents of persistent organic pollutants such as PCBs and DDTs as well as high levels of mercury and cadmium in the liver, and high levels of nonylphenol in bile. Significant alterations in some biochemical markers (cytochrome P450 system, phase II activities, methallothionein) to pollutants exposure were also observed [48].

7.1 Biogeochemistry at the Ebro Delta

Specific biogeochemical characteristics define the Ebro Delta. This area receives the impacts of the activities on the whole river basin. Before the dam construction its sedimentary dynamics was mainly fluvial, while nowadays it is mostly driven by the sea effects because of the limited supply of sediments to the Delta [49, 50]. The Ebro River is a highly regulated fluvial system, and this drastically affects the sediment transport. The analysis of discharge patterns and sediment transport before and after the large dams of Mequinenza, Ribaroja, and Flix during flood events clearly shows a reduction of sediment transport in the lower Ebro River. The dams captured over the 95% of the fine sediment carried in suspension in the river channel (from 0.5 g/L to 0.05 g/L of total solid mean concentrations before and after the dams, respectively), preventing it from reaching the lowermost reaches of the river and the delta plain [27]. At the same time, the marine salt wedge occasionally penetrates 25 km upstream, especially in summer, and disappears during the high flow episodes of spring. Nowadays, the lower high flood frequency as well as the flow regulation favors the persistence of the saline wedge [51, 52], decreasing oxygen content down to anoxia [53]. Therefore, the lowest stretch of the Ebro River (about 42 km) behaves mostly as a highly stratified estuary with a salt wedge, and determines saltwater intrusion problems both for aquifers and crops [54, 55]. The lower sediment input affects the physical consistency of the deltaic sediments, and is associated with the current regression of the Delta. Further effects at the Ebro delta due to increases in sea level are suspected [56].

7.2 Biogeochemistry at the Endorheic Saline Lakes

Located in the middle Ebro River basin, the arid zone of Los Monegros is characterized by the presence of several hypersaline inland lakes. The hydrology of these saline lakes is mainly regulated by high evapotranspiration (1,000–1,500 mm/year), low rainfall (300–350 mm/year), runoff, irrigation returns, and groundwater flow. The latter represents an important source of solutes due to dissolution of the carbonate and evaporitic rocks of tertiary formation [57]. These depressions are developed in Miocene lacustrine strata, formed by karstic processes acting on the underlying limestone and gyprock [58, 59]. Salinity reaches values >30 g/L mainly due to the ions Ca, Na, Mg, SO_4, and Cl [60, 61]. These saline lakes undergo fluctuations at different time scales, difficulting its management [62]. Different time scales (annual, decadal) exert their influence in the phytoplankton and zooplankton community dynamics of these lakes, which are mainly related to fluctuations in water level and salinity, strongly correlated with meteorological events [62]. Anoxic conditions have been registered to occur periodically at the bottom layers. These habitats are highly sensitive to anthropic effects such as the extension of irrigation on the expansion of arable agriculture, which have been produced in the Monegros area for the period 1984–1997 [63]. In these semi-arid areas the analysis of irrigation and drainage management and their effects on the loading of salts is important for the control of on-site and off-site salinity effects of irrigated agriculture [64].

Acknowledgments The manuscript was partially funded by Consolider-Ingenio SCARCE (CSD2009-00065). The Confederación Hidrográfica del Ebro (CHE) provided long term physical and chemical data.

References

1. Santisteban JI, Schulte L (2007) Fluvial networks of the Iberian Peninsula: a chronological framework. Quaternary Sci Rev 26:2738–2757
2. Garcia-Castellanos D, Verges J, Gaspar-Escribano J, Cloetingh S (2003) Interplay between tectonics, climate, and fluvial transport during the Cenozoic evolution of the Ebro Basin (NE Iberia) – art. no. 2347. J Geophys Res Solid Earth 108:2347–2347
3. Jones SJ, Frostick LE, Astin TR (1999) Climatic and tectonic controls on fluvial incision and aggradation in the Spanish Pyrenees. J Geol Soc 156:761–769
4. Arenas C, Pardo G (1999) Latest Oligocene-Late Miocene lacustrine systems of the north-central part of the Ebro Basin (Spain): sedimentary facies model and palaeogeographic synthesis. Palaeogeogr Palaeoclimatol Palaeoecol 151:127–148
5. Perez A, Luzon A, Roc AC, Soria AR, Mayayo MJ, Sanchez JA (2002) Sedimentary facies distribution and genesis of a recent carbonate-rich saline lake: Gallocanta Lake, Iberian Chain, NE Spain. Sediment Geol 148:185–202
6. Schutt B (2000) Holocene paleohydrology of playa lakes in northern and central Spain: a reconstruction based on the mineral composition of lacustrine sediments. Quatern Int 73–4:7–27
7. Sanchez JA, Coloma P, Perez A (1999) Sedimentary processes related to the groundwater flows from the Mesozoic Carbonate Aquifer of the Iberian Chain in the Tertiary Ebro Basin, northeast Spain. Sediment Geol 129:201–213
8. Gutierrez F, Orti F, Gutierrez M, Perez-Gonzalez A, Benito G, Gracia FJ, Duran JJ (2002) Paleosubsidence and active subsidence due to evaporite dissolution in Spain. Carbonates Evaporites 17:121–133
9. Gutierrez F, Orti F, Gutierrez M, Perez-Gonzalez A, Benito G, Prieto JG, Valsero JJD (2001) The stratigraphical record and activity of evaporite dissolution subsidence in Spain. Carbonates Evaporites 16:46–70
10. Benito G, Perez-Gonzalez A, Gutierrez F, Machado MJ (1998) River response to Quaternary subsidence due to evaporite solution (Gallego River, Ebro Basin, Spain). Geomorphology 22:243–263
11. Valero-Garces BL, Gonzalez-Samperiz P, Navas A, Machin J, Delgado-Huertas A, Pena-Monne JL, Sancho-Marcen C, Stevenson T, Davis B (2004) Paleohydrological fluctuations and steppe vegetation during the last glacial maximum in the central Ebro valley (NE Spain). Quatern Int 122:43–55
12. Gutierrez-Santolalla F, Gutierrez-Elorza M, Marin C, Maldonado C, Younger PL (2005) Subsidence hazard avoidance based on geomorphological mapping in the Ebro River valley mantled evaporite karst terrain (NE Spain). Environ Geol 48:370–383
13. Molina Holgado P (2002) Análisis y comparación de la vegetación de las riberas de los ríos Ebro, Tajo y Jarama. Madrid, Ediciones de la Universidad Autónoma de Madrid
14. Ribera I, Blasco-Zumeta J (1998) Biogeographical links between steppe insects in the Monegros region (Aragon, NE Spain), the eastern Mediterranean, and central Asia. J Biogeogr 25:969–986
15. Confederación Hidrográfica del Ebro (2009) http://www.chebro.es
16. Confederació Hidrográfica del Ebro (2005) http://www.chebro.es. WFD Report
17. Sterk G, Lopez MV, Arrue JL (1999) Saltation transport on a silt loam soil in northeast Spain. Land Degrad Dev 10:545–554

18. Martínez-Cob A, Tejero-Juste M (2004) A wind-based qualitative calibration of the Hargreaves ET0 estimation equation in semiarid regions. Agric Water Manage 64:251–264
19. Val R, Niñerola D, Armengol J, Dolz J (2003) Incidencia de los embalses en el régimen térmico del río. El caso del tramo final del río Ebro. Limnetica 22:85–92
20. Sanchez-Cabeza JA, Masque P, Martinez-Alonso M, Mir J, Esteve I (1999) Pb-210 atmospheric flux and growth rates of a microbial mat from the northwestern Mediterranean Sea (Ebro River Delta). Environ Sci Technol 33:3711–3715
21. Rodo X, Giralt S, Burjachs F, Comin FA, Tenorio RG, Julia R (2002) High-resolution saline lake sediments as enhanced tools for relating proxy paleolake records to recent climatic data series. Sediment Geol 148:203–220
22. Novoa M (1984) Precipitaciones y avenidas extraordinarias en Catalunya. Ponencias y comunicaciones de las Jornadas de Trabajo sobre Inestabilidad de laderas en el Pirineo 1:1–15, Barcelona
23. Lopez-Moreno JI, Beniston M, Garcia-Ruiz JM (2008) Environmental change and water management in the Pyrenees: Facts and future perspectives for Mediterranean mountains. Global Planet Change 61:300–312
24. Gallart F, Llorens P (2004) Observations on land cover changes and the headwaters of the Ebro catchment, water resources in Iberian Peninsula. Phys Chem Earth 29:769–773
25. Gallart F, Llorens P (2002) La cubierta forestal de la cuenca del Ebro: Caracterización espacio-temporal y afección en la disminución de recursos hídricos. Confederación Hidrográfica del Ebro, Zaragoza, Unpublished report
26. Lopez-Moreno JI, Begueria S, Garcia-Ruiz JM (2002) Influence of the Yesa reservoir on floods of the Aragon River, central Spanish Pyrenees. Hydrol Earth Syst Sci 6:753–762
27. Vericat D, Batalla RJ (2005) Sediment transport in a highly regulated fluvial system during two consecutive floods (lower Ebro River, NE Iberian Peninsula). Earth Surf Process Landforms 30:385–402
28. Batalla RJ, Gomez CM, Kondolf GM (2004) Reservoir-induced hydrological changes in the Ebro River basin (NE Spain). J Hydrol 290:117–136
29. González-Sampériz P, Utrilla P, Mazo C, Valero-Garcés B, Sopena MC, Morellón M, Sebastián M, Moreno A, Martínez-Bea M (2009) Patterns of human occupation during the early Holocene in the Central Ebro Basin (NE Spain) in response to the 8.2 ka climatic event. Quatern Res 71:121–132
30. García MA, Moreno MC (2000) Los aprovechamientos en la Cuenca del Ebro: Afección en el regimen hidrológico fluvial, Internal Technical Report no. 2000-PH-24.1, Confederación Hidrográfica del Ebro, Zaragoza, 83pp
31. Lassaletta L, García-Gómez H, Gimeno BS, Rovira JV (2009) Agriculture-induced increase in nitrate concentrations in stream waters of a large Mediterranean catchment over 25 years (1981–2005). Sci Total Environ 407:6034–6043
32. García-Garizábal I (2009) Causapé Valenzuela J, Abrahão R (2009) Evolution of the efficiency and agro-environmental impact of a traditional irrigation land in the middle Ebro Valley (2001–2007). Span J Agric Res 7:465–473
33. Dise NB, Rothwell JJ, Gauci V, van der Salm C, de Vries W (2009) Predicting dissolved inorganic nitrogen in European forests using two independent databases. Sci Total Environ 407:1798–808
34. Torrecilla NJ, Galve JP, Zaera LG, Retamar JF, Alvarez AN (2005) Nutrient sources and dynamics in a mediterranean fluvial regime (Ebro river, NE Spain) and their implications for water management. J Hydrol 304:166–182
35. Sabater S, Artigas J, Durán C, Pardos M, Romaní AM, Tornés E, Ylla I (2008) Longitudinal development of chlorophyll and phytoplankton assemblages in a regulated large river (the Ebro River). Sci Total Environ 404:196–206
36. Roura M (2004) Incidence of the Mequinensa reservoir in the transport of suspended solids and in the water quality of the River Ebro. Ph D dissertation, Universitat de Barcelona

37. Tisseuil C, Wade AJ, Tudesque L (2008) Modeling the stream water nitrate dynamics in a 60,000-km2 European catchment, the Garonne, southwest France. J Environ Qual 37:2155–2169
38. Causape J, Quilez D, Aragues R (2004) Assessment of irrigation and environmental quality at the hydrological basin level – II. Salt and nitrate loads in irrigation return flows. Agric Water Manage 70:211–228
39. Bouza-Deaño R, Ternero-Rodríguez M, Fernández-Espinosa AJ (2008) Trend study and assessment of surface water quality in the Ebro River (Spain). J Hydrol 361:227–239
40. Ibáñez C, Prat N, Durán C, Pardos M, Munné A, Andreu R (2008) Changes in dissolved nutrients in the lower Ebro river: causes and consequences. Limnetica 27:131–142
41. Oscoz J, Durán C, Pardos M, Gil J, Viamonte A (2008) Historical evolution of the biological water quality in the Ebro Basin (Spain) (1990–2005). Limnetica 27:119–130
42. Lacorte S, Raldua D, Martinez E, Navarro A, Diez S, Bayona JM, Barcelo D (2006) Pilot survey of a broad range of priority pollutants in sediment and fish from the Ebro river basin (NE Spain). Environ Pollut 140(3):471–482
43. Fernandez MA, Alonso C, Gonzalez MJ, Hernandez LM (1999) Occurrence of organochlorine insecticides, PCBs and PCB congeners in waters and sediments of the Ebro river (Spain). Chemosphere 38:33–43
44. Eljarrat E, De La Cal A, Raldua D, Duran C, Barcelo D (2004) Occurrence and bioavailability of polybrominated diphenyl ethers and hexabromocyclododecane in sediment and fish from the Cinca River, a tributary of the Ebro River (Spain). Environ Sci Technol 38:2603–2608
45. Eljarrat E, de la Cal A, Raldua D, Duran C, Barcelo D (2005) Brominated flame retardants in Alburnus alburnus from Cinca River Basin (Spain). Environ Pollut 133:501–508
46. Gascon J, Salau JS, Oubina A, Barcelo D (1998) Monitoring of organonitrogen pesticides in the Ebro river. Preliminary loadings estimates. Analyst 123:941–945
47. Lavado R, Thibaut R, Raldua D, Martin R, Porte C (2004) First evidence of endocrine disruption in feral carp from the Ebro River. Toxicol Appl Pharmacol 196:247–257
48. Lavado R, Urena R, Martin-Skilton R, Torreblanca A, del Ramo J, Raldua D, Porte C (2006) The combined use of chemical and biochemical markers to assess water quality along the Ebro River. Environ Pollut 139:330–339
49. Jimenez JA, Sanchez-Arcilla A, Valdemoro HI, Gracia V, Nieto F (1997) Processes reshaping the Ebro delta. Mar Geol 144:59–79
50. Maidana MA, Naudin JJ, Espino M, Garcia MA, Sanchez-Arcilla A (2002) Feasibility and usefulness of steady-state calculations of the mean circulation in the vicinity of the Ebro mouth. Model tests against field data. Cont Shelf Res 22:229–245
51. Durand N, Fiandrino A, Fraunie P, Ouillon S, Forget P, Naudin JJ (2002) Suspended matter dispersion in the Ebro ROFI: an integrated approach. Cont Shelf Res 22:267–284
52. Ibanez C, Saldana J, Prat N (1999) A model to determine the advective circulation in a three layer, salt wedge estuary: Application to the Ebre River estuary. Estuar Coast Shelf Sci 48:271–279
53. Muñoz I (1998) C, N y P en la parte baja del río Ebro y en los canales de riego del delta. Oecología Aquatica 11:23–53
54. Muñoz I, Prat N (1989) Effects of river regulation on lower river Ebro. Regul Rivers: Res Manage 3:345–354
55. Sierra JP, Sanchez-Arcilla A, Figueras PA, Del Rio JG, Rassmussen EK, Mosso C (2004) Effects of discharge reductions on salt wedge dynamics of the Ebro River. River Res Appl 20:61–77
56. Sánchez-Arcilla A, Jiménez JA, Valdemoro HI, Gracia V (2008) Implications of climatic change on Spanish Mediterranean Low-Lying Coasts: The Ebro Delta Case. J Coast Res 24:306–316
57. Valero-Garces BL, Delgado-Huertas A, Navas A, Machín J, González-Sampériz P, Kelts K (2000) Quaternary palaeohydrological evolution of a playa lake: Salada Mediana, central Ebro Basin, Spain. Sedimentology 47:1135–1156

58. Castañeda C, Herrero J, Casterad MA (2005) Facies identification within the playa-lakes of the Monegros desert, Spain, from field and satellite data. Catena 63:39–63
59. Gracia FJ, Gutierrez F, Gutierrez M (2002) Origin and evolution of the Gallocanta polje (Iberian Range, NE Spain). Zeitz Geomorf 46:245–262
60. Corzo A, Luzon A, Mayayo MJ, van Bergeijk SA, Mata P, de Lomas JG (2005) Carbonate mineralogy along a biogeochemical gradient in recent lacustrine sediments of Gallocanta Lake (Spain). Geomicrobiol J 22:283–298
61. de Wit R, Falcon LI, Charpy-Roubaud C (2005) Heterotrophic dinitrogen fixation (acetylene reduction) in phosphate-fertilised Microcoleus chthonoplastes microbial mat from the hypersaline inland lake 'la Salada de Chiprana' (NE Spain). Hydrobiologia 534:245–253
62. Comin FA, Cabrera M, Rodo X (1999) Saline lakes: integrating ecology into their management future. Hydrobiologia 396:241–251
63. Koch M (2000) Geological controls of land degradation as detected by remote sensing: a case study in Los Monegros, north-east Spain. Int J Remote Sens 21:457–473
64. Tedeschi A, Beltran A, Aragues R (2001) Irrigation management and hydrosalinity balance in a semi-arid area of the middle Ebro river basin (Spain). Agric Water Manage 49:31–50
65. Prats J, Armengol J, Marcé R, Sánchez-Juny M, Dolz J (2010) Dams and reservoirs in the lower Ebro river and its effects on the river thermal cycle. Hdb Env Chem DOI 10.1007/698_2010_68
66. Batalla RJ, Veircat D (2010) Hydrology and sediment transport. Hdb Env Chem DOI 10.1007/698_2010_67
67. Zambrano-Bigiarini M, Majone B, Bellin A, Isabella Bovolo C, Blenkinsop S, Fowler HJ (2010) Hydrological impacts of climate change on the Ebro River basin. Hdb Env Chem DOI 10.1007/698_2010_85

Hydrology and Sediment Transport

Ramon J. Batalla and Damià Vericat

Abstract This chapter presents a summary of the most notable trends of river hydrology and sediment transport observed in the Ebro basin. Measurement efforts of fluvial process have concentrated in the past decade in the lowermost part of the river downstream from major dams. This chapter emphasises the results obtained in this part of the catchment. Flow and sedimentary regimes of the Ebro are profoundly marked by the human activity. Reservoirs and land use changes have systematically altered the pattern of water and sediment yield, and the associated river processes (i.e. channel morphodynamics) along the twentieth century. Data indicate that runoff and especially magnitude of frequent floods have been reduced all over the basin; moreover, sediment supply has diminished due to extensive afforestation of catchment mountainous headwaters, while dams trap most of sediment that still circulating in the drainage network. Overall, sediment yield in the Ebro basin is estimated to be less than 2% of the original load at the beginning of the twentieth century. An average of 0.45×10^6 tones of sediment per year (60% in suspension and 40% as bedload) have been measured during the intensive monitoring period 2002–2004 at the control section of Móra d'Ebre (in the lowermost part of the basin downstream from dams), and further corroborated by data obtained between 2005 and 2008. Sediment deficit is mostly evident in the lower reaches of the river mainstem but also in some of its main tributaries. However, the channel is still active from the sedimentary point of view showing a net export of sediment during competent floods, a fact that produces channel deepening and incision; at the

R.J. Batalla (✉)
Department of Environment and Soil Sciences, University of Lleida, 25198 Lleida, Catalonia, Spain
Forestry and Technology Center of Catalonia, 25280 Solsona, Catalonia, Spain
e-mail: rbatalla@macs.udl.cat

D. Vericat
Forestry and Technology Center of Catalonia, 25280 Solsona, Catalonia, Spain
Institute of Geography and Earth Sciences, Aberystwyth University, Ceredigion, SY23 3DB, Wales, UK

D. Barceló and M. Petrovic (eds.), *The Ebro River Basin*,
Hdb Env Chem (2011) 13: 21–46, DOI 10.1007/698_2010_67,

same time, though, riverbed armouring occurs, a fact that progressively stabilises the channel facilitating, for instance, a massive colonisation of the riverbed by aquatic vegetation (macrophytes).

Keywords Dams, Flow regime, River Ebro, River hydrology, Sediment transport

Contents

1 River Hydrology

The hydrology of a river is defined primarily by its flow regime. The flow regime is characterised by a complex range of interrelated flow elements, such as total runoff (i.e. water yield), mean discharge, frequency and magnitude of flood events, baseflows, shape of hydrographs, and seasonal and interannual variability, between many others. The flow regime controls many of physical and ecological aspects of river form and processes, including sediment and nutrient loads [1]. Dams, in particular, alter the downstream flow regime of rivers (e.g. [2, 3]). The resultant hydrological alterations may include changes in flood frequency and magnitude, reduction in overall flow, changes in baseflows, and flow fluctuations as a consequence of the altered timing of releases. These alterations have, consequently, a wide range of effects on riverine ecology (e.g. [3–7]). Floods are responsible for regular energy inputs that subsequently shake and disturb channel morphology and sedimentology which, in turn, control river habitat conditions. Floods in regulated rivers will tend to be reduced more in dry years and early in the season, when reservoir levels are lower and a major storage capacity is available. Moreover, it can be expected that reservoir effects are more pronounced in drier climates because of greater storage needs and greater likelihood that the reservoir will be drawn-down when floods enter. Because channel form and river ecology in dryland rivers will be adapted to highly variable flow regimes, dam-induced reduction in flow variability is likely to have a relatively larger effect on river ecology, both through reductions in high flows (reducing disturbance) and often increasing baseflows (i.e. making these environments more suitable for exotic species not adapted to seasonal drying cycles). The degree of regulation depends basically on the regional hydroclimatological conditions and water demands and uses. Rivers in humid environments tend to be less regulated than their dryland counterparts. The Ebro and, in general, other large rivers in the Iberian Peninsula are the most regulated in Europe, reaching a regulation capacity of almost twice the annual runoff in some of their main tributaries (e.g. Noguera Ribargorçana, NE Ebro basin).

The Ebro basin is located in the northeast of the Iberian Peninsula and drains an area of around 85,000 km^2 (Fig. 1). Mean annual precipitation in the basin is 650 mm, but varies from over 2,000 mm in the summits of the Pyrenean Range to less than 300 mm in the dry Central Depression. Mean annual discharge in Tortosa (the most downstream gauging station on the Ebro, see Fig. 1) is 435 m^3/s (period 1913–2008, operated by the Ebro Water Authorities, hereafter CHE) Mean annual runoff for the same period is around 158 mm (i.e. 13,700 hm^3, where 1 $hm^3 = 1 \times 10^6$ m^3), and has varied from a minimum of 50 mm (i.e. 4,284 hm^3) to a maximum of 353 mm (i.e. 30,800 hm^3). The maximum historical peak flow in Tortosa is estimated at 12,000 m^3/s, which occurred in 1907 following a major rainfall storm [8]. The Ebro basin was progressively impounded during the second half of the twentieth century, with 67% of the storage capacity accomplished between 1950 and 1975, when 5,200 hm^3 of water was effectively impounded. The 24 largest reservoirs with capacity between 50 and 500 hm^3 have a total storage capacity of 4,200 hm^3. This value represents more than half the total basin storage capacity and it is equivalent to 30% of the total annual runoff in the basin's outlet. These numbers show that dams in the Ebro play an important role in regulating flows and, consequently, in altering flow regime and associated processes and dynamics. Over the almost 200 reservoirs in the basin, just three of them may store more than 500 hm^3: Ebro and Mequinenza at the upstream and downstream ends of the River Ebro respectively, and Canelles in the Noguera Ribagorçana (Fig. 1). According to works by García and Moreno, MIMAM [9, 10] only the tributaries draining from the headwaters still have natural flow regimes. The largest complex of reservoirs is located in the lower reaches of the river and it is formed by Mequinenza (constructed 1966, with a capacity of 1,534 hm^3), Riba-roja (1969, 207 hm^3) and Flix dams (1948, 11 hm^3), altogether collecting water from 97% of the basin area (Fig. 1). The most important tributary downstream from the Flix Dam is the Siurana River which is itself heavily regulated by three dams since the 1980s and profoundly altered by gravel mining activities since the 1990s (Fig. 1). Overall, rivers in the basin have experienced significant reductions in flood magnitude downstream from dams. Reductions between 21 and 85% for frequent floods (i.e. 2-year flood) were reported by Batalla et al. [11], while the magnitude of the 10–25-year floods has been reduced by 25% on average. This chapter will present the main hydrological trends of the Ebro basin, particularly in its lower reaches, and will summarise the effects of regulation on flow regime.

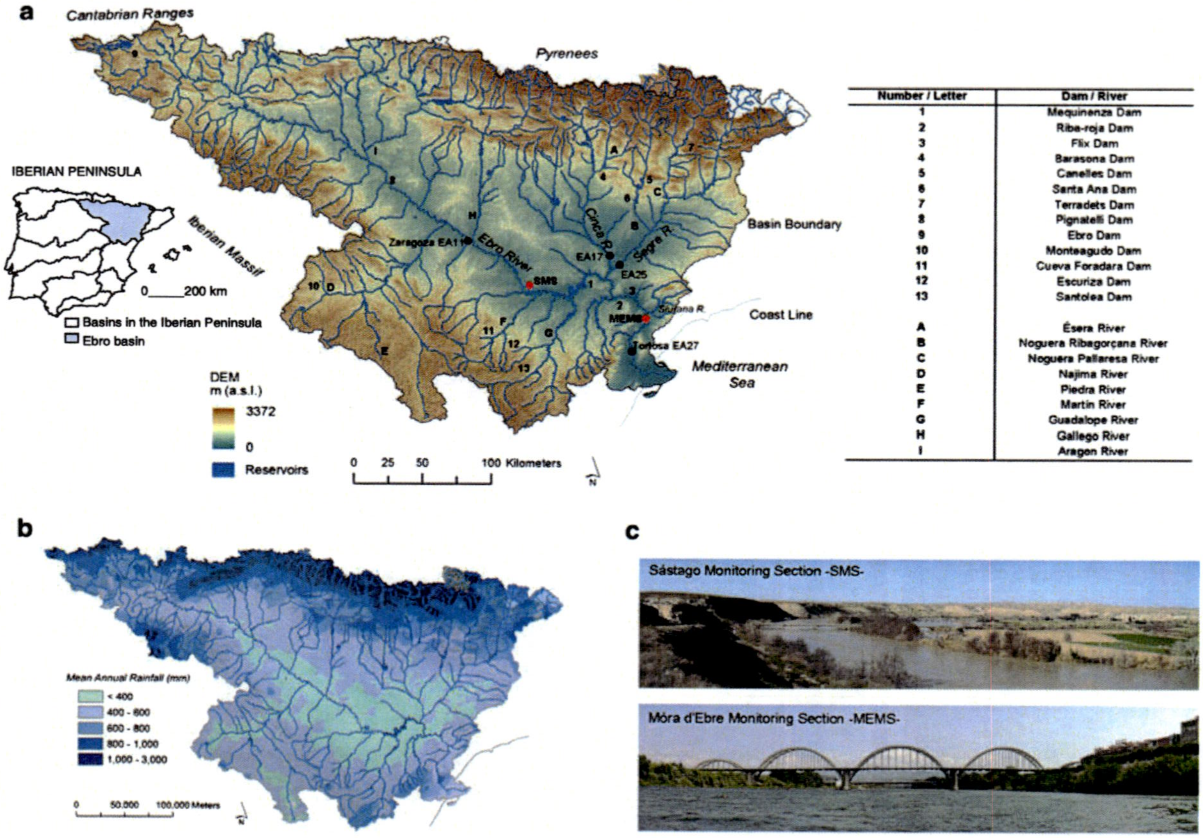

Number / Letter	Dam / River
1	Mequinenza Dam
2	Riba-roja Dam
3	Flix Dam
4	Barasona Dam
5	Canelles Dam
6	Santa Ana Dam
7	Terradets Dam
8	Pignatelli Dam
9	Ebro Dam
10	Monteagudo Dam
11	Cueva Foradara Dam
12	Escuriza Dam
13	Santolea Dam
A	Ésera River
B	Noguera Ribagorçana River
C	Noguera Pallaresa River
D	Najima River
E	Piedra River
F	Martin River
G	Guadalope River
H	Gallego River
I	Aragon River

Fig. 1 (**a**) Digital Elevation Model of the Ebro Basin showing the principal drainage channel network and reservoirs (see *labels* and *numbers* in the table). A figure indicating the location of the basin in the Iberian Peninsula is also presented. (**b**) Mean annual rainfall in the Ebro basin. (**c**) Panoramic views of the two Monitoring Sections where a sediment transport sampling programme has been carried out since 2002 (see location in (**a**)) by the University of Lleida

2 Fluvial Sediment Transport

Rivers carry water but also sediments from continents to oceans, being the responsible for the equilibrium between fluvial and marine processes in deltas and beaches. River channel morphology is considered to be, over the long-time scale, maintained in a dynamic quasi-equilibrium, whereby sediment exported from a given reach is almost equal to the supply from upstream [2]. River sediments form the morphological and sedimentological structure that support physical habitat for a plethora of riverine species, from invertebrates to fish and aquatic vegetation. Dams disrupt the continuity of water and sediment transfer, which in turn affect river morphology and ecology (e.g. [2, 3, 12]). Generally speaking, the effects of an individual reservoir depend upon its position in the river system, the time that has been in operation, reservoirs' depth and surface area, its purpose (e.g. irrigation, hydroelectric generation, flood control) and dam operating rules, as well as regional geographic factors such as climate [13]. As it has been introduced in the last section, dams diminish the frequency and magnitude of floods events that are responsible for the majority of sediment transfer in rivers. At the same time, dams also affect sediment transport processes along the river corridor by trapping the majority of bedload (i.e. coarse material that typically move in contact with the riverbed) and variable quantities of suspended load (i.e. fine particles that travel in suspension in the water column). As a consequence, dams, especially large ones in which water resides for long time, release *clear-water* to downstream reaches, which may induce geomorphic effects such as armouring [2], incision or channel widening [7]. One of the most acknowledged examples of the effects of sediment trapping at delta environments is the River Nile, downstream Aswan Dam. In the lowermost reaches of the Nile, the annual suspended sediment load was reduced from 100×10^6 t/y to almost zero [14]. This lack of nourishment has produced rates of erosion in the Nile Delta of 150 m/y. The changes in flow and sediment transport regimes, and morpho-sedimentary characteristics of regulated rivers are also often accompanied by alterations to river's ecosystem [5]. For example, the vegetation cover in and along channels downstream from the dams generally increases [2, 15], a fact that in turn changes channel hydraulics, modifying energy dissipation and, consequently, altering sediment transport processes and associated dynamics. In addition, the modified regime favours the spread of species with life history characteristics atypical of the pre-dam environment, including non-native species, resulting in altered species composition and vegetation dynamics [16].

Sedimentation in reservoirs is usually cited as a relevant cause for sediment disequilibrium in river systems. As indicated before, more than 180 reservoirs were constructed during the twentieth century in the Ebro basin, impounding water but also trapping an important part of the sediment transferred from upstream reaches. Using the same average sedimentation rate than the one estimated for global Spain [17], and taking into account that reservoirs in the Ebro are on average 50 years old, mean annual reduction of reservoir capacity in this basin is estimated at around 0.2%/y. This value would give a total annual sedimentation of ca. 15 hm^3 [17].

The same annual sedimentation rate can be obtained from data reported by [18] for 17 reservoirs representing 50% of the total regulation capacity of the Ebro basin. Sediment retained in reservoirs along the Ebro basin is composed mainly by silt (62%), followed by clay (25%) and sand (13%) [19]. Some reservoirs are already full of sediment (e.g. Pignatelli in the Ebro mainstem, constructed in 1790 with an original capacity of 1 hm^3, and Escuriza in the Martín River constructed in 1890, with an original capacity of 6 hm^3) (see Fig. 1). In others, sedimentation has been recognised as a socio-economic and environmental problems because threatens water quality and the use of infrastructures associated to them (e.g. Terradets on the Noguera Pallaressa, constructed in 1953, with an actual capacity of 8 hm^3 from an original 23 hm^3, and Barasona in the Ésera constructed in 1931 with an original capacity of 71 hm^3, and then regrowth in 1970) (see Fig. 1).

Besides dams, afforestation is considered to play an important role in the reduction of the delivery of sediment from headwaters to lowlands. In the case of the Ebro, land abandonment of mountainous areas (i.e. Pyrenees, Fig. 1), mostly during the second half of the twentieth century has led to re-vegetation of upland areas, reducing runoff and soil erosion; hence, sediment contribution to the drainage network, which ultimately would reach the coastline and the delta. Gallart and Llorens [20] reported a decrease of the water yield in the Ebro headwaters caused by the land use change, mainly related to an increase of the forest cover. Effects of land use changes in the sediment production zone propagate to the lowermost reaches of the basin that is the sedimentation zone. Assuming a long-term direct linear relation between runoff and sediment delivery, the reduction of sediment transfer due to changes in land use may account for the 30% of the current sediment deficit. As indicated previously, magnitude and frequency of floods have also been reduced [11], in turn reducing the river's transport competence and capacity and the role of riverchannel as sediment contributor (i.e. source). Many reaches that were anciently active gravel-bed channels have been massively colonised by vegetation and occupied by agricultural fields, reducing its potential to be scoured and to supply sediment to the downstream reaches. Together with the hydrological analysis, this chapter will also summarise the existing data on sediment transport in the Ebro basin, with special emphasis on the sediment dynamics in the lower most reach of the catchment where the largest dam complex is located (Fig. 1) and where the most complete dataset on total sediment transport exists.

3 Data Collection and Processes

3.1 Hydrology

Flow records in the Ebro basin are available on the Ebro Water Authorities web page (i.e. *Confederación Hidrográfica del Ebro*-CHE). The CHE is a government agency established in 1926 to manage the water resources in the Ebro River basin (http://www.oph.chebro.es). Batalla et al. [11] explored the main trends of the hydrology of the catchment as the bases to identify gauges that would reflect hydrologic effects of reservoirs by virtue of their locations and years of operation. Data sets were screened and used only series with enough data to represent pre and post-dam river hydrology. They introduced the Impoundment Ratio index (IR) as an indicator of the degree to which reservoirs could potentially change flows. The IR expresses the ratio between reservoir's capacity and mean annual runoff as a dimensionless decimal fraction. The available values were for total reservoir (i.e. including dead storage). Although calculating the IR with values of total reservoir's storage could lead to overestimating the potential hydrologic effect of the dams, these were the only data available at the scale of the entire basin. Authors then examined relations between IR and the selected hydrologic variables, with special emphasis on the effects of dams on flood magnitude and frequency. The reason of this was the interest on the intimate relation between floods, sediment transport and associated fluvial processes.

3.2 Sediment Transport

Data on sediment transport in the Ebro basin is very scarce. Besides some historical studies in the lowermost reaches of the catchment (e.g. [21]), almost at the delta plain, and few others undertaken during the 1990s (see Table 1), only the work undertaken by the Fluvial Dynamics group of the University of Lleida (hereafter UdL) between 2002 and 2008 in the lower part of the river supply reliable data on the total sediment load and changes in sediment transport and channel morpho-sedimentary characteristics upstream and downstream from large dams.

3.2.1 Sediment Yield in the Ebro Basin

Historical sediment yield in the whole catchment can be roughly estimated from bathymetrical records undertaken during the twentieth century in different reservoirs distributed all along the entire Ebro basin [22, 23]. Specifically, the sediment yield presented in this chapter has been estimated from rates of reservoir sedimentation in eighteen large dams (i.e. height >15 m) [23]. Their capacity represents 48% of the total reservoir capacity in the basin, so, owing to their location, characteristics and size, we consider that they represent well the average Ebro basin hydro-climatic conditions. Unpublished reservoir sedimentation data from bathymetric surveys (ranging from 13 to 74 years apart) have been supplied by the CEDEX (*Centro de Estudios y Experimentación de Obras Públicas*, Spanish Ministry of Public Works) and, to our knowledge, is the only available information. A majority of sedimentation records start in the 1930s when the first important phase of dam construction occurred in Spain; a second group of data start in the 1950s and 1960s and represent reservoirs that were built up during the big

Table 1 Summary of sediment yield estimations in the lower River Ebro (modified from [55])

Year	Sediment yield (10^3 t/y)	Reference
1877	30,000[a]	[48][k]
1900	15,000[a, b]	[46, 47]
1900	1,000–1,500[c]	[56]
1944	22,000[a]	[40]
1950–1975	400[d]	[56]
1961–1963	2,200[a]	[57][k]
1964	8,700[a]	[58][k]
1975–2000	170[d]	[56]
1976–1982	320[a]	[58][k]
1976–1990	260[e]	[18]
1983–1986	150[a]	[59][k]
1986–1987	130[a]	[60][k]
1988–1990	120[a]	[21]
1998–1999	500[f]	[61]
1998–1999	30[g]	[61]
2002–2004	455[h]	[25]
2002–2004	1,650[i]	[25]
2005–2008	216[j]	l

[a]Total load downstream Riba-roja Dam (Fig. 1)
[b]Contribution of the Ebro to its delta at the beginning of twentieth century
[c]Bedload, considering the 10% of the total load estimated by Bayerri [46] and Nelson [47]
[d]Bedload transport capacity in Tortosa (Fig. 1)
[e]Suspended load downstream Flix Dam (Fig. 1)
[f]Suspended load upstream from Sástago (Fig. 1, SMS)
[g]Suspended load below Mequinenza Dam (Fig. 1)
[h]Mean annual load in Móra d'Ebre (Fig. 1, MEMS) during 2002–2004. The 60 and 40% of this was transported as suspended and bedload, respectively
[i]Mean annual load in Sástago (Fig. 1, SMS) during 2002–2004. Almost all of this was transported as suspended load
[j]Mean annual suspended load in Móra d'Ebre (Fig. 1, MEMS) during 2005–2008
[k]Source: [40]
[l]Data presented in this chapter. Unpublished data

expansion of dam construction in the country. Many records reach the 1990s. Annual sedimentation (hm^3/y) has been converted to sediment yield, using the average sediment density of 1.1 t/m^3, estimated by Sanz-Montero et al. [18] from sediment stored in reservoirs in the basin using the Miller approach [24].

3.2.2 Sediment Load in the Lower Ebro

The work that the UdL research group carries out in the lower Ebro river is based on a sediment sampling programme designed to determine the sediment transport regime upstream and downstream from the large Mequinenza, Riba-roja and Flix reservoir complex. The sampling programme included monitoring of suspended and bedload transport at the Sástago Monitoring Section (hereafter SMS, channel width 110 m, between 2002 and 2004) located upstream from the dams; and at the

Móra d'Ebre Monitoring Section (hereafter MEMS, channel width 160 m, between 2002 and 2008) located downstream from the dams (Fig. 1). Special attention was devoted to the measurement of sediment transport during floods although routine samplings have also been performed during low flow conditions. The sampling programme was established to obtain reliable data on sediment transport to assess the magnitude of the changes to sediment supply in the river, to discern the relative contribution of baseflows, naturals floods and flushing flows (i.e. dam generated and controlled floods), to model the transit of sediment to the downstream reaches and, finally, to inform ongoing restoration projects in the river and in its delta plain. A complete description of sediment sampling techniques is given in [25]. Worth mentioning here is the fact that a flushing flows programme is being developed in the lower Ebro to control the excess of macrophytes since 2002 [26, 27]. Flushing flows hydrographs are designed based on riverbed material entrainment calculations. Hydrographs are generated and controlled combining releases from Mequinenza, Riba-roja and Flix dams. Results from this programme are satisfactory in terms of macrophytes removal in areas close to the Flix Dam. The artificial releases constitute so far the unique restoration practice that is systematically carried out in the lower Ebro, constituting a first, but pioneering step, towards the renaturalization of the river's flow regime.

Calculation of suspended sediment load at the two monitoring sections has been based on the analysis of depth integrated water and sediment samples. Samples are regularly obtained during floods, in which most sediment load is transported, and sparsely collected during low flows (i.e. below mean discharge). The whole range of discharges for the study period has been sampled. Samples are obtained at a single vertical section by means of cable-suspended depth-integrating samplers (i.e. 28 kg US DH74 and 12 kg US DH59, Fig. 2a, b, respectively). The sampling vertical was located at the centre of both monitored sections, the same vertical where sporadic hydraulic measurements are done. Around 1 litre of water was collected in every sample. Samples were carried to the laboratory and filtered using 1.2-μm cellulose filters. Annual suspended sediment yield was calculated from load-rating relations between discharge (Q in m^3/s) and suspended sediment concentration (SSC in mg/l) using the Flow Duration Curve method [28]. Annual statistically significant relations were used to estimate suspended sediment concentrations, and thus loads, during periods or discharges for which measurements were unavailable. Since sediment-rating curves are based on instantaneous measurements of discharge, they possibly underestimate the sediment loads [28], especially for the very high flows. In the case of the lower Ebro River, unsampled discharges are equalled or exceeded less than 0.5%; thus, the suspended load was probably not underestimated (for more details on precision of measurements and calculations see [25]).

Bedload was sampled during competent flows at the same vertical than suspended sediment. Bedload analysis has been based upon 215 samples, 145 during 2002–2003 and 70 during 2003–2004. At SMS we used a 29-kg cable-suspended Helley–Smith sampler with a 76-mm intake and an expansion ratio (i.e. ratio of nozzle exit area to entrance area) of 3.22 (Fig. 2c). Bedload was measured at

Fig. 2 Cable suspended sediment samplers in operation at MEMS (Fig. 1), lower Ebro, during floods: (**a**) US DH-74 depth-integrating suspended sediment sampler; (**b**) US DH-59 depth-integrating suspended sediment sampler; (**c**) 152 mm intake Helley–Smith sampler; (**d**) 76 mm intake Helley–Smith sampler. See text and [25] and [31] for operational details

MEMS by means of a 76-kg cable suspended Helley–Smith sampler with a 152-mm intake and an expansion ratio of 3.22 (Fig. 2d). The bedload samplers were operated from a bridge using a manual crane and an automatic crane at SMS and MEMS, respectively. In order to keep sampling efficiency as high as possible, sampling time did not exceed 5 min; thus preventing the sampler bag being filled to more than 50% of its capacity [29, 30]. Samples were collected and taken to the laboratory, where they were dried, sieved and weighed to obtain the total mass and the grain size distribution. Vericat et al. [31] analysed the efficiency of the Helley–Smith samplers in the Ebro. They concluded that its efficiency will be controlled by the size of the sampler's intake in relation to the largest particles potentially apt to be entrained during competent flows. Samplers with intakes smaller than five times the largest particles may underestimate the instantaneous bedload rates and grain size distribution, subsequently, may interfere sediment yield and texture estimates based on that data. Despite being a fundamental element to characterise riverbed dynamics in this and other catchments, we are not aware that other systematic further measurements on bedload exist in the Ebro basin besides those reported by Vericat and Batalla [25] and subsequent publications (e.g. [32, 33]).

Changes in river's load modify the channel sedimentary characteristics and dynamics. In the case of the lower Ebro, high magnitude floods that pass the dams have typically greater transport capacity than the amount of sediment supplied and keep sufficient competence to entrain and move most sediment sizes from the riverbed; the main effect is hence channel incision. During smaller floods the bed shear stress is less than the critical stress needed to entrain the largest particles of the bed surface but sufficient to move the finer material; under such conditions, the surface becomes coarser and, consequently, an armour layer develops. The degree of armouring of a river bed influences bed load transport rates and grain size distributions for instance, reducing the amount and sizes of sediment transported [34]. Vericat et al. [35] analysed riverbed sediment dynamics in the lower Ebro by means of systematic sampling of surface and subsurface bed materials in exposed morphological equivalent sediment areas in a 28-km reach between the Flix Dam and MEMS. The surface layer was characterised using the pebble count method [36, 37]. Around 3,800 particles were measured. The surface layer was also sampled using the area-by-weight method [38]. This method allows the determination of the percentage of the fine material (i.e. particles finer than 8 mm) that is underestimated by the pebble count method. The subsurface material was sampled using the volumetric method. The largest particle in the subsurface layer did not exceed 1% of the sample weight (i.e. sampling efficiency). Almost 1,400 kg of material were sampled. Area-by-weight and volumetric samples were sieved at 1/2 phi-intervals and weighed in the field. Samples containing wet material were taken to the laboratory and dried prior to sieving and weighing to obtain the grain size distribution.

4 Hydrology

4.1 Runoff Characteristics

Most runoff is generated in the northeast part of the basin, where the main tributaries draining the Pyrenean region are located (Fig. 1). The rivers Aragon, Gállego, and especially Cinca and Segre generate almost of the mean annual water yield in the catchment (i.e. 10,700 hm^3); the rest of the runoff comes from small basins located in the Cantabrian region at the north-western catchment headwaters

(23%) and rivers draining the Iberian Massif (5%) (Fig. 1). Distribution of the water yield follows a clear rainfall gradient from the Pyrenean to the Iberian Ranges, by virtue of the concentration of precipitation in the northern part of the basin (Fig. 1). According to the Ebro Basin Water Management Plan, rivers draining the Cantabrian region and the Western Pyrenees show a pluvio-oceanic regime, which is characterised by a regular flow along the year with maximum values in winter and minimum during summer. The Central and the Eastern Pyrenees displays a nivo-pluvial regime owing to the snow cover during winter and the snowmelt in spring. It is characterised by a main flow peak in spring (May–June) and a smaller peak in autumn, with two baseflow periods in winter and summer. In the southern part of the basin the oceanic and high mountain influences disappear and flow regime show dryland characteristics. The regime in this region is defined as pluvial-Mediterranean and it is characterised by a marked unstable flow with a maximum in autumn (http://www.chebro.es).

4.2 Changes in Runoff

Batalla et al. [11] analysed 23 flow series of regulated rivers and four in non-impounded ones distributed along the Ebro catchment. Results indicated that annual runoff significantly decreased after dams in eleven rivers and increased in two of them. In contrast, in ten rivers annual runoff did not significantly change after dams were closed. All regulated rivers in the dry southern half of the basin showed significant reduction (i.e. reduction between 5 and 70%), while in the northern and Atlantic zones only 40% of rivers showed significant reduction. Although there is a certain degree of variability between periods and regions, there is no statistical evidence that rainfall decreased during the twentieth century in any of the regions of the Ebro basin [39]; in general, annual oscillations are smaller than 1% and they are compensated between decades and regions. Observed trends in runoff increase and decrease may thus respond to particular subcatchment characteristics (i.e. temporal distribution of rainfall), dam operation (i.e. evaporation is likely to happen more intensively in reservoirs located dry areas), and water uses (i.e. agriculture is generally pointed out as the main source of water losses).

Integrating the entire Ebro basin, the flow series from Tortosa gauging station (A27- CHE, where Ai represents the code of the gauging station operated by CHE) display a marked reduction in annual runoff along the twentieth century (Fig. 3a). Mean annual runoff in the period 1970–2008 (taken here as the post-dam period, i.e., in 1969 the dam complex Mequineza-Riba-roja was closed and already in operation) was 40% lower than during the pre-dam period. Worth to remember at this stage that reservoir capacity for the whole basin in 1960 was 2,000 hm^3 with IR = 0.15 and over 6,000 hm^3 by 1975 with IR nearly 0.50 (i.e. therefore, in 1975 the dams in operation in the whole basin had a potential water storage capacity equivalent to half of the annual runoff at the outlet of the basin). In turn, the main tributary of the Ebro, the River Segre at the lowermost station of Serós (A25-CHE), also displays a

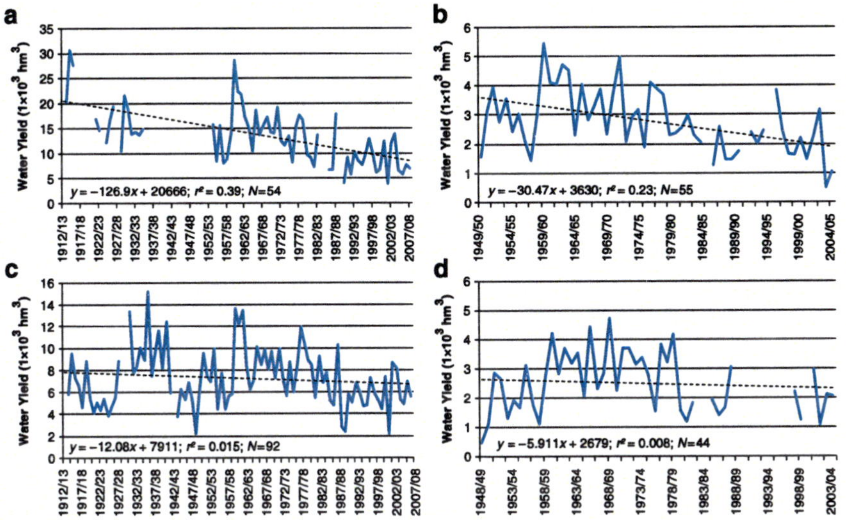

Fig. 3 Long-term annual runoff in key hydrological locations along the basin (**a**) Ebro in Tortosa, (**b**) Segre in Serós, (**c**) Ebro in Zaragoza and (**d**) Cinca in Fraga. See Fig. 1 for geographical references

remarkable reduction in the mean annual runoff since the 1950s (−25%) (Fig. 3b). It is important to mention that the 1960s decade was especially wet, concentrating the largest rainfall amounts registered along the century (see Fig. 3a, b); however, the runoff reduction in the Segre and in the Ebro in Tortosa is appreciable at the long-term perspective (century). In contrast, the Ebro main stem in Zaragoza (A11-CHE, Fig. 3c) and the River Cinca in Fraga (A17-CHE, Fig. 3d) show a much less pronounced reduction in runoff that is even not statistically significant. It appears thus that the reduction of runoff in the lower reaches of the Ebro (downstream the Riba-roja Reservoir) may be essentially related to the reduction in the Segre; the overall reduction may be complemented by a less notable reduction in the Ebro mainstem upstream form Mequinenza -in Zaragoza- (−12%) and in the second most important tributary, the River Cinca (−10%). Overall, water in most reservoirs is effectively used on an annual cycle, as the IR index lower than one indicates [11]. Therefore, water store over several years cannot be directly pointed out as long-term reason to explain the runoff reduction downstream some reservoirs. In contrast, according to [40] 22% of the decrease of mean annual runoff can be attributable to reservoir evaporation and 75% to irrigation losses. Collier et al. [41] found similar results in rivers of the Western US. Reduction in water yield is remarkable and corroborates results reported by Gallart and Llorens [20] for some Pyrenean tributaries. They attributed runoff reduction to afforestation that has massively occurred in the Pyrenees during the second half of the century.

Batalla et al. [11] assessed the changes produced by reservoirs on monthly flows by means of a correlation coefficient calculated for each flow series ($\Phi_{pre;post}$, where

$1 \geq \Phi \geq -1$); the coefficient is obtained by dividing the covariance of the pre- and post-dam average values of the particular data set by the product of their standard deviations, thus providing a variable to indicate the degree of impact of the impoundment on runoff (i.e. negative values indicate an inversion of the flow regime). The index is as follows: $\Phi_{\text{pre;post}} = \text{cov}(x,y)/(\sigma_x \times \sigma_y)$, where x and y represent the pre and post-dam average values for the analysed monthly data set. The correlation coefficient index varied greatly. For instance, there was virtually no change on monthly flows in the upper Noguera Pallaresa, with a correlation coefficient $\Phi_{\text{pre;post}}$ of 0.99. In contrast, the seasonal flow regime was inverted in the river below the Ebro Dam ($\Phi_{\text{pre;post}} = -0.84$) and the Piedra River ($\Phi_{\text{pre;post}} = -0.42$). Figure 4 shows some examples of monthly changes due to dam operation in the Ebro basin and a summary of the results. Overall, $\Phi_{\text{pre;post}}$ averaged 0.65 for the 34 analysed data series, suggesting a moderate impact of dams on the monthly regime of the basin. Rivers monthly regimes are most affected in the Mediterranean region of the catchment (mean $\Phi_{\text{pre;post}} = 0.19$) in comparison with the northern (Pyrenean) and the Atlantic counterparts (mean $\Phi_{\text{pre;post}} = 0.82$), with the exception of the Noguera Ribargorçana, one of the most regulated rivers in Europe, whose monthly regime has disappeared ($\Phi_{\text{pre;post}} = 0.07$, i.e. flow is steady most of the time). Similar results were found for flood magnitude (see results of the flood analysis in the text below, [11]). A given percentage of regulation produces greater change in the drier part of the basin than in the more humid zones. The reason for this is not entirely clear. However, it may be hypothesised that reservoirs in drier parts are rarely full, facilitating the retention of an important part of the incoming water when floods arrive; in contrast, in the more humid parts, reservoirs are typically more full, thus, having less capacity to store water when floods occur. Consequently, downstream environmental effects of dams in Mediterranean-climate rivers are likely to be more pronounced.

Most flow records showed a pronounced reduction in flood frequency and magnitude as a consequence of dam operation. The ratio between post-dam and pre-dam flood values ($\delta_{-\text{i}} = Q_{\text{i}-\text{post}}/Q_{\text{i}-\text{pre}}$, where Q_{i} indicates the flood magnitude for a given recurrence i of the flow series pre and post dams) averaged 0.65 (i.e. 35% decrease) for the Q_2 (2-year return period flood); 0.67 for the Q_{10}; and 0.59 for the Q_{25}, but the reduction varied greatly. For instance, small floods (Q_2 and Q_{10}) downstream of the Ebro dam (Ebro River), Canelles and Santa Ana dams (Noguera Ribagorçana), Cueva Foradada Dam (Martín), Santolea Dam (Guadalope) and Moteagudo Dam (Najima) were the most affected, with average δ values of 0.30 for Q_2 and 0.40 for Q_{10} (i.e. decreases of 70 and 60%, respectively). Williams and Wolman [2] for rivers of the Western United States and Kondolf and Matthews [42] for the Sacramento-San Joaquin River system in California reported similar results. The River Guatizalema showed an extreme reduction ($\delta = 0.13$ for Q_2 and 0.22 for Q_{10}), but is an unusual case because of the multiple diversions in the reach. Few of the rivers showed little change in flood magnitude (for details see [11]). In general, flood frequency and magnitude decreased with increasing IR. Trends are evident despite considerable scatter (Fig. 5). The best-fit relations ($p < 0.05$) between δ and IR for the studied return intervals have similar slopes, suggesting a similar degree of

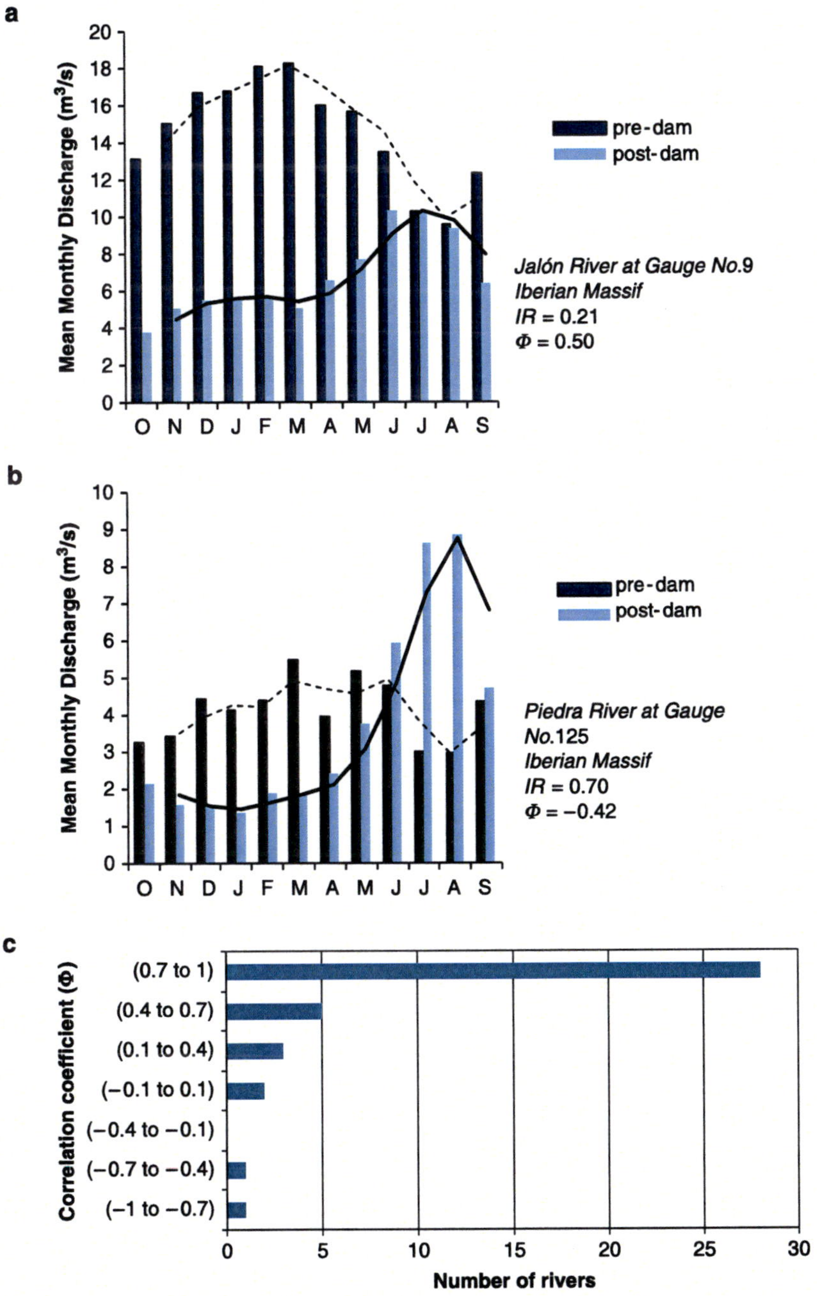

Fig. 4 Changes in the monthly pattern of two selected catchments in the Ebro basin: (**a**) the Jalon River showing moderate impact after dam operation, and (**b**) the Piedra River in which natural flow distribution has completely changed after dam closure (lines indicate the running means model of the respective data sets). (**c**) summary of analysed rivers (see [11] for more examples)

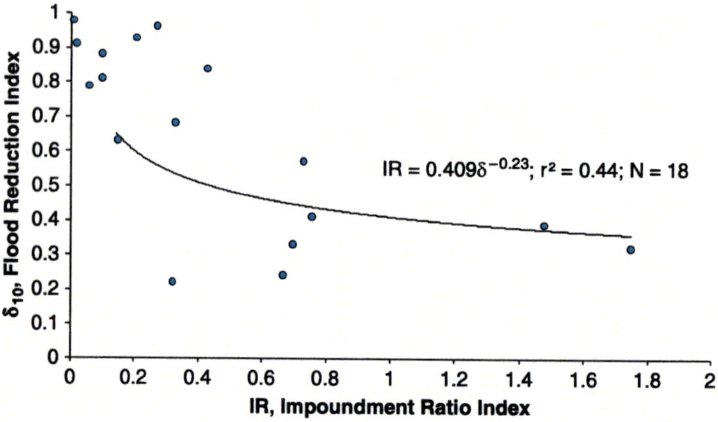

Fig. 5 Flood reduction in the Ebro basin in relation to the degree of impoundment (IR). Flood reduction is expressed by δ as the ratio between the post and the pre-dam flood of a given return period Q_i; for this particular case the 10-year flood has been selected. See [11] for more examples

alteration to each of the studied recurrence intervals. Similar to the results reported for the monthly pattern, the change in flood magnitude produced by a given level of impoundment varied regionally. Reduction in floods was again least for the wettest parts of the Ebro Basin (West-Central Pyrennean and Atlantic zones) and greatest in the dry southern Mediterranean area.

5 Sediment Transport

5.1 Sediment Yield

Available reservoir sedimentation data has been used to estimate the (minimum) sediment yield in the subcatchments upstream from them. Bathymetric data indicate that the mean reservoir capacity loss is 7.7% in 43 years (i.e. average time from reservoir construction to the last bathymetrical survey), equating to mean loss of 0.3% per year. Considering that these reservoirs have an impoundment capacity of 48% of the total, the total sedimentation in the Ebro basin occurring during the twentieth century would account for a total of 670×10^6 tonnes. Considering than the majority of the sediment trapped in the reservoirs is transported in suspension, sediment yield averages 400 t/km^2/y. This plots in the upper range of those reported for basins in the Mediterranean region (e.g. [43–45]).

Reservoir sedimentation data allow estimating sediment loads of regulated catchments, although it is important to note that this sedimentation represents a fraction of the true yield since reservoir sediment trapping efficiency is usually lower than 100%. In the case of the Ebro basin, a reduction of sediment supply from the catchments might be also expected as a result of the afforestation process that took place following land abandonment in the mountains during the second half of the twentieth century. We estimate a contemporary natural sediment load (i.e. 1900s) for the whole basin of 19 × 10^6 t/y. Our estimate comes from the sum of the annual sediment yields in the basin subcatchments, and the sediment deposited in the Mequinenza and Riba-roja reservoirs, that includes the sediment generated in small rivers draining the Central Depression and the sediment that would not be trapped in the upstream reservoirs. The value is in the range of the value reported by Bayerri [46] at the time that dam construction began. Similar estimations were presented by Nelson [47]. However, our result is substantially lower than the one reported by Gorría [48] (i.e. 30 × 10^6 t/y) when the conditions in the basin were different, and closer to the historical sediment load (i.e. the mountain ranges were highly populated and agriculture was the predominant land use all over the basin). The natural load in the Ebro is in the range of the annual estimates in basins such as the Rhone and Po, with similar drainage area but with three times more annual runoff [49].

5.2 *Sediment Load in the Lower Ebro*

5.2.1 Suspended Load

As we have indicated in the previous sections, there is little information on sediment transport in the River Ebro, and most of it is concentrated in the lower reaches of the catchment, downstream the reservoir complex of Mequinenza-Riba-roja-Flix (Fig. 1). We present here a summary of the results obtained from direct measurements on suspended and bedload obtained by the UdL group during the period 2002–2004 upstream (i.e. SMS) and downstream (i.e. MEMS) the reservoir complex, together with the data collected downstream from the dams during the period 2005–2008 (in the later case referred only to suspended sediment). A hydrological context is provided for each of the sediment transport data sets.

The first sampling period 2002–2004 show contrasting hydrological characteristics, although both years are representative of the flow regime of the post-dam period (>1970). The hydrological year 2002–2003 experienced high-magnitude low frequency floods (i.e. Q_{10}, 2,500 m^3/s); meanwhile, the year 2003–2004 floods were smaller but more frequent (Q_{annual}, 1,000 m^3/s) (see [25] for more details). Suspended sediment concentrations at the sampling sites SMS and MEMS (Fig. 1) showed a positive and statistically significant relation with discharge (Q) ($p<0.01$) for the two study years, 2002–2003 and 2003–2004 (Fig. 6a, b). Although both load rating curves present certain degree of scatter, the positive relation of these indicates that the increment of the suspended load is related to the increase of discharge during floods (i.e. hydraulically dependent). The mean suspended concentration at SMS (estimated as the mean of the measured concentrations) was 530 mg/l for a

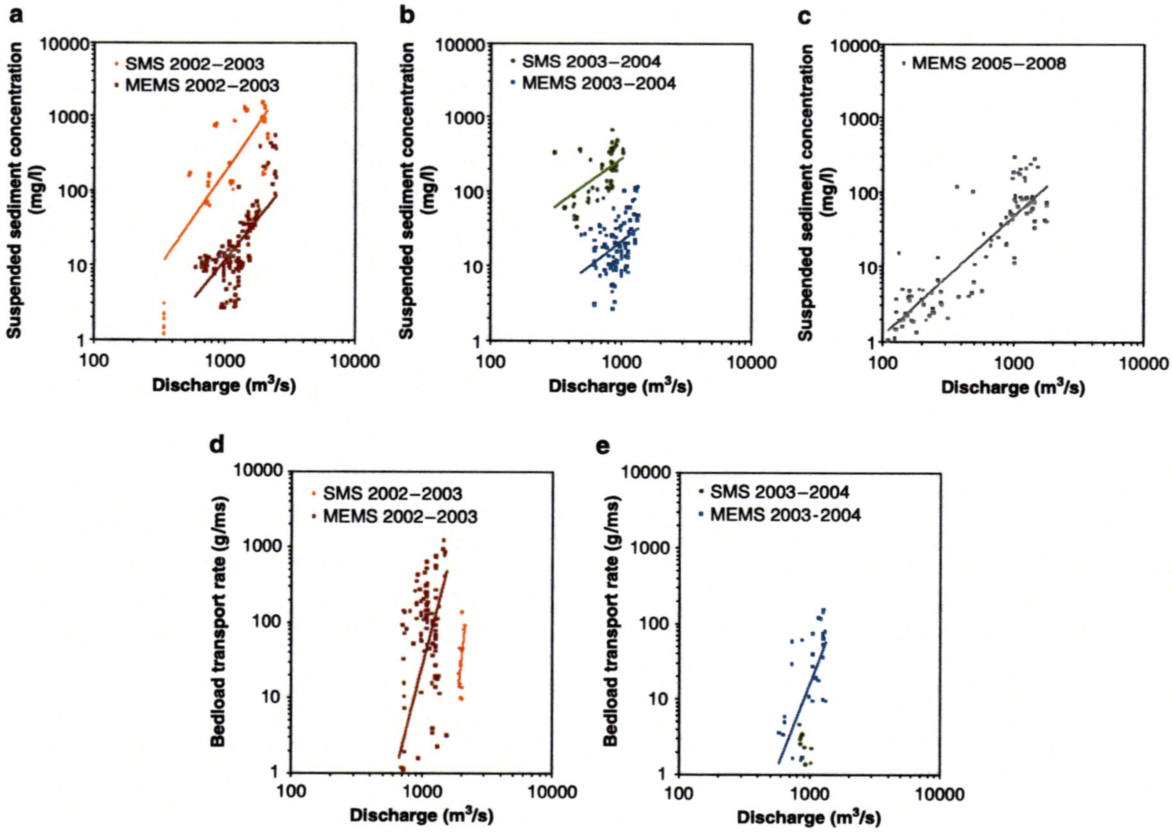

Fig. 6 Suspended sediment transport relations obtained at SMS (upstream from Mequinenza Reservoir) and MEMS (downstream from the Flix Dam) during the period (**a**) 2002–2003 and (**b**) 2003–2004. For location details see Fig. 1. (**c**) Suspended sediment transport relation obtained at MEMS (downstream from the Flix Dam) during the period 2005–2008. For location details see Fig. 1. Bedload transport relations obtained at SMS and MEMS during the period (**d**) 2002–2003 and (**e**) 2003–2004. Note that statically significant models are presented for all the relations (excepted at SMS in (**e**), see text for details) as a reference and to identify general trends

mean sampled discharge of 1,280 m^3/s in 2002–2003, and 215 mg/l for a mean sampled discharge of 725 m^3/s in 2003–2004. The maximum concentration was recorded during 2002–2003 (ca. 1.5 g/l) under discharges close to 2,000 m^3/s. The mean annual suspended sediment load passing SMS and entering the Mequinenza Reservoir was calculated at around 2.3×10^6 t ($\pm0.32 \times 10^6$ t) for the year 2002–2003 and 0.97×10^6 t ($\pm0.07 \times 10^6$ t) for the year 2003–2004, which represent specific sediment yields of approximately 47 t/km^2/y and 19 t/km^2/y. These values are lower than others reported for smaller basins in the Mediterranean region [43, 50–53]. In the case of MEMS, the mean suspended concentration (estimated as the mean of the measured concentrations) was 38 mg/l for a mean sampled discharge of 1,284 m^3/s in 2002–2003, and 27 mg/l for a mean sampled discharge of 910 m^3/s in 2003–2004. The maximum concentrations were recorded during 2002–2003 (0.55 g/l) under discharges close to 2,440 m^3/s. As in the case of SMS, the degree of scatter is very high, indicating notable temporal variability. This again can be related mainly to seasonal effects [54]. Mean and maximum concentrations are one order of magnitude lower than those obtained at SMS under similar discharges (Fig. 6a, b). This fact reflects the influence of the reservoirs in trapping fine sediment that otherwise would circulate downstream. The mean annual suspended sediment load passing MEMS was calculated at around 0.26×10^6 t ($\pm0.06 \times 10^6$ t) for the year 2002–2003 and 0.29×10^6 t ($\pm0.02 \times 10^6$ t) for the year 2003–2004, which represents a specific sediment yield of approximately 3 t/km^2/y and 3.4 t/km^2/y, respectively. These values are much lower than the ones obtained at SMS, pointing out the important role of the 120-km chain of reservoirs in trapping the river's solid load in its lowermost reaches. One of the consequences of sediment trapping is that the river channel downstream from dams acts as the main sediment source during high flows (see bedload section for discussion). Overall, during the period 2002–2004 a total of 3.27×10^6 t circulated through the Sástago measuring site (SMS), while in Móra d'Ebre (MEMS) the total suspended load was 0.55×10^6 t.

Field data obtained at MEMS for the period 2005–2008 corroborates the positive relation between discharge and suspended sediment load. The relation is statistically significant, but the scatter remains high, with variations that may reach one order of magnitude for the same range of discharges (Fig. 6c). Concentrations were generally low, with values oscillating between 1 mg/l during baseflows to 300 mg/l during floods. The mean concentration was 50 mg/l for a sampled discharge of 800 m^3/s. As previously indicated we have used the rating curve method [28] to estimate the annual load for the period. The total load for the study period (i.e. three hydrological years) was 0.65×10^6 t, (i.e. mean of 0.19×10^6 t/a, ranging from 0.065×10^6 t in the dry 2005–2006 year, to 0.29×10^6 t and 0.3×10^6 t attained during the relative wet 2006–2007 and 2007–2008, respectively). Results are of the same order, although slightly lower than those reported by Vericat and Batalla [25] for the period 2002–2004 estimated at the same control section; water yield during that period was 13,870 hm^3/y (very close to the mean for the post-dam), while it attained only 8,440 hm^3/y during the period 2005–2008.

Sediment load does not distribute uniformly through time. The transport of sediment is less constant and sporadic compared to the runoff, since floods typically

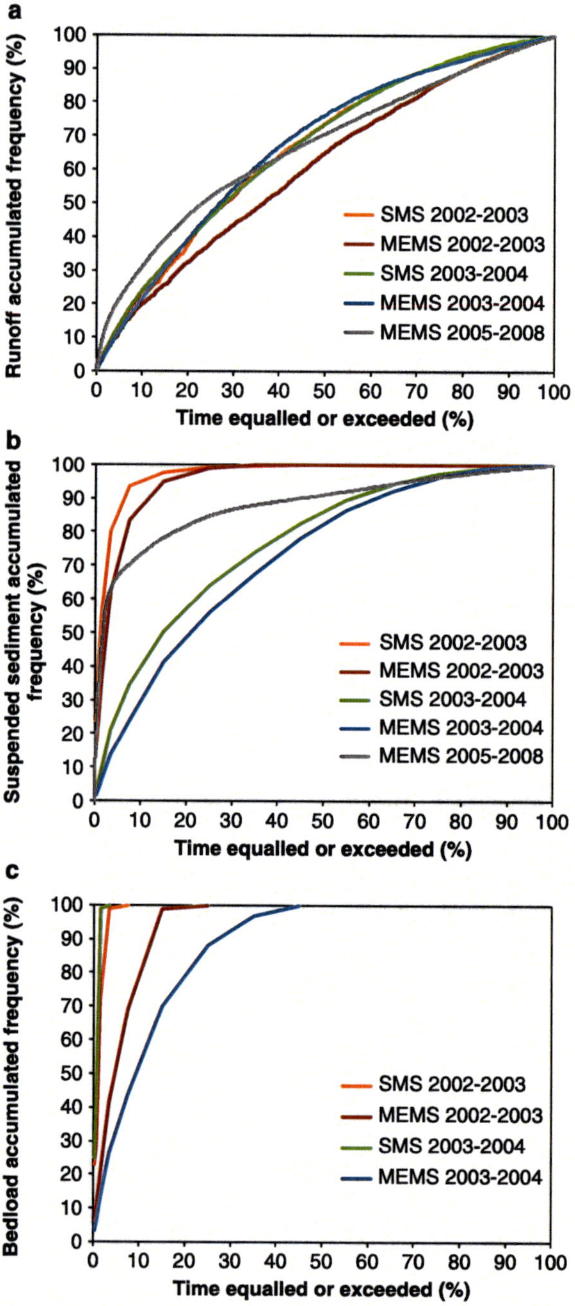

Fig. 7 Duration curves for (**a**) runoff, (**b**) suspended sediment load and (**c**) bedload in the lower Ebro River for the period 2002–2008

accounts for the transport of most of the particulate load. In the case of the Ebro at MEMS suspended load was transported in much less time during 2002–2003 (e.g. 10% of time carried around 99% of the load) than during 2003–2004 (10% of time carried 45% of the load) (Fig. 7). This fact is a reflection of the different hydrological behaviour between study years one and two: (1) during 2002–2003 several floods occurred and they were responsible for most of the load (e.g. 50% of the load was transported by discharges equalled or exceeded only 2% of the time); (2) during 2003–2004 sediment transport was more constant through time and relatively frequent discharges achieved most of the transport (e.g. 50% of the load was transported by discharges equalled or exceeded 20% of the time). During the period 2005–2008, 35% of the suspended load was transported in just 1% of the time ($Q > 1{,}500$ m^3/s); further, the 80% of the load was transported during the 10% of the time ($Q > 500$ m^3/s), while 20% of the time ($Q > 380$ m^3/s) transported the 90% of the particulate load [25].

5.2.2 Bedload and Riverbed Dynamics

Bedload rates at SMS and MEMS showed a positive and statistically significant relation with discharge for the study period 2002–2004 (Fig. 6d, e). The exception was SMS where competent discharge were rarely exceeded in the hydrological year 2003–2004. This fact limited the elaboration of a statistically significant rating curve for that year. Only samples fluctuating around entrainment were collected, although they now showing a clear transport pattern; hence, precluding any significant relation with discharge. The degree of scatter of the rating curves is, overall, very high, indicating expected fluctuations in gravel transport and possible bias related to operational effects caused by a misalignment of the sampler during raising and lowering. The mean bedload rate (estimated as the mean of the measured rates) at SMS was 41 g/sm for a mean sampled discharge of $Q = 2{,}012$ m^3/s in 2002–2003, and 2 g/sm for a mean sampled discharge of 894 m^3/s in 2003–2004, which represents a daily transport of 389 and 19 t/day. Median bedload size ranged from 21 to 40 mm in 2002–2003 and from 11 to 22 mm during 2003–2004, which represents D_{60} and D_{84} of the bed grain size distribution (in absence of a surface layer) for the sizes collected in the first year, and D_{40} and D_{60} for the sizes collected in the second year. The mean annual bedload carried downstream from SMS and entering the Mequinenza Reservoir was calculated at around 15,300 t (±4,440 t for the year 2002–2003) and 155 t for the year 2003–2004, which represents specific sediment yields of approximately 0.3 and 0.003 t/km^2/y, respectively. The statistically significant relations between discharge and bedload transport rate at MEMS show a high degree of scatter as in the case of SMS (Fig. 6d, e). This indicates the importance of bedload variability and possible operational effects (sampler uncertainty) (see [31]). Mean bedload rate (estimated as the mean of the measured rates) was 146 g/sm for a mean sampled discharge of 1,050 m^3/s in 2002–2003 and 68 g/sm for a mean sampled discharge of 1,020 m^3/s in 2003–2004. For a given discharge, bedload rates were, on average, double during the first study year what they were during the second. Maximum rates were recorded during 2002–2003 (1,200 g/sm) under discharges close to 1,500 m^3/s. Maximum rates are an order of magnitude higher than those obtained at SMS under higher discharges (135 g/sm under 2,025 m^3/s). This fact clearly reflects the different bedload transport in each section: (1) small bedload in a wandering low gradient channel upstream SMS and (2) higher bedload in a *hungry water* [7] river system below dams. The D_{50} in samples collected during 2002–2003 varied from 1 to 72 mm. During 2003–2004 the D_{50} varied from 4 to 44 mm. The upper limits (72 and 44 mm) correspond to D_{75} and D_{50} of the bed surface grain size distribution at MEMS, respectively. The mean annual bedload yield at the downstream section of MEMS was calculated at around 0.28×10^6 t ($\pm 0.30 \times 10^6$ t, reflecting the high variability for the year 2002–2003) and 0.08×10^6 t ($\pm 0.04 \times 10^6$ t) for the year 2003–2004, which represents a specific sediment yield of approximately 3.4 t/km^2/y, and 0.9 t/km^2/y, respectively. Overall, during the period 2002–2004 a total of 3.28×10^6 t circulated through the SMS measuring site, almost all of it in suspension; on the meantime, the transport in MEMS was 0.91×10^6 t, around half in suspension and half as bedload, reflecting again the important role of the riverchannel as the main sediment supplier in the absence of sediment coming from upstream.

Bedload shows a different pattern between the 2 years (Fig. 7), owing to different hydrology (i.e. number and magnitude of floods occurred during 2002–2003 were substantially higher than during 2003–2004). Bedload was transported over a greater duration in 2002–2003 (e.g. 3% of time carried 95% of the load) than in 2003–2004 (1% of time carried 95% of the load) at SMS. Critical discharge (865 m^3/s, estimated by means of the Shields, 1936 equation) was equalled or exceeded 7% of the time in 2002–2003 and only 1% of the time in 2003–2004 [25]. In the case of MEMS, bedload was transported in much less time during 2002–2003 (e.g. 10% of time carried around 75% of the load) than during 2003–2004 (10% of time carried 50% of the load).

Dams trap more than 90% of the suspended load and the totality of bedload in the Ebro. In the absence of sediment replacement from upstream, the riverbed acts as the unique sediment source during competent floods, a fact that causes incision and/or armouring depending on flood magnitude and grain-size of riverbed sediments [35]. The surface, subsurface, and bed load grain size distribution constitute the bases for the analysis of bed-armouring dynamics. The initial (summer 2002) mean armouring ratio ($D_{50-\text{surface}}/D_{50-\text{subsurface}}$) was 2.3 for the 28-km study reach downstream the Flix Dam, with maximum values reaching 4.4. During high magnitude floods in the winter of 2002–2003 (Q_8), the armour broke up and released fine subsurface sediments, increasing bedload transport and, consequently, creating generalised incision (e.g. 60 mm of riverbed degradation was observed in the reach between Flix and Mora d'Ebre after the succession of 2002–2003 flood events). Most grain size classes were entrained and transported increasing the sediment deficit in the reach. The mean armouring ratio decreased to 1.9. During the low magnitude floods that occurred during 2003–2004 (i.e. around Q_{annual} and Q_2), the coarsest fractions (64 mm) did not take part in the bedload while finer particles were winnowed, thus surface deposits coarsened and the coarse armour layer re-established, a fact that reduced bedload transport and minimised incision (i.e. the mean armouring ratio increased to 2.3 and incision was almost negligible all over the reach, especially in the river reach immediately below the dam, where the supply and transport of bed material appeared to be in balance). Further downstream the transport of finer classes was higher than their supply from upstream, a phenomenon that progressively reduced their availability in the riverbed surface, hence the armour layer re-established at the end of 2003–2004 winter floods [35].

6 Summary and Final Remarks

This chapter has summarised the main hydrological and sediment transport trends observed in the Ebro basin during the twentieth century. Most data concentrates in the lower reach of the catchment and in recent years and, therefore, this area has driven most of the attention of the work. The Ebro flow and sedimentary regimes are profoundly marked by the human action in the entire basin. Changes in land uses and especially dam construction have change the historical pattern of water and sediment yield, and the associated physical processes (i.e. riverchannel sedimentary structure and morphology).

Overall runoff and flood magnitude have been reduced; meanwhile reservoirs were continuously trapping most of the sediment circulating in the drainage network at the same time that catchment headwaters reduced their sediment supply due to extensive afforestation. Sediment deficit is evident in the lower reaches of the Ebro mainstem and in some of its main tributaries (e.g. rivers Segre, Noguera Ribargorçana and Gállego, between many others; in the case of the latest deficit worsens due to gravel mining). However, the river mainstem is still active from the sedimentary point of view in many of its reaches, so a new equilibrium has not reach yet. For instance, under present conditions the riverbed and channel of the lower Ebro River continues exporting sediment during competent floods, causing sustained degradation (i.e. net export of alluvial sediment, channel deepening and incision).

Information on sediment transport and riverbed characteristics and dynamics has been of use to design experimental flows (i.e. flushing flows [27]) to control the excess of macrophytes, and it is being used to model the sediment load at the reach scale, a practice that might be interesting to design experiments that combine artificial sediment feeding and flow releases that helps ameliorating the sediment disequilibrium in the river and between the river and its coastline.

Acknowledgements The bulk of this research has been carried out within the framework of a series the research projects funded by the Spanish Ministry of Education and Science 2002–2009 (REN2001-0840-C02-01/HID, CGL2005-06989-C02-02/HID, CGL2006-11679-C02-01/HID, CGL2009-09770 BTE) and by the research contract "RiskFlix: *Encomienda de gestión* to study the polluted sediments in the Flix Reservoir" 2006–2008, funded by the Spanish Ministry of Environment and the Catalan Water Agency. Special thanks are due to the Ebro Water Authorities – CHE and Endesa SA for the collaborative support during the course of the investigations in the Ebro.

References

1. Poff NL, Allan JD, Bain MB, Karr JR, Prestegaard KL, Richter BD, Sparks RE, Stromberg JC (1997) The natural flow regime: a paradigm for river conservation and restoration. Bioscience 47(11):769–784
2. Williams GP, Wolman MG (1984) Downstream effects of dams in alluvial rivers, US Geological Survey, Professional Paper, 1286
3. Petts GE (1984) Impounded rivers: perspectives for ecological management. Wiley, New York
4. Ward JV, Stanford JA (1979) The ecology of regulated streams. Plenum Press, New York
5. Ligon FK, Dietrich WE, Trush WJ (1995) Downstream ecological effects of dams, a geomorphic perspective. Bioscience 45(3):183–192
6. Ward JV, Stanford JA (1995) Ecological connectivity in alluvial river ecosystems and its disruption by flow regulation. Regulated Rivers: Res Manag 11:105–119
7. Kondolf GM (1997) Hungry water: effects of dams and gravel mining on river channels. Environ Manag 21(4):533–551
8. Novoa M (1984) Precipitaciones y avenidas extraordinarias en Catalunya. Ponencias y comunicaciones de las Jornadas de Trabajo sobre Inestabilidad de laderas en el Pirineo, 1:1–15, Barcelona
9. García MA, Moreno MC (2000) Los aprovechamientos en la Cuenca del Ebro: Afección en el regimen hidrológico fluvial, Internal Technical Report no.2000-PH-24.1, Confederación Hidrográfica del Ebro, Zaragoza, p 83
10. MIMAM, Ministerio de Medio Ambiente (2000) Libro Blanco del Agua en España, Chapter 3, La situación actual y los problemas existents y previsibles
11. Batalla RJ, Kondolf GM, Gomez CM (2004) Reservoir-induced hydrological changes in the Ebro River basin, NE Spain. J Hydrol 290:117–136
12. Richter BD, Baugartuer JV, Powell J, Braun DP (1995) A method for assessing hydrologic alteration within ecosystems. Conserv Biol 10(4):1163–1174
13. Ahearn DS, Sheibley RW, Dahlgren RA (2005) Effects of river regulation on water quality in the lower Mokelumne River, California. River Res Appl 21:651–670
14. Walling DE, Fang D (2003) Recent trends in the suspended sediment loads of the world's rivers. Glob Planet Change 39:111–126
15. Batalla RJ, Vericat D, Martínez TI (2006) River-channel changes downstream from dams in the lower Ebro River. Z Geomorphol Suppl B 143:1–14
16. Cowell CM, Dyer JM (2002) Vegetation development in a modified riparian environment: human imprints on an Allegheny River Wilderness. Ann Assoc Am Geogr 92:189–202
17. Batalla RJ (2003) Sediment deficit in rivers caused by dams and instream gravel mining. A review with examples from NE Spain. Cuaternario y Geomorfología 17(3–4):79–91
18. Sanz-Montero ME, Avendaño-Salas C, Cobo-Ryan R (1999) Influencia de los embalses en el transporte de sedimentos hasta el río Ebro (España). Proceedings of the congress on hydrological and geochemical processes in large-scale river basins. HIBAM, Shahin
19. Sanz-Montero ME, Cobo-Rayán R, Gómez-Montaña JL, Avendaño-Salas C (1998) Composición de los sedimentos acumulados en embalses españoles. Ingeniería del Agua 5(4):21–28
20. Gallart F, Llorens P (2004) Observations on land cover changes and water resources in the headwaters of the Ebro catchment, Iberian Peninsula. Phys Chem Earth 29:769–773

21. Guillén J, Palanques A (1992) Sediment dynamics and hydrodynamics in the lower course of a river highly regulated by dams: the Ebro River. Sedimentology 39:567–579
22. Avendaño C, Cobo R, Sanz ME, Gómez JL, (1997) Capacity situation in Spanish reservoirs. I.C.O.L.D. Proceedings of the nineteenth congress on large dams 74(52):849–862
23. Batalla RJ, Vericat D (2010) An appraisal of the contemporary sediment yield in the Ebro basin. J Soils Sediments (in revision)
24. Miller CR (1953) Determination of the unit weight of sediment for use in sediment volume computations. US Bureau of Reclamation, Dept. of Interior, Denver, CO
25. Vericat D, Batalla RJ (2006) Sediment transport in a large impounded river: the lower Ebro, NE Iberian Peninsula. Geomorphology 79:72–92
26. Palau A, Batalla RJ, Rosico E, Meseguer A, Vericat D (2004) Management of water level and design of flushing floods for environmental river maintenance downstream of the Riba-Roja Reservoir (Lower Ebro River, NE Spain). Proceedings of the International Conference HYDRO 2004: a new era for hydropower, Porto, Portugal, 18–21 October 2004
27. Batalla RJ, Vericat D (2009) Hydrological and sediment transport dynamics of flushing flows: implications for management in large mediterranean rivers. River Res Appl 25:297–314
28. Walling DE (1984) Dissolved loads and their measurements. In: Hadley RF, Walling DE (eds) Erosion and sediment yield: some methods of measurements and modeling. Geo Books, London, 111–177
29. Emmett WW (1980) A field calibration of the sediment trapping characteristics of the Helley–Smith bedload sampler. US Geological Survey Professional Paper, 1139
30. Habersack HM, Laronne JB (2001) Bed load texture in an alpine gravel bed river. Water Resour Res 37(12):3359–3370
31. Vericat D, Church M, Batalla RJ (2006) Bedload bias: comparison of measurements obtained using two (76 and 152 mm) Helley–Smith samplers in a gravel-bed river. Water Resour Res 42(1–13):W01402
32. Vericat D, Batalla RJ (2007) Fractional bedload during small floods in a regulated gravel-bed river. Z für Geomorphol 51(2):227–240
33. Vericat D, Batalla RJ, Garcia C (2008) Bed-material mobility in a large river below dams. Geodinamica Acta 21(1–2):3–10
34. Parker G, Sutherland AJ (1990) Fluvial armour. J Hydraulic Res 28:529–544
35. Vericat D, Batalla RJ, Garcia C (2006) Breakup and reestablishment of the armour layer in a highly regulated large gravel-bed river: the lower Ebro. Geomorphology 76:122–136
36. Wolman MG (1954) A method of sampling coarse bed material. Am Geophys Union Trans 35:951–956
37. Rice S, Church M (1996) Sampling surficial fluvial gravels: the precision of size distribution percentile estimates. J Sedimentary Res 66(3):654–665
38. Kellerhals R, Bray DI (1971) Sampling procedures for coarse fluvial sediments. J Hydraulics Division ASCE97(HY8):1165–1180
39. García MA (2000) Las precipitaciones en la cuenca del Ebro: caracterización de la evolución espacio-temporal y análisis de tendencias, Internal Technical Report no.2001-PH-14.1, Confederación Hidrográfica del Ebro, Zaragoza, p 118
40. Ibañez C, Prat N, Canicio A (1996) Changes in the hydrology and sediment transport produced by large dams on the Lower Ebro River and its estuary. Regulated Rivers: Res Manag 12:51–62
41. Collier M, Webb RH, Schmidt JC (1996) Dams and rivers: a primer on the downstream effects of dams. US Geological Survey Circular 1126:94
42. Kondolf GM, Matthews WVG (1993) Management of coarse sediment in regulated rivers of California. University of California Water Resources Center, Riverside, Report No. 80
43. Inbar M (1992) Rates of fluvial erosion in basins with a Mediterranean type climate. Catena 19:393–409

44. de Vente J, Poesen J, Bazzoffi P, Van Rompaey A, Verstraeten G (2006) Predicting catchment sediment yield in Mediterranean environments: the importance of sediment sources and connectivity in Italian drainage basins. Earth Surf Processes Landforms 31:1017–1034
45. de Vente J, Poesen J, Arabkhedri M, Verstraeten G (2007) The sediment delivery problem revisited. Progress Phys Geogr 31(2):155–178
46. Bayerri E (1934–1935) Historia de Tortosa y su comarca. Imprenta Moderna de Alguerri, Tortosa
47. Nelson CH (1990) Post Messinian deposition rates and estimated river loads in the Ebro sedimentary system. In: Nelson CH, Maldonado A (eds) Marine geology of the Ebro continental margin. Geology 95:395–418
48. Gorría H (1877) Desecación de las marismas y terrenos pantanosos denominados de Los Alfaques. Imprenta La Giralda, Madrid
49. Jansen P, Van Bendegom L, Van den Berg J, De Vries M, Zanen A (1979) Principles of river engineering. Pitman, London, 509 p
50. Walling DE, Webb BW (1983) Patterns of sediment yield. In: Gregory KJ (ed) Background to Paleohydrology. Wiley, Chichester, pp 69–100
51. Lvovich MI, Karasik GY, Bratseva NL, Medvedeva GP, Maleshko AV (1991) Contemporary intensity of the world land intracontinental erosion. USSR Academy of Sciences, Moscow, Russia
52. Batalla RJ, Sala M, Werritty A (1995) Sediment budget focused in solid material transport in a subhumid Mediterranean drainage-basin. Z Geomorphol 39(2):249–264
53. Walling DE, Webb BW (1996) Erosion and sediment yield: a global overview. Proceedings of the IAHS symposium on erosion and sediment yield: global and regional perspectives, vol 236. IAHS Red Book, Exeter, pp 3–19
54. Vericat D, Roura M, Dolz J, Batalla RJ (2009) Suspended sediment dynamics in a highly regulated river: the lower Ebro (NE Iberian Peninsula). Geophysical research abstracts, 11, SRef-ID: EGU2009-5991. European Geosciences Union General Assembly, Viena, Austria, 19–24 April, 2009
55. Vericat D (2005) Sediment transport and processes in a highly regulated river. The lower Ebro, NE Iberian Peninsula. Universitat de Lleida, Lleida. PhD Thesis. Digital publication on the TDX online service. ISBN: L-1221-2006/ 84-690-0835-2, p 320
56. Vericat D, Batalla RJ (2004) Efectos de las presas en la dinámica fluvial del curso bajo del río Ebro. Cuaternario y Gemorfología 18(1–2):37–50
57. Catalán JG (1969) Química del agua. Blume, Barcelona
58. Varela J, Gallardo A, López de Velasco A (1986) Retención de los sólidos por los embalses de Mequinenza y Ribarroja. In: Mariño M (ed) Efectos sobre los aportes al Delta del Ebro. In El Sistema Integrado del Ebro. Gráficas Hermes, Madrid, pp 203–219
59. Palanques A (1987) Dinámica sedimentaria, meneralogía y microcontaminantes inorgánicos de las suspensiones y de los sedimentos superficiales en el margen continental del Ebro. PhD Thesis, University of Barcelona
60. Muñoz I (1990) Limnologia de la part baixa del riu Ebre I els canals de reg: Els factors físico químics, el fitoplancton i els macroinvertebrats bentònics. PhD thesis, Department of Biology, University of Barcelona
61. Roura M (2004) Incidència de l'embassament de Mequinensa en el transport de sòlids en suspensió i la qualitat de l'aigua del riu Ebre. PhD Thesis, Facultat de Biologia, Universitat de Barcelona, Barcelona, pp 145

Hydrological Impacts of Climate Change on the Ebro River Basin

Mauricio Zambrano-Bigiarini, Bruno Majone, Alberto Bellin, C. Isabella Bovolo, Stephen Blenkinsop, and H. J. Fowler

Abstract This work presents hydrological simulations on the Ebro River basin (Spain), using both control (1961–1990) and future (2071–2100) climate scenarios, to investigate the effect of climate change on water resources. Using the Soil and Water Assessment Tool hydrological model, simulations were carried out in four subcatchments representative of typical situations within the basin. Model parameters were identified using sensitivity analysis and long-term calibration procedures, which enabled the historical behaviour of the catchments to be reproduced. Following validation, the parameters were used to simulate the effects of climate change on future streamflows.

Bias-corrected daily time series of precipitation and mean temperature from two regional climate models (RCMs), using the same medium-high SRES A2 emissions scenario, were used as drivers of the hydrological simulations during the future scenarios. Important annual and seasonal differences in the projected future precipitation and air temperature fields were observed among the RCMs. However, the two models project an overall increase in the mean annual temperature accompanied by a reduction in the annual precipitation, with the strongest differences with respect to the control period observed during the summer season.

When these changes were used to project future streamflows, a general decrease was observed in the streamflows at the outlet of the selected catchments. Changes in streamflows were in general agreement with the projections of daily precipitation and temperature fields, with the largest drop in predicted monthly streamflows for

M. Zambrano-Bigiarini (✉), B. Majone, and A. Bellin
Department of Civil and Environmental Engineering, University of Trento, Via Mesiano 77, I-38123 Trento, Italy
e-mail: mauricio.zambrano@ing.unitn.it

C.I. Bovolo, S. Blenkinsop, and H.J. Fowler
School of Civil Engineering and Geosciences, University of Newcastle upon Tyne, Newcastle NE1 7RU UK

D. Barceló and M. Petrovic (eds.), *The Ebro River Basin*,
Hdb Env Chem (2011) 13: 47–75, DOI 10.1007/698_2010_85,

the subcatchments with the lowest aridity index, and seasonal differences that appears to be related to the elevation range of the subcatchments.

Keywords Climate change, Hydrological impact, Hydrological modelling, PRUDENCE, SWAT

Contents

1 Introduction

Among the most significant impacts of climate change on our society, those related to regional water availability and frequency/intensity of extreme events may convey implications on a variety of activities, from agriculture and energy production to flood control, emphasizing the necessity of better understanding how those changes in global climate will affect local water resources [1]. According to the Intergovernmental Panel on Climate Change [2], almost the entire Europe will probably be negatively affected by the projected climate, including a higher risk of droughts, floods, and erosion. In this context, southern Europe is likely to suffer the important alterations, with an expected decrease in annual runoff ranging from 0 to 23% by 2020s, and from 6 to 36% by 2070s, along with a decrease by up to 80% of low summer flows [2], making the risk of droughts particularly important. Moreover, water withdrawals are also expected to increase in southern Europe, amplifying the risks associated with climate change, being the Mediterranean (Spain and Portugal) the region more exposed to drought risk [2].

This chapter is organized as follows. We first present a short description of the criteria used for selecting the data used for driving the hydrological simulations at the basin scale. Subsequently, we briefly describe Soil and Water Assessment Tool (SWAT), the hydrological model adopted for this study, and the setup thereof. Later on, we continue with a brief review of the main spatio-temporal patterns of climate, for both the control period 1961–1990 and the future climate scenarios for the period 2071–2100. Finally, the main results of the hydrological simulations on four subcatchments are discussed for both the control period and the two selected future climate scenarios.

2 Study Area

Figure 1 shows the location of the Ebro River basin and the study area selected for the hydrological simulations. Topography exerts a continental effect over the Mediterranean climate in a large area of the basin, with semi-arid condition in its centre. Annual precipitation on the entire basin equals 622 mm/year, from 1920 to 2002, ranging from a minimum of 452 mm/year to a maximum of 840 mm/year [38]. Mean annual temperature over the period 1961–1990 is 12.2°C, with mean winter minimum and summer maximum temperatures of 4.8°C and 20.5°C, respectively. The study area selected for the hydrological simulations corresponds to the western part of the Ebro River basin (red polygon in Fig. 1), with a total area of 42,000 km^2. The elevation ranges from 185 to 2,875 m a.s.l., with an average value of 790 m a.s.l. and 50% of the area below 770 m a.s.l.

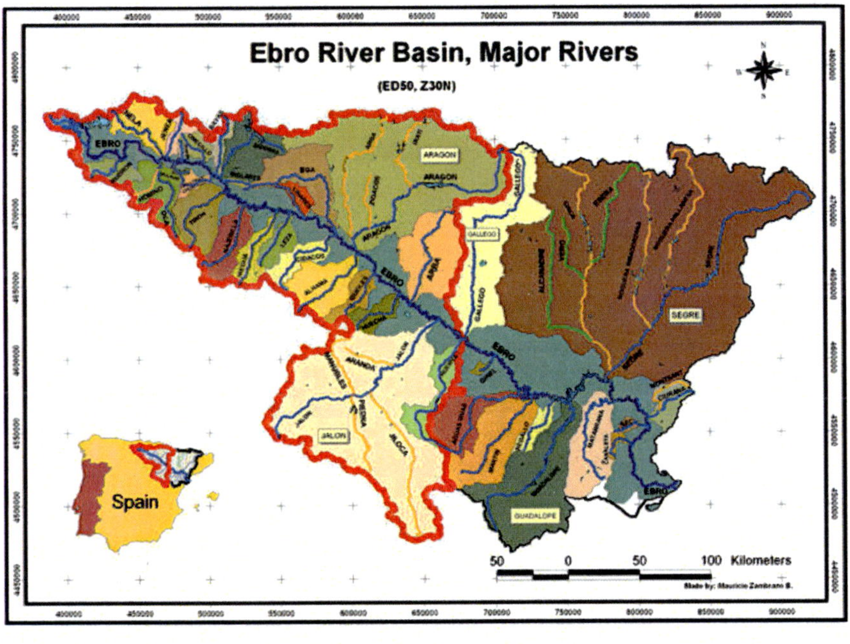

Fig. 1 Location of the Ebro River basin. The *red polygon* shows the extent of the study area selected for the hydrological simulations, with a total surface of 42,000 km^2. The *inset* shows the location of the Ebro River basin in the Iberian Peninsula

3 SWAT Model

Hydrological simulations were carried out using the version 2005 of the SWAT hydrological model [3–6]. SWAT has been widely used to model hydrological and biogeochemical processes at the catchment scale [7–9], and its hydrological modules have been tested in watersheds covering a wide range of spatial scales, climates and hydrogeologic conditions [7, 10, 11]. SWAT is a physically based, semi-distributed model developed to simulate continuous-time processes and streamflows with a high level of spatial detail, as obtained by dividing the main basin into subcatchments connected through the stream network. Each subcatchment can be further subdivided into homogeneous Hydrological Response Units (HRUs), to reproduce the combinations of land use and soil type representative of the subcatchment [4–6]. SWAT operates at a daily time step and is quite broad in perspective since, according to Arnold et al. [4], it is designed to simulate water, sediment, and agricultural chemical transport in large ungauged basins, and to evaluate the effect of different management scenarios on watershed's hydrology, as well as the effects of point and non-point source pollution. A complete description of the model components can be found in Arnold et al. [4] and Neitsch et al. [5, 6].

4 Model Setup

Spatial information needed for watershed delineation was obtained from a digital elevation model (DEM) with a resolution of 60 m, which was re-sampled from the original DEM, with spatial resolution of 20 m, provided by the "Confederación Hidrográfica del Ebro" (CHE, Hydrological Confederation of Ebro River), public office responsible for the administration and control of the Ebro River basin. After a preliminary analysis, the study area was divided into 120 different subcatchments, as shown in Fig. 2, taking into account the location of streamgauges and important hydraulic infrastructures (e.g. reservoirs and channels). The input files required for running the hydrological simulations were constructed using the ArcView GIS interface for SWAT 2005 (AVSWAT-X, [12, 13]), including the drainage area, overland field slope and length, channels slope and length, and this information was used at the subcatchment level. It should be noted that the results are presented only for four catchments that capture the characteristic response of other catchments with similar climate or elevation.

Land use maps (Fig. 3) were provided by the CHE, for the years 1984, 1991 and 1995, at a 1:100,000 scale, and the land use cover of the year 1984 was selected as representative for the simulations referring to the control period (1961–1990). Table 1 shows the six major land use classes identified within the study area. Information on soil types provided by the "Oficina de Planificación Hidrológica" (OPH) of CHE was used to assign the soil properties required by SWAT 2005. The original 136 lithological classes were first aggregated into 23 main classes, as

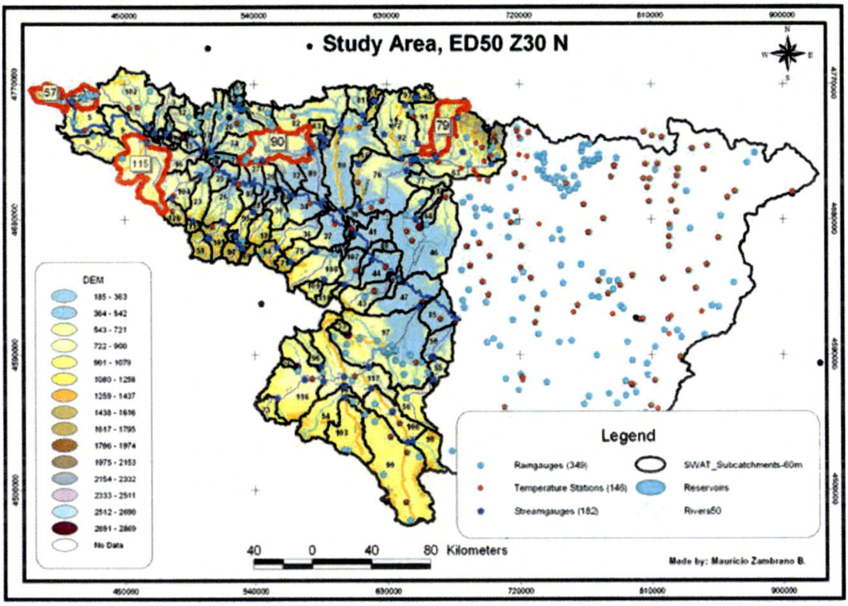

Fig. 2 Study area, digital elevation model and subcatchments used for the hydrological simulations, along with the location of the gauging stations of precipitation, air temperature and streamflows. *Yellow polygons* indicate the location of the four subcatchments selected for carrying out the hydrological simulations during the control and the future scenarios described in Sect. 6

shown in Fig. 4, and then a first guess of soil properties was obtained through a correspondence with the soil types included in the SWAT database.

Daily precipitation at 1,569 stations, daily mean temperature at 859 stations and daily mean streamflow at 318 stations, unevenly sampled during the period 1900–2004, were provided for the entire Ebro River basin by the OPH. Due to the fact that the available datasets varied in length and quality, with several missing data unevenly distributed in time and space, only gauging stations with an amount of daily information during the control period 1961–1990 above a given threshold were retained for the hydrological simulations. In particular, a threshold of 70% was selected for precipitation, whereas 65% was used for temperature and streamflow, leading to 349, 146 and 182 gauging stations of precipitation, air temperature and streamflow, respectively. The location of the selected gauging stations is shown in Fig. 2.

SWAT requires daily maximum and minimum air temperature at all the selected temperature stations, data that were not available during the control period 1961–1990. The missing values were reconstructed using linear relationships between the maximum/minimum daily air temperature and the elevation and daily mean air temperature of 34 stations of the HidroEbro database, provided by the CHE for the period 1 January 2003 to 16 October 2006. These relationships

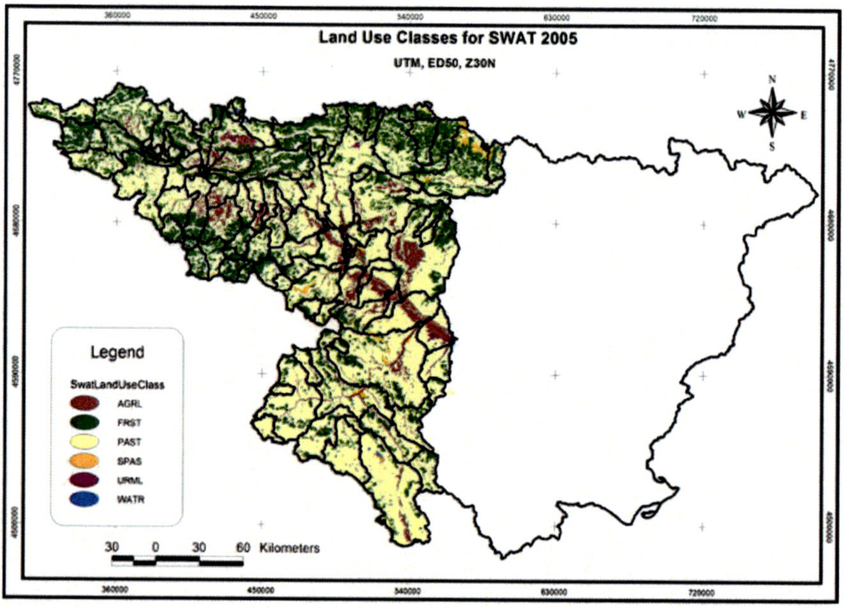

Fig. 3 Land use classification based on the information for the year 1984 at 1:100,000 scale. Legend refers to the classification reported in Table 1

Table 1 Land use classification adopted for the hydrological simulations

ID	SWAT land cover/plant class	% Basin area	Original name
SPAS	Summer pasture	0.91	Roquedo
FRST	Forest-mixed	26.12	Bosques
PAST	Pasture	64.48	Suelo Desnudo
URML	Urban, medium density	0.42	Suelo Artificial
WATR	Water	0.03	Lagos Interiores
AGRL	Agricultural land-generic	8.04	Regadios

were then used for generating daily time series of maximum and minimum air temperature in each one of the 146 gauging stations selected for the hydrological simulations, both during the control period 1961–1990 and the future time-slice 2071–2100, under the hypothesis that climate change does not affect the established relationships between maximum/minimum and mean daily temperatures.

4.1 *Daily Mean Precipitation on Sub-watersheds*

Precipitation is a key driver of the hydrological cycle and should be accurately reproduced in simulations, since a wrong representation of precipitation

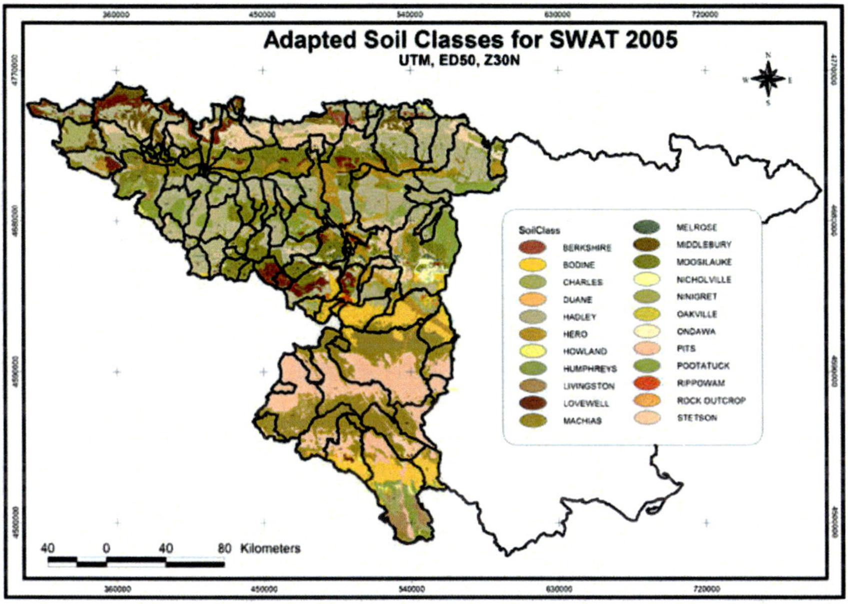

Fig. 4 Soil types classification adopted in the present study, based on soil information provided by CHE aggregated into 23 main classes, which soil properties are then assigned by comparison with similar soils of the SWAT database

volumes over a catchment can lead to significant bias in the simulations. SWAT allows for reproducing spatial variability of precipitation by assigning spatially uniform daily precipitation to each subcatchment. This means that the detail with which spatial variability is reproduced within the catchment depends essentially on the number and size of subcatchments, and therefore is controlled by the user. To distribute the precipitation measured at the selected 349 raingauges over the 120 subcatchments of the study area (see the previous section), spatial interpolation was used to compute a uniform daily precipitation for each subcatchment. The computations were performed using the *gstat* [14] and *sp* [15] packages of *R* [16].

The procedure for computing the average daily precipitation over each subcatchment can be summarized as follows: (1) the study area is subdivided into square cells of 1 km^2; (2) the daily precipitation is computed at the centre of each cell using inverse distance weighted (IDW) interpolation considering the 40 closest raingauge stations; and (3) a spatially uniform daily precipitation for each of the 120 subcatchments is then obtained by averaging the values generated in all the cells belonging to each subcatchment. Steps (1) and (3) are repeated for all the days within the control and future climate scenarios.

4.2 *Reservoirs*

Within the study area there are several infrastructures that substantially modify the natural flow regime, as described in Zambrano-Bigiarini et al. [17]. Consequently, the effects of reservoirs on streamflow should be carefully evaluated and modelled. Unfortunately, the influence of reservoirs on the hydrological regime is often difficult to predict, because their effect spans a wide range of time scales (from daily to weekly, monthly and annual), and most importantly it changes with time [18]. After recognizing the practical impossibility to obtain data describing how the reservoirs have been actually operated at the daily scale, we decided to implement simplified monthly operational rules. These rules are specific for each reservoir, since they depend on its own particular characteristics and management. Subsequently, these operational rules were implemented into SWAT considering a monthly target release-storage approach for a controlled reservoir [5], in which the water is released as a function of a desired target storage in the reservoir and a minimum and maximum allowed discharges. The implementation of the monthly target release-storage approach requires specifying monthly target volumes to be preserved and daily maximum and minimum releases for each month, which are kept the same during all the simulations performed in both control and future periods. Operational rules for the simulated reservoirs were obtained by analysing the available time series of storage and deliveries, which were downloaded from the website of the Centro de Estudios y Experimentación de Obras Públicas (CEDEX, Centre of Studies and Experimentation of Public Works) on http://hercules.cedex.es/anuarioaforos/, for the period 1 January 1961 to 31 December 1990. To properly implement these rules, we introduced a modification in the SWAT source code, which prevents that the volume of water stored into the reservoir exceeds its maximum capacity.

5 Projected Changes in Climate

5.1 *Control Period (1961–1990)*

Before analysing the expected changes in climate, knowledge about the historical climate is needed for a better understanding of the significance of the ongoing changes. Following the criterion described in Sect. 4, only gauging stations with less than 30% and 35% of missing daily data for precipitation and air temperature, respectively, were used in the analysis of the climatic characteristics during the control period 1961–1990. In this period, the mean annual precipitation on the entire Ebro River basin, computed by averaging the precipitation at the centre of the square cells (which are obtained as described in Sect. 4), amounts to 544.6 mm/year, against the value of 608.90 mm/year obtained by simply averaging the values recorded at the 349 raingauges (Table 2). The same computation

Table 2 Annual and seasonal averages of precipitation and air temperature for the control period 1961–1990

Season	Precipitation (mm)		Mean temperature (°C)	
	Average over stations	Average over the cells	Average over stations	Average over the cells
Annual	608.9	544.6	12.2	12.7
Winter (DJF)	145.6	127.3	4.8	5.3
Spring (MAM)	171.0	156.3	10.7	11.1
Summer (JJA)	121.4	109.5	20.5	20.9
Autumn (SON)	170.9	151.5	13.1	13.5

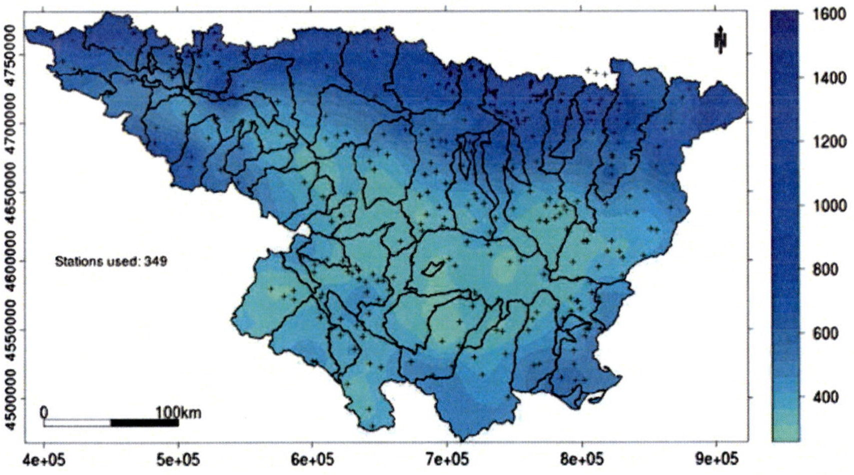

Fig. 5 Spatial distribution of the average annual precipitation during the control period 1961–1990, computed using ordinary kriging interpolation with cells of 1 km^2 and 40 nearest neighbours. *Crosses* represent the location of the gauging stations

repeated for the mean daily temperature leads to a mean annual temperature of 12.7°C and 12.2°C (Table 2), when computed averaging the interpolated values of each cell and the values measured at the raingauges, respectively. The spatial distribution of the average annual precipitation and the average of the annual air temperature for the control period 1961–1990, both obtained by ordinary kriging interpolation, are shown in Figs. 5 and 6, respectively.

5.2 *Future Scenarios (2071–2100)*

To take into account some of the uncertainties involved in the projections of climate change, two future climate scenarios for the entire Ebro River basin were selected

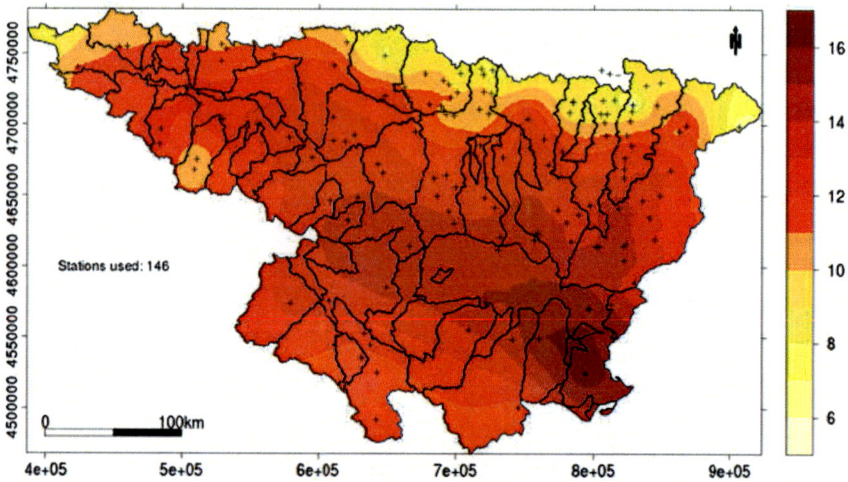

Fig. 6 Spatial distribution of the average annual air temperature during the control period 1961–1990, computed using ordinary kriging interpolation with cells of 1 km^2 and 40 nearest neighbours. *Crosses* represent the location of the gauging stations

Table 3 Regional climate models (RCMs) adopted in the present study for the control (CRTL) and future (SCEN) periods

RCM	INST	Driving GCM	Days per month	CTRL	SCEN	CTRL RCM filename	SCEN RCM filename
HIRHAM_H	DMI	HadAM3H A2	30	1961–1990	2071–2100	HC1	HS1
RCAO_E	SMHI	ECHAM4/ OPYCA2	30	1961–1990	2071–2100	MPICTL	MPIA2

The table shows the originating institution (INST, where *DMI* Danish Meteorological Institute, *SMHI* Swedish Meteorological and Hydrological Institute), the driving global climate model (GCM) and the model's filename of the RCMs

from the red set of the EU FP5 PRUDENCE (Prediction of Regional scenarios and Uncertainties for Defining EuropeaN Climate change risks and Effects) project [19], as described in Bovolo et al. [20, 39]. These two scenarios, shown in Table 3, correspond to the outputs of two different RCMs run for the time-slice 2071–2100 and the same medium-high SRES A2 emissions scenario [21], and they were selected because they are representative of climate conditions derived from the two general circulation models (GCMs) used as boundary conditions of the PRUDENCE experiment.

Despite the spatial resolution of RCM outputs is much higher than the one of the driving GCMs, additional correction is usually required to overcome biassed representation of observed climate, mainly due to problems in conceptualization, discreatization and spatial averaging within grid cells [22]. A common practice for correcting outputs of climate models is to apply a statistical downscaling technique to obtain point-scaled climate variables (a comprehensive review of downscaling techniques with focus on hydrological applications can be found in Fowler et al. [23]).

In this work, a simple bias-correction method was used to downscale daily precipitation and air temperature fields from the (large) grid-cell scale of the two regional climate models (RCMs) to the point scale of the corresponding gauging stations, as described in Bovolo et al. [39]. This bias-correction method use monthly correction factors to modify both the absolute magnitude and the seasonality of the modelled values, based on the relationship between observed values of precipitation and air temperature and the corresponding grid-cell values during the control period (see [24, 39]), to allow the modelled monthly means to match the observed monthly averages during the control period 1961–1990. Under the assumption that the same model biases that apply during the control period will apply also in the future, the computed bias-correction factors are then used to downscale the outputs of the RCMs for the time-slice 2071–2100, additively in the case of temperature and multiplicatively for precipitation (see [24] for more details).

To apply the bias-corrected method using monthly mean statistics, the time series of the observed data should be long enough to provide significant statistics of the 30-year time control period, and the period of observations should be well represented in the models control period. Following this approach, missing values in the daily time series of precipitation and air temperature described in Sect. 4 for the control period 1961–1990 were infilled using a modified version of the IDW algorithm, proposed by Teegavarapu and Chandramouli [25], where the spatial distance is replaced by the Pearson's product-moment coefficient of correlation between the daily time series of the stations. The original procedure was slightly modified to consider for interpolations only the four closest stations (in terms of the coefficient of correlation between the daily time series) to each target station, instead of all the available stations, as proposed by Teegavarapu and Chandramouli [25]. A leave-one-out cross-validation procedure was used for assessing the goodness-of-fit between the interpolated values and the observed ones. The overall mean errors resulted of 0.011 mm/day and 0.30°C/day for precipitation and air temperature, respectively.

Table 4 shows the projected anomalies of annual and seasonal precipitation and air temperature for the Ebro River, whereas Figs. 7–12 show the spatial variation of these anomalies, computed using ordinary kriging. Anomalies are computed as the difference between projected bias-corrected values for the climate scenarios (January 2071 to December 2100) and the corresponding values observed during the control period (January 1961 to December 1990), and they can be viewed as expected values about which uncertainties of different origin exist. Table 4 shows that both RCMs predict a reduction in the mean annual precipitation, accompanied by an increase in the mean annual temperature with respect to the control period. In particular, the RCAO_E model projects a reduction of 21.8% for the mean annual precipitation and an increase of +6.3°C for the mean annual temperature.

Table 4 Anomalies of annual and seasonal precipitation and air temperature of the Ebro River basin, for the future period 2071–2100 with respect to the control period 1961–1990

RCM	Anomaly of mean precipitation (%)					Anomaly of mean temperature (°C)				
	Annual	DJF	MAM	JJA	SON	Annual	DJF	MAM	JJA	SON
DMI.HS1 (HIRHAM_H)	−3.8	21.2	−14.1	−31.6	4.0	4.2	3.3	3.3	5.6	4.4
SMHI.MPIA2 (RCAO_E)	−21.8	−2.8	−35.1	−35.3	−16.7	6.3	4.2	6.2	8.8	5.5
Average	−12.8	+9.2	−24.6	−33.5	−6.4	+5.3	+3.8	+4.8	+7.2	+5.0

Values computed by averaging over the corresponding gauging stations

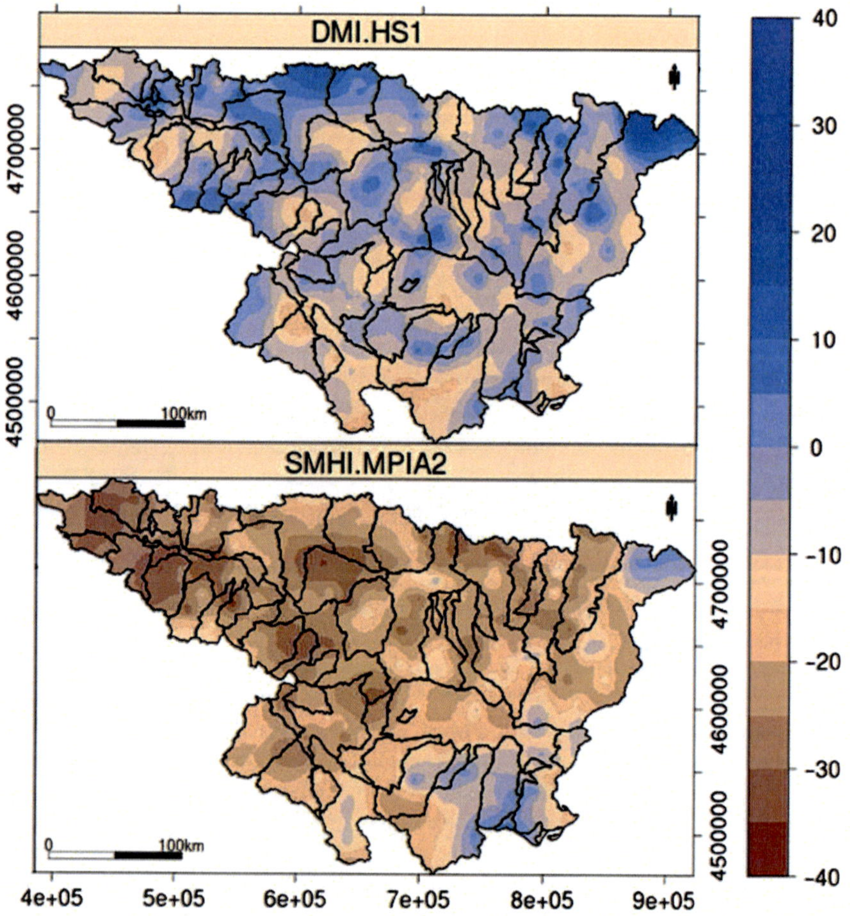

Fig. 7 Spatial distribution of the anomalies of annual precipitation, projected by the two RCMs: HIRHAM_H (DMI.HS1) and RCAO_E (SMHI.MPIA2)

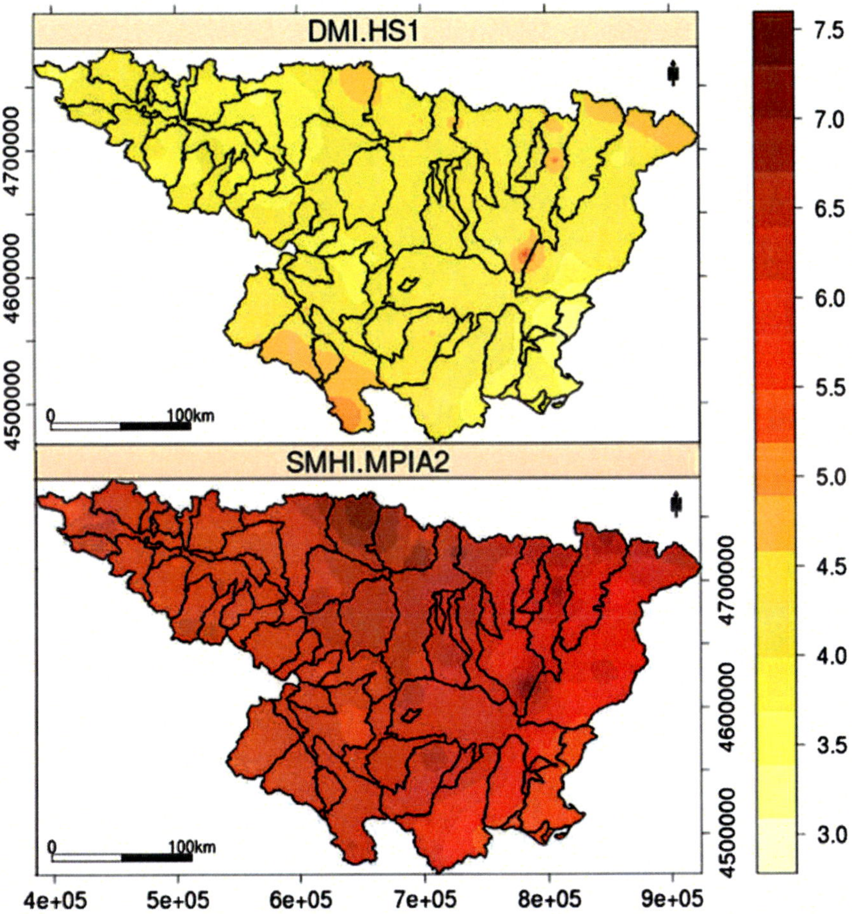

Fig. 8 Spatial distribution of the anomalies annual temperature, projected by the two RCMs: HIRHAM_H (DMI.HS1) and RCAO_E (SMHI.MPIA2)

Expected changes are less dramatic in the HIRHAM_H model, which projects a reduction of 3.8% for the mean annual precipitation and an increase of 4.2°C for the mean annual temperature. While the two RCMs predict quite different reductions of the mean annual precipitation, with HIRAM_H showing only a slight reduction, both models agree in a dramatic increment of the annual temperature for the time-slice 2071–2100, with RCAO_E projecting the largest increase with respect to the control period.

Additional considerations can be drawn if we look at the seasonal distribution of the projected changes in climate, as shown in Table 4. In particular, both RCMs project an important reduction of the precipitation field in both the spring and

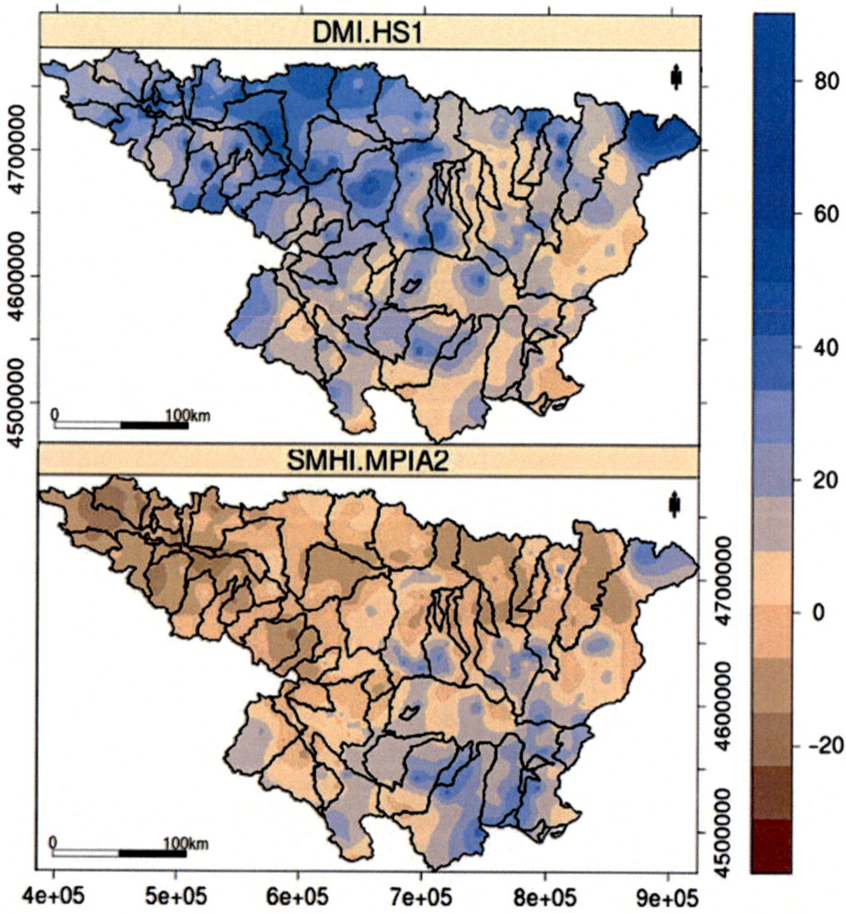

Fig. 9 Spatial distribution of the anomalies of winter (DJF) precipitation, projected by the two RCMs: HIRHAM_H (DMI.HS1) and RCAO_E (SMHI.MPIA2)

summer seasons, with the largest decrease during summer (31.6% and 35.3% for HIRHAM_H and RCAO_E, respectively). These strong seasonal reductions are only partially compensated by higher winter and autumn precipitations in the HIRHAM_H model (21.2% and 4.0%, respectively), whereas the RCAO_E model project smaller reductions in the winter and autumn precipitation (2.8% and 16.7%, respectively), compared to spring and summer.

Considering now the air temperature, we note that both RCMs project an increase in all the seasonal temperatures, with the largest increase obtained during the summer (5.6°C and 8.8°C for the HIRHAM_H and RCAO_E, respectively), which confirm the findings of previous studies conducted in the Iberian peninsula

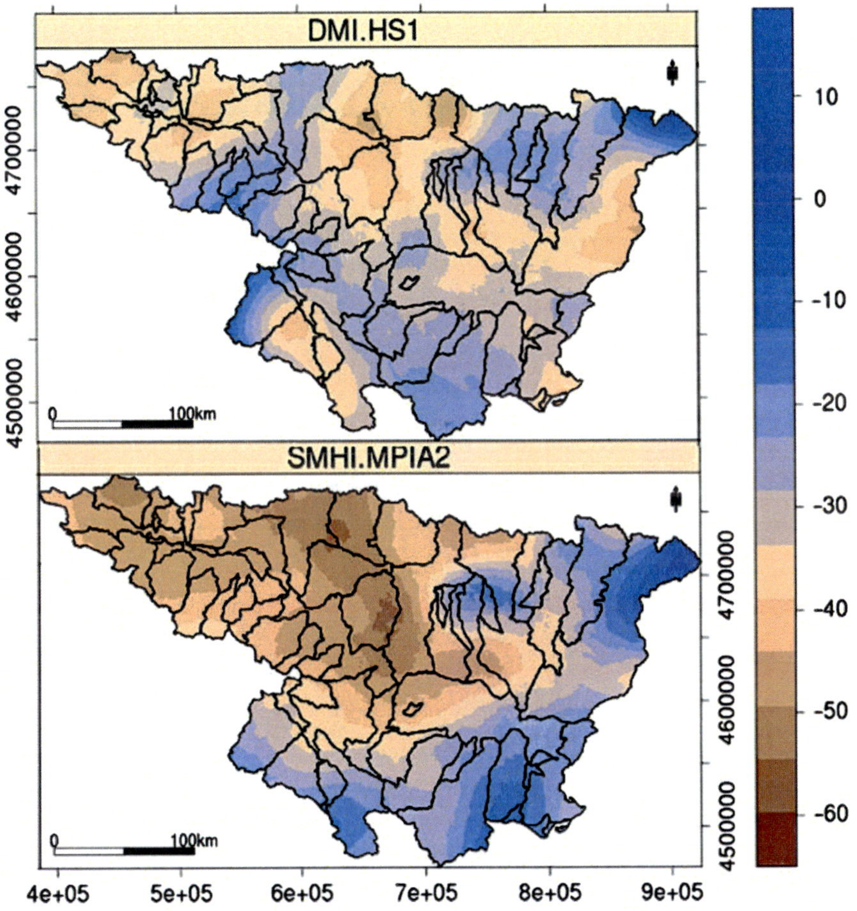

Fig. 10 Spatial distribution of the anomalies of summer (JJA) precipitation, projected by the two RCMs: HIRHAM_H (DMI.HS1) and RCAO_E (SMHI.MPIA2)

[26–28]. The aforementioned seasonal changes of the meteorological driving forces may have important effects on the water availability of the Ebro River, because the largest projected reductions of precipitation are mainly concentrated during the seasons with the highest water demand for irrigation.

Figures 9–12 show the spatial distribution of the anomalies of winter and summer precipitation and air temperature projected by the selected two climate scenarios. It is possible to observe that projected changes show large spatial variations within the Ebro River basin, with the largest differences for the precipitation field. In addition, in some places one RCM projects an increase whereas the other projects a decrease in the seasonal precipitation (e.g. winter precipitation on the north-western part of the catchment).

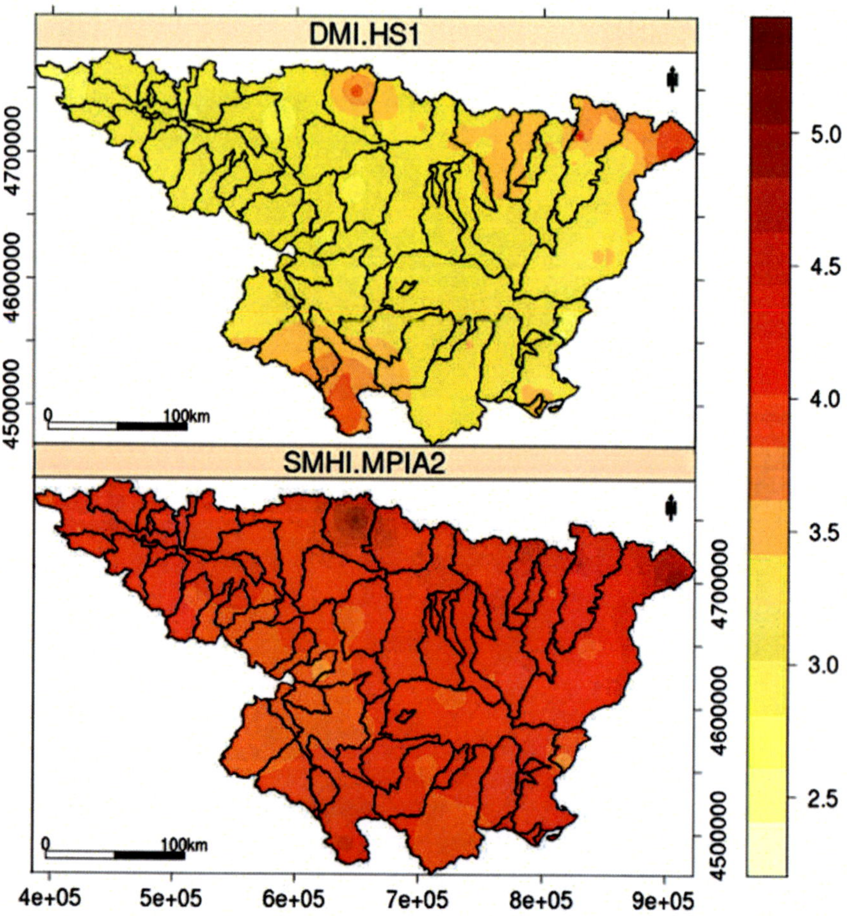

Fig. 11 Spatial distribution of the anomalies of winter (DJF) temperature, projected by the two RCMs: HIRHAM_H (DMI.HS1) and RCAO_E (SMHI.MPIA2)

6 Hydrological Modelling Results

As mentioned in Sect. 4, four subcatchments were selected for carrying out the hydrological simulations within the Ebro River basin. The selected catchments are characterized by different annual precipitation (from 685.7 to 1063.6 mm/year). Although the selected subcatchments show similar aridity index [29], they were selected as representative of catchments with different elevation and climate regime for testing the capabilities of the modelling framework, with the main purpose of testing the proposed approach in most of the typical situations within the Ebro river basin. The location and main characteristics of these subcatchments are presented

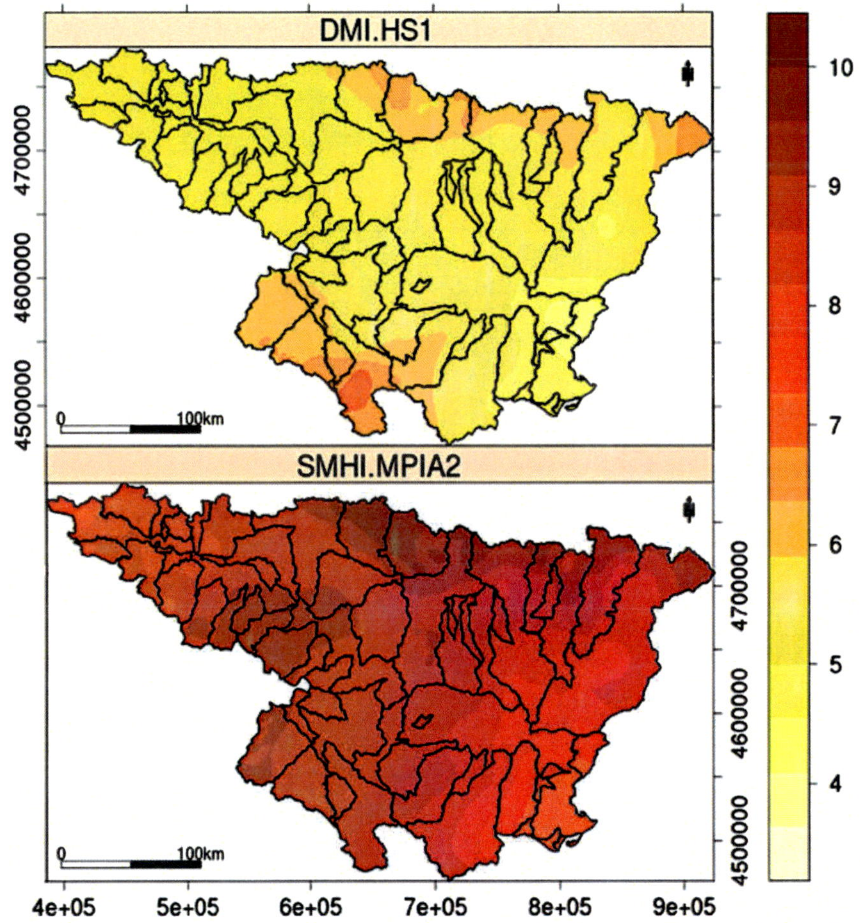

Fig. 12 Spatial distribution of the anomalies of summer (JJA) temperature, projected by the two RCMs: HIRHAM_H (DMI.HS1) and RCAO_E (SMHI.MPIA2)

in Fig. 2 and Table 5, respectively, and a short description of each one of them is presented below:

1. *Subcatchment 057 (Ebro River)*: This catchment is located at the headwaters of the Ebro River basin, and it feeds the Embalse del Ebro reservoir, the largest of the basin (E801, built in 1952, with 62.50 km^2 of maximum surface area and a capacity of 540×10^6 m^3). Because of its size, this reservoir plays a major role in downstream streamflows.
2. *Subcatchment 079 (Aragon River)*: Located in the north-western part of the study area, with 1,919 m of elevation range (from 503 to 2,422 m a.s.l.). This catchment undergoes important snow effects during the year.

Table 5 Main characteristics of the selected four subcatchments

Subb. ID	Draining area (km^2)	Elevation range (m a.s.l.)	Annual precipitation (mm)	Aridity index[a]	ID Q station	Station name	Q data period	No. missing daily data	Mean annual streamflow (m^3/s)
057+ E801[b]	461.59	824–2,173	874.2	1.45	Q801 (CEDEX)	Embalse del Ebro	1958–2004	26	11.29
079	509.10	503–2,422	1063.6	1.64	Q063	Esca en Sigues	1931–2002	15	11.44
090	808.07	420–1,424	811.3	1.21	Q071	Ega en Estella	1931–2002	7	12.51
115	1040.34	570–1,250	685.7	1.02	Q093	Oca en Oña	1959–2002	480	5.03

[a]Average annual precipitation has been estimated for the period 1961–1990 following the approach presented in Sect. 4.1, while average annual potential evapotranspiration (PET) was computed by averaging the annual PET estimates provided by CHE at several meteorological stations during the same period

[b]Measured streamflows correspond to the observed deliveries of reservoir E801 (Embalse del Ebro), which is feeded by subcatchment 057

3. *Subcatchment 090 (Ega River)*: Located in the central part of the study area, with 1,004 m of elevation range (from 420 to 1,424 m a.s.l.).
4. *Subcatchment 115 (Homino River)*: Located in the north-eastern part of the catchment, with 680 m of elevation range (from 570 to 1,250 m a.s.l.). It is the catchment with the smallest annual precipitation of the four selected catchments.

Gauging stations Q063, Q071 and Q093 are only slightly disturbed [30], and therefore no withdrawals were incorporated in the simulations of their corresponding catchments.

6.1 Hydrological Simulations During Control Period (1961–1990)

Parameters subject to calibration within SWAT were selected after a preliminary sensitivity analysis and literature review, to partially compensate for the inadequacy of the initial values assumed for some of them (especially those related to soil type), model structure and other sources of uncertainty. A detailed description about the SWAT parameters can be found in [5, 6], while a brief description of the selected parameters is provided next:

1. *CN2*: Initial Soil Conservation Service (SCS) runoff curve number for moisture condition II. The SCS curve number is a function of the soil's permeability, land use and antecedent soil water conditions.
2. *ALPHA_BF* (days): Baseflow alpha factor. The baseflow recession constant is a direct index of groundwater flow response to changes in recharge. This parameter can vary between 0 and 1, with values in the range 0.1–0.3 for a land with slow response to recharge, and 0.9–1.0 for a land with a rapid response.
3. *GWQMN* (mm H_2O): Threshold depth of water in the shallow aquifer required for return flow to occur. Groundwater flow to the reach is allowed only if the depth of water in the shallow aquifer is equal to or greater than GWQMN.
4. *RCHRG_DP*: Deep aquifer percolation fraction. It represents the fraction of percolation from the root zone which recharges the deep aquifer. This parameter may vary between 0 and 1.
5. *SOL_Z* (mm): Depth from soil surface to the bottom of a layer. Since all the subcatchments were conceptualized as single layer units, SOL_Z is equal to SOLZMX (mm), the maximum rooting depth of soil profile.
6. *SOL_AWC* (mm/mm): Available water capacity of the soil layer. The available water in the soil is calculated by subtracting the water content at the permanent wilting point from that at field capacity: *SOL_AWC= FC – WP*.
7. *SOL_K* (mm/h): Saturated hydraulic conductivity.

The first year of the control period (1961) was used as spin-up for all the simulations. Later, the four subcatchments were calibrated during the period January 1962–December 1983, and verified during the remaining part of the control period

(1984–1990), to allow using the calibrated parameters during the future climate scenarios. Performance statistics for both calibration and verification periods were computed for the monthly streamflows, while simulations were carried out at daily time step.

Subcatchment 057 was simulated in a different way, because it feeds the reservoir E801 (Embalse del Ebro), which is considered completely within the subcatchment. The only streamflows available for being used during the calibration and verification periods were the deliveries of the reservoir, preventing to use those measurements as representative of the catchment. Therefore, subcatchment 057 and reservoir E801 were simulated as a single system, using the measured daily deliveries and stored volumes of the reservoir as target variables to be matched by the simulations. The procedure for this system can be summarized as follows:

- Definition of operational rules for the reservoir E801 (those relating stored volume with deliveries, to be implemented in SWAT), which were not available, and therefore were computed during the calibration period (1962–1983) using the observed monthly deliveries and stored volumes.
- Reconstruction of the streamflows entering the reservoir during the calibration period, using the water balance equation used by SWAT for the reservoirs:

$$V = V_{\text{stored}} + V_{\text{flowin}} + V_{\text{pcp}} - V_{\text{flowout}} - V_{\text{evap}} - V_{\text{seep}}, \tag{1}$$

where V is the volume of water in the reservoir at the end of the day (Mm^3); V_{stored} is the volume of water stored in the reservoir at the beginning of the day (Mm^3); V_{flowin} is the volume of water entering the reservoir during the day (Mm^3), which was assumed equal to the runoff produced by subcatchment 057; V_{pcp} is the volume of precipitation falling on the water body during the day (Mm^3); V_{flowout} is the volume of water delivered by the reservoir during the day (Mm^3); V_{evap} is the volume of water removed from the reservoir by evaporation during the day (Mm^3) and V_{seep} is the volume of water lost from the reservoir by seepage (Mm^3), which was assumed to be equal to zero. For each day, V, V_{stored} and V_{flowout} were known, whereas V_{pcp} was computed knowing the daily average precipitation over the subcatchment and the surface area corresponding to the stored volume of water, and V_{evap} was computed by approximating the evaporation from the reservoir with 0.6 times the potential evapotranspiration computed by SWAT for the subcatchment multiplied by the surface area corresponding to the stored volume of water. We decided to approximate evaporation from the reservoir with potential evapotranspiration because of the small influence of this term in the above balance equation.

- Calibration of the hydrological parameters of subcatchment 057, using the previously computed streamflow entering the reservoir during the calibration period as runoff at the outlet of the subcatchment.
- Computation of the simulated deliveries and stored volumes of the reservoir during the verification period (1984–1990), by running SWAT with the same

parameter values and operational rules obtained during calibration. Computed deliveries and stored volumes were then compared to the measured ones.

Based on previous recommendations [31], a combination of graphical techniques and error index statistics was used for evaluating the goodness-of-fit between the simulated and observed streamflow values, both during the calibration and validation period. The used statistics were: the mean error (ME), the percent bias (PBIAS, [32]) and the Nash–Sutcliffe efficiency (NSeff, [33]):

$$\mathrm{ME} = \frac{1}{N}\sum_{i=1}^{N}\left(Q_i^{\mathrm{sim}} - Q_i^{\mathrm{obs}}\right), \tag{2}$$

$$\mathrm{PBIAS} = \frac{\sum_{i=1}^{N}\left(Q_i^{\mathrm{sim}} - Q_i^{\mathrm{obs}}\right) \times 100}{\sum_{i=1}^{N} Q_i^{\mathrm{obs}}}, \tag{3}$$

$$\mathrm{NSeff} = 1 - \frac{\sum_{i=1}^{N}\left(Q_i^{\mathrm{obs}} - Q_i^{\mathrm{sim}}\right)^2}{\sum_{i=1}^{N}\left(Q_i^{\mathrm{obs}} - \bar{Q}^{\mathrm{obs}}\right)^2}, \tag{4}$$

where N is the total number of observations, Q_i^{obs} and Q_i^{sim} represent the i-th observed and simulated values, respectively, and $\bar{Q}^{\mathrm{obs}}$ is the mean of the observed streamflows. In particular, PBIAS measures the average tendency of the simulated data to be larger or smaller than their observed counterparts, whereas NSeff measures the fraction of the variance of the observed flows explained by the modelled values. In this study, we assumed that modelled streamflows are behavioural [34] and thus can be accepted as suitable simulators of the system, when NSeff > 0.50, PBIAS is within $\pm 25\%$ and ME is close to zero [31]. The resulting goodness-of-fit indexes, for both calibration and validation period, are shown in Table 6, whereas a graphical comparison between the simulated and observed streamflows is presented, as an example, in Fig. 13 for subcatchment 090.

Since for almost all the subcatchments the NSeff was larger than 0.5, and the PBIAS was within $\pm 11\%$ during the validation period, the calibrated parameters were considered satisfactory for the use in the future climate scenarios. It is worth noting that the performance of the simulated deliveries and stored volumes of the reservoir E801 suffered a large decrease during the validation period. This change in the skill of the model in reproducing the corresponding observed values is very likely due to the Ebro–Besaya water transfer, which started in 1982 with a reversible nature, for assuring the water supply in the Torrelavega community. This water transfer has a capacity of 4 m^3/s, with 49.9 Mm^3/year transferred and

Table 6 Goodness-of-fit indexes for the monthly calibrations and verifications

Subcatchment ID	Calibration			Validation		
	ME	PBIAS (%)	NSeff (–)	ME	PBIAS (%)	NSeff (–)
079	−1.90 (m^3/s)	−15.8	0.65	−0.98 (m^3/s)	−10.8	0.67
090	−2.24 (m^3/s)	−16.7	0.74	0.06 (m^3/s)	0.7	0.82
115	−0.50 (m^3/s)	−8.9	0.68	−0.07 (m^3/s)	−1.8	0.71
E801 – Qout	0.15 (m^3/s)	1.2	0.83	2.51 (m^3/s)	31.3	0.58
E801 – Vol	−4.06 (Mm^3)	−1.1	0.77	34.64 (Mm^3)	12.7	0.57

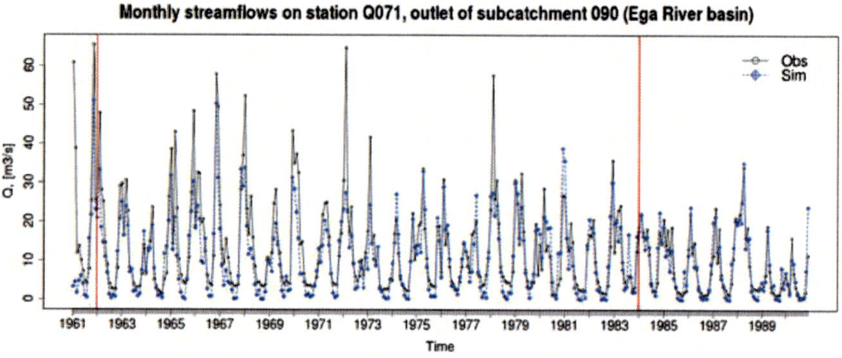

Fig. 13 Monthly observed and simulated streamflows on station Q071 (Ega en Estella, ID 090) during the calibration (1962–1983) and validation period (1984–1990). Observed values are represented by *black dots* and a *continuous black line*, whereas simulated values are represented by *blue squares* with a *dotted blue line*. *Vertical red lines* separate the warming-up, calibration and validation periods, respectively

23.7 Mm^3/year returned, between the Ebro and the Saja-Besaya River basins [35]. This water transfer was only introduced during the future scenarios, with a constant consumptive water transfer to the Saja-Besaya basin of 2.183 Mm^3/month.

6.2 *Future Scenarios (2071–2100)*

The hydrological simulations for the future time-slice 2071–2100 were performed using the bias-corrected daily time series derived from the two different climate scenarios described in Table 4 as external forcing, and the hydrological parameters obtained during the calibration and verification period described in Sect. 6.1. One operational modification was made to the simulation of subcatchment 057, where a withdrawal corresponding to the Ebro–Besaya water transfer, described in Sect. 6.1, was allowed to remove water outside the catchment. This water transfer was introduced only for the simulations of the future scenarios, while the operational rules of reservoir E801 (Embalse del Ebro) were kept the same as in the control period. A spin-up period of 1 year was used for the hydrological simulations of the future scenarios, to obtain a correct initialization of all the fluxes within each catchment.

The streamflows expected for the future scenarios at the outlet of the four selected catchments are summarized with monthly flow duration curves (FDCs, [36]), and they are compared with the FDCs corresponding to the streamflows observed during the control period, as shown in Fig. 14. To characterize the information content of a FDC, we followed the criterion proposed by Yilmaz et al. [37], where the curve is divided into three segments corresponding to different flow magnitudes: (1) a high-flow portion (0–0.2 exceedance probability) that represents the catchment response to large precipitation events; (2) a medium-flow portion (0.2–0.7 exceedance probability) representing flows controlled by moderate precipitation events coupled to medium-term baseflow; and (3) a low-flow segment (0.7–1.0 exceedance probability) representing a catchment response dominated by long-term baseflow during extended dry periods.

To gain a better understanding of the annual and seasonal effects of the two future climate scenarios considered in the present analysis, three types of monthly flow duration curves are computed: one with all the monthly streamflows (left column), another only with the winter (DJF) monthly streamflows (centre column), and final one with the summer (JJA) streamflows only (right column).

The left column of Fig. 14 shows that the two future climate scenarios project an overall reduction in the monthly streamflow in the four catchments. The projected decrease is larger for the streamflows derived from the RCAO_E RCM, than those derived from HIRHAM_H RCM. This reduction in monthly flows is in agreement with the reduction of annual precipitation and the rise of the annual mean air temperature projected by both RCMs, as discussed in Sect. 5.2, where RCAO_E RCM proved to be more severe than HIRHAM_H. In particular, subcatchment 115 experiences a much larger decrease in monthly medium and high flows, in comparison to the other three subcatchments, which seems to indicate that catchments located at lower altitude may suffer a larger decrease of monthly flows than those located at higher elevations.

The central column of Fig. 14 depicts the effect of the two future climate scenarios on the winter (DJF) monthly streamflows. Catchments 079, 090 and 115 show a substantial reduction of high and low flows, with a larger decrease projected by the RCAO_E model. However, in subcatchment 115 these reductions are, again, much larger than those suffered by the other subcatchments. On the other hand, changes predicted for the subcatchment 057 (feeding catchment of the reservoir E801) by the two RCMs are opposed, with HIRHAM_H projecting an increase in medium and low flows, whereas RCAO_E predicts a general decrease in the winter monthly flows.

The right column of Fig. 14 shows the projected changes for the summer (JJA) streamflows. Subcatchment 079, located in the northern part of the basin and with the largest elevation range, shows a decrease in low flows accompanied by an increase in the medium and high flows, with the largest increments predicted by the

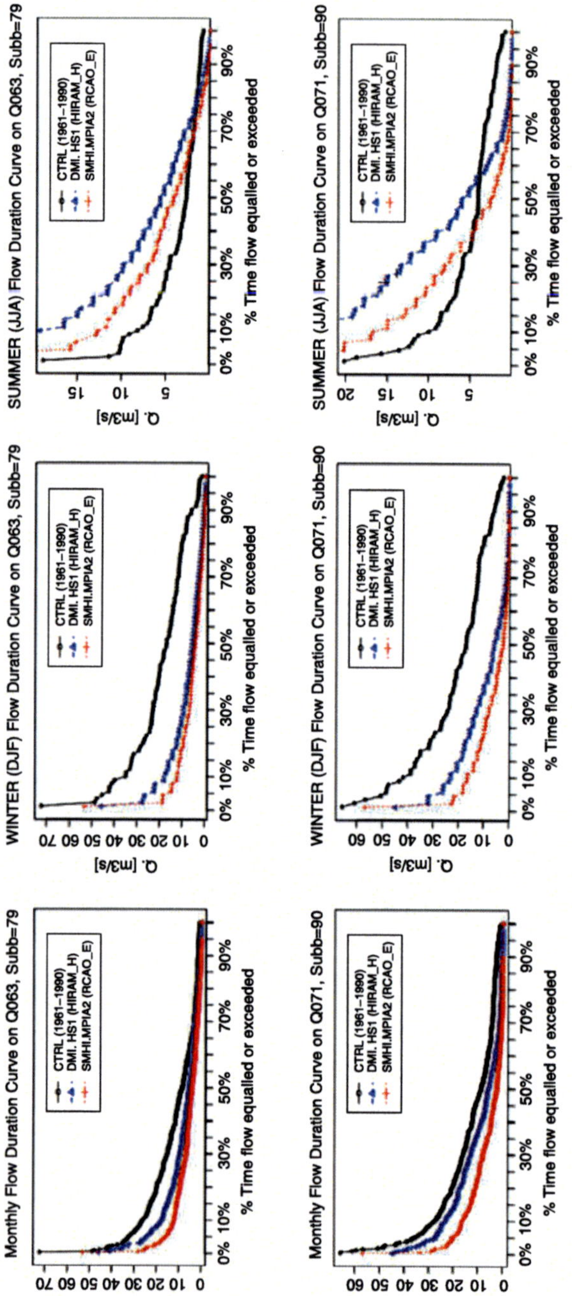

Fig. 14 (Continued)

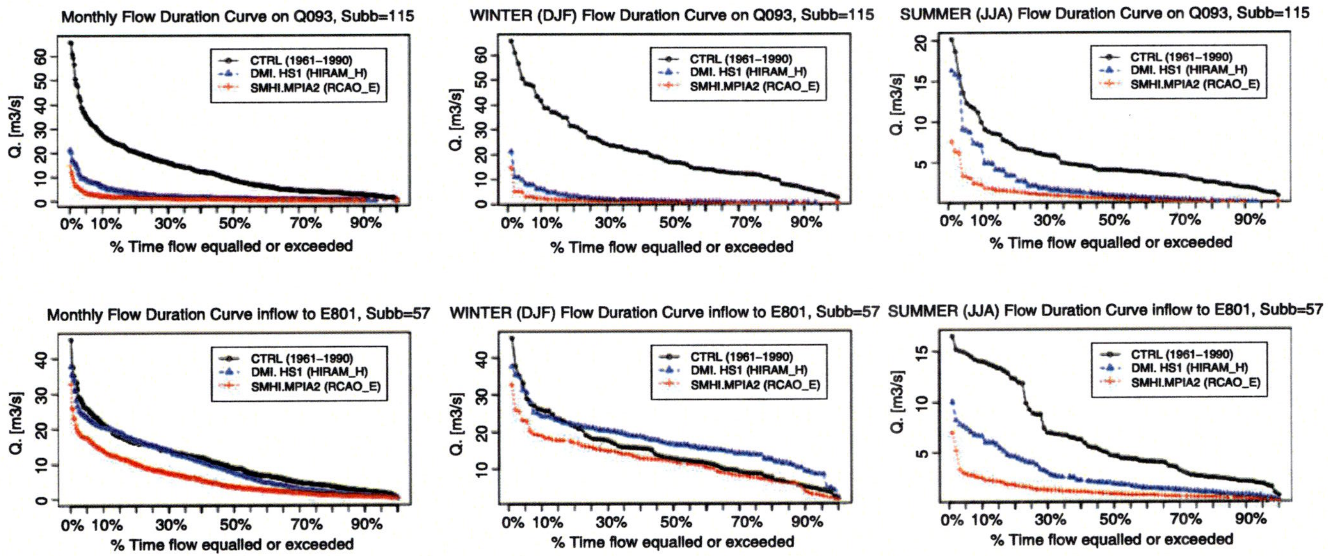

Fig. 14 Monthly flow duration curves for the subcatchments 057, 079, 090 and 115. *Black lines* refer to the flow duration curves obtained by considering the observed values during the control period 1961–1990, while the *blue* and *red lines* refer to the streamflows derived from the two RCMs HIRHAM_H and RCAO_E, respectively. The *left column* represent the flow duration curves computed with all the monthly values during the period 2071–2100; the *central column* was computed with the monthly values belonging to the winter season (DJF); and the *right column* represent the flow duration curve computed with the monthly values belonging to the summer season (JJA). Note that vertical scales are different

HIRHAM_H model. For subcatchment 090, a decrease in medium and low flows is projected by both RCMs, whereas high flows are projected to increase by both RCMs, and again the largest increments are obtained with the HIRHAM_H model. This shows that the two subcatchments located at medium and high elevations show similar patterns during summer. According to both RCMs, subcatchments 115 and 057 will experience a general decrease in the summer streamflow, with the largest decreases projected by the RCAO_E model.

7 Conclusions

This study presents expected changes in precipitation and air temperature for the Ebro River Basin by the end of this century (2071–2100), along with hydrological simulations conducted on four subcatchments thereof, for both the control period (1961–1990) and two future climate scenarios (2071–2100), to investigate the effect of climate change on water resources. Hydrological parameters that allowed a good reproduction of the monthly catchments' response during the control period were selected after a sensitivity analysis conducted by SWAT, and then obtained through calibration, performed on the period 1962–1983, followed by verification on the remaining part of the time series (1984–1990). Subsequently, bias-corrected daily time series of precipitation and mean air temperature derived from two RCMs (HIRHAM_H and RCAO_E) were used as climatic drivers for the hydrological simulations, during the future time-slice 2071–2100. The two RCMs provided scenarios with substantially different spatial and temporal patterns of daily precipitation and mean air temperature. Annual and seasonal analyses of projected streamflows suggest an important role of these climate modifications in the streamflows at the outlet of the four selected catchments. In fact, the projected flow reduction is – in general – much more severe for RCAO_E than HIRHAM_H.

For what concerns seasonal projections, both RCMs project a reduction in precipitation during the spring (14.1% and 35.1% for HIRHAM_H and RCAO_E, respectively) and summer seasons (by 31.6% and 35.3% for HIRHAM_H and RCAO_E, respectively). Likewise, both RCMs predict an increase in all the seasonal temperatures, with the largest increase projected for summer (+5.6°C and +8.8°C for the HIRHAM_H and RCAO_E, respectively).

In general, projected changes in daily precipitation and temperature resulted in a substantial reduction of the streamflow at the outlet of the four analysed catchments. Changes in streamflows were in general agreement with those of daily precipitation and air temperature, since the drop in projected monthly streamflows were larger when RCAO_E was used as driver, and the drop was larger for the catchment with the lowest aridity index (115, Homino River). For the time-slice 2071–2100, our simulations project a substantial decrease in the winter streamflow at all the four catchments. In summer, projections indicate a reduction in low flows accompanied by an increase in high flows for catchments located at higher elevations, whereas the catchment located at the lower elevation is expected to suffer a general reduction of streamflows.

This work is a first step towards a better understanding of the likely effects of climate change on water resources of the Ebro River basin (uncertainties coming from hydrological parameterisation and future climate scenarios are subject of ongoing research), which is instrumental to guide the choice of suitable mitigation and adaptation policies.

Acknowledgements The authors would like to thank the Oficina de Planificación Hidrológica de la Confederación Hidrográfica del Ebro (Hydrological Planning Office of the Hydrological Confederation of Ebro River) for the collaboration during the data collection process. In particular, many thanks to José Losada for his help with the GIS information, and to Rogelio Galván for providing daily time series. Additionally, we acknowledge financial support by the European Union FP6 Integrated Project Aquaterra (project no. GOCE 505428).

References

1. Xu C (1999) From GCMs to river flow: a review of downscaling methods and hydrologic modelling approaches. Prog Phys Geogr 23(2):229–249. doi:10.1177/030913339902300204
2. Alcamo J, Moreno JM, Novaky B, Bindi M, Corobov R, Devoy RJN, Giannakopoulos C, Martin E, Olesen JE, Shvidenko A (2007) Europe. In: Parry ML, Canziani OF, Palutikof JP, van der Linden PJ, Hanson CE (eds) Climate change 2007: impacts, adaptation and vulnerability. Contribution of Working Group II to the Fourth Assessment Report of the Intergovernmental Panel on Climate Change. Cambridge University Press, Cambridge, UK, pp 541–580
3. Arnold JG, Fohrer N (2005) SWAT2000: current capabilities and research opportunities in applied watershed modeling. Hydrol Processes 19(3):563–572. doi:10.1002/hyp. 5611
4. Arnold JG, Srinivasan R, Muttiah RS, Williams JR (1998) Large area hydrologic modeling and assessment part I: model development. J Am Water Resour Assoc 34(1):73–78
5. Neitsch SL, Arnold JG, Kiniry JR, Williams JR (2005) Soil and water assessment tool theoretical documentation Version 2005. Grassland, Soil and Water Research Laboratory; Agricultural Research Service, 808 East Blackland Road, Temple, Texas; Blackland Research Center; Texas Agricultural Experiment Station 720 East Blackland Road, Temple, Texas 76502, USA. Available at: www.brc.tamus.edu/swat/doc.html
6. Neitsch SL, Arnold JG, Kiniry JR, Srinivasan R, Williams JR (2005) Soil and water assessment tool input/output file documentation, Version 2005. Grassland, Soil and Water Research Laboratory; Agricultural Research Service 808 East Blackland Road, Temple, Texas; Blackland Research Center; Texas Agricultural Experiment Station 720 East Blackland Road, Temple, Texas 76502, USA. Available at: www.brc.tamus.edu/swat/doc.html
7. Arnold JG, Srinivasan R, Muttiah RS, Allen PM (1999) Continental scale simulation of the hydrologic balance. J Am Water Resour Assoc 35(5):1037–1051
8. Borah DK, Bera M (2004) Watershed scale hydrologic and nonpoint source pollution models: review of applications. Trans ASAE 47(3):789–803
9. Gassman PW, Reyes MR, Green CH, Arnold JG (2007) The soil and water assessment tool: historical development, applications and future research directions. Trans ASABE 50(4):1211–1250
10. Hao FH, Zhang XS, Yang ZF (2004) A distributed non-point source pollution model: calibration and validation in the Yellow River Basin. J Environ Sci 16(4):646–650
11. Van Liew MW, Garbrecht J (2003) Hydrologic simulation of the Little Washita River experimental watershed using SWAT. J Am Water Resour Assoc 39(2):413–426

12. Di Luzio M, Srinivasan R, Arnold JG (2002) Integration of watershed tools and SWAT model into BASINS. J Am Water Resour Assoc 38(4):1127–1141
13. Di Luzio M, Srinivasan R, Arnold JG, Neitsch SL (2002) Soil and water assessment tool. ArcView GIS Interface Manual: Version 2000, GSWRL Report 02-03, BRC Report 02-07. Texas Water Resources Institute TR-193, College Station, TX. 346p. Available at: www.brc.tamus.edu/swat/downloads/doc/swatav2000.pdf
14. Pebesma EJ (2004) Multivariable geostatistics in S: the gstat package. Comput Geosci 30:683–691
15. Pebesma EJ, Bivand RS (2005) Classes and methods for spatial data in R. R News 5(2). Available at: http://cran.r-project.org/doc/Rnews/
16. R Development Core Team (2009) R: A language and environment for statistical computing. R Foundation for Statistical Computing, Vienna, Austria. ISBN 3-900051-07-0. Available at: http://www.R-project.org
17. Zambrano-Bigiarini M, Majone B, Bellin A, Botter G, Nicotina L, Zanetti S, Alessi E, Marani M, Rinaldo A (2008) GEOTRANSF applications to the Ebro River basin (analysis of data consistency and conceptual model). AquaTerra deliverable C3.5
18. Zolezzi G, Bellin A, Bruno MC, Maiolini B, Siviglia A (2009) Assessing hydrological alterations at multiple temporal scales: Adige River, Italy. Water Resour 45(12):W12421. doi:10.1029/2008WR007266
19. Christensen JH, Carter TR, Rummukainen M, Amanatidis G (2007) Evaluating the performance and utility of climate models: the PRUDENCE project. Clim Change 81(1):1–6. doi:10.1007/s10584-006-9211-6
20. Bovolo CI, Blenkinsop S, Fowler HJ, Zambrano-Bigiarini M, Bellin A (2009) Production of climate change scenarios for large catchments. AquaTerra deliverable H1.13
21. Nakićenović N, Alcamo J, Davis G, de Vries HJM, Fenhann J, Gaffin S, Gregory K, Grubler A, Jung TY, Kram T, La Rovere EL, Michaelis L, Mori S, Morita T, Papper W, Pitcher H, Price L, Riahi K, Roehrl A, Rogner HH, Sankovski A, Schlesinger M, Shukla P, Smith S, Swart R, van Rooijen S, Victor N, Dadi Z (2000) Emissions scenarios. A special report of Working Group III of the Intergovernmental Panel on Climate Change. Cambridge University Press, Cambridge, UK
22. Teutschbein C, Seibert J (2010) Regional climate models for hydrological impact studies at the catchment scale: a review of recent model strategies. Geogr Compass 4(7):834–860
23. Fowler HJ, Blenkinsop S, Tebaldi C (2007) Linking climate change modelling to impacts studies: recent advances in downscaling techniques for hydrological modelling. Int J Climatol 27:1547–1578. doi:10.1002/joc.1556
24. Fowler HJ, Kilsby CG (2007) Using regional climate model data to simulate historical and future river flows in northwest England. Clim Change 80:337–367
25. Teegavarapu RSV, Chandramouli V (2005) Improved weighting methods, deterministic and stochastic data-driven models for estimation of missing precipitation records. J Hydrol 312:191–206
26. Bürger CM, Kolditz O, Fowler HJ, Blenkinsop S (2007) Future climate scenarios and rainfall-runoff modelling in the Upper Gállego catchment (Spain). Environ Pollut 148:842–854
27. Christensen JH, Christensen OB (2007) A summary of the PRUDENCE model projections of changes in European climate by the end of this century. Clim Change 81(S1):7–30
28. Giorgi F, Lionello P (2008) Climate change projections for the Mediterranean region. Glob Planet Change 63(2–3):90–104. doi:10.1016/j.gloplacha.2007.09.005
29. United Nations Environment Programme (UNEP) (1992) World atlas of desertification. Edward Arnold, London
30. CHE (2008) Confederación Hidrográfica del Ebro, Oficina de Planificación Hidrológica Caracterización de las Alteraciones del Régimen Hidrológico Sufridas en las estaciones de Aforos de la Cuenca del Ebro [last visited December 2008]. Available at: http://oph.chebro.es/documentacion/TablaResumenCaudalEA/AlteracionesEA.htm

31. Moriasi DN, Arnold JG, Van Liew MW, Bingner RL, Harmel RD, Veith TL (2007) Model evaluation guidelines for systematic quantification of accuracy in watershed simulations. Trans ASABE 50(3):885–900
32. Yapo PO, Gupta HV, Sorooshian S (1996) Automatic calibration of conceptual rainfall-runoff models: sensitivity to calibration data. J Hydrol 181(1–4):23–48
33. Nash JE, Sutcliffe JV (1970) River flow forecasting through conceptual models part I – a discussion of principles. J Hydrol 10(3):282–290
34. Beven KJ, Binley A (1992) The future of distributed models – model calibration and uncertainty prediction. Hydrol Processes 6(3):279–298
35. CHE (2000) Confederación Hidrográfica del Ebro, Oficina de Planificación Hidrológica Los Aprovechamientos en la Cuenca del Ebro: Afección en el Régimen Hidrológico Fluvial [Withdrawals in the Ebro River Basin: Impact on the Hydrological Fluvial Regime] PH-24.I
36. Vogel R, Fennessey NM (1994) Flow duration curves I: A new interpretation and confidence intervals. ASCE, J Water Resour Plan Manage 120(4):485–504
37. Yilmaz KK, Gupta HV, Wagener T (2008) A process-based diagnostic approach to model evaluation: application to the NWS distributed hydrologic model. Water Resour Res 44(9): W09417. doi:10.1029/2007WR006716
38. Ministerio del Medio Ambiente y Confederación Hidrográfica del Ebro (2007) Plan Especial de actuación en situaciones de alerta y eventual sequía en la cuenca hidrográfica del Ebro [special plan of action in situations of alert and possible droughts on the Ebro river basin], Tech.rep. Available at http://www.mma.es/secciones/acm/aguas_continent_zonas_asoc/ons/planes_sequia_isas/pdf/EBRO_cap00.pdf
39. Bovolo CI, Blenkinsop S, Majone B, Zambrano-Bigiarini M, Fowler HJ, Bellin A, Burton A, Barceló D, Grathwohl P, Barth JAC (2010) Climate change, water resources and pollution in the Ebro basin: Towards an integrated approach. Hdb Env Chem. doi 10.1007/698_2010_86

Dams and Reservoirs in the Lower Ebro River and Its Effects on the River Thermal Cycle

Jordi Prats, Joan Armengol, Rafael Marcé, Martí Sánchez-Juny, and Josep Dolz

Abstract River regulation can cause direct and indirect alterations in the river thermal cycle, which in turn may affect biological processes. In the lower Ebro River, two weirs (Ascó and Xerta) and a system of three dams (Mequinensa, Riba-roja, Flix) can be found. The weir of Ascó is used for the derivation of water to be used in the cooling system of a nuclear plant, which is returned later to the river and causing an increase of 3°C in the river water temperature. Instead, the weir of Xerta is used for the derivation of water for irrigation which most probably results in a diminution of the thermal inertia of the flowing water mass. On the other side, the system of reservoirs of Mequinensa, Riba-roja and Flix have a seasonal effect on water temperature. Water exiting the system of reservoirs is cooler than the river water entering them in spring and summer and is warmer in autumn and winter. Also, water temperature variability is reduced both in the daily and annual timescales. The reservoir of Riba-roja, receiving the contribution of two main affluents (Ebro and Segre), presents a most interesting hydrodynamic behaviour, that neither is typical lacustrine nor is that of a river. It has a variable pattern of circulation: in winter and early spring, the water column is mixed; in late spring and most of summer, the Segre River water flows above the Ebro River water; and in the rest of summer and most of autumn, the circulation pattern is inverse to the previous one, with the Ebro River water flowing above the Segre River water.

J. Prats, M. Sánchez-Juny, and J. Dolz
Fluvial Dynamics and Hydrological Engineering (FLUMEN), Civil Engineering School, Technical University of Catalonia, C. Jordi Girona 1-3, 08034 Barcelona, Spain

J. Armengol (✉)
Faculty of Biology, Fluvial Dynamics and Hydrological Engineering (FLUMEN), University of Barcelona, Av. Diagonal 645, 08028 Barcelona, Spain
e-mail: jarmengol@ub.edu

R. Marcé
Faculty of Biology, Fluvial Dynamics and Hydrological Engineering (FLUMEN), University of Barcelona, Av. Diagonal 645 08028 Barcelona, Spain
Catalan Institute for Water Research (ICRA), Scientific and Technological Park of the University of Girona, 17003 Girona, Spain

D. Barceló and M. Petrovic (eds.), *The Ebro River Basin*,
Hdb Env Chem (2011) 13: 77–95, DOI 10.1007/698_2010_68,

Keywords Dam, Reservoir, River regulation, Water temperature

Contents

1 Introduction

The water temperature is a variable of great importance for freshwater living organisms. It triggers egg eclosion and determines the biological cycle. It influences growth rate, food rate and metabolic activity. At the community level, alterations in water temperature can induce modifications in the composition of the community by exclusive competition or by making the environment no longer suitable for certain species. Mankind can alter water temperature in rivers in many ways through: reservoirs, thermal effluents, flow derivations, etc. Because of the intensity of the alteration and because of their great number, one of the most usual and altering ways of doing it are reservoirs. In an atmosphere of general concern for the climate change, the careful and accurate study of this subject is even more important. For example, Cid et al. [1] attribute the change in the life cycle of *Ephoron virgo* in the Ebro River to an increase in water temperature and Muñoz and Prat [2] relate the uniformity in the macroinvertebrate composition of the lower Ebro River to the reduced annual range caused by the reservoirs.

1.1 Thermal Alterations Caused by Reservoirs

Water temperature modification is an important consequence of river regulation. We can consider that there are two types of impacts on the thermal regime: direct and indirect alterations. The direct alterations are those in which there is a direct modification of the temperature of water, mainly through the release of water at a temperature different from the expectable in natural conditions. The indirect alterations of the thermal regime are those that alter factors that affect the processes of heat exchange between water and the environment.

Also, the alterations can be classified depending on their intensity. The more intense or frequent the intensity, the easier to predict the consequences. When the alterations are slighter or less frequent, it may occur that the natural factors are more important than the alteration itself [3]. The effect produced by the reservoirs depends on different factors, like the size of the reservoir, the residence time, the stability of the thermal stratification and the withdrawal depth. Moreover, there is a certain interannual variability in the magnitude as well as in the timing of the alterations [4]. Among these factors, the most important one is the depth at which water is released.

In reservoirs with discharge of hypolimnetic water, the effects on the water temperature have been widely studied. If there is stratification, water temperature downstream from the dam is higher in winter, colder in summer, the daily and annual thermal amplitude is reduced and the maximum annual temperature is delayed [5–7]. Recently, it has been observed that hypolimnetic discharges can reduce the variability of the temperature of the water in reduced time scales, in the range of days to weeks [8]. The low temperatures in summer can modify the composition of the fluvial community, but they can also bring the river to a previous stage of the river continuum [9, 10].

As an example of the alterations produced by this type of reservoirs, we quote the case of the lake Powell in the river Colorado. After the construction of this reservoir, there was a change in the thermal regime, mainly consisting of a decrease of the mean annual temperature by 10–15°C, with water colder in summer and warmer in winter. While before the construction of the reservoir the water temperature along the year fluctuated between 0 and 29.5°C, afterwards it only varied between 6 and 15°C [11, 12].

Another example is the Keepit Dam in the river Namoi (Australia). The alterations produced by this dam on water temperature were studied by Preece and Jones [13]. This dam has a storage capacity of 423 hm^3, a maximum depth of 40 m and water is released through a fixed-level intake structure with the inlet located at 24 m bellow full supply level. The mean discharge of the river Namoi downstream of the dam is circa 7 m^3/s. Just downstream from the dam, a 5°C decrease of the annual maximum temperature was observed, and a delay of the same of 22 days. As the distance increased, the natural thermal regime was recovered: at 100 km downstream from the dam, the differences in water temperature with respect to the situation former to the construction of the dam were less than 1°C.

By contrast, in the case of dams with discharge of surface water the spilled water is warm in summer and cold in winter in temperate latitudes [14] and they normally produce an increase of the annual thermal amplitude [15].

Lessard and Hayes [16] studied the impact caused by ten small dams of this kind in rivers of Michigan. They found a warming of water in summer (up to 5°C) in all the cases except one, in which there was a cooling by 1°C. Moreover, no return to measured upstream water temperature values was observed downstream from the dam.

The indirect alterations of the thermal regime are those produced on factors that affect the processes of heat exchange between water and the environment. We have already discussed the direct alterations caused by the reservoirs, but the management of these structures can also cause indirect alterations by means of the alteration of the flow regime. The reservoirs with hydroelectric uses often cause sudden fluctuations of the water level with daily periodicity [17]. On the other side, the reservoirs designed for the irrigation of culture fields make the flow increase in summer and make it diminish in winter and in some cases summer flows can be even higher than the winter ones [18, 19]. In other cases, the regulation of the flows causes some reduced summer flows and, consequently, a wider daily range of water temperatures. On the other hand, the regulation also causes the reduction in the magnitude of the underground flow, reducing the assimilative capacity of the river in front of variations of the water temperature [20].

1.2 Thermal Behaviour of the Ebro River

In order to determine the effect of the reservoirs on the Ebro thermal regime, it is necessary to know the thermal behaviour and characteristics of the river. In the case of rivers with high discharge, the knowledge of the local behaviour and characteristics is important, as well as having a more general perspective. In such rivers, the water temperature at a given point may depend on the river conditions several kilometres upstream (water temperature, water temperature alterations and heat exchange processes).

1.2.1 Ebro River Basin

Margalef [21] used data from the Water Quality Control Network (Red de Control de la Calidad del Agua, COCA) for the years 1972–1974 to draw the maps of the water temperature in the Spanish rivers for the months of January and July (Fig. 1). In January, in the high parts of the Ebro River basin and a great part of its contributing streams in the left margin the water temperature was lower than 5°C. In the medium and low stretches of the river and part of the tributaries of the right margin, the mean water temperature was between 5 and 9°C. In summer, however, the water temperature of the head waters of the basin in the Pyrenees was found to be below 15°C. In the central axis of the river, the temperature was between 20 and 25°C. And in the rivers placed to the South of the main axis of the Ebro River, the mean water temperature was above 25°C.

Alberto and Arrúe [22] also used data from the COCA (years 1972–1982) and data provided by private companies to study the thermal regime of streams in the Ebro River basin. Using data from the thermal power plant at Escatrón from years

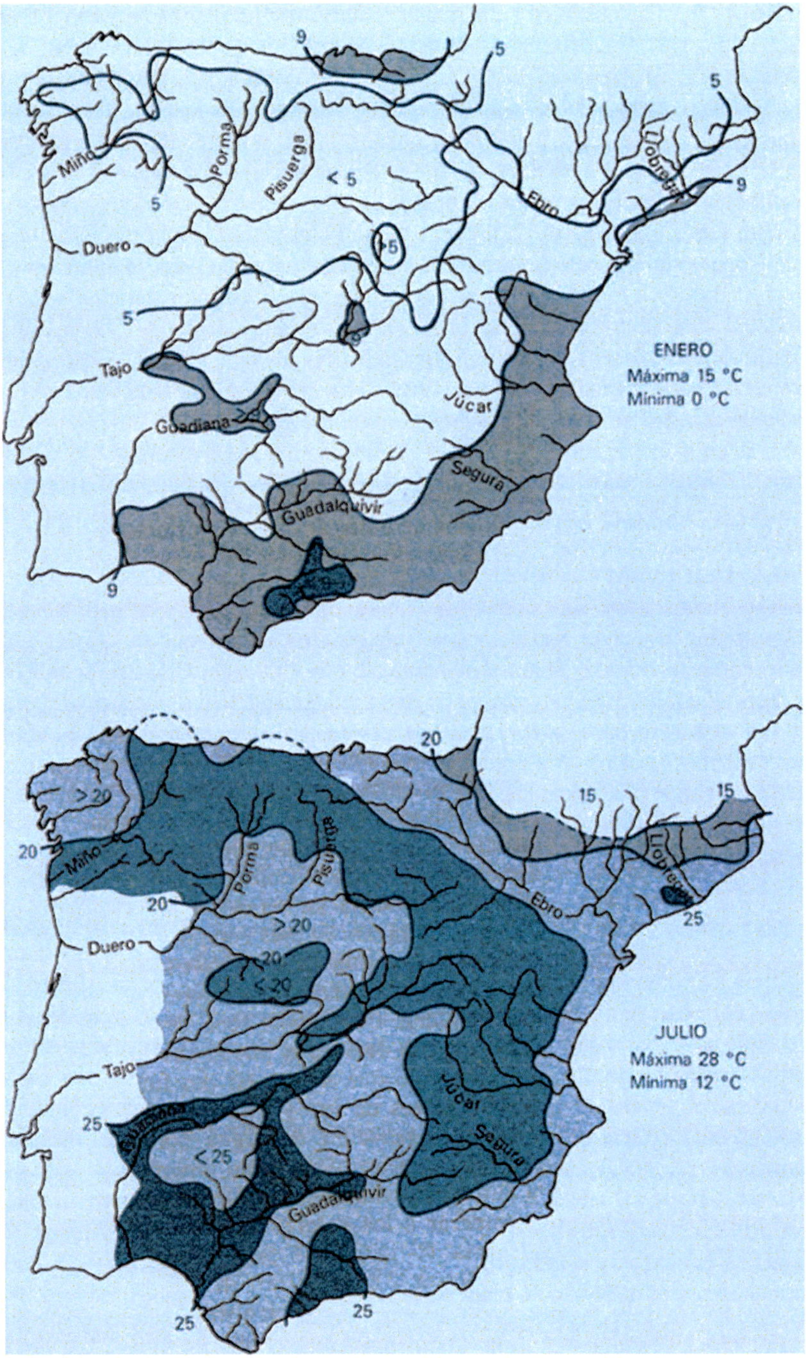

Fig. 1 Water temperature of the Spanish rivers
Source: [21]

1949 to 1978, they observed a dial water temperature oscillation between 1.5 and 2.5°C at this site. The maximum oscillation occurred in summer and the minimum in November. At the same place and despite the considerable distance from the river source (around 640 km), the origin of the flowing water had an effect on water temperature. When high flows due to snow melt occur, a decrease in water temperature can be observed. However, in the case of high flows caused by rainfall the effect on water temperature is not so clear.

Also, Arrúe and Alberto [22] obtained good correlations between mean annual water temperatures, mean annual water temperature ranges and heights for the rivers of the Ebro River basin. Comparing the actual water temperature with the estimated water temperature they observed that some rivers (rivers Jalón, Jiloca, Martín) showed higher temperatures than expected. However, although the authors formulated some hypothesis for such alterations (reservoirs, thermal sources) the available data did not suffice to isolate the actual cause.

There are some examples of studies dealing with water temperature alterations in the Ebro River basin in the scientific literature. They include different types of alterations such as those caused by power plants [22–26] and reservoirs [22, 26–30].

The effect caused by the nuclear power plant of Santa María de Garoña on the water temperature was studied by Alberto and Arrúe [22] with data from the 1970s. Using the river water for the refrigeration of the nuclear power plant produced an increase in water temperature by 2–3°C. According to the authors, the thermal equilibrium between water and the atmosphere did not recover before 150 km, at Mendavia. However, the use of water for refrigeration by the thermal power plant of Escatrón has not an appreciable effect on the river water temperature [31]. It should be noted that the power plant has the right to use a maximum of 9.1 m^3/s of the river water for refrigeration, which is small in relation to the discharge of the river at this point (mean annual discharge of about 180 m^3/s in the years 1998–1999).

Regarding the effect of reservoirs on water temperature, both heating and cooling effects have been observed. Comparing mean annual water temperatures upstream and downstream from different reservoirs of the river basin, Alberto and Arrúe [22] observed an increase in water temperature. Instead, García de Jalón et al. [27] observed a decrease in water temperature in summer because of a small capacity reservoir at the River Cimca upstream from Ainsa. The low water temperatures induced a change in the composition of the downstream community. The observed effects of the reservoirs of the lower Ebro River are discussed below.

In the long term, mean annual water temperature at Escatrón, in the medium Ebro River course, shows an increasing trend, as demonstrated by Alberto and Arrúe [22] for the period 1955–1978 and Prats et al. [32] for the period 1955–2000. During the period 1955–2000, mean annual water temperature increased by 2.3°C. This increase seems to be related to an increase in air temperature and a decrease in discharge. Also, Alberto and Arrúe [22] suggested it was due to the cumulative effects of reservoirs, urban wastewater, power plants and irrigation.

1.2.2 Lower Ebro River

Miravall [33] offers a historical description of the meteorological phenomena registered at Tortosa, in the lower Ebro River course (Fig. 2). Among others, his study included the floods and freezings of the Ebro River at Tortosa since the fifteenth century (Table 1). Notably, in most occasions the freezings of the river occurred in December or January. From the analysis of these data, it can also be deduced that the freezings had a periodicity of about 30 years in the period between 1442 and 1891. The fact that no more freezings have occurred since then can be related to the end of the Little Ice Age, which lasted from the fourteenth to the nineteenth century approximately.

The water temperature of the lower Ebro River was measured monthly by Muñoz [34] between Xerta and the mouth of the river in the years 1986–1987 (Fig. 3). Her measures showed a clear annual cycle, with minima of 8–10°C in January–February and maxima of 24–29°C in summer. In addition, the annual thermal range was around 16–18°C. The measurements of Val [29] show a similar water temperature cycle in the lower Ebro River downstream from the nuclear plant of Ascó. However, in the stretch between the reservoir of Flix and the nuclear power plant of Ascó the mean annual water temperature is about 3°C lower than downstream from the nuclear plant [26].

The heat exchanges between the river water and the environment were studied in the stretch between Flix and the power plant of Ascó by Dolz et al. [28] in the

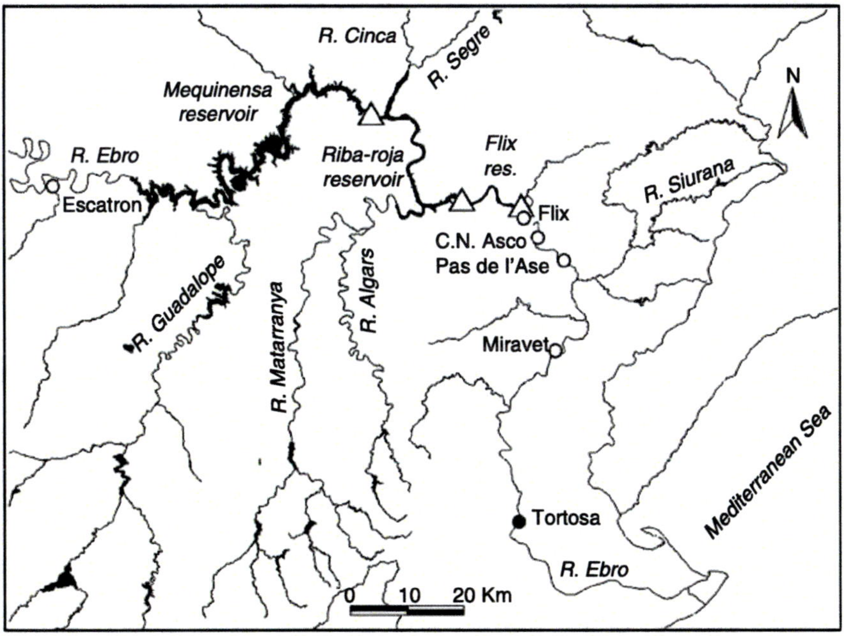

Fig. 2 Lower Ebro River. The reservoirs are indicated by *triangles* and the towns by *circles*

Table 1 Chronology of the Ebro River freezings at Tortosa since the fifteenth till the twentieth century

Chronology of the freezings of the Ebro River (fifteenth to twentieth centuries)	
End of December 1442	11 January 1709
1–2 January 1447	1712
12 December 1506	11 January 1766
29 December 1573	10 January 1784
1590	29 January 1788
30 December 1624	7 January 1789
1649	13 December 1829
11 January 1694	17 December 1891

Source: [33]

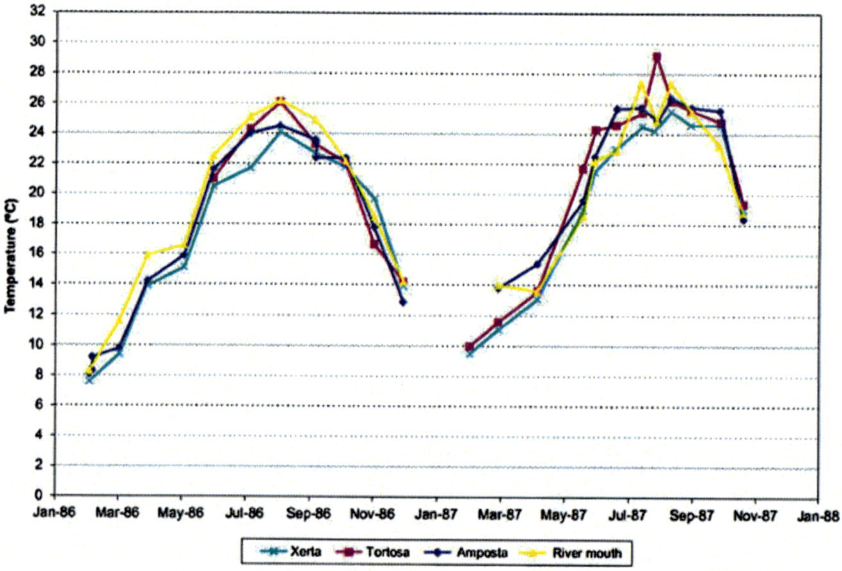

Fig. 3 Water temperature in the lower Ebro River in the years 1986–1987
Source: [34]

summer of 1990, and by Val [29] in the summer of 1998, 1999 and 2000. The different components of the heat balance in the summer of 2000 took the following mean daily values [29]:

- The solar radiation had a certain variability, although in general it took values around 200–300 W/m^2
- The long-wave incoming radiation was rather constant with values of 350–400 W/m^2
- The emitted long-wave radiation was very constant and maintained in values of −410 W/m^2
- The evaporation varied widely taking values between 0 and −150 W/m^2

- The exchange of sensible heat between the water and the atmosphere was limited and always below 50 W/m^2
- The exchange of heat between the water and the riverbed was supposed to be negligible at the dial scale

In the whole, the mean daily heat gain of the river was about 200 W/m^2.

2 Dams and Weirs in the Lower Ebro River

2.1 Weirs

In the Ebro River, hydraulic engineering works have been carried out since the Middle Ages. The first important work was the weir of Xerta, of Arabic origin. The weir of Xerta is located at some 50 km downstream from the hydroelectric power plant of Flix. The function of the weir of Xerta is the derivation of water to the right and left lower Ebro river channels, which bring water to irrigate the rice fields of the Ebro Delta. The communities that exploit these channels are entitled to extract up to 31 and 17 m^3/s of the river water respectively. No thermal alterations caused by the weir are known. However, the reduction of discharge in the river because of the derivation for irrigation should result in an increased daily water temperature range [35].

In the river stretch between Flix and Tortosa, there is another weir, the weir of Ascó. It is placed 5 km downstream from Flix, by the nuclear power station of Ascó. Its main function is to derive part of the flow of the river to the inflow channel of the refrigeration system of the nuclear power station. The power plant has been assured water right to 72.30 m^3/s of the river water by the Water Authority of the Ebro River. Water is returned to the river at some hundreds of metres downstream from the weir. Water released by the nuclear power plant is warmer than the river water and after mixing it produces an increase of mean annual water temperature of about 3°C. However, the actual effect has a certain variability as it depends on the energy production (which is rather constant) and on the discharge of the river; the higher the discharge, the lower the thermal effect. Limnos [24] and Ibàñez [25] studied the effects of the nuclear plant effluent on the biological communities of the Ebro River. The results were not conclusive as most of the species of the river are eurithermal and the upper tolerance limit was not exceeded in the study period, although according to Limnos [24] some species (*Dugesia*, *Hydropsyche exocellata*) might have been favoured by the higher water temperatures.

2.2 Dams

The lower Ebro River is under the influence of the system of reservoirs of Mequinensa, Riba-roja and Flix with the respective capacity or volume storage of 1,500 × 10^6, 210 × 10^6 and 11 × 10^6 m^3. These reservoirs regulate the

Table 2 Technical data of the reservoirs of Mequinensa, Riba-roja and Flix

	Mequinensa	Riba-roja	Flix
Storage capacity (hm^3)	1,534	210	11
Surface (ha)	7,540	2,152	320
Length (km)	100	35	13
Maximum level (masl)	121	70	41.1
Minimum level (masl)	70	40	25
Maximum depth (m)	62	34	16
Mean depth (m)	20	9.8	3.4
Spillway (masl)	106	59.5	45
Bottom outlet (masl)	87	43.3	–
Hydroelectric intake (masl)	75.2	41.4	
Additional outlet (masl)	60	40.0	–
Mean residence time	72 days	6 days	7.5 h

downstream flow of the river, with the purpose of avoiding floods and assuring the necessary flow for the main water concessions in the lower course of the Ebro River: the nuclear power station of Ascó concession and that of the channels for the irrigation of the Ebro Delta. The technical data of the reservoirs of Mequinensa, Riba-roja and Flix is shown in Table 2.

Mequinensa is a huge reservoir, 100 km in length and having more than 1,500 hm^3 of storage capacity. It has many meanders. The dam is a gravity one and the dam axis is straight. It was finished in 1966. It is owned by the Spanish power generating company ENDESA. It has four groups of hydroelectric production with a turbine flow of 150 m^3/s per group and a total power of 324 MW. The main tributary of the reservoir is the Ebro River, even though it also receives the scarce contributions of the rivers Guadalope and Martín. During the ordinary management of the reservoir, only the hydroelectric intake, the same for the four groups, and the overflow are used. The hydroelectric intake in summer takes water from the hypolimnion if the level of the reservoir is high or medium. In this case, there is a risk of releasing anoxic water. If the reservoir level is low, water is taken from the metalimnion [36]. The residence time is low, between 1 and 2.5 months [36–38]. The concentration of nutrients is high and the reservoir is considered eutrophic or meso-eutrophic [36–39].

The reservoir of Riba-roja, owned by the company ENDESA, is located in the Ebro River, between the reservoirs of Mequinensa and Flix. The construction of the dam – a gravity, straight axis one – was completed in 1969. The reservoir of Riba-roja is a valley meandering reservoir. Its storage capacity is 210 hm^3, its mean depth is 9.7 m and the maximum depth is 34 m. Its main use is power generation and has four groups with capacity to propel 225 m^3/s for each one and a total power of 263 MW. Other uses include fishing, navigation, and the supply of drinking water and irrigation for the town of Riba-roja. The reservoir of Riba-roja receives contributions of water from the reservoir of Mequinensa, and from the rivers Segre, Cinca and Matarranya.

The residence time is very low, less than 1 month [36], and it can even attain 5 days during high flows [38]. The trophic state of the reservoir can change from year to year, from mesotrophic conditions [36, 40] to eutrophic ones [38, 39].

The dam of Flix, located at the beginning of a very closed meander, is a gravity one. The hydroelectric power plant of Flix, property of ENDESA, is located near the dam, and it has a power generation of 42.5 MW and the capacity of using up to 400 m^3/s. The hydroelectric intake consists of a river diversion tunnel that cuts through the meander, so that it releases water at the end of it. When the flow is lower than 400 m^3/s, the river water passes directly through this tunnel without going through the meander. When the water level rises downstream from the hydroelectric power plant, water can even climb up the meander. However, when the river flow is higher than 400 m^3/s, a part of the flow is released through the overflow and travels through the meander, until it meets the water released from the hydroelectric power plant.

The reservoir of Flix has little storage capacity in relation to the other two reservoirs, and the residence time is low, only some hours [31, 40, 41]. In consequence, its behaviour is not the typical of a reservoir, but neither typical of a river. Instead, its behaviour is that of a river held by a great weir (i.e. a run-off-the-river reservoir).

3 Hydrodynamic Behaviour of the Reservoirs

The water temperature and thermal behaviour of the reservoirs of Mequinensa and Riba-roja have been dealt with in different studies and reports. These include water quality controls commissioned by the Ebro River Authority, the CHE [36–39], some study by the Catalan Water Authority, the ACA [42], several degree and PhD thesis at the Technical School of Civil Engineering of Barcelona [43–48] and research projects aimed to study the dispersion of the zebra mussel *Dreissena polymorpha* [40, 41]. However, the reservoir with the most information available regarding its thermal behaviour is the reservoir of Riba-roja as it has received an important research effort because of the invasion of zebra mussel.

3.1 Mequinensa

Like many other reservoirs in the temperate regions, Mequinensa is monomictic. The thermal stratification begins in spring, intensifies and attains its maximum in summer. In the autumn the water column mixes and water temperature is uniform in the vertical dimension in winter. The summer stratification is more intense close to the dam [39]. In this area, during the stratification period the surface temperatures can attain 24–27°C, while at the bottom remain around 14–16°C [36–39].

The thermocline is located at about 14–25 m deep [38, 39]. In the hypolimnion, the stratification can produce oxygen depletion and anoxic water with H_2S [36, 38, 39]. In winter the minimum water temperature can be around 5–9°C [38, 43]. The surface water temperature of the reservoir can be estimated as [28, 43]

$$T_s = 16.3 + 8.6\cos[2\pi(t + 135)/365], \tag{1}$$

where t is the ith day of the year.

Roura [46] modelled the thermal and hydrodynamic behaviour of the reservoir using the computer model DYRESM. According to the simulation results, the stratification initiates at the end of June and reaches its maximum at the end of September. Since then, the intensity of the stratification weakens, so that in January the water column is completely mixed.

3.2 Riba-roja

One of the main characteristics of the Riba-roja reservoir is that it receives the contribution of two inflows of similar order of magnitude. On one side, it receives the outflow from the reservoir of Mequinensa. On the other side, it receives water from the river Segre, which also receives water from the Cinca River just before entering the reservoir. For a great part of the year, the discharge of the two inflows is similar, although in winter and spring the inflow from the Ebro River is usually much higher (Fig. 4). The water temperature of the inflows for the years 1997–1999 can be seen in Fig. 5. Given the low residence time, the characteristics of the two main inflows determine the hydrodynamic behaviour of the reservoir.

The reservoir of Riba-roja presents a weak stratification in summer, so that surface water temperatures are 24–27°C and 18–21°C at the bottom [36, 39]. The thermocline is situated between 14 and 21 m deep [38, 39]. During this time of the year, the water entering the reservoir of Riba-roja from the reservoir of Mequinensa is colder than the water coming from the Segre River. In consequence, the denser Ebro River water sinks and flows along the bottom of the reservoir while the Segre River water floats and flows along the surface [40, 41]. The deep water can be anoxic [36, 38, 39]. In winter, the water temperature is uniform in the water column and the minimum values are around 11°C [38].

The thermal and hydrodynamic behaviour of the reservoir has been modelled by Gonzalez [47] and Salgado [48] using the DYRESM model. These works have dealt with the study of the variables necessary for the simulation [47] and the calibration of the parameters of the model [48]. On the basis of their work, it has been possible to determine a complex hydrodynamic behaviour mainly driven by the temperature and discharge of the inflows. A numerical simulation experiment with a conservative tracer was designed by considering the salinity of the Segre River to be 0 psu and that of the Ebro River 0.1 psu. It must be said that the real river salinity is low enough not to affect the results of the modelling. The results of this numerical

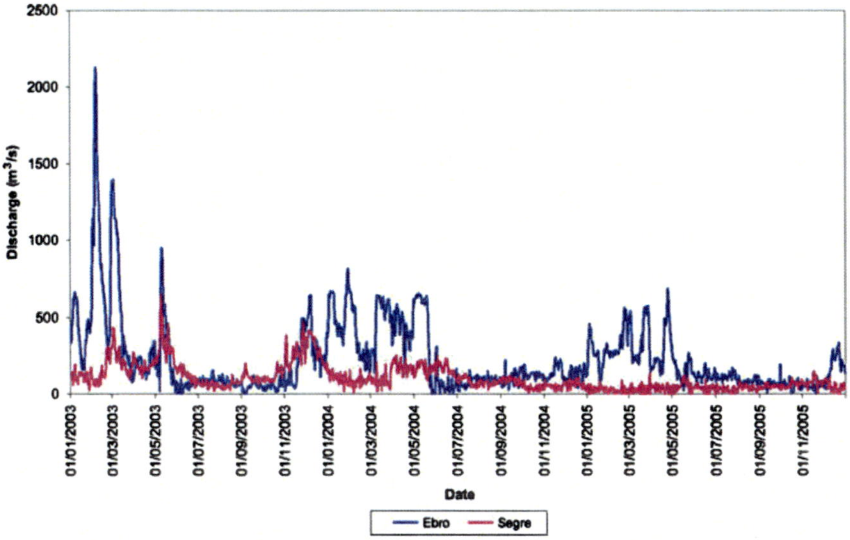

Fig. 4 Inflows to the reservoir of Riba-roja (2003–2005)
Source: ENDESA

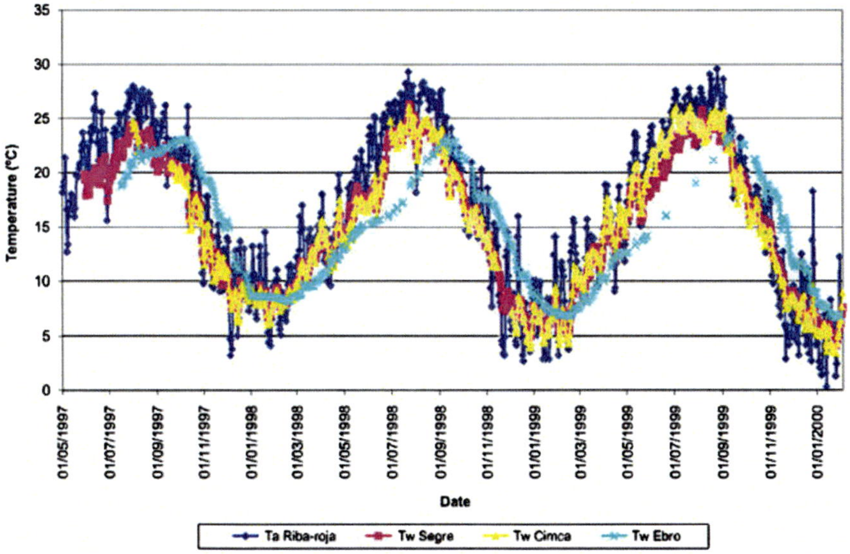

Fig. 5 Mean daily water temperature (Tw) of the inflows to the reservoir of Riba-roja and air temperature (Ta). Years 1997–1999
Source: [29]

experiment can be seen in Fig. 6. At the beginning of the year, from January to March, the water column is mixed and most of the water comes from the Ebro River. Since April, after the spring floods, it can be seen how the water of the Segre

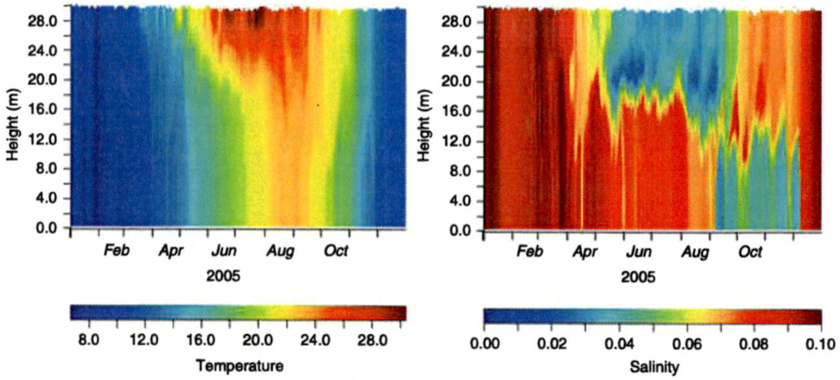

Fig. 6 Numerical simulation experiment with tracer. The Ebro River water is indicated by water with salinity equal to 0.1 psu and that of the Segre River equal to 0 psu

River circulates on the surface. The water from the Segre is warmer than that of the Ebro River at this moment of the year. In consequence, it is less dense and floats above. During September, the pattern of circulation switches: the Ebro River water is now warmer than the Segre River water and flows on the surface, although the mixing of the two layers is a bit more intense. The situation remains like this until December, when the winter high flows make the water column uniform again.

The summer pattern of circulation with the Segre River flowing above the Ebro River water has been pointed as providing optimal epilimnetic conditions for the growth of the zebra mussel [40].

4 Downstream Thermal Effects of the Reservoirs

Ward [49], Barnes and Minshall [5] and Ward [6] describe the typical effects of deep-release reservoirs on downstream water temperature: higher temperatures in winter, lower temperatures in summer, reduced daily and annual thermal amplitude and displaced annual maximum and minimum water temperature. These effects have been observed when studying the reservoir of Mequinensa. While just upstream from the reservoir, at Escatrón, the annual thermal amplitude is about 19°C, downstream from the reservoir it is reduced to about 15°C and the annual maxima occur about 20 days later. Also, the effect on water temperature varies along the year on a cyclical basis. In November, the mean monthly water temperature at Escatrón can be as much as 6°C higher than that of the water exiting the reservoir. On the other hand, it can be as much as 4°C lower in May.

In the reservoir of Riba-roja, water from the Segre River somewhat mitigates the effect of the reservoir of Mequinensa, although the impact of reservoirs downstream of the system of reservoirs is still remarkable, as demonstrated by

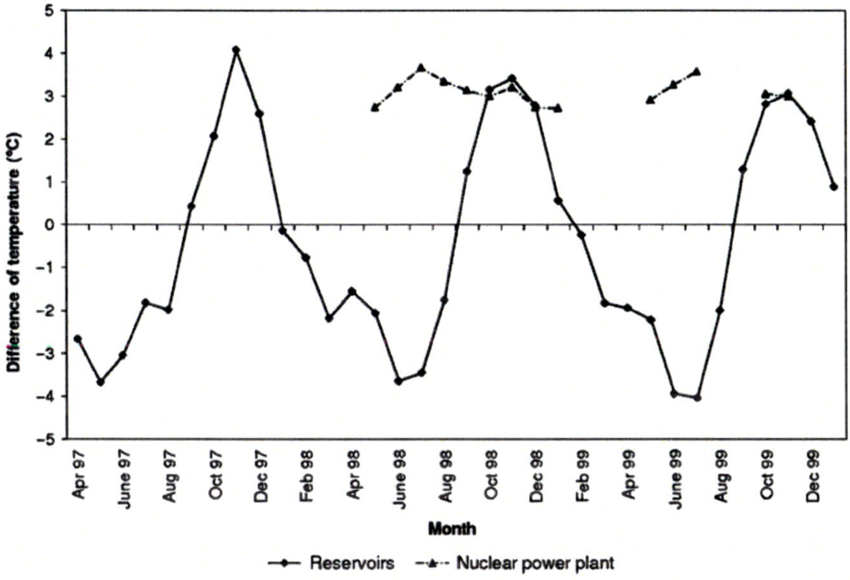

Fig. 7 Water temperature variation caused by the system of reservoirs of Mequinensa, Riba-roja and Flix and by the nuclear power plant of Ascó

Dolz et al. [28], Val [29] and Prats et al. [26, 50] (Fig. 7). During summer, the water leaving the system of reservoirs was colder than the water entering the system at Escatrón, with a maximum difference of mean monthly water temperature of 3.5–4°C. During winter, the situation reversed and the outflow from the reservoirs was warmer than the inflowing water; the increase in temperature could be as high as 3–3.5°C.

About 5 km downstream from the system of reservoirs, there is the nuclear power plant of Ascó. In general, it produces an increase in water temperature of about 3°C. In consequence, it cancels out the cooling effect of the reservoirs in summer. However, in winter it sums up to the warming effect.

5 Concluding Remarks

Like water temperature, water quality downstream from reservoirs depends on the hydrodynamic processes that take place in the reservoirs. In the case of the Ebro River, the effect of the reservoir of Mequinensa on water quality was studied by Roura [46]. According to her results, this reservoir acted as a treatment plant that improved the quality of the river water. Also, the hydrodynamics of the reservoir depends on upstream water temperature, discharge and meteorological variables. Such forcing parameters change over the years and, consequently, the water quality should change in accordance. Water temperature at Escatrón, at the entrance of the reservoir of Mequinensa, has been observed to increase by 2.3°C in the period 1955–2000 [32] paired up with an increasing trend in air temperature of 0.03–0.07°C/year over the last 3 decades of the twentieth century in the Ebro River basin and in the Iberian Peninsula [51–53]. Mean river discharge has decreased by 5,760 hm^3 in the period 1940–1997 [54] and precipitation has diminished by 4% in the period 1947–1999 [55]. How these changes have affected water quality has not been studied to our knowledge. In the present situation of concern for climate change effects, further and wider studies on the hydrodynamics and chemical behaviour of the Mequinensa reservoir could help in taking future reservoir management decisions.

The special importance of studying the reservoir of Riba-roja should also be noted. It is a paradigmatic example of a reservoir that does not behave as a typical model because of its particular pattern of circulation. Although one of the models most extensively used to explain the behaviour and ecology of reservoirs is the monomictic lake model, the necessity of taking into account other kinds of models has been stated [56]. The model has been, for many years, a valuable tool of interpretation of the physical and chemical processes that take place in reservoirs. It is still a perfectly valid tool in many cases. However, as it has been shown in the case of the Riba-roja reservoir, more complex hydrodynamic models may be applicable. The behaviour of this reservoir is not typical lacustrine and cannot be assimilated to that of a river. Instead, it has an intermediate behaviour. In the study of these behaviours, the use of numerical modelling can be very useful. Furthermore, in the case of the reservoir of Riba-roja, fouled by the zebra mussel, a detailed knowledge of its hydrodynamic behaviour is even more important as it has been seen that it can influence the distribution of the invading organism in the reservoir [40].

Acknowledgements The authors thank Antoni Palau and ENDESA for their assistance and the Confederación Hidrográfica del Ebro for the data provided. The authors also thank the Center for Water Research of the University of Western Australia for providing the DYRESM model. Part of the data presented here was the result of projects CGL2004-05503-C02-01/02/HID and GL2008-06377-C02-01/02 funded by the *Programa de Recursos Hídricos del Plan Nacional de Investigación y Desarrollo*. One of the authors has benefitted from an FPI grant from the *Programa de Recursos Hídricos del Plan Nacional de Investigación y Desarrollo* and the European Social Fund.

References

1. Cid N, Ibáñez C, Prat N (2008) Life history and production of the burrowing mayfly Ephoron virgo (Olivier, 1791) (Ephemeroptera: Polymitarcyidae) in the lower Ebro river: a comparison after 18 years. Aquatic Insects 30:163–178
2. Muñoz I, Prat N (1994) Macroinvertebrates in the lower Ebro River (NE Spain). Hydrobiologia 286:64–78

3. Prat N, Ward JV (1994) The tamed river. In: Margalef R (ed) Limnology now: a paradigm of planetary problems. Elsevier, Amsterdam, pp 219–236
4. Webb BW, Walling DE (1993) Temporal variability in the impact of river regulation on thermal regime and some biological implications. Freshw Biol 29:167–182
5. Barnes JR, Minshall GW (1983) Stream ecology: application and testing of general theory. Plenum, New York, p 399
6. Ward JV (1985) Thermal characteristics of running waters. Hydrobiologia 125:31–46
7. García de Jalón D (1996) Impactos de las modificaciones del régimen térmico en las comunidades fluviales. In: Dolz J, Puertas J, Aguado A, Agulló L (eds) Efectos térmicos en presas y embalses. Colegio de Ingenieros de Caminos Canales y Puertos, Madrid, pp 95–107
8. Steel EA, Lange IA (2007) Using wavelet analysis to detect changes in water temperature regimes at multiple scales: effects of multi-purpose dams in the Willamette River basin. River Res Appl 23:351–359
9. Ward JV, Stanford FR (1983) The serial discontinuity concept of lotic ecosystems. In: Fontaine TD, Bartell SM (eds) Dynamics of lotic ecosystems. Ann Arbor Science, Ann Arbor, pp 29–42
10. Sabater F, Armengol J, Sabater S (1989) Measuring discontinuities in the Ter River. Regul Rivers Res Manage 3:133–142
11. Stanford JA, Ward JV (1986) The Colorado River System. In: Davies BR, Walker KF (eds) The ecology of river systems. Dr. Junk Publishers, Dordrecht, p 793
12. Ward JV, Zimmermann HJ, Cline LD (1986) Lotic zoobenthos of the Colorado system. In: Davies BR, Walker KF (eds) The ecology of river systems. Dr. Junk Publishers, Dordrecht, p 793
13. Preece RM, Jones HA (2002) The effect of Keepit Dam on the temperature regime of the Namoi River, Australia. River Res Appl 18:397–414. doi:10.1002/rra.686
14. Wotton RS (1995) Temperature and lake-outlet communities. J Therm Biol 20:121–125
15. Allan JD (1995) Stream ecology. Structure and function of running waters. Chapman & Hall, London
16. Lessard JL, Hayes DB (2003) Effects of elevated water temperature on fish and macroinvertebrate communities below small dams. River Res Appl 19:721–732. doi:10.1002/rra.713
17. Prats J, Dolz J, Armengol J (2009) Variabilidad temporal en el comportamiento hidráulico del curso inferior del río Ebro. Ingeniería del Agua 16(4):259–272
18. CHE (2000) Los aprovechamientos en la Cuenca del Ebro: afección en el régimen hidrológico fluvial. Oficina de planificación hidrológica, CHE, Zaragoza, p 83
19. Batalla RJ, Gómez CM, Kondolf GM (2004) Reservoir-induced hydrological changes in the Ebro River basin (NE Spain). J Hydrol 290:117–136
20. Poole GC, Berman CH (2001) An ecological perspective on in-stream temperature: natural heat dynamics and mechanisms of human-caused thermal degradation. Environ Manage 27:787–802
21. Margalef R (1983) Limnología. Editorial Omega, Barcelona, p 1010
22. Alberto F, Arrúe JL (1986) Anomalías térmicas en algunos tramos de la red hidrográfica del Ebro. Anales de la Estación Experimental Aula Dei 18:91–113
23. Departament d'Enginyeria Hidràulica, Marítima i Ambiental (DEHMA) (1990) Estudio del comportamiento térmico del río Ebro en el tramo C. N. Ascó – Pas de l'Ase. Technical University of Catalonia, Barcelona, p 44
24. Limnos (1997) Estudi dels efectes de l'abocament tèrmic de la central nuclear d'Ascó sobre les comunitats biològiques. Junta de Sanejament, Barcelona, p 50
25. Ibàñez C (1998) Anàlisi i conclusions dels resultats de l'estudi dels efectes de l'abocament tèrmic de la Central d'Ascó sobre les comunitats biològiques. Junta de Sanejament, Barcelona, p 22
26. Prats J, Val R, Armengol J, Dolz J (2004) Regulation in the lower Ebro River: thermal regime alterations. In: Proceedings of the fifth international symposium on ecohydraulics. Aquatic habitats: analysis & restoration, Madrid, Spain, 12–17 September 2004, pp 873–878

27. Garcia de Jalon D, Montes C, Barcelo E, Casado C, Menes F (1988) Effects of hydroelectric scheme on fluvial ecosystems within the Spanish Pyrenees. Regul Rivers Res Manage 2:479–491
28. Dolz J, Puertas J, Herrero E (1994) Water temperature alteration downstream from a reservoir. The Ebro river case (Spain). Commission International de Grands Barrages 69, pp 211–225
29. Val R (2003) Incidencia de los embalses en el comportamiento térmico del río Ebro. Caso del sistema de embalses Mequinenza-Ribarroja-Flix en el río Ebro. Doctoral thesis, Technical University of Catalonia, Barcelona
30. Val R, Niñerola D, Armengol J, Dolz J (2003) Incidencia de los embalses en el régimen térmico del río. El caso del tramo final del río Ebro. Limnetica 23:85–93
31. Limnos (1998) Estudio de la calidad ecológica integral del río Ebro (on line). Available at: http://oph.chebro.es/DOCUMENTACION/Calidad/estudios/calecolebro1998/index.htm
32. Prats J, Val R, Armengol J, Dolz J (2007) A methodological approach to the reconstruction of the 1949–2000 water temperature series in the Ebro River at Escatrón. Limnetica 26:293–306
33. Miravall R (1997) Flagells naturals sobre Tortosa. Riuades, gelades, aiguats i sequeres. Ventades i terratrèmols. Columna Tresmall, Barcelona, p 191
34. Muñoz I (1990) Limnologia de la part baixa del riu Ebre i dels canals de reg: els factors físico-químics, el fitoplàncton i els macroinvertebrats bentònics. Doctoral thesis, University of Barcelona, Barcelona
35. Meier W, Bonjour C, Wüest A, Reichert P (2003) Modeling the effect of water diversion on the temperature of mountain streams. J Environ Eng 129:755–764. doi:10.1061/(ASCE)0733-9372(2003) 129:8(755)
36. Limnos (1996) Diagnóstico y gestión ambiental de embalses en el ámbito de la Cuenca Hidrográfica del Ebro. Confederación Hidrográfica del Ebro, Zaragoza
37. Infraestructura y Ecología (2003) Seguimiento de la calidad de las aguas en embalses de zonas sensibles. Confederación Hidrográfica del Ebro, Zaragoza
38. Infraeco (2006) Ejecución de trabajos relacionados con los requisitos de la Directiva Marco (2000/60/CE) en el ámbito de la Confederación Hidrográfica del Ebro referidos a: elaboración del registro de zonas protegidas, determinación del potencial ecológico de los embalses, desarrollo de programas específicos de investigación. Embalse de Ribarroja. Confederación Hidrográfica del Ebro, Zaragoza, p 44
39. URS (2002) Actualización limnológica de embalses. Confederación Hidrográfica de Ebro, Zaragoza
40. Navarro E, Bacardit M, Caputo L, Palau T, Armengol J (2006) Limnological characterization and flow patterns of a three-coupled reservoir system and their influence on *Dreissena polymorpha* populations and settlement during the stratification period. Lake Reservoir Manage 22:293–302
41. Armengol J, Bacardit M, Caputo L, Gallegos MA, Navarro E (2003) Ecologia aquàtica dels embassaments de Mequinensa, Ribarroja i Flix. University of Barcelona, Department of Ecology, p 27
42. Agència Catalana de l'Aigua (ACA) (2003) Caracterització i propostes d'estudi dels embassaments catalans segons la Directiva 2000/60/CE del Parlament Europeu. Departament de Medi Ambient, Barcelona, p 212
43. Miguel JJ (1993) Análisis de la estratificación térmica del embalse de Mequinenza. Degree thesis, Technical University of Catalonia
44. Perucho X (1994) Estudi del fenomen d'estratificació tèrmica de l'aigua i aplicació a diversos embassaments. Degree thesis, Technical University of Catalonia
45. Martí B (1997) Estudio de la estratificación térmica de los embalses aplicado al caso de Mequinenza. Degree thesis, Technical University of Catalonia
46. Roura M (2004) Incidència de l'embassament de Mequinensa en el transport de sòlids en suspensió I la qualitat de l'aigua del riu Ebre. PhD thesis, Faculty of Biology, Un iversity of Barcelona
47. González O (2007) Modelització de la hidrodinàmica d'un embassament. Aplicació al cas de l'embassament de Riba-roja al riu Ebre. Degree thesis, Technical University of Catalonia

48. Salgado A (2008) Optimització de paràmetres del model hidrodinàmic de l'embassament de Riba-roja al riu Ebre. Degree thesis, Technical University of Catalonia
49. Ward JV (1974) A temperature-stressed stream ecosystem below a hypolimnial release mountain reservoir. Archiv für Hydrobiologie 74:247–275
50. Prats J, Val R, Armengol J, Dolz J (2010) Temporal variability in the thermal regime of the lower Ebro River (Spain) and alteration due to anthropogenic factors. Journal of Hydrology 387:105–118
51. Seguí J (2003) Análisis de la serie de temperature del Observatorio del Ebro 1894-2002. Observatori de l'Ebre, Roquetes, Spain, p 83
52. Martín Vide J (2005) Factors geogràfics, regionalització climàtica, i tendencies de les series climàtiques a Catalunya. In: Llebot JE, Queralt A, Rodó J (eds) Informe sobre el canvi climatic a Catalunya. Departament de la Presidència, Consell Assessor per al Desenvolupament Sostenible de Catalunya (CADS), Barcelona, pp 81–111
53. Moreno JM (coord) (2005) Evaluación preliminar de los impactos en España por efecto del cambio climático. Ministerio de Medio Ambiente, Madrid, Spain, p 822
54. Gallart F, Llorens P (2001) Water resources and environmental change in Spain. A key issue for sustainable integrated catchment management. Cuadernos de Investigación Geográfica 27:7–16
55. Ayala-Carcedo FJ (2001) Impactos del cambio climático sobre los recursos hídricos en España y viabilidad del Plan Hidrológico 2000. In: Arrojo P (ed) El Plan Hidrológico Nacional a Debate. Bakeaz, Bilbao, Spain, pp 51–56
66. Marcé R, Moreno-Ostos E, López P, Armengol J (2008) The role of allochtonous inputs of dissolved organic carbonon the hypolimnetic oxygen content of reservoirs. Ecosystems 11:1035–1053

Dissolved Fluxes of the Ebro River Basin (Spain): Impact of Main Lithologies and Role of Tributaries

Emmanuelle Petelet-Giraud and Philippe Négrel

Abstract The aim of this study was to evaluate, over more than 20 years, the export fluxes for dissolved loads on the catchment scale of the Ebro River basin in Spain. Data are compiled from the databank of the *Confederación Hidrográfica del Ebro* (CHE). The spatial and temporal distribution of daily discharges, physico-chemical parameters and chemical data covering the last two decades (1981–2003) were investigated in the Ebro Basin on five monitoring stations along the Ebro River (Mendavia, Castejon, Zaragoza, Sastago and Tortosa), as well as six stations at the outlet of the main tributaries (Arga, Aragon, Gallego, Jalon, Cinca and Segre). The dissolved load of the rivers at the Ebro Basin scale was characterised through the EC, total dissolved solids (TDS) and the major elements chemical data. The surface water can be classified into three main categories, a clear dominance of Ca–SO_4 water type, a Ca–HCO_3 type mainly encountered in the upper part of the basin and some data presenting a Na–Cl water type. The TDS values are highly variable, both in time and in space, in the range 390–1,360 mg L^{-1}. The dissolved exportations to the Mediterranean Sea and the relative contribution of the different tributaries were calculated. The Ebro basin in its upper part (upstream Mendavia) contributes around 22.4% of the total exported flux near the outlet (Tortosa) over the studied period. The tributaries that mainly contribute to the total exported load are the Cinca and Segre (19% and 17% respectively). The Aragon, Gallego and Jalon contributions are very low, often less than 5% of the total exported flux. The specific TDS flux at the outlet of the Ebro is 70 ± 23 t km^{-2} $year^{-1}$ and 108 ± 24 t km^{-2} $year^{-1}$ upstream in Mendavia while the highest chemical erosion rate was calculated for the Arga with 251 ± 55 t km^{-2} $year^{-1}$. The dissolved export fluxes represent the major export from the Ebro basin, and the respective contribution of carbonate and evaporite (gypsum) with respect to the TDS was then calculated using the major element concentrations and discharge data. In the

E. Petelet-Giraud (✉) and P. Négrel
BRGM, Avenue C. Guillemin, BP 36009, 45060 Orléans Cedex 02, France
e-mail: e.petelet@brgm.fr

D. Barceló and M. Petrovic (eds.), *The Ebro River Basin*,
Hdb Env Chem (2011) 13: 97–120, DOI 10.1007/698_2010_91,

upper part of the Ebro Basin, carbonate weathering is dominant compared to gypsum weathering while downstream the dissolved exportations are dominated by carbonate weathering. For the tributaries, most of them are dominated by evaporite weathering. The exportation rates at the outlet of the watershed shows dissolved exportation derived from gypsum weathering, that are about 1.5 times that derived from carbonate weathering.

Keywords Carbonate dissolution, Dissolved flux, Dissolved load, Ebro River basin, Evaporite dissolution, Long-term fluxes

Contents

1 Introduction

Over the last decades, numerous studies dedicated to weathering fluxes assessment have been carried out on major world rivers. Purposes were multiple, from CO_2 consumption rates estimates to global erosion rates (chemical and mechanical) from continents to oceans. Main factors controlling the erosion processes (climate, lithologies and tectonic) were studied to define general erosion laws [1–8]. The impact of growing anthropogenic pressures (industry, agriculture and domestic inputs) on chemistry of river waters versus the natural contribution derived from weathering of various lithologies was also investigated [9–11]. Knowledge of the fluxes for present-day weathering (determination of inputs and outputs as well as the controlling parameters) is needed to understand the past and to predict the future, in terms of both water quality and ecosystem evolution, especially in delta areas.

In the Ebro River basin, it has been established that the human impact through dam construction has drastically changed the natural basin functioning in terms of transported loads to the Mediterranean Sea. Nowadays, dissolved transport dominates the solid one [30], suspended matter (SPM) being massively trapped in the dams along the Ebro course and its main tributaries [12, 13]. This study thus focuses on the dissolved exportation rates at the Ebro River basin scale, using five monitoring stations along the Ebro River and six stations from the outlet of the main tributaries, and to assess the relative contributions of the main lithologies (carbonates and evaporites) to the total dissolved load at the outlet of the Ebro Basin.

2 The Ebro River Basin

The Ebro River (Fig. 1) is located in north-eastern Spain, it rises near the Atlantic coast in the Cantabrian Mountains. It flows into the western Mediterranean Sea through several large cities and agricultural, mining and industrial areas. The river is 928 km long, has an 85,530 km^2 drainage basin and is the largest Spanish fluvial system. There are 21 main tributaries that feed the Ebro main stream, 10 flowing from the Northern part, i.e. the left bank, and 11 from the Southern part, i.e. the right bank. The relative contribution to the total runoff (approximately $6{,}837.10^6$ m^3) of the Southern area is only about 5% [13]. The Ebro River flow regime is regulated by more than 187 dams, with a total capacity equivalent to 57% of the total mean annual runoff, inducing a reduction in flood magnitude, and a reduction of the variability of mean daily flows due to storage of winter floods and increased

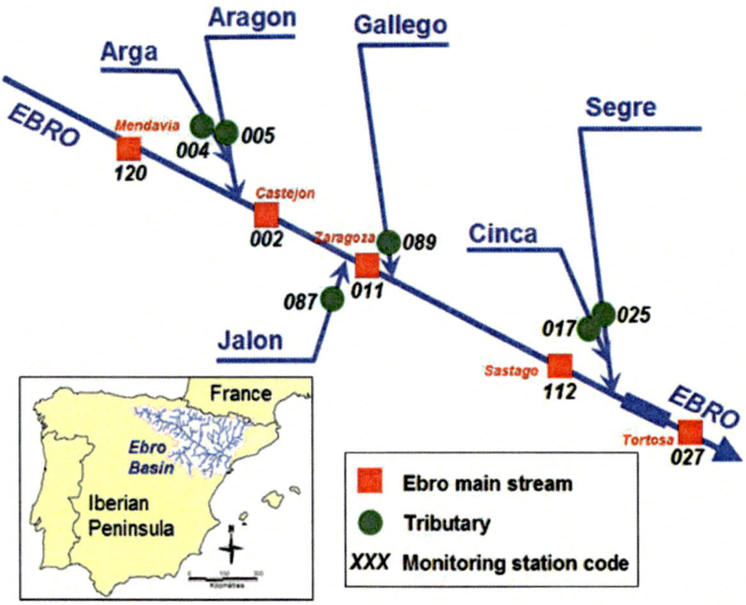

Fig. 1 Location map of the Ebro basin in Spain and scheme of the Ebro River basin with the monitoring stations (name and code) selected for this study

baseflows in summer for irrigation [13]. Monthly water discharge is quite irregular, with a significant decrease over the last century, mainly attributed to increased water use for human activities (agricultural irrigation, reservoirs, electricity production and domestic consumption). Irrigation is responsible for an important hydraulic deficit, with an average of around 300 $m^3\ s^{-1}$ taken from the river. The Ebro natural flow between 1914 and 1935 was around 590 $m^3\ s^{-1}$, whereas in the last few decades (1960–1990) it was 430 $m^3\ s^{-1}$, which is a 28% reduction [12]. The Ebro River is one of the largest contributors of freshwater into the Mediterranean Sea. The river ends in the Ebro Delta, one of the most important wetlands in Europe, covering an area of about 320 km^2 of sediments with swamps and a coastal lagoon system, which are valuable in terms of natural resources and related economic activities [14].

The catchment area is heterogeneous in terms of geology, topography and land use, but the most heterogeneous parameter is the climatology. The Ebro Basin is divided into four main climatic areas, showing different degrees of runoff impoundment: the Atlantic headwaters, with average annual precipitation of about 900 mm, the west-central Pyrenees (about 950 mm), the eastern Pyrenees (about 800 mm) and the southern Mediterranean zone (about 500 mm). More generally, the mean annual precipitation varies from 2,000 mm in the Pyrenees to less than 400 mm in the arid interior, e.g. annual precipitation averages 324 mm in Zaragoza, where a potential evapotranspiration of more than 1,000 mm has been calculated [13].

The Ebro River basin covers several geological terrains: (1) The sedimentary and metamorphic Palaeozoic terrains of the Iberian and Pyrenees mountains, the igneous terrains of these two areas and the Priorat Massif. (2) The sedimentary Mesozoic terrains of the Iberian Chain. The Iberian Chain is an intraplate mountain range that occupies the north-eastern part of the Iberian Peninsula and forms the southern margin of the Ebro Basin. It is made up of rocks that cover most of the Phanerozoic stratigraphic record; however, Jurassic and Cretaceous carbonated formations predominate in the Iberian Chain. (3) The sedimentary Tertiary terrains of the Pre-Pyrenees and the Ebro Valley. (4) The lower part of the Ebro River flows across a Neogene depression, bends through the Catalan Chain and the Baix Ebre Valley and flows across delta plain deposits [15].

3 Data Collection: Chemistry and Hydrology

The Ebro headwater tributaries still have a natural flow regime [16], whereas the fluvial regimes of the main tributaries reaching the Ebro mainstream in the central and lower part of the catchment (Segre, Cinca, Gallego and Aragon rivers) are slightly to moderately altered.

The Centro de *Estudios Hidrográficos del Centro de Estudios y Experimentación de Obras Públicas* (CEH-CEDEX, http://hercules.cedex.es) provides the daily discharges for 275 gauging stations in the Ebro watershed. The CHEBRO website (http://www.chebro.es/) gathers the available chemical data (Ca, Na, K, Mg, Cl, SO_4, NO_3, HCO_3) and physico-chemical parameters (T, electrical conductivity EC, pH) together with the discharge value (*W*) corresponding to the sampling date. For this study, a selection of monitoring station was done to characterise the main sub-catchments and several sections of the Ebro main stream. In that way, six monitoring stations were selected at the outlet of the main tributaries (Fig. 1), five from the left bank with the highest drained volumes: the Arga (station n°004) and Aragon (n°005) before their confluence and the confluence in the Ebro main stream, the Gallego (n°089), and the Cinca (n°017) and Segre (n°025) before their confluence and the confluence in the Ebro main stream. The Jalon (n°087) was selected as the main tributary of the Southern area on the right bank. A total of five monitoring stations were selected along the Ebro main stream (Fig. 1), located between the confluence of the main tributaries to evidence their impact on both the chemistry and the hydrology of the Ebro River. Up- to downstream, the following Ebro monitoring stations were selected: Mendavia (n°120), Castejon (n°002), Zaragoza (n°011), Sastago (n°112) and Tortosa (n°027). The available chemical data cover the last few decades beginning in 1981. For this study, data covering the period 1981–2003 were selected. In the same way, the daily discharge at each sampling station was extracted for the same period. For both chemical and hydrological data, the dataset is not complete and there are missing data for some periods up to some years. The mean monthly discharges are reported in Fig. 2, for the Ebro River stations and the tributaries. The mean monthly discharge at Mendavia (n°120) concerns the period 1948–2006 for a total of 699 months, while at Tortosa

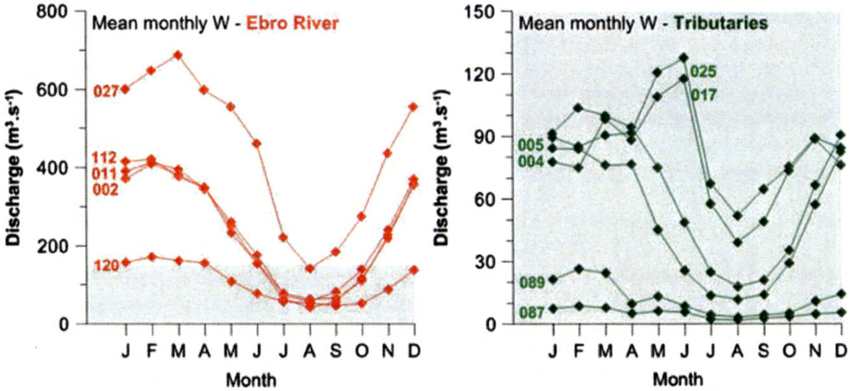

Fig. 2 Evolution of the mean monthly discharge ($m^3\ s^{-1}$). *Left*: along the Ebro River for five monitoring stations: 120: Mendavia (1948–2006), 002: Castejon (1928–2006), 011: Zaragoza (1912–2006), 112: Sastago (1945–1997) and 027: Tortosa (1912–2006). *Right*: for the selected tributaries: 004: Arga (1912–2006), 005: Aragon (1912–2006), 017: Cinca (1928–2003), 025: Segre (1925–2006), 087: Jalon (1934–2006), 089: Gallego (1973–2006). Data are not available for all hydrological years
Source: CEH-CEDEX, http://hercules.cedex.es

(n°027) there are two periods of records, one between 1912 and 1935 and the other between 1953 and 2006 for a total of 916 months. The two curves are roughly similar with high flow between December and May and a low flow period between July and October. It is worth noting that the mean monthly discharges for the Ebro intermediate station (n°112, 011 and 002) are very similar. The Jalon (n°087) and Gallego (n°089) tributaries presents the lowest mean monthly discharges (<25 m^3 s^{-1}), the Arga (n°004) and Aragon (n°005) present similar profiles than the Ebro main stream, whereas the Cinca (n°017) and Segre (n°025) present the highest picks of discharge later in the springtime (May and June) reflecting the snowmelt from the Pyrenees.

For the Ebro main stream, a total of 164 sets of complete data (major elements, EC, T, discharge) from 1980 to 2003 were selected at Mendavia station, 85 sets of complete data in Castejon, 244 in Zaragoza, 43 in Sastago and 203 in Tortosa. For the tributaries, a total of 43 sets of complete data were selected at the Arba and Aragon stations, 45 from the Jalon, 42 from the Gallego, 47 from the Cinca and 43 from the Segre.

4 Dissolved Loads in the Ebro River Basin

4.1 Physico-Chemical Parameters

Over the investigated period (1980–2003), the Ebro River presents a general increasing trend of EC from up- to downstream (Fig. 3a), the mean EC in Mendavia being EC = 646 ± 156 μS cm^{-1} ($n = 264$) and EC = 911 ± 227 μS cm^{-1}

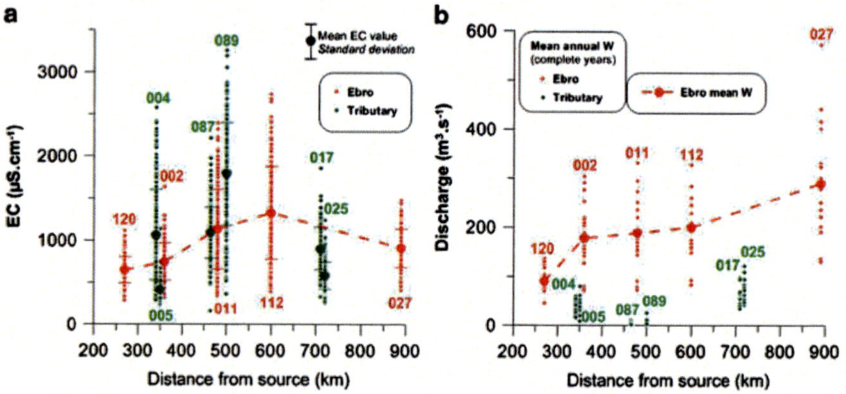

Fig. 3 (**a**) Electrical conductivity (EC, μS cm^{-1}) monitored for the period 1981–2003 in all stations. (**b**) Mean annual discharge (W, m^3 s^{-1}) at each station, only complete years (365 daily measurements) are reported. The Ebro mean discharge is also reported. All the data are reported versus the distance from the source of the Ebro River

(n = 264) downstream in Tortosa. A pick of EC is observed in Sastago with a mean EC = 1,329 ± 551 μS cm^{-1} (n = 255), after the confluence of the Gallego the tributary with the highest mean EC (1,789 ± 609 μS cm^{-1}, n = 296). Such values are typical of carbonate and evaporate weathering according to Meybeck [17] classification from French unpolluted rivers. The other tributaries present extremely variable mean EC value, the lowest value being measured in the Aragon (EC = 409 ± 102 μS cm^{-1}, n = 256). The pH of all the river water presents small seasonal variations, ranging from 7.0 to 8.7, with a mean value of 7.9 ± 0.3, there is no difference between the Ebro main stream and the tributaries. The mean annual temperature increases from up- to downstream the Ebro River, T = 13.9 ± 5.6°C in Mendavia to 17.6 ± 5.7°C in Tortosa with large variations along the hydrological cycle from 2.5 up to 30°C. The mean discharge over the studied period (calculations on complete years) increases from 89 to 289 m^3 s^{-1} from Mendavia to Tortosa, respectively (Fig. 3b). The year to year discharge mean discharge is highly variable, especially in Tortosa with a minimum of 131 m^3 s^{-1} during the 2001–2002 (October to September) hydrological cycle, and a maximum of 573 m^3 s^{-1} in 1987–1988. These values reflect the general trend observed at the basin scale during very dry and wet hydrological years. The Ebro in Tortosa presents a strong and continuous reduction of its water discharge over the last decades (CEH-CEDEX database), commonly attributed to reservoirs and anthropogenic water use. However, Ludwig et al. [18] calculated that more than half of the reduction was related to climate change, and the rest to anthropogenic water extraction.

4.2 Chemical Data and Total Dissolved Solids

Most of the data from the complete set at each station do not show a substantial charge imbalance, considering that the cationic composition should be balanced by the anionic composition. The magnitude of the imbalance is best seen by using the Normalised Inorganic Charge Balance (NICB = $\{\Sigma^- - \Sigma^+/\Sigma^- + \Sigma^+\}$ 100). Most points plot within the ±5% field with a mean NICB close to 0% and a standard deviation of 2% for all the Ebro monitoring stations and the tributaries.

The chemistry of the Ebro River basin surface water can be classified using ternary diagrams (Piper diagrams, not shown). The main characteristics of the surface water of the Ebro River can be mainly summarised as Ca–SO_4 water type, some data plot in the Ca–HCO_3 field, they reflect water draining the upper part of the basin. In the Ebro surface waters, Ca varies from 30% up to 80% of the Σ^+ while Mg fluctuates from 5% up to 40% of the Σ^+. The total concentration major cations ($\Sigma^+ = Ca^{2+} + Mg^{2+} + Na^+ + K^+$) ranges from 3.4 to 11.5 meq L^{-1} with a mean value of 7.4 ± 1.6 meq L^{-1} upstream in Mendavia, from 5.9 to 17.9 5 meq L^{-1} with a mean value of 10.8 ± 2.5 meq L^{-1} downstream in Tortosa with a pick in Sastago (mean value Σ^+ = 16.0 ± 6.6 meq L^{-1}). These data are clearly higher than the average for world rivers ($\Sigma^+ = 1.47$ meq L^{-1} [4, 19]). Such characteristics (calcium dominance and $\Sigma^+ > 4$ meq L^{-1}) and the water type can be related to the drained lithologies, since the weathering of evaporites and carbonates should lead to a Ca–SO_4 water type with high Σ^+ [17, 20]. The main tributaries present mostly Ca–SO_4 facies at their outlet, except the Aragon presenting Ca–HCO_3 characteristics and the Arga presenting alternatively Ca–HCO_3 (winter), Ca–SO_4 and Na–Cl (summer) water types. Water quality may also be affected by anthropogenic factors, such as increased water abstraction from the Ebro, changes in water demand patterns, growing pressure from farming and agricultural activities (increasing discharge of nutrients and pesticides), changes in agricultural practices and land cover, and industrial development.

The total dissolved solids (TDS) is calculated as the sum of the dissolved major ions. Along the Ebro River, the mean TDS over the studied period ranges from 521 ± 106 mg L^{-1} in Mendavia ($n = 164$) to 735 ± 165 mg L^{-1} in Tortosa, with a pick of $1{,}085 \pm 430$ mg L^{-1} in the Sastago station. On the other hand, the mean TDS of the tributaries ranges from 386 ± 96 mg L^{-1} at the outlet of the Aragon to $1{,}359 \pm 460$ mg L^{-1} in the Gallego. Compared to the World Rivers, the Ebro Basin surface waters present high TDS values, attributed to the drained lithologies (carbonates and evaporates). Contributors to the Mediterranean Sea from the French Mediterranean side present lower mean TDS, the Rhone River with a mean value

Table 1 Relations between TDS and EC, and EC and W at each monitoring station

Ebro main stream	Tributary	TDS (mg L^{-1}) versus EC (µS cm^{-1})	EC (µS cm^{-1}) versus W (m^3 s^{-1})
120-Mendavia		TDS = (0.798 ± 0.07)× EC; $R^2 = 0.72$	EC = (1,463.424 ± 89.695)× $W^{(-0.196 \pm 0.015)}$; $R^2 = 0.41$
	005-Aragon	TDS = (0.860 ± 0.016)× EC; $R^2 = 0.75$	EC = (581.586 ± 23.699)× $W^{(-0.115 \pm 0.013)}$; $R^2 = 0.35$
	004-Arga	TDS = (0.713 ± 0.011)× EC; $R^2 = 0.92$	EC = (2,842.634 ± 195.580)× $W^{(-0.320 \pm 0.025)}$; $R^2 = 0.46$
002-Castejon		TDS = (0.797 ± 0.010)× EC; $R^2 = 0.77$	EC = (2,411.011 ± 136.623)× $W^{(-0.256 \pm 0.013)}$; $R^2 = 0.67$
	087-Jalon	TDS = (0.838 ± 0.012)× EC; $R^2 = 0.86$	No relation: mean value EC = 1,085.471 ± 297.725
011-Zaragoza		TDS = (0.781 ± 0.005)× EC; $R^2 = 0.92$	EC = (8,444.876 ± 558.438)× $W^{(-0.442 \pm 0.016)}$; $R^2 = 0.80$
	089-Gallego	TDS = (0.786 ± 0.009)× EC; $R^2 = 0.94$	EC = (3,310.504 ± 150.001)× $W^{(-0.365 \pm 0.021)}$; $R^2 = 0.77$
112-Sastago		TDS = (0.790 ± 0.014)× EC; $R^2 = 0.91$	EC = (6,264.687 ± 386.030)× $W^{(-0.337 \pm 0.015)}$; $R^2 = 0.72$
	025-Segre	TDS = (0.805 ± 0.011)× EC; $R^2 = 0.88$	EC = (1,073.318 ± 70.380)× $W^{(-0.158 \pm 0.017)}$; $R^2 = 0.26$
	017-Cinca	TDS = (0.806 ± 0.013)× EC; $R^2 = 0.80$	EC = (2,416.045 ± 234.836)× $W^{(-0.273 \pm 0.027)}$; $R^2 = 0.48$
027-Tortosa		TDS = (0.774 ± 0.005)× EC; $R^2 = 0.8$	EC = (1,873.582 ± 210.9)× $W^{(-0.137 \pm 0.022)}$; $R^2 = 0.15$

TDS total dissolved solids (mg L^{-1}), *EC* electrical conductivity (µS cm^{-1}), *W* river discharge (m^3 s^{-1})

of 339 mg L^{-1} [4] using the databank of Meybeck and Ragu [19] and the Hérault River with a mean TDS of 374 mg L^{-1} (RMC Water Agency, and Petelet-Giraud unpublished data). The Nile River has a mean TDS value of 388 mg $L^{-1}$and the Po River a mean TDS of 353 mg L^{-1} [4].

The EC (μS cm^{-1}) and the TDS (mg L^{-1}) both reflect the water ionic content, i.e. the dissolved load also called water salinity. The EC, easily obtained compared to chemical data, is thus widely documented in the CHEBRO database ($n = 2{,}860$ versus 999 complete major element analyses). These two parameters (EC and TDS) are linked by a linear relation: TDS (mg L^{-1}) $= b$ EC (μS cm^{-1}), with a mean b factor: $0.54 < b < 0.96$ according to water types and range of salinity [21, 22]. The linear relations between TDS and EC were calculated for each monitoring station, b factor ranging from 0.713 (Arga) and 0.86 (Aragon), whereas the Ebro River stations present less variability (0.774–0.798) with R^2 always better than 0.72, all the relations are summarised in Table 1. These relations are very similar to that defined for the whole Ebro basin with $b = 0.81$ [23].

5 Dissolved Fluxes in the Ebro River Basin

5.1 *Dissolved Load Variation with the River Discharges*

The annual exported dissolved load at each monitoring station corresponds to the instantaneous TDS concentration in the river multiplied by the instantaneous discharge of the river. As mentioned above, chemical analyses are not systematically done (in most of the cases one every 6 months), but EC is more often measured (mainly on a monthly basis). The good correlations between EC and TDS at each station allow to assess the TDS when only the EC is available. First, the co-variations between daily discharge (W) and punctual measured EC were investigated, the best correlations being obtained with power fit equations ($EC = aW^b$). Figure 4 presents the best and worst power relations for the Ebro River and the main tributaries. Along the Ebro main stream, the relations between these two parameters present R^2 coefficient between 0.41 in Mendavia and 0.80 in Zaragoza. Nevertheless, R^2 coefficient is only 0.15 in Tortosa ($n = 245$), reflecting the lower influence of the water flow on the EC downstream and highlights the buffer role of dam on the EC (resulting from mixing of Ebro waters of different dates in the Mequinenza reservoir, and the Ebro and Segre in Ribarroja and Flix dams). At stations where flows are regulated by the presence of reservoirs, EC responds to the concentrations of the flows of the previous months that mix in the reservoirs; Relations between monthly values of EC and values of previous months average monthly flow and instantaneous flows at the time of sampling of previous months were tested [24]. These relationships were especially tested in the Ebro at Tortosa, resulting in improved regressions between EC and W, confirming the main role of reservoirs. Considering the outlet stations of the main tributaries, the R^2 coefficient

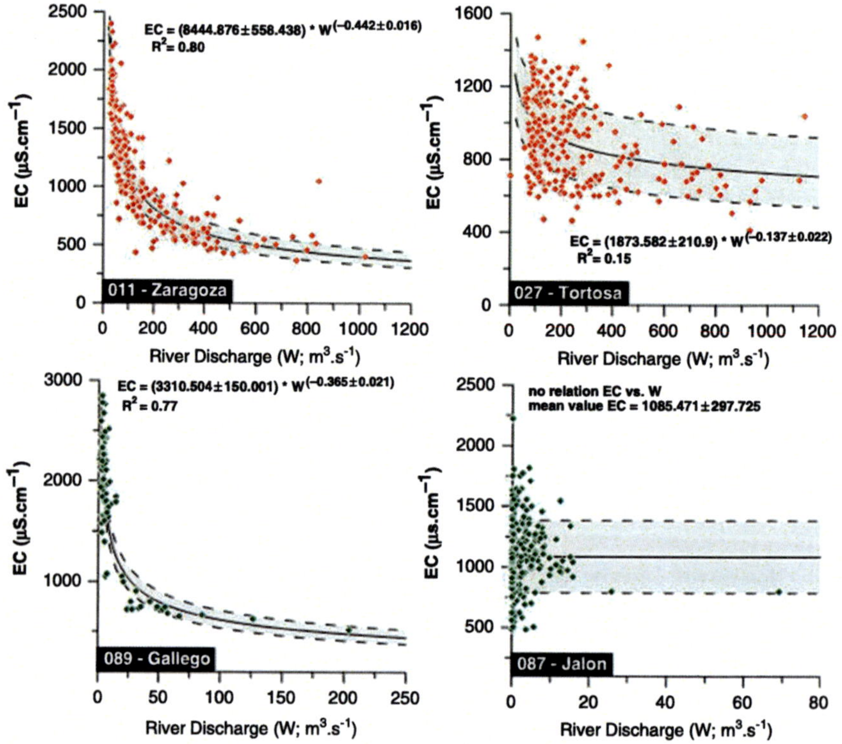

Fig. 4 Relation between EC (μS cm^{-1}) and river discharge (W, m^3 s^{-1}) with the power relation (EC = $a \times W^b$) linking both parameters. Best and worth relation are illustrated for the Ebro (011, 027) stations and the tributaries (089, 087)

is very variable, from 0.01 (Jalon) to 0.77 (Gallego), and independent of the river salinity, for instance for the Jalon outlet station, drawdowns for irrigation control the Jalon River flow in its lower part leading to the worst W versus EC relation of the basin (Fig. 4).

5.2 *Dissolved Load Fluxes at the Ebro Basin Scale*

Most of the studies on large word rivers dedicated to weathering fluxes (dissolved and/or particulate) are based on short-term monitoring often on a limited number of stations, resulting in incomplete time series and low sampling frequency to reflect long-term and episodic fluxes ([1] for instance). Only few studies have been devoted to long-term monitoring, to verify the temporal variability of the average characteristics, their trends, and the representativeness of short-term investigations [8, 25]; some studies also focus on extreme events (e.g. floods) to assess their impact on the general exportation rates [8, 26–29]. Dissolved fluxes are established from the EC and chemical data available in the CHEBRO database for the period 1980–2003, and the daily discharges from the CEH-CEDEX database. In this study, we consider five stations along the Ebro main stream and six tributaries, mainly located on the left bank (Fig. 1). As illustrated above, relations established between EC and discharge at each station (Table 1) are used to calculate the daily mean EC from the daily mean discharge of the river. The calculated ECs are then converted into a calculated TDS value according to the EC-TDS relation (Table 1), except for the Jalon where no relation exists between EC and Discharge, the mean EC value is then used. Finally, annual flux of dissolved material can be calculated as the sum of daily fluxes (F = TDS $\times$ W).

The mean annual flux of the Ebro main stream over the studied period 1980–2003 (calculated for complete years, i.e. when daily discharge are available for each day of the hydrological year) increase from Mendavia to Tortosa reflecting the increase of both parameters TDS and W (Fig. 5a, b). In Mendavia, the mean annual TDS flux is $1.30 \pm 0.28 \times 10^6$ t year^{-1} ($n = 21$) very similar to that calculated by Négrel et al. [30] based on the mean interannual discharges of the Ebro River at Mendavia (1.6×10^6 t year^{-1}). At Tortosa, the annual dissolved flux reaches a mean value of $5.86 \pm 1.93 \times 10^6$ t year^{-1} ($n = 19$) in very good agreement with that calculated by Négrel et al. ([30], 5.9×10^6 t year^{-1}). These data are consistent with the calculated for other hydrological periods: 6.67×10^6 t year^{-1} for the 1973–2004 hydrological period (30 full years within a period of 32 years [24]), 6.70×10^6 t year^{-1} for the hydrologic years 1974, 1975 and 1977 [31] and 6.95×10^6 t year^{-1} for the period 1975–1990 [32]. The mean dissolved exported flux in Tortosa is 4.5 times higher than that observed in Mendavia. The annual variability of dissolved exportations at the outlet of the Ebro catchment is very high, from 3.05×10^6 t year^{-1} for the dry hydrological year 2001–2002 up to 10.80×10^6 t year^{-1} during wet hydrological year 1987–1988, i.e. a 3.5 factor.

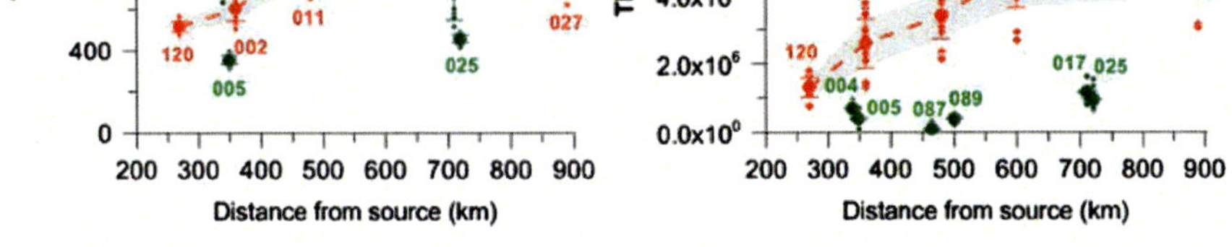

Fig. 5 Mean annual TDS (**a**) expressed in mg L^{-1} and TDS Flux (**b**) expressed in t year^{-1}, together with mean value and standard deviation at each monitoring station. All the data are reported versus the distance from the source of the Ebro River

To assess the robustness of fluxes calculations, we systematically test the worst favourable conditions for each step of the flux calculation, by adding and subtracting the standard deviation (SD) value to the mean value in both equations EC = $(a \pm SD_a)\ W^{(c \pm SDc)}$ and TDS = $(b \pm SD_b)$ EC. The resulting minimum and maximum annual dissolved fluxes are compared to the interannual variability of the mean dissolved fluxes; it appears that the extreme values calculated are within the range of the interannual variability. In that way, the extreme case of the Tortosa station, where there is a poor EC versus W relation (Fig. 4), gives the following results: Minimum mean annual TDS flux is 4.45×10^6 t year^{-1} and the maximum mean annual TDS flux is 7.48×10^6 t year^{-1} that fall within the range defined by the interannual variability during the period 1980–2003 ($5.86 \pm 1.93 \times 10^6$ t year^{-1}). The mean annual dissolved exportations from the Ebro tributaries vary from $0.08 \pm 0.07 \times 10^6$ t year^{-1} at the outlet of the Jalon, up to $1.16 \pm 0.27 \times 10^6$ t year^{-1} for the Cinca (Fig. 5b). In spite of its high TDS mean value, the Ebro at its outlet in Tortosa presents relatively limited exportation rates compared to the largest word rivers and main Mediterranean River. The Rhone River, because of its high discharge, has a mean exported dissolved flux of 18.3×10^6 t year^{-1}, about three times higher than the Ebro TDS flux. The Po and the Nile export 16.5×10^6 t year^{-1} and 32.2×10^6 t year^{-1} of dissolved load, respectively [4], using the databank of Meybeck and Ragu [19], whereas the Hérault River has a mean TDS flux of 0.56×10^6 t year^{-1}.

The relative proportions of the total exported dissolved flux at Tortosa from the main tributaries are illustrated in Fig. 6. The Ebro basin in its upper part (upstream Mendavia station) contributes to 16.5–28% of the total exported flux in Tortosa with a mean contribution of 22.4% over the studied period. The Cinca is the tributary that mainly contributes to the total exported load at Tortosa with a mean

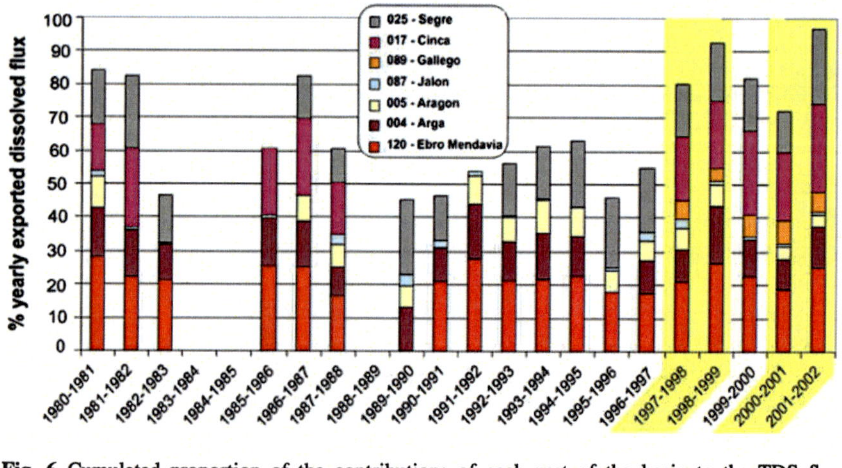

Fig. 6 Cumulated proportion of the contributions of each part of the basin to the TDS flux measured in Tortosa. Note that data are not available for all the hydrological years. Years with data for all stations are highlighted in *yellow*

contribution of 19% (14–26.5%). The Segre is the second main contributor with a mean contribution of 16.8% (10.3–22.4%). In the upper basin, the Arga mean contribution is 11.7%, whereas the Aragon, Gallego and Jalon contributions are very low, each representing 6.5% and less than 5% of the total exported flux for the two latest ones. When all the contributors to the total flux in Tortosa are available for a same year, the sum of the tributary fluxes added to the flux in Mendavia represents more than 80% of the exported dissolved flux at the outlet of the Ebro. When only one or two minor contributors is missing, the sum can also reach more than 80%. However, all the major contributors to the Ebro discharge (Cinca, Segre, Aragon + Arga) present mean TDS lower than that of the Ebro at Tortosa, suggesting possible important contributions of dissolved material from small tributaries and also uncontrolled flow (especially groundwater).

The specific TDS fluxes reflect the TDS fluxes by surface unit (t km^{-2} $year^{-1}$) and represent the chemical denudation rate of a catchment if the anthropogenic and impact are neglected. The specific TDS flux at the outlet of the Ebro in Tortosa in 70 ± 23 t km^{-2} $year^{-1}$ over the studied period (Fig. 7) is very consistent with previous calculations (73.8 t km^{-2} $year^{-1}$ [24]). The Ebro upstream in Mendavia presents a slightly higher value with 108 ± 24 t km^{-2} $year^{-1}$. Most of the main tributaries present specific TDS fluxes in the same range of value, except the Jalon with a very small exportation per unit area (9 ± 7 t km^{-2} $year^{-1}$) and the Arga in the upper part with a high specific exportation flux (251 ± 55 t km^{-2} $year^{-1}$), the latter having a variable water type along the hydrological cycle (Ca–HCO_3, Ca–SO_4 and Na–Cl). Compared to the main Mediterranean Rivers, the Ebro has a specific exportation rate about 2.7 times lower than that of the Rhone River (191 t km^{-2} $year^{-1}$) and 3.4 times lower than that of the Po River (236 t km^{-2} $year^{-1}$), but about six times higher than the Nile River, according to Gaillardet et al. [4].

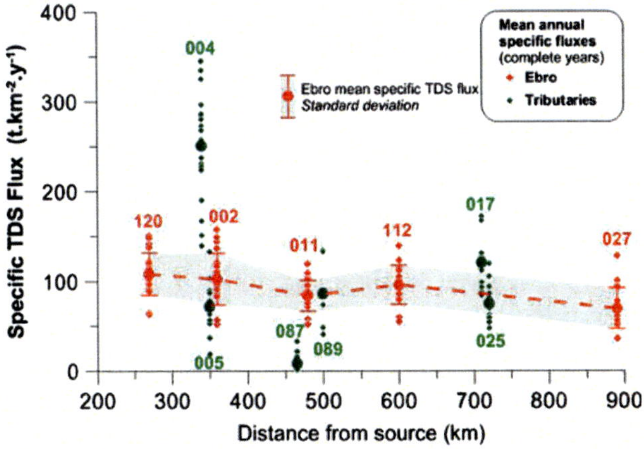

Fig. 7 Specific TDS flux expressed in t km^{-2} $year^{-1}$, together with mean value, and standard deviation for Ebro monitoring stations. Data are reported versus the distance from the source of the Ebro River

6 Impact of Lithologies on the Ebro Basin Chemistry

6.1 Influence of Evaporite and Carbonate Dissolution

Water types in the Ebro River basin can be schematically classified into three categories based on Piper diagrams, most of the waters show a Ca–SO_4 nature, and a few of them Ca–HCO_3 and Na–Cl types. Upstream in Mendavia, the Pearson correlation coefficients between the major dissolved ions and the TDS show that the TDS is mostly controlled at the first order by SO_4, Na and Cl concentrations (Pearson's R in the range 0.82–0.90). Downstream in the Tortosa Ebro station, SO_4, Na and Cl also have the highest effect on the TDS, with Pearson's R coefficients close to 0.96–0.98 but a higher effect by Ca is clearly demonstrated with R coefficients of 0.92. For both Mendavia and Tortoasa Stations, NO_3 and HCO_3 have a limited effect on the TDS, with R coefficients lower than 0.5. The Sastago Ebro station, with the highest TDS values, also presents a main control of TDS by SO_4, Na and Cl with Pearson's R^2 coefficients between 0.97 and 0.98, and by Ca $R = 0.93$. Concerning the main tributaries, the Arga TDS is mostly controlled at the first order by Cl and Na (Pearson's $R = 0.98$) and by SO_4 and Ca (Pearson's $R = 0.89$). The TDS of the Aragon is mainly controlled by Na and SO_4 ($R = 0.95$) and Cl ($R = 0.92$). The Gallego, with the highest dissolved contents, presents a main control of TDS by Cl, SO_4 and Na with Pearson's R coefficients between 0.98 and 0.95, respectively, and in a less extent by Ca $R = 0.86$. The TDS of the Cinca is mainly controlled by Na ($R = 0.98$), SO_4 ($R = 97$) and Cl (0.96). Finally, for the Segre, TDS is controlled by SO_4 and Ca (Pearson's $R = 0.97$), Na ($R = 0.95$), Mg ($R = 0.94$) and HCO_3 and Cl ($R = 0.92$ and 0.9, respectively).

Cl–Na binary relation (not shown) presents a wide range of contents (0.5–13 mmol L^{-1} for the Ebro stations; 0.3–17 mmol L^{-1} for the main tributaries). Ebro stations present mean Na/Cl molar ratios close to 1 (0.9–1.1), i.e. reflecting NaCl (i.e. halite) dissolution (Na=Cl). Concerning the tributaries, the Arga, Jalon and Gallego also present Na/Cl molar ratio close to 1, suggesting that Na and Cl originate from halite dissolution. However, some other tributaries present clearly higher Na/Cl molar ratio, the Aragon in the upper part of the catchment present a mean Na/Cl = 1.5 with a low dissolved content compared to other waters of the Ebro basin. This Na excess may originate from Na–aluminosilicates weathering [33]. The same scenario could explain the Na/Cl = 1.38 for the Segre waters, whereas for the Cinca, with a mean Na/Cl ratio of 1.28 but with a clearly higher dissolved content, the Na excess could originate from ion exchange reaction [34] resulting in the release of Na adsorbed on clays and replaced by Ca.

Ca and SO_4 contents in surface waters from the Ebro Basin also show large variations (both between 0.5 and 7 mmol L^{-1}) and plot below the seawater dilution line in an SO_4 versus Ca diagram (Fig. 8). There is a clear evolution of the Ca and SO_4 contents along the Ebro from up to downstream, with decreasing Ca/SO_4 molar ratios mean values from 2.19 ± 0.70 in Mendavia to 1.28 ± 0.49 in Sastago and finally a slight increase (1.40 ± 0.33) in Tortosa. In the middle part of the basin, for

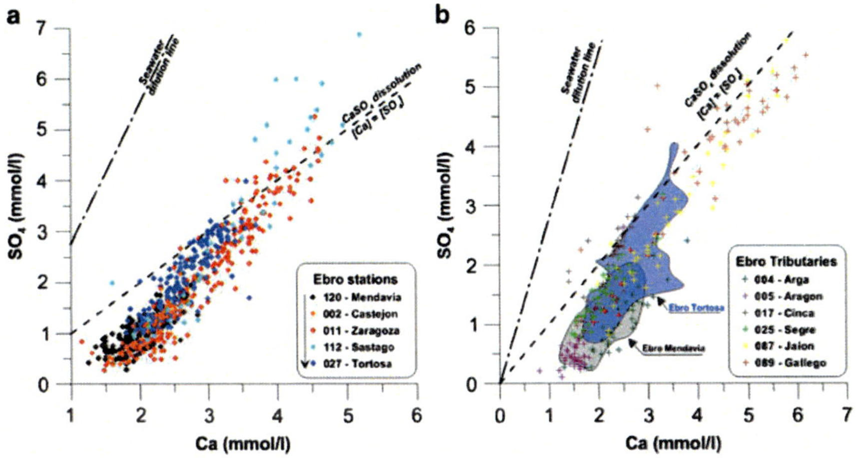

Fig. 8 Relationships between Ca^{2+} and SO_4^{2-} (mmol L^{-1}), in Ebro River waters (**a**) and tributaries (**b**)

Zaragoza and Sastago stations, individual samples define a trend from Ca excess ($Ca/SO_4 > 1$) towards the gypsum dissolution line ($Ca=SO_4$) when dissolved contents increase (Fig. 8a), so that both ions are mainly controlled by evaporite (gypsum) dissolution. In Sastago, some samples present a Ca deficit when compared to SO_4 that could result from calcite precipitation, this process is supported by a supersaturated calcite saturation index ($SI_{calcite}$ between 0.5 and 1.4, calculated using the PHREEQC geochemical code [35]) especially with the highest TDS values. For the Ebro River samples, Ca and SO_4 seem to be controlled by gypsum dissolution when dissolved content is high, but another lithology is required for low dissolved content when $Ca/SO_4 \gg 1$ (molar ratio). As demonstrated by Négrel et al. [30], carbonate weathering plays an important role in the Ebro river basin, as a Ca supplementary source, especially for low dissolved contents in the upper part of the basin. This is also well illustrated by the Aragon water sample that presents Ca contents close to half that of HCO_3, typical of carbonate dissolution. The other tributaries present the same trends observed for the Ebro stations (Fig. 8b) with the Gallego and Sastago samples, presenting the highest dissolved contents, plotting close to the gypsum dissolution line.

6.2 *Dissolved Contents: Evaporite and Carbonate Contribution*

Ibanez et al. [12] and Batalla et al. [13] demonstrated the impact of the numerous dams in the basin on the decreasing exportations of SPM at the outlet of the Ebro River in the Mediterranean Sea, so that dissolved exportations (TDS flux) become dominant [30]. Based on the calculated dissolved fluxes along the Ebro River and from the main tributaries, the contribution of carbonate and evaporite weathering to the dissolved fluxes can thus be assessed. Weathering processes have been summarised by Tardy et al. [36]. Evaporite dissolution reactions are congruent and produce dissolved ions in a similar proportion as in rocks. For example, gypsum or anhydrite dissolution follows the reaction below without CO_2 consumption:

$$CaSO_4 = Ca^{2+} + {SO_4}^{2-} \tag{1}$$

The carbonate weathered mineral is calcium/magnesium and the reaction consumes one CO_2 from the atmosphere and liberates one CO_2 coming from the matrix. The main weathering equation is:

$$(Ca, Mg)CO_3 + CO_2 + H_2O = \left(Ca^{2+}, Mg^{2+}\right) + 2{HCO_3}^{-} \tag{2}$$

Thus, half of the ${HCO_3}^{-}$ released by carbonate weathering originates from atmospheric CO_2.

The estimation of the carbonate and evaporite weathering rate requires the quantification of the elements originating from the two components. Roy et al. [10] calculated the carbonate weathering rate (TDS originating from carbonate dissolution) as follows:

$$TDS_{carb} = Ca^{2+} + Mg^{2+} + 1/2{HCO_3}^{-} \tag{3}$$

Similarly, the evaporite weathering rate was defined as follows:

$$TDS_{evap} = Ca^{2+} + K^{+} + Na^{+} + Cl^{-} + {SO_4}^{2-} \tag{4}$$

However, Cl content can be influenced by anthropogenic and atmospheric inputs and Na by silicate weathering [4, 10, 11, 37, 38] and thus the calculation of TDS_{evap} is limited in the Ebro Basin, since the computation of Cl from pollution and Na from silicate weathering cannot be easily estimated while the input from the atmophere can be neglected according to Avila and Alarcon [39].

In the Central Ebro Basin, the gypsum unit was deposited in a large saline lake–saline mud flat complex related to distal areas of large alluvial fans during the Lower Miocene. These deposits are related to the shallow lacustrine environment combining carbonate sedimentation and evaporite precipitation [40]. The evaporite succession in the Ebro Basin consists mainly of gypsum ($CaSO_4$) with halite (NaCl), thenardite (Na_2SO_4), bloedite ($Na_2Mg(SO_4)_2$) and glauberite ($Na_2Ca(SO_4)_2$) often present, but thenardite generally dominates [41]. We thus calculate a $TDS_{evap(GYP)}$, only corresponding to the weathering of gypsum:

$$TDS_{evap(GYP)} = Ca^{2+} + {SO_4}^{2-} \tag{5}$$

We thus assume that all SO_4 originates from the evaporite dissolution (gypsum and other SO_4-bearing minerals). Anthropogenic sulphates can be present, but up to now it has not been possible to separate and quantify between both sources. Ca is likely to be influenced by the weathering of both carbonate and evaporite and the proportion of each component should be determined. For the determination of Ca originating from evaporite weathering, we assume that all SO_4^{2-} originates from evaporite weathering (mainly gypsum) and thus Ca_{carb} is deducted as follows:

$$Ca_{carb} = Ca^{2+}{}_{T} - SO_{4T}^{2-} \tag{6}$$

where Ca_T and SO_{4T} represent the total Ca and SO_4 measured in the river water.

Then we calculate the proportion of Ca originating from evaporite (gypsum) weathering ($Ca_{evap(GYP)}$) as follows:

$$Ca_{evap(GYP)} = Ca^{2+}{}_{T} - Ca_{carb} \tag{7}$$

assuming that no other major input of Ca occurs (fertilisers, silicate weathering). Finally, TDS_{carb} and $TDS_{evap(GYP)}$ are calculated according to (3) and (5) with their respective Ca value, for all the monitoring stations.

In Mendavia, the TDS derived from carbonate weathering (TDS_{carb}) represents about 30% (min. 18% and max. 47%) of the total dissolved load and the TDS derived from evaporate (gypsum) weathering ($TDS_{evap(GYP)}$) represents about 25% (min. 11%; max. 38%) of the total TDS, resulting in a residual TDS of 45% of the total TDS, representing the dissolved load originating from other minerals weathering (evaporites, silicates) and anthropogenic inputs, atmospheric inputs being considered negligible. It is worth noting that the relative contribution of TDS_{carb} is higher than that of $TDS_{evap(GYP)}$ for total TDS lower than about 550–600 mg L^{-1}, and that the relative contribution of $TDS_{evap(GYP)}$ is higher than that of TDS_{carb} when total TDS is higher about 550–600 mg L^{-1}. Downstream in Tortosa, the contribution of TDS_{carb} decreases to a mean value of 21%, whereas the $TDS_{evap(GYP)}$ increases up to a mean value of 36%, the residual TDS remaining stable with about 43% of the total TDS. At the Ebro River scale, for the three upper monitoring stations (Mendavia, Castejon and Zaragoza) $TDS_{carb} > TDS_{evap(GYP)}$ up to a total TDS around 500–600 mg L^{-1}, and for higher total TDS $TDS_{evap(GYP)}$ becomes dominant. For the downstream monitoring stations (Sastago and Tortosa), the relative contribution of TDS_{carb} and $TDS_{evap(GYP)}$ is roughly the same for the lowest total TDS and then the proportion of $TDS_{evap(GYP)}$ becomes dominant in comparison with TDS_{carb}. Concerning the tributaries, the Arga and Aragon located upstream present the same behaviour than the upper Ebro stations, whereas all the other tributaries (Jalon, Gallego, Segre and Cinca) have similar behaviour than the Ebro stations located downstream of the river basin. Figure 9 illustrates the amount of TDS derived from carbonate and evaporite (gypsum) dissolution versus the total TDS. For all the monitoring stations, the dissolved part derived from gypsum dissolution ($TDS_{evap(GYP)}$) increases with the increasing total TDS, with a slope

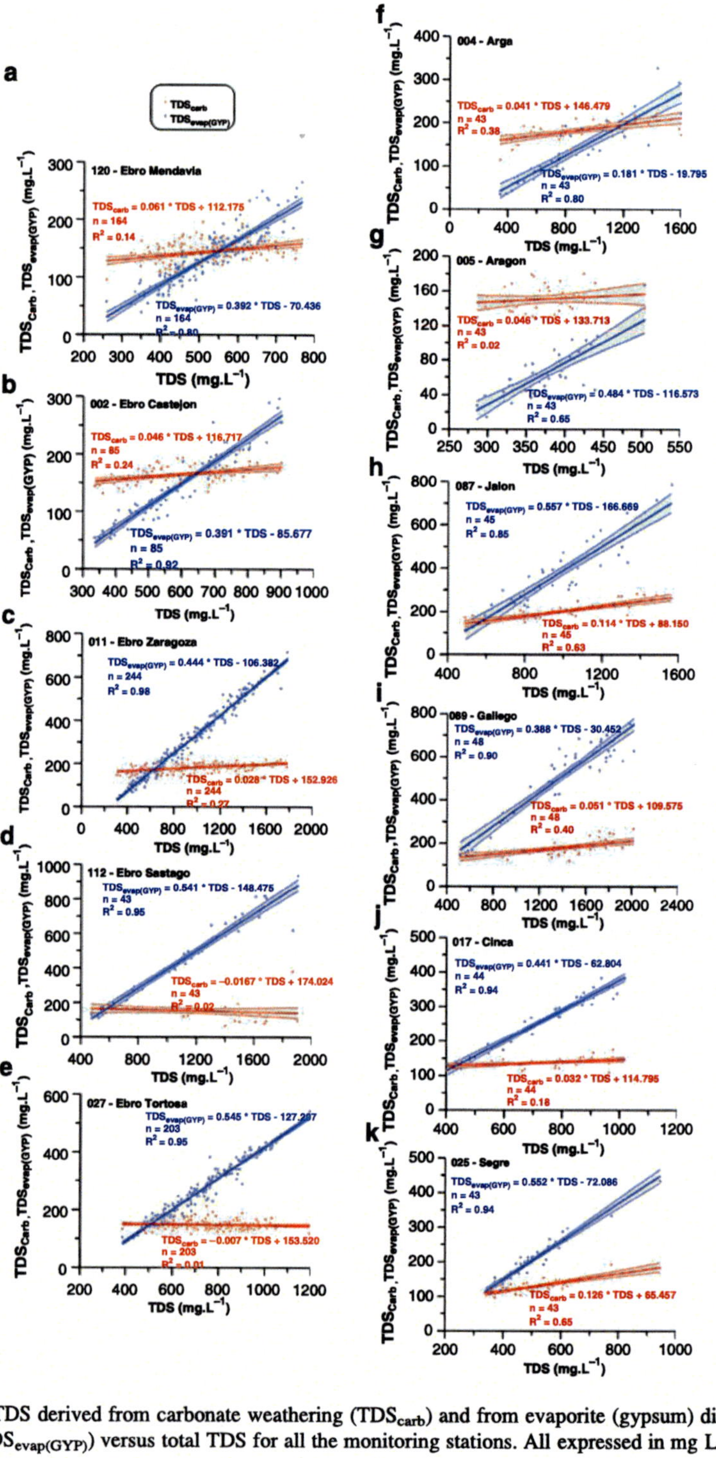

Fig. 9 TDS derived from carbonate weathering (TDS_{carb}) and from evaporite (gypsum) dissolution ($TDS_{evap(GYP)}$) versus total TDS for all the monitoring stations. All expressed in mg L^{-1}

between 0.4 and 0.55, except for the Arga where the slope is only 0.18 and reflect a $TDS_{evap(GYP)}$ proportion between 10% and 20% of the total dissolved load. On the other hand, the TDS_{carb} is relatively constant around 150 ± 20 mg L^{-1} whatever the total TDS, which vary from 300 up to 1,900 mg L^{-1}. This implies that the dissolved content derived from carbonate dissolution (TDS_{carb}) is almost independent of the amount of water draining carbonates within the basin, in other words the dissolved content resulting from carbonate dissolution reaches a plateau resulting from the inability of water to weather more Ca, Mg and HCO_3 from carbonate. This is fully supported by the saturation index computations showing that the water samples at the whole Ebro Basin scale are at least in equilibrium with calcite, or even supersaturated ($SI_{calcite} = 0 \pm 0.5$ up to 1.3; calculated using the PHREEQC geochemical code [35]). Ca, Mg and HCO_3 contents in the rivers are controlled by calcite and dolomite. This phenomenon was already observed in the middle part of the Loire River [11] and in the Hérault River [42]. Thus, it appears that the dissolved content of the Ebro and the tributaries of the lower of the basin is only very partly controlled by carbonate weathering and that evaporite dissolution, especially through gypsum, plays a major role.

6.3 *Dissolved Fluxes from Evaporite and Carbonate Dissolution*

Based on linear correlations identified in Fig. 9, it is then easy to calculate the dissolved exported load from carbonate (TDS_{carb} flux) and evaporite ($TDS_{evap(GYP)}$ flux), from the total dissolved exported load (TDS flux). As for total TDS flux calculated above, the calculations are only done for complete years (each daily discharge value available). As already observed for the total TDS fluxes, the TDS derived from carbonate and evaporite (gypsum) dissolution, respectively, presents large inter-annual variations, because of the main control by the rivers discharges. Figure 10 presents the TDS_{carb} flux versus $TDS_{evap(GYP)}$ flux for each monitoring station and for all the available years. Ebro stations from the upper part of the basin (Mendavia and Castejon), the dissolved exportations resulting from carbonate weathering clearly dominate that from evaporite weathering, especially for wet hydrological year, and exportation rate are similar for dry years. The two upper tributaries (Arga and Aragon) also present the same behaviour (exportations from carbonates are higher than exportation from gypsum). Thus in the upper part of the Ebro Basin, carbonate weathering is dominant in comparison with gypsum weathering. In the Middle part of the basin, the Ebro monitoring station in Zaragoza, the dissolved exportations are dominated by carbonate weathering except during dry hydrological years, where exportations are dominated by gypsum dissolution. The dissolved fluxes from the two tributaries draining the middle part of the basin (Gallego and Jalon) are dominated by evaporite weathering. Finally, in the lower part of the Ebro Basin, the Segre and Cinca are also largely controlled by evaporite dissolution and clearly impact the exportation rates at the outlet of the watershed in

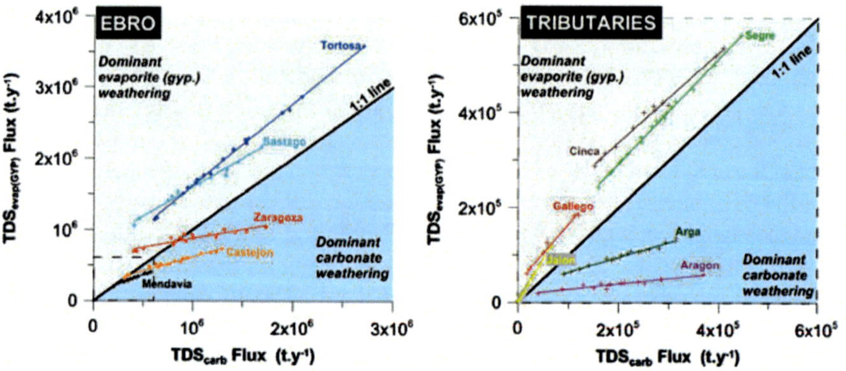

Fig. 10 $TDS_{evap(GYP)}$ versus TDS_{carb} for all the available complete hydrological years over the studied period. Ebro stations (*left*), tributaries (*right*). Data expressed in t year^{-1}

Tortosa, where dissolved exportation derived from gypsum weathering is about 1.5 times derived from carbonate weathering.

The mean exportation rate from carbonate in Mendavia is $0.39 \pm 0.10 \times 10^6$ t year^{-1}, and $1.36 \pm 0.52 \times 10^6$ t year^{-1} in Tortosa, while the mean exportation rate from gypsum is roughly similar in Mendavia with $0.31 \pm 0.06 \times 10^6$ t year^{-1}, but $2.03 \pm 0.60 \times 10^6$ t year^{-1} in Tortosa, similar to the gypsum transport calculated by Navas [43] with 2.95×10^6 t year^{-1}.

With the specific export rate for carbonate and evaporite (gypsum), we can calculate the specific weathering rate per surface unit for these two lithologies. In Mendavia, the specific weathering rate for carbonate and evaporite (gypsum) is similar with 30 ± 12 t km^{-2} year^{-1} and 24 ± 9 t km^{-2} year^{-1}, respectively. In Tortosa, the specific weathering rate for evaporite (gypsum) dominates, with a value of 21 ± 11 t km^{-2} year^{-1}, while the specific weathering rate for carbonate is 14 ± 8 t km^{-2} year^{-1}. The latter is lower than the value of the specific weathering rate for carbonate determined by Ollivier et al. [8] for the Rhone River (89 t km^{-2} year^{-1}), by Roy et al. [10] for the Seine River (45 t km^{-2} year^{-1}) and by Grosbois et al. [11] for the Loire River (47 and 23 t km^{-2} year^{-1} at Orleans and Tours, respectively; the latter being 150 km downstream). For the Loire River, the difference between the two locations was explained through Ca^{2+} and HCO_3^- consumption processes, such as the formation of authigenic calcite [11, 44].

7 Conclusions

The main purpose of this work was to estimate, over a long time period (more than 20 years), the export fluxes for dissolved loads on the catchment scale of the Ebro River basin. Data were compiled from the databank of the *Confederación Hidrográfica del Ebro* (CHE), the Spanish government agency in charge of managing water resources within the Ebro River basin (http://www.chebro.es/). Five monitoring stations were selected along the Ebro River (Mendavia, Castejon, Zaragoza, Sastago and Tortosa), as well as six stations at the outlet of the main tributaries (Arga, Aragon, Gallego, Jalon, Cinca and Segre). Data compilation of daily discharges, physico-chemical parameters and chemical data covers the last two decades (1981–2003), all the calculations were done for complete hydrological years (October to September) for each monitoring station.

The first objective was to characterise the dissolved load of the rivers at the Ebro Basin scale through the EC, TDS and the major elements chemical data. Surface water can be classified into three main categories, a clear dominance of Ca–SO_4 water type, then a Ca–HCO_3 type mainly encountered in the upper part of the basin, and some data presenting a Na–Cl water type. The TDS values are highly variable, both in time and in space, with mean value 521 ± 106 mg L^{-1} in Mendavia to 735 ± 165 mg L^{-1} in Tortosa, with a pick of 1,085 ± 430 mg L^{-1} in the Sastago before the confluence of the Segre and Cinca tributaries. Mean TDS of the tributaries ranges from 386 ± 96 mg L^{-1} in the Aragon to 1,359 ± 460 mg L^{-1} in the Gallego.

The second objective was to calculate dissolved exportations to the Mediterranean Sea and the relative contribution of the different tributaries. EC being the parameter measured with the highest frequency compared to TDS, it was used to define the relation between the dissolved river content and the river discharge. Then the linear correlation between TDS and EC was used to calculate the TDS exportations. TDS fluxes are highly variable from year to year, being mainly controlled by the river discharge. In Mendavia, the mean annual TDS flux is estimated to be 1.30 ± 0.28 × 10^6 t $year^{-1}$ at Tortosa, the annual dissolved flux reaches a mean value of 5.86 ± 1.93 × 10^6 t $year^{-1}$. The Ebro basin in its upper part (upstream Mendavia station) contributes to about 22.4% (mean value) of the total exported flux in Tortosa over the studied period. The Cinca and Segre are the tributaries that mainly contribute to the total exported load at Tortosa with a mean contribution of 19% and 17%, respectively. The Aragon, Gallego and Jalon contributions are very low, often less than 5% of the total exported flux. The specific TDS flux at the outlet of the Ebro in Tortosa is 70 ± 23 t km^{-2} $year^{-1}$, whereas in Mendavia it is slightly higher with 108 ± 24 t km^{-2} $year^{-1}$. The highest chemical erosion rate was calculated for the Arga with 251 ± 55 t km^{-2} $year^{-1}$.

Given that the dissolved export fluxes represent the major export from the basin [30], the third objective was to calculate the respective contribution of carbonate and evaporite (gypsum) with respect to the TDS. It appears that the TDS, on the whole catchment scale, is mostly controlled by SO_4, Na and Ca concentrations, reflecting the preferential contribution of both evaporite and carbonate dissolution during surficial chemical weathering. The chemical weathering fluxes relative to evaporite (gypsum) and carbonate contribution were estimated from major element concentrations and discharge data. In the upper part of the Ebro Basin, carbonate weathering is dominant compared to gypsum weathering. Downstream in Zaragoza, the dissolved exportations are dominated by carbonate weathering except during dry hydrological years, where exportations from gypsum dissolution become dominant. In the middle part of the basin, Gallego and Jalon tribuaries are dominated by evaporite weathering, as well as the Segre and Cinca that are largely controlled by evaporite dissolution and that impact the exportation rates at the outlet of the watershed in Tortosa, where dissolved exportation derived from gypsum weathering is about 1.5 times derived from carbonate weathering.

Acknowledgements This work was funded by the BRGM Research Division and by the European Union (integrated project AquaTerra, Project 505428) under the thematic priority "Sustainable Development, Global Change and Ecosystems (FP6) Water Cycle and Soil Related Aspects". The views expressed are purely those of the authors and may not be regarded in any circumstances as stating the official position of the European Commission. The authors acknowledge the support on the information received from the *Agencia Catalana de l'Aigua* and the *Confederación Hidrográfica del Ebro* (CHE) and from D. Barcelo (CSIC, Department of Environmental Chemistry, Barcelona, Spain). Thanks are due to the BRGM Translation Service for proofreading the English.

References

1. Stallard RF, Edmond JM (1983) Geochemistry of the Amazon. 2. The influence of geology and weathering environment on the dissolved load. J Geophys Res 88(C14):9671–9688
2. White AF, Blum AE (1995) Effects of climate on chemical weathering in watersheds. Geochim Cosmochim Acta 59:1729–1747
3. Louvat P, Allègre CJ (1997) Present denudation rates on the island of Réunion determined by river geochemistry: Basalt weathering and mass budget between chemical and mechanical erosions. Geochim Cosmochim Acta 61(17):3645–3669
4. Gaillardet J, Dupre B, Louvat P, Allegre CJ (1999) Global silicate weathering and CO_2 consumption rates deduced from the chemistry of large rivers. Chem Geol 159(1–4):3–30
5. Dalai TK, Krishnaswami S, Sarin MM (2002) Major ion chemistry in the headwaters of the Yamuna river system: chemical weathering, its temperature dependence and CO_2 consumption in the Himalaya. Geochim Cosmochim Acta 66(19):3397–3416
6. Jacobson AD, Blum JD, Chamberlain CP, Craw D, Koons PO (2003) Climatic and tectonic controls on chemical weathering in the New Zealand Southern Alps. Geochim Cosmochim Acta 67(1):29–46
7. Wu L, Huh Y, Qin J, Du G, Van Der Lee S (2005) Chemical weathering in the Upper Huang He (Yellow River) draining the eastern Qinghai-Tibet Plateau. Geochim Cosmochim Acta 69(22):5279–5294
8. Ollivier P, Hamelin B, Radakovitch O (2010) Seasonal variations of physical and chemical erosion: a three-year survey of the Rhone River (France). Geochim Cosmochim Acta 74(3):907–927
9. Négrel Ph, Allègre CJ, Dupré B, Lewin E (1993) Erosion sources determined by inversion of major and trace element ratios and strontium isotopic ratios in river water: the Congo Basin case. Earth Planet Sci Lett 120:59–76
10. Roy S, Gaillardet J, Allègre CJ (1999) Geochemistry of dissolved and suspended loads of the Seine River, France: anthropogenic impact, carbonate and silicate weathering. Geochim Cosmochim Acta 63:1277–1292
11. Grosbois C, Négrel Ph, Grimaud D, Fouillac C (2001) An overview of dissolved and suspended matter fluxes in the Loire river basin: natural and anthropogenic inputs. Aquat Geochem 7:81–105
12. Ibáñez C, Prat N, Cancio A (1996) Changes in the hydrology and sediment transport produced by large dams on the lower Ebro River and its estuary. Regul Riv 12:51–62

13. Batalla RJ, Gomez CM, Kondolf GM (2004) Reservoir-induced hydrological changes in the Ebro River Basin, NE Spain. J Hydrol 290:117–136
14. Vericat D, Batalla RJ (2006) Sediment transport in a large impounded river: the lower Ebro, NE Iberian Peninsula. Geomorphology 79:72–92
15. Guillén J, Palanques A (1992) Sediment dynamics and hydrodynamics in the lower course of a river highly regulated by dams: the Ebro River. Sedimentology 39:567–579
16. García MA, Moreno MC (2000) Los aprovechamientos en la Cuenca del Ebro: Afección en el regimen hidrológico fluvial. Internal Technical Report no. 2000-PH-24.1, Confederación Hidrográfica del Ebro, Zaragoza, 83 pp
17. Meybeck M (1986) Composition des ruisseaux non pollués de France. Sci Géol Bull 39:3–77
18. Ludwig W, Dumont E, Meybeck M, Heussner S (2009) River discharges of water and nutrients to the Mediterranean and Black Sea: major drivers for ecosystem changes during past and future decades? Prog Oceanogr 80(3–4):199–217
19. Meybeck M, Ragu A (1996) River Discharges to the Oceans. An assessment of suspended solids, major ions, and nutrients. Environment Information and Assessment Rpt., United Nations Environment Programme. Nairobi, 250 p
20. Durand S, Chabaux F, Rihs S, Duringer P, Elsass P (2005) U isotope ratios as tracers of groundwater inputs into surface waters: example of the Upper Rhine hydrosystem. Chem Geol 220:1–19
21. Hem JD (1985) Study and interpretation of the chemical characteristics of natural water, 3rd edn. US Govt Printing Office (USGS Water Supply Paper 2254), Washington, pp 66–69
22. McNeil VH, Cox ME (2000) Relationship between conductivity and analysed composition in a large set of natural surface-water samples, Queensland, Australia. Environ Geol 39 (12):1325–1333
23. Interlab (2006) Caracterizacion de la salinidad de la aguas superficiales de la cuenca del Ebro. Analisis de sus origenes potenciales y diseño de un modelo de seguimiento. Informe final, Diciembre 2006. CHE Ministerio de Medio Ambiente, 208 p
24. CHE (2006) Caracterización de la calidad de las aguas superficiales y control de los retornos del riego en la cuenca del Ebro. Confederación Hidrográfica del Ebro, Dirección General de Obras Hidráulicas y Calidad de las Aguas, 194 p
25. Tipper ET, Bickle MJ, Galy A, West AJ, PomiFs C, Chapman HJ (2006) The short term climatic sensitivity of carbonate and silicate weathering fluxes: insight from seasonal variations in river chemistry. Geochim Cosmochim Acta 70(11):2737–2754
26. Ben Othman D, Luck JM, Tournoud MG (1997) Geochemistry and water dynamics: application to short time-scale flood phenomena in a small Mediterranean catchment. I- Alkalis, alkali-earths and Sr isotopes. Chem Geol 140:9–28
27. Serrat P, Ludwig W, Navarro B, Blazi J-L (2001) Variabilité spatio-temporelle des flux de matières en suspension d'un fleuve côtier méditerranéen: la Têt (France). C R Acad Sci Paris 333:389–397
28. Petelet-Giraud E, Négrel P (2007) Apport de matières en suspension à la Méditerranée: le rôle des crues dans le bilan des exportations du bassin versant de l'Hérault. Techniques Sciences Méthodes 10:55–69
29. Salles C, Tournoud MG, Chu Y (2008) Estimating nutrient and sediment flood loads in a small Mediterranean river. Hydrol Process 22(2):242–253
30. Négrel P, Roy S, Petelet-Giraud E, Millot R, Brenot A (2007) Long-term fluxes of dissolved and suspended matter in the Ebro River Basin (Spain). J Hydrol 342(3–4):249–260
31. Alberto F, Navas A (1986) La participación de los yesos en la salinización de las aguas superficiales de la cuenca del Ebro. II. Contribución relativa de los iones SO_4^{2-} y Ca^{2+} a la salinidad. Anales de Aula Dei 18(1–2):19–29
32. Navas A, Machín J (1995) Salinidad de las aguas superficiales de la cuenca del Ebro. In: Cobelas M, Cabrera F (eds) La calidad de las aguas continentales españolas: estado actual e investigación. Geofroma Ediciones, Logroño, pp 223–231

33. Meybeck M (1987) Global chemical weathering of surficial rocks estimated from river dissolved loads. Am J Sci 287:401–428
34. Cerling TE, Pederson BL, Von Damm KL (1989) Sodium-calcium ion exchange in the weathering of shales: implications for global weathering budgets. Geology 17:552–554
35. Parkhurst DL, Appelo CAJ (1999) User's guide to PHREEQC (version 2) – a computer program for speciation, batchreaction, one-dimensional transport, and inverse geochemical calculations: U.S. Geological Survey Water-Resources Investigations Report 99-4259, 312 pp
36. Tardy Y, Bustillo V, Boeglin JL (2004) Geochemistry applied to the watershed survey: hydrograph separation, erosion and soil dynamics. A case study: the basin of the Niger River, Africa. Appl Geochem 19:469–518
37. Grosbois C, Négrel Ph, Fouillac C, Grimaud D (2000) Dissolved load of the Loire River: chemical and isotopic characterization. Chem Geol 170:179–201
38. Millot R, Gaillardet J, Dupré B, Allègre CJ (2002) The global control of silicate weathering rates and the coupling with physical erosion: new insights from rivers of the Canadian Shield. Earth Planet Sci Lett 196:83–98
39. Avilla A, Alarcon M (1999) Relationship between precipitation chemistry and meteorological situations at a rural site in NE Spain. Atmos Environ 33:1663–1677
40. Benito G, Gutiérrez F, Pérez-González A, Machado MJ (2000) Geomorphological and sedimentological features in quaternary fluvial systems affected by solution-induced subsidence (Ebro Basin, NE-Spain). Geomorphology 33:209–224
41. Valero-Garcès BL, Delgado-Huertas A, Navas A, Machin J, Gonzalez-Sampèriz P, Kelts K (2000) Quaternary palaeohydrological evolution of a playa lake: Salada Mediana, central Ebro Basin, Spain. Sedimentology 47:1135–1156
42. Petelet-Giraud E, Négrel P (2007) Geochemical flood deconvolution in a Mediterranean catchment (Herault, France) by Sr isotopes, major and trace elements. J Hydrol 337 (1–2):224–241
43. Navas A (1991) The pattern of gypsum transport in the Ebro River network. Catena 18:45–49
44. Négrel Ph, Grosbois C (1999) Changes in chemical and $^{87}Sr/^{86}Sr$ signature distribution patterns of suspended matter and bed sediments in the upper Loire river basin (France). Chem Geol 156:231–249

Aquatic and Riparian Biodiversity in the Ebro Watershed: Prospects and Threats

Sergi Sabater, Isabel Muñoz, Joan Artigas, Anna M. Romaní, Maricarmen Pérez, and Concha Duran

Abstract This chapter provides a state-of-the art of the who is who in the biological communities of the Ebro river, as well as considerations about the threats imposed by habitat deterioration, eutrophication, pollution and species invasions, as well as the current knowledge gaps that still exist. The chapter is structured in three main sections describing the structure and composition of aquatic biological communities (phytoplankton, macroinvertebrates and fish communities), the problematic of invasive species and other hazards, and the management and conservation actions related to ecosystem health. The longitudinal patterns of planktonic chlorophyll concentration did not follow a regular pattern of increase in the Ebro River. Planktonic chlorophyll sharply decreased after the large reservoir system located in the lower reaches. Those changes were not solely explained by nutrient availability, but by alterations in the hydrological regime due to flow regulation. The lower water residence time in this section favors the growth of submerged macrophytes. Zebra mussels (*Dreissenia polymorpha*) invasion shows a massive proliferation in the lower part of the river. Biodiversity and density of macroinvertebrates decrease in the regulated areas of the river. Macroinvertebrate communities

---

S. Sabater (✉)
Institute of Aquatic Ecology, University of Girona, Girona, Spain
Catalan Institute for Water Research (ICRA), Girona, Spain
e-mail: sergi.sabater@udg.edu

J. Artigas
Catalan Institute for Water Research (ICRA), Girona, Spain

I. Muñoz
Department of Ecology, University of Barcelona, Barcelona, Spain

M. Pérez
Calle Busot, 6-1, 46007 Valencia, Spain

C. Duran
Confederación Hidrográfica del Ebro, Zaragoza, Spain

A.M. Romaní
Institute of Aquatic Ecology, University of Girona, Girona, Spain

D. Barceló and M. Petrovic (eds.), *The Ebro River Basin*,
Hdb Env Chem (2011) 13: 121–138, DOI 10.1007/698_2010_69,

in the lower part of the river are mostly filter feeders (e.g., *Hydropsyche* spp., *Ephoron virgo*). Nevertheless, the biological quality based on macroinvertebrates has improved during the last decade in the mid and low reaches of the tributaries. Stable populations of the critically endangered giant European freshwater pearl mussel (*Margaritifera auricularia*) occur in the middle Ebro section. Fish populations have been deeply altered, allochthonous fish species accounting for the 42% of the total fish species in the Ebro. Most native fish species are considered as vulnerable or threatened. Although a recent bioassessment indicates "good" and "very good" status in the 70–77% of the Ebro tributaries, caution must be taken on the potential hazards (e.g., global change, climate change) affecting the Ebro system, especially in the prospect of global change. Lower precipitation regime and higher water abstraction in rivers of the Mediterranean region such as the Ebro could contribute to the complexity loss of habitats, loss of biodiversity, and loss of ecosystem functions (e.g., nutrient recycling). The combined higher residence flow and increased nutrient concentrations in these scenarios could further enhance phytoplankton growth.

Keywords Ebro River, Global change, Invasive species, Longitudinal patterns, Management, Phytoplankton, Reservoirs

Contents

1 Introduction

Large rivers are characterized by a high complexity in habitats and biological compartments. While in small rivers the activity in the water column is biologically and biogeochemically minor, in large rivers plankton is a major contributor in any balance of organic matter and energy. The hydrological and geomorphological dynamics of large rivers configure the existence of habitats associated (i.e., hydrologically connected) to the main channel. The floodplain, i.e., the area that periodically may be flooded by the river due to peak flows, enclose a rich variety of habitats. These include oxbow lakes (mostly derived from abandoned meanders), secondary channels, and dense riparian strips [1]. Within the channel itself, rapidly flowing areas are regularly interspersed with more quiet zones where waters have longer residence time. Sediments cover extensive sections of the river bed and are areas of relevant biogeochemical activity. Macrophytes may occur in sediment-rich areas, and are themselves a relevant biological compartment that host associated flora and fauna. Fish account for a rich community in terms of diversity and biomass, larger in rivers of well-preserved heterogeneity [2]. The Ebro as a large river shows this variety of habitats, and encompasses a large set of influences. While it shows a remarkable Mediterranean regime in extensive areas of the river network [3], it is also highly regulated and accounts for high human pressure, since its original character has been historically altered, and its natural habitat and biological communities appear as patchy and irregularly preserved.

The Ebro watershed has a large surface area (85,000 km^2) and a complex drainage network (a total of 347 streams). The Ebro is a relatively well known river from the point of view of the biological communities composition. Studies exist on the aquatic fauna and flora [3]. However, the functional activity of the river is largely unknown, mostly because of its complexity and associated technical difficulties. This chapter provides a state-of-the art of the who is who in the biological communities in the river, as well as considerations about the threats imposed by habitat deterioration, eutrophication, pollution and species invasions, and indicates the current gaps that still exist in the knowledge of the river.

2 Phytoplankton Biomass and Communities: Longitudinal and Interannual Variations

The phytoplankton development in large river systems occurs in spite of unfavorable hydrological and hydrodynamic conditions, if certain conditions are accomplished. High discharge (or low water residence time) affects the planktonic development, and causes that rivers show lower planktonic densities or chlorophyll densities than lacustrine systems of similar nutrient conditions [4]. Development of river phytoplankton is constrained by the advective losses that characterize flowing waters, i.e., the drift of the populations and the decrease of the population growth rates [5]. Further, light limitation derived from higher water turbidity may limit photosynthesis and prevent phytoplankton development [5]. Though these constraints are applicable to large sections of the Ebro, a large development of phytoplankton biomass occurs in the river, mostly in the period from early summer to early autumn.

Few studies have addressed the dynamics of suspended and attached primary producers in the Ebro basin. Previous data on phytoplankton assemblages are available only for the Ebro delta [6], and for phytoplankton occurring in the lower course of the river [7, 8]. However, there is no information about other areas in the catchment, and an overall perspective of the longitudinal patterns of plankton in the system cannot be attained. This lack of data is not exclusive to the Ebro but also applies to other Mediterranean river systems. To date, the only data available for this region are those reported by [9, 10] on the river Lot (France; [9, 10]), on the river Po (Italy; [11]), by on the river Ter [12], and more recently on rivers in Greece [13, 14].

Sabater et al. [3] performed chemical and planktonic analyses at 31–43 sampling sites scattered along the main course of the Ebro (Fig. 1). Twenty-five sites covered from the upper reach down to the Ebro Reservoir to the reservoirs of Flix, Mequinenza and Ribaroja, while the other six were downstream up to 30 km from the river mouth. Samples were collected in June and October 2005 and 2006. Other surveys were completed in 2008 and 2009. Physical, chemical, and biological analyses were performed at all sites and were later related to the phytoplankton biomass.

Planktonic chlorophyll increased from the upper reaches of the main section to the meandering zone of the Ebro Depression, and sharply decreased after the large reservoirs located in the lower reaches. On the basis of the chlorophyll concentration, the river Ebro can be defined as eutrophic [15]. Plankton chlorophyll levels ranged between 10 and 17 μg/L from the headwaters to the city of Zaragoza, and then increased up to ~60 μg/L in the meander plain (Fig. 2a). Suspended chlorophyll in the Ebro ranged between 10 and 30 μg/L in most of the sites

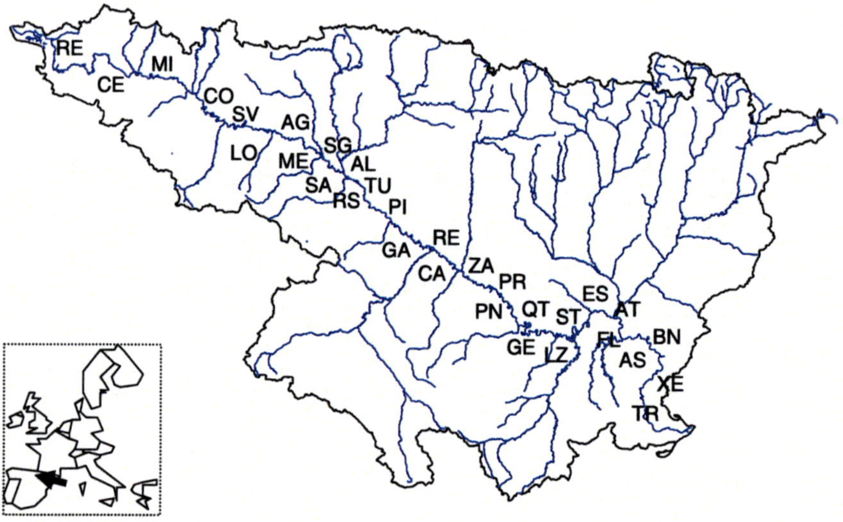

Fig. 1 Basin of the Ebro river in which the main course and the most relevant tributaries are represented. The sampling sites selected for planktonic chlorophyll analyses are marked with acronyms [Reinosa (RE), Cereceda (CE), Miranda de Ebro (MI), Conchas de Haro (CO), San Vicente de la Sonsierra (SV), Logroño (LO), Agoncillo (AG), Mendavia (ME), San Adrián (SA), Sartaguda (SG), Rincón de Soto (RS), Alfaro (AL), Tudela (TU), Pignatelli (PI), Gallur (GA), Remolinos (RE), Cabañas de Ebro (CA), Zaragoza (ZA), Presa Pina (PR), Pina de Ebro (PN), Quinto (QT), Gelsa (GE), La Zaida (LZ), Sástago (ST), Escatrón (ES), Almatret (AT), Flix (FL), Ascó (AS), Benifallet (BN), Xerta (XE), Tortosa (TR)]

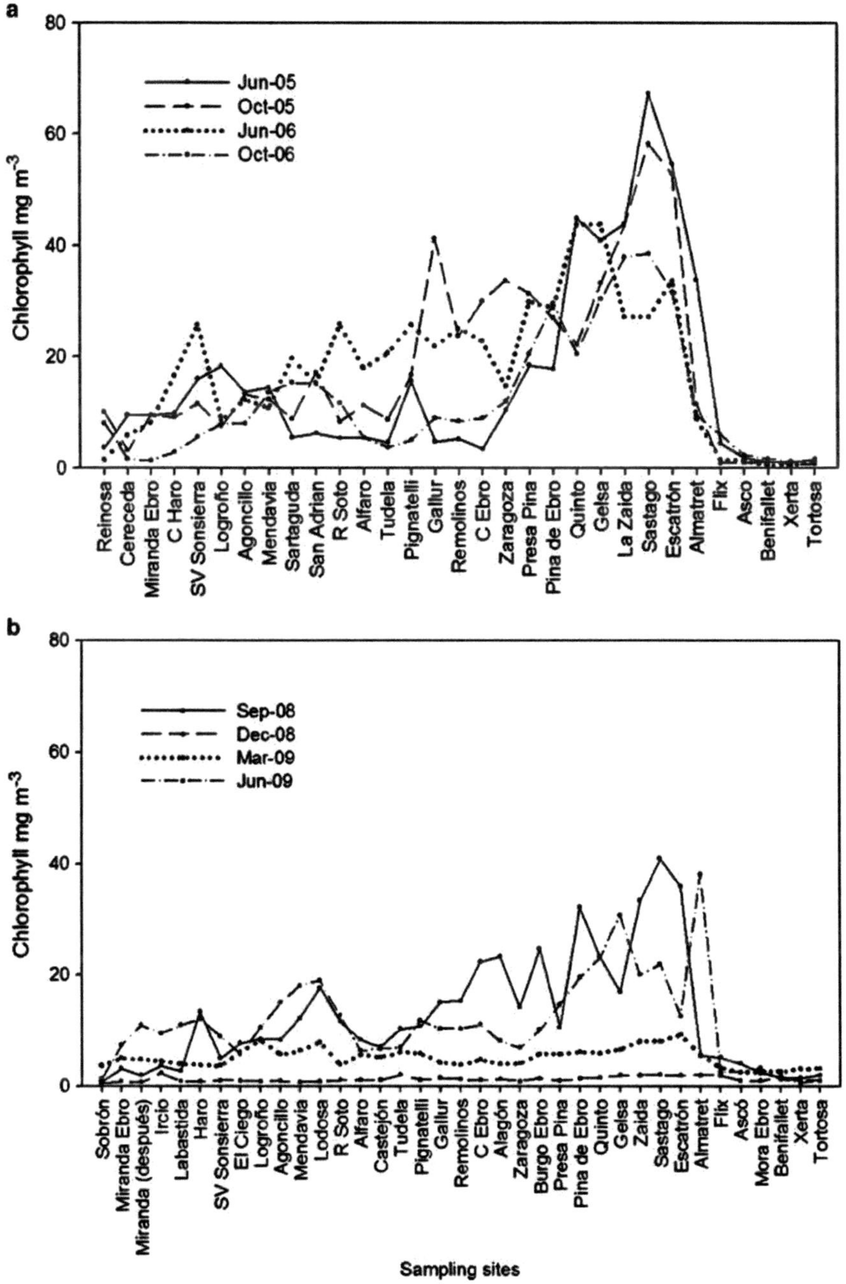

Fig. 2 Planktonic chlorophyll concentration along the main course of the Ebro river during the periods 2005–2006 (**a**) and 2008–2009 (**b**). Values are means of the chlorophyll samples taken in triplicate at each sampling time and in the different sampling sites

upstream of the reservoirs. The lowest planktonic chlorophyll values were registered in the upper part of the river, though concentrations below 10 μg/L were recorded only in the very first sites of the river. Recordings between 20 and 30 μg/L occurred in most sites located in the mid section. Concentrations increased further downstream in the meandering zone (up to 75 μg/L from Quinto to Escatrón, last site before the reservoirs). A remarkable decrease in chlorophyll concentrations (5–10 μg/L) was observed in the lower section, after the reservoirs. It is likely that processes occurring in the reservoirs (e.g., grazing, cell settlement) and the usual hypolimnetic outlet cause this abrupt interruption in the chlorophyll trend. Sestonic chlorophyll in the Ebro downstream the reservoirs was close to the values recorded in the upper section. The OD430/OD665 ratio of the planktonic pigment extracts generally showed low values (within the range 1.5–2.5), indicating the dominance of active chlorophyll in relation to other photosynthetic pigments. Slightly higher values (approaching 3) occurred after the reservoirs, indicating a certain physiological stress in chlorophyll-containing organisms.

Although the detailed longitudinal pattern of planktonic chlorophyll has been observed during 5 years of surveys, intensive surveys in 2008–2009 showed that this chlorophyll pattern is subjected to seasonality (Fig. 2b). Chlorophyll concentration strongly decreases in winter (e.g., December 2008) and flooding episodes tend to homogenize values along the river, e.g., in March 2009. During low to moderate flows the river reaches remarkable nutrient concentrations and higher chlorophyll concentration.

The longitudinal patterns of chlorophyll concentration did not follow a regular pattern of increase. The longitudinal trends in unregulated lowland rivers range from linear to second-order polynomial patterns depending on the periods of the year [16]. Hence, the expected pattern of planktonic chlorophyll in large rivers is of a downstream increase, associated with time available for plankton to develop and the corresponding increase in available nutrients downstream, followed by a decrease as the river deepen and the transparency lowers [17, 18]. In the Ebro, however, the longitudinal patterns of nutrients and chlorophyll did not closely match this model. Moreover, the presence of three large reservoirs in the lower part of the river further separated the respective longitudinal patterns.

There was no longitudinal coincidence between nutrient concentration and chlorophyll in the Ebro. This reinforces the role that hydraulics – and perhaps other factors – could exert on the development of phytoplankton [19, 20]. In some areas of the Ebro River, though, the effects of nutrients can reinforce the phytoplankton dynamics which is associated to hydraulics. The concentrations of nitrogen and phosphorus along the Ebro are related to the entrance of diffuse and local inputs [21], but also to the water quality management in the river. Nitrates were usually higher in areas of the river with the highest planktonic chlorophyll. However, phosphates were completely depleted where chlorophyll registered the highest concentrations. The co-occurrence of suspended chlorophyll with nitrates is maxima in the middle section of the river, and indicates that this nutrient was never limiting for phytoplankton development. Higher chlorophyll concentrations occurred in the meandering zone (e.g., Quinto, Gelsa, Sástago and Escatrón), where the high nitrates related to the agricultural inputs co-occurred with the low water velocity of the water to favor phytoplankton development. The estimated water travel time in this area of the river is amongst the lowest occurring in the main section of the river. Under these circumstances, the phytoplankton growth is favored. The observed depletion of phosphorus is commonly observed in large eutrophic rivers during mass phytoplankton growth [22]. An obvious consequence is that the arrival of higher P concentrations to the meandering area of the Ebro could further enhance phytoplankton biomass in this section of the river. The location and magnitude of the reservoirs of Mequinenza, Ribaroja, and Flix complicate the longitudinal pattern of planktonic chlorophyll in the lower Ebro, and interfere in the already different hydrology in that part of the river. Water travel time in the last 100 km of the river has been derived from the recent estimates [23] and approaches ca. 2 days under low flow. This is a short time to allow river plankton growth [19] that possibly minimizes the relevance of the inoculum provided by the reservoirs on the riverine phytoplankton dynamics [20].

Low chlorophyll concentrations in the lower section clearly diverge from those recorded during the late eighties. Historical values of chlorophyll-*a* in the lower section of the river [7] ranged from 5 to 46 μg/L, with maximum values in spring and summer (20–45 μg/L) and lowest values in winter (5–12 μg/L). The comparable periods to those of the present study reached 20–45 μg/L, while they currently range between 2 and 5 μg/L. This change could be attributed to several factors. The nutrient inputs in the lower section have changed over the last 20 years. The average phosphate concentrations in this part have decreased (from 0.08–0.27 mg/L P-PO_4 in 1980–2004 down to 0.02–0.06 mg/L in 2004–2005; data from CHE). This decrease is related to the construction of waste water treatment plants in the mid and lower parts of the river. There has been a certain decrease of suspended solids in the lower section along the last 10 years (CHE data available at www.chebro.es). Possibly associated to this decrease in particulate material (that allows higher light availability) there has been a large increase in macrophytes in the lower section, which at times cover >40% of the water surface. Further, zebra mussels (*Dreissena polymorpha*) in the Mequinensa and Ribarroja reservoirs [24] may contribute to deplete the inoculum from the reservoirs to the lower part of the river.

The Ebro sustains a diverse phytoplankton flora. Surveys being carried out in the last 5 years have allowed determining that up to 171 taxa occur in the main section of the river. The phytoplankton community is dominated by centric diatoms and single-cell and colonial chlorophytes, cyanobacteria being also present. Diatoms characterize more turbulent conditions (autumn), while chlorophytes are dominant in more lentic conditions, which are common in spring-summer. Amongst the diatoms, the centric diatoms are the most diverse [8, 25]. There is a large group of small and medium size centric diatoms of the genera *Cyclotella* and *Stephanodiscus*, while *Skeletonema potamos* is common in the medium and lower part of the river (Fig. 3). Among the green algae, *Scenedesmus* and *Desmodesmus* are the most abundant [26]. *Planktothrix* is the most common cyanobacteria throughout the year, probably deriving from the reservoirs. After heavy rains benthic diatoms (from the sediments and plants) appear in the water column, while the common

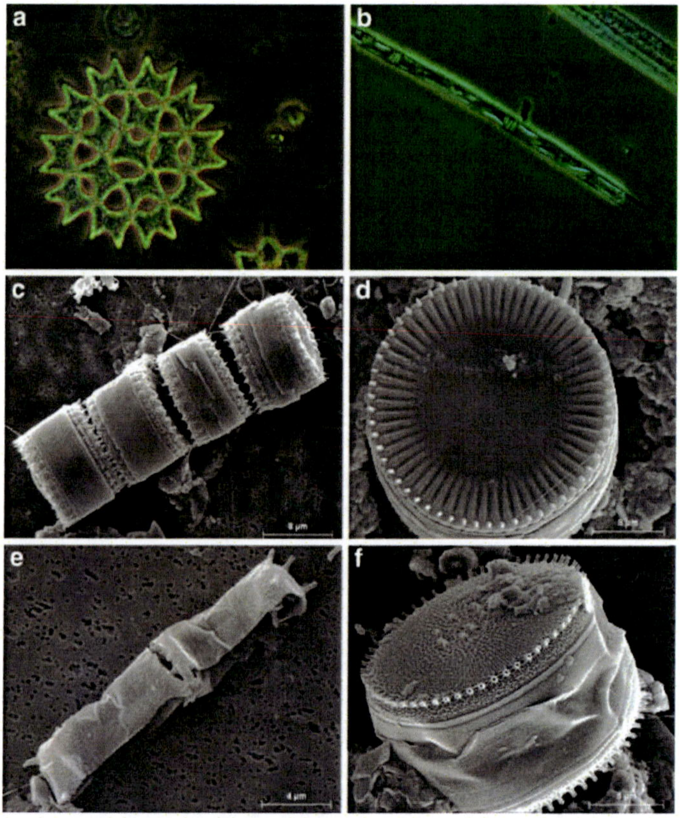

Fig 3 Images of phytoplankton cells from the Ebro river obtained by means of epifluorescence microscopy and scanning electron microscopy. Images correspond to the species *Pediastrum duplex* (**a**), *Aulacoseira granulata* (**b**), *Stephanodiscus* cf. *neoastraea* (**c**), *Cyclotella meneghininana* (**d**), *Skeletonema potamos* (**e**), and *Thalassiosira weissflogii* (**f**)

phytoplankton cells are washed away. Maximal cell densities (33,970 cells/mL in Escatrón) were analogous to those recorded in large eutrophic river systems [27]. However, minimum abundance of 4 cells/mL may occur downstream of the reservoirs (Benifallet). Planktonic chlorophyll is significantly related to cell density ($r = 0.447, p = 0.0001$).

The ordination of 23 taxa or groups of taxa in relation to the environmental predictors was examined by means of multivariate analyses. Nutrients (nitrates, total phosphate, and ammonia), conductivity, total suspended solids and water flow were retained by the redundancy analysis (RDA) (Fig. 4). The first RDA axis (RDA1; 16.6% of the variance) showed a strong correlation between nutrients and suspended solids and (negatively) water flow. Large centric diatoms and *Scenedesmus* sp.pl. best defined this situation, which was characteristic of the meandering zone, mostly in autumn. Other taxa (*Coelastrum microporum*,

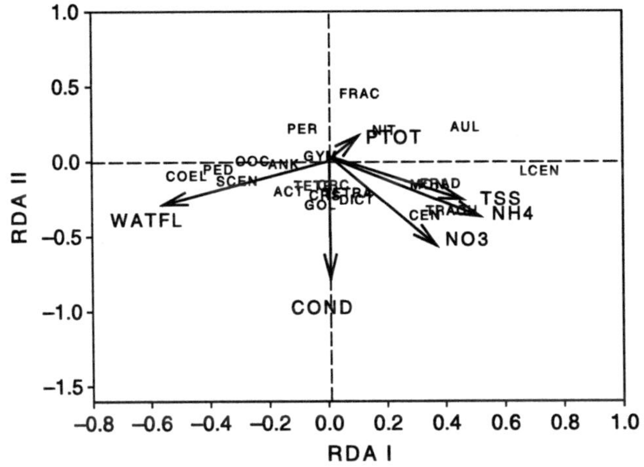

Fig. 4 Redundance analysis (RDA) on phytoplankton communities collected during the surveys of 2005–2006 and their relationship with the environmental variables, such as the Total Phosphorus (PTOT), Water Flow (WATFL), Total Suspended Solids (TSS), ammonia concentration (NH4), nitrate concentration (NO3), and conductivity (COND). Acronyms were also used for phytoplankton taxa representation

Pediastrum sp.pl.; Fig. 3) were probably washed off the reservoirs, since they were particularly abundant in sites below the Flix, Mequinensa and Ribaroja reservoirs. The second RDA axis (RDA2) explained a low proportion of the variance (4%), and was mostly negatively related to water conductivity. Samples associated with high conductivity were those from the mid section of the river, characterized by several Chlorococcales (*Actinastrum hantzschii*, *Golenkinia radiata*, *Tetrastrum* sp. pl.), a community that contrasted with those in the upper part, characterized by the abundance of *Fragilaria crotonensis*.

Differences in chlorophyll concentrations between present and past records in the lower part of the Ebro do not correspond with significant changes in the distribution of phytoplankton assemblages. The general trends in the distribution of phytoplankton communities appear to be consistent with those reported in previous surveys, at least in the lower part of the river. Centric diatoms such as *Aulacoseira granulata*, *Cyclotella* sp. and *Stephanodiscus* sp. were dominant in autumn, spring, and early summer 1989–1990, while *Scenedesmus* sp., *Coelastrum* sp., and *Pediastrum* sp. were most abundant in the summer of that period [7].

3 Phytobentos and Macrophytes

Algae colonize both natural (cobbles and sediments, macrophytes and wood) and artificial substrata (walls, infrastructures). Light attenuation limits the development of the algal community to the shallow areas of the river. Benthic chlorophyll in the main channel is on average 266 mg/m^2 in summer and 196 mg/m^2 in autumn [28]. Benthic chlorophyll was over 100 mg Chl-*a*/m^2 in most of the sites and campaigns (average concentration of 167 mg Chl-*a*/m^2), therefore characteristic of eutrophic rivers. The highest concentration of benthic chlorophyll was observed in the mid section of the Ebro (Agoncillo; 781.7 mg Chl-*a*/m^2) in June 2005, and the lowest also in the same area (Gallur, vicinity of Zaragoza; 7 mg Chl-*a*/m^2 in October 2006). The concentration of benthic chlorophyll in sites below the Flix, Mequinensa and Ribaroja reservoirs was generally high (50–350 mg Chl-*a*/m^2), in contrast to the low values in the planktonic compartment. A surface transformation of the planktonic chlorophyll values in the section downstream the reservoirs provide equivalent values of 10–12 mg Chl-*a*/m^2. However, the benthic compartment is less relevant than the planktonic compartment in terms of organic matter and energy transfers.

Regarding the benthic diatom communities, *Achnanthes minutissima* and *Achnanthes biasolettiana* are the dominant taxa of a low diversity community in the streams and rivers (Erro, Gállego, headwaters of the Ebro) headwaters, where fast water current and low nutrient content are characteristic. Epilithic diatoms in the upper Segre are dominated by *Achnanthidium subatomus*, *Diatoma mesodon*, *Cymbella silesiaca*, *Fragilaria arcus*, *F. capucina*, *Gomphonema pumilum*, *Meridion circulare*, and *Nitzschia pura* [29]. *Amphora pediculus* and *Cocconeis placentula* become dominant when waters are slower and with higher mineral content (Guadalopillo, Cambrones, Cortiella). In middle and lower courses, with high mineral and nutrient contents a more abundant diatom flora made up of *Navicula cryptotenella*, *Navicula subminuscula*, *Navicula minima*, and *Nitzschia inconspicua* characterized the sites. This was common in the main section of the Ebro, as well as in the main reaches of the large Ebro tributaries.

In temporary saline lakes occurring in endorheic areas of the Ebro Depression, the phototrophic community is composed of planktonic (*Dunaliella* sp., *Aphanothece* sp.) and benthic organisms (*Hantzschia amphyoxis*) [30]. The permanent Lake Salada de Chiprana (78 g/L salinity on average) is covered by microbial mats of *Microcoleus chthonoplastes* (~30 μg Chl-*a*/cm^2) and the charophyte *Lamprothamnium papulosum* [31]. Microbial mats at the Ebro delta are composed of three pigmented layers of phototrophic organisms: an upper brown layer of *Lyngbya aestuarii* and diatoms, an intermediate green layer of the cyanobacterium *M. chthonoplastes*, and an underlying pink layer of purple sulfur bacteria [32].

Macrophytes are common in the lower river below the reservoirs. Floating algae and macrophytes in summer may cover >40% of the water surface in the lower river. The fast growth of macrophytes has relevant hydraulic effects that can lead to remarkable increases in the water level. These macrophytes also favor the massive development of blackflies (*Simulidae erythrocephalum*) [33].

The most abundant macrophytes are *Potamogeton pectinatus*, *Potamogeton crispus*, *Potamogeton densus*, *Ceratophyllum demersum*, *Myriophyllum spicatum*, and *Lemna gibba* [34]. *P. pectinatus* L., *M. spicatum* L., and *C. demersum* L. are the most abundant species in this section. *P. pectinatus* has the highest success, since is a rooted macrophyte with remarkable ability for phosphorus uptake, a fast growth, and highly resistant to shear stress related to water flow fluctuations. Most of the massive macrophyte growth occurs in summer when water is at its lowest flow level, water temperature increases, and the low dilution (and associated higher nutrient concentration) allows for a massive growth [35]. The problem with macrophytes has been managed through controlled flood events [36]. Artificial floods are periodically produced in order to decrease the abundance of macrophytes. For instance, induced floods in March, May, and November 2006 consisted of peak flows of 1,350 m^3/s (up to 6–7 times the basal flow in those periods) during short periods of time.

At the Ebro delta, *P. pectinatus* is mainly found in the least brackish areas, while *Ruppia cirrhosa* inhabits transitional zones between freshwater and seawater. Mixed stands of *Zostera noltii*, *R. cirrhosa*, and the floating macroalga *Chaetomorpha linum* develop in saline areas [37].

4 Riparian Vegetation

Riparian vegetation is deeply altered in the Ebro river basin. In the middle section area, where the river builds up a large floodplain, the riparian vegetation presently covers only 4.5% of the area, while this amounted up to ca. 40% in the 1950s [38]. Several factors contributed to this decrease. The agricultural development occupied the fertile floodplain areas, and dykes were built to stabilize the banks. Large floods affecting floodplain areas (such as the one occurring in 1961, [38]) triggered the construction of several bank defenses in the 1960s and 1980s. Elimination of secondary arms (or filling them up with debris) and simplification of meanders constrained the natural hydrological dynamics of the river. Obviously, these alterations constrained that during high discharge periods waters could occupy the flood plain. The possibility of the riparian vegetation to develop, and that well-structured riparian forests occur is therefore limited.

In spite of the loss of the riparian habitat, still well preserved riparian forests are found at the Aragón, Arga, Irati, Cinca, Segre and Gállego tributaries. At the main Ebro river channel, some well preserved riparian forests (*Sotos*) are observed at the headwaters and in short reaches at the Rioja, Navarra and Aragón. Riparian tree species having rapid growth and being well-adapted to water level fluctuations are widespread (*Salix alba*, *Populus alba*, *Populus nigra*, *Tamarix africana*, *Tamarix gallica*). Forests of *Alnus glutinosa* occur in small headwater areas in the northern catchment. Reed beds and herbs also cover the deposits of sediment materials accumulating in the river banks and bars. In these areas opportunistic species such as *Polygonum lapathifolium*, *P. persicaria*, *Xanthium echinatum*, and *Paspalum paspalodes* are very common.

5 Invertebrate Communities

The term *macroinvertebrate* refers to the larger invertebrate fauna, defined as those retained by a 500-μm mesh net. This is the most known and studied invertebrate group in aquatic ecosystems. Invertebrate community in rivers generally constitutes an important component of the benthos. Invertebrates are associated with all the different substrata of the river bed (bedrock, cobbles, sediment, leaves, wood, roots, and aquatic vegetation). Their role in the energy transference between compartments is a fundamental link in the food web, since they intersect between the organic matter resources (algae, leaf litter, detritus) they use, and fishes that feed on them. They are ubiquitous in rivers and are represented by a high number of species. Changes in water quality and in microhabitat structure remarkably affect their distribution and abundance.

The macroinvertebrate community showed a high diversity at the headwaters of the main channel and tributaries of the Ebro River. This relevance is related with higher water quality [39] and suitable habitat characteristics in these sites. Biodiversity and density of macroinvertebrates decrease in the regulated areas of the river [40]. Nevertheless, the biological quality based on macroinvertebrates has improved during the last decade in the mid and low reaches of the tributaries, except for the Jalón river [41].

The Ebro river forms in its floodplain a complex mosaic of interconnected aquatic and terrestrial patches with different environmental and hydrological features. Such variety of habitats is at the base of the high biological macroinvertebrate diversity [42] in the area. In the middle Ebro river floodplain, the most important factor associated with macroinvertebrate assemblage was the hydrological connectivity with the river channel [43]. Wetland restoration and conservation enhance biological diversity and aids in the functional and ecological recovery of the riverine landscape. Some experiences [43] showed the presence of new macroinvertebrate species (e.g., *Trithemis annulata*, *Coenagrion scitulum*) in new and restored wetlands.

The Mediterranean influence is mostly obvious in the tributaries entering the Ebro from the South. The macroinvertebrate taxa (*Physella acuta*, *Perla marginata*, *Hydroptila insubrica* and *Hydropsyche instabilis*) are adapted to avoid the impact of floods and droughts [44]. Macroinvertebrate communities in the lower part of the river are mostly filter feeders (e.g., *Hydropsyche* spp., *Ephoron virgo*). *E. virgo* is an indicator of the good ecological quality of the river, and it is abundant in gravel and sand habitats [45]. This mayfly decreased its density dramatically in most of the European rivers due to contamination. In those rivers were the quality was improved, this insect returns to its previous density. Where light reaches the river bottom, stones and boulders are covered by grazers such as the gastropods *Melanopsis* sp. and *Theodoxus fluviatilis* [46]. Euryhaline species such as the polychaete *Ficopotamus* and the trichopteran *Ecnomus* are abundant nearby the Ebro delta [46].

Up to four naiad species have been found in the main river (Ebro) and in the artificial channels and ditches. These are *Margaritifera auricularia*, *Potomida litoralis*, *Unio mancus*, and *Anodonta* sp. The Critically Endangered (IUNC) giant European freshwater pearl mussel (*M. auricularia*), present in the lower part of the river, has declined dramatically since the early twentieth century [47]. The only stable population is located in a channel (Canal Imperial) fed by water of the mid Ebro River. This is by far the largest European population of this species. A secondary population is located in Canal de Tauste [48] in well oxygenated, subsaline and mesotrophic waters with high values of calcium (150 mg/L), and gravel and pebbles substrata. Their life cycle is closely linked to the presence of different fish species. Mussel reproductive cycle includes the glochidia infection of host fish. Habitat disturbance, lost of host fish species and low water quality are the most important factors that determine the decline of this species [49].

6 Fish Communities

Most of the native fish species of the Ebro are vulnerable (such as *Anguilla anguilla*, *Petromyzon marinus*, *Cobitis calderoni*, and *Cobitis paludica*), while others are threatened (*Aphanius iberus*, *Valencia hispanica*, *Gasterosteus gymnurus*, *Salaria fluviatilis*, *Acipenser sturio*) (CH [50]). The Ebro catchment encompasses a biogeographic zone where southern and northern fishes occur, making assemblages vulnerable to introduced and invasive species such as *Micropterus salmoides*, *Sander lucioperca*, *Gambusia holbrooki*, *Esox lucius*, *Ameiurus melas*, and the large *Silurus glanis* (CH [50]). Dam construction has also caused an increase in fish density, especially in the smaller species.

The Ebro catchment has the highest richness of autochthonous limnetic species in the Iberian Peninsula [51]. *Salaria (blennius) fluviatilis*, is common at the main channel and tributaries, and *Chondrostoma toxostoma* only occurs at the tributaries. *Leuciscus cephalus* is a fish of narrow distribution, found only at the Ebro river basin and other small catchments of NE Spain. Salmonids are common at the headwaters: one species is native *Salmo trutta* and two introduced others, *Salmo gairdneri* and *Salvelinus fontinalis*. Part of the autocthonous species are endemic, for example the cyprinids *Barbus graellsii* and *Barbus haasi*, the cobitid *C. calderoni* with scattered and scarce populations [52] and *A. iberus*.

The fish community in the lower Ebro (including its delta) is the result of the interaction between the freshwater origin species and those coming from marine and adapted to environment fluctuations. Up to 50 species colonize the Ebro delta [53]. In the fluvial reaches, there are species with a high faunistical interest, such as *S. fluviatilis* that is endangered in this area. This section of the river is one of the few rivers in the Spanish Mediterranean where migratory species are present: *Alosa fallax*, *Alosa alosa*, *P. marinus*, and *A. anguilla*. Eel fishery is of great importance in the delta area. Amongst the delta species, distinguished by their faunistic value, are *A. iberus* and *V. hispanica*. The former with a Mediterranean distribution, colonize waters with variability in salinity and temperature and it is present at the river mouth, coastal lagoons and salt marshes of the Ebro delta. *V. hispanica* is present mainly at the irrigation channels and coastal lagoons. Its habitat degradation and the introduction of *G. holbrooki* with which it shared habitat, have diminished significantly its population.

In the Ebro River, the allochthonous species constitutes the 42% of the total fish species. In the last 50 years, this number has tripled. The effects of this introduction are more severe in the Ebro due to the relatively low fish diversity and the high grade of endemism. In addition, there are few predatory species, so the introduction of carnivorous species has had a major impact on the original communities [53]. Some of these species are clearly in expansion or long-established: *S. glanis*, *S. lucioperca*, *Alburnus alburnus*, *G. holbrooki*, *E. lucius*. The introduced Asian cyprinid *Pseudorasbora parva* has recently been recorded in the delta [54], as well as *Misgurnus anguillicaudatus* [55] and *Fundulus heteroclitus* [56].

7 Invasive Species and Other Biological Hazards

Different invasive species have been recorded in different waterbodies of the River Ebro. The European catfish (*S. glanis*), the largest fish in European freshwaters, was introduced in the lower Ebro in 1974 in the reservoirs of Mequinenza and Ribarroja, for angling purposes [57, 58]. Introducing the catfish came together with the introduction of yearlings of another invasive species, the common bleak (*A. alburnus*), used as a living bait for the former. Further, waters transporting the baits or catfish could be the vector of other invasive species, highly problematic in the Ebro, the zebra mussel (*D. polymorpha*). Larvae of this mussel were able to survive in the container waters and infect the same areas where the catfish was introduced. The species is currently progressing upstream from the Mequinenza and Ribarroja reservoirs to many lacustrine and fluvial areas in the river, where it is considered as a pest. The presence of the mussel was first detected in 2001 [24] and ever since has been increasing its occupation area. *Dreissena* is a very active filterer that grows massively attached to any hard substrata in the river, including infrastructures as well as other benthic species. *Dreissena* has caused extensive damage to the aquatic ecosystems and directly competes with sensitive, or endemic species, such as the bivalve *M. auricularia*. The massive growth of *Dreissena* has produced remarkable economic costs (CHE [50]). *Dreissena* increases water transparency through its filtering activity that favors light penetration and the higher production of benthic primary producers. Their presence is associated to the helminth *Phyllodistomum folium*, which infects the zebra mussels, and represents a new invasive species in the Ebro river basin [59]. The physical changes induced by the presence of *Dreissena* favors the massive growth of macrophytes in extensive areas of the lower Ebro, as detailed above [60]. Recently, the presence of the Asian invasive bivalve *Corbicula fluminea* (Spanish Society of Malacology, 2003) in the lower part of the Ebro (and even in the city of Zaragoza) has added to the other alterations.

Other organisms have accidentally reached the Ebro waters. This is the case of the North American freshwater Planarian *Dugesia tigrina*. The yellowbelly slider Florida turtle (*Trachemys scripta*) and the American crayfish (*Procambarus clarkii*) have displaced native species wherever they progress. The latter is a vector of fungal infections (aphanomycosis) that threats the survival of the autochthonous European crayfish. Also, the diatom *Didymosphenia geminata* is an invader in the Ebro basin, mostly in middle reaches of tributaries [61]. The aquatic fern *Azolla filiculoides* has also reached several sites in the Ebro [62].

8 Management and Conservation

The economic use of water for irrigation and reservoir construction is the main environmental disturbance in the Ebro catchment, and has altered the flow regime except in upper tributaries. The Ebro has 187 reservoirs impounding 57% of the mean annual runoff. All dams were constructed during the twentieth century with 67% of the reservoir capacity built in 1950–1975. Three reservoirs have >500 Mm^3 in capacity (Ebro, Mequinenza and Canyelles). The Hydrologic Basin Authority (est. 1926) of the Ebro (CH Ebro) was the first organization for managing Spanish rivers. The first objective was to organize irrigation for agriculture. Today, it is responsible for the control of catchment master plans based on the Water Framework Directive. A recent bioassessment indicates "good" and "very good" status of 70–77% of the examined rivers (CHE [50]).

Hydraulic structures in the river (dams, reservoirs, canals) add to the habitat complexity occurring in the river and would cause an optimal scenario for phytoplankton growth under more prolonged low flow situations. Further, low water flows are associated with higher nutrient concentrations which would reinforce phytoplankton mass outgrowth, in particular in the shallow and slow-moving areas of the river. Given the predictions concerning rainfall in the Iberian Peninsula [63], it is therefore essential to increase the control of nutrients from both diffuse and local inputs to the river, particularly in the meandering section upstream of the reservoirs, and specifically regarding phosphorus. On the basis of our findings, we consider that the combined higher residence flow and increased nutrient concentrations in these scenarios could further enhance phytoplankton growth. If this occurs, the river Ebro would be further away from attaining an acceptable ecological status.

References

1. Petts GE, Amoros C (1996) Fluvial hydrosystems. Chapman & Hall, London, p 307
2. Allan JD, Castillo MM (2007) Stream ecology. Structure and function of running waters, 2nd edn. Springer, Dordrecht, p 435
3. Sabater S, Feio MJ, Graça MAS, Muñoz I, Romaní A (2009) The Iberian rivers. In: Tockner K, Robinson C, Uehlinger U (eds) Rivers of Europe. Elsevier, Amsterdam, pp 113–150

4. Søballe DM, Kimmel BL (1987) A large-scale comparison of factors influencing phytoplankton abundance in rivers, lakes and impoundments. Ecology 68:1943–1954
5. Reynolds CS (2006) Ecology of phytoplankton. Cambridge Univ. Press, Cambridge, p 550
6. Comin FA (1984) Caracteristicas fisicas y químicas y fitoplancton de las lagunas costeras, Buda, Encañizada y Tancada (Delta del Ebro). Oecologia Aquatica 7:79–162
7. Sabater S, Muñoz I (1990) Successional dynamics of the phytoplankton in the lower part of the river Ebro. J Plankton Res 12:573–592
8. Sabater S, Klee R (1990) Observaciones en diatomeas centrales del fitoplancton del río Ebro, con especial interés en algunas pequeñas Cyclotella. Diatom Res 5:141–154
9. Capblanc J, Dauta A (1978) Phytoplancton et production primaire de la riviere Lot. Ann Limnol 14:85–112
10. Dauta A (1975) Étude du phytoplankton du Lot. Ann Limnol 11:219–238
11. Andreoli C, Fricano G (1981) Variazione stagionali del fitoplancton del tratto medio del Po interessato dalla construenda centrale termonucleare di Caorso (Piacenza). Riv Idrobiol 20:139–152
12. Sabater S (1990) Phytoplankton composition in a Mediterranean river: the Ter (Spain). Limnetica 6:47–55
13. Montesanto B, Tryfon E (1999) Phytoplankton community structure in the drainage network of a Mediterranean river system (Aliakmon, Greece). Int Rev Hydrobiol 84:451–468
14. Montesanto B, Ziller S, Danielidis D, Economou-Amilli A (2000) Phytoplankton community structure in the lower reaches of a Mediterranean river (Aliakmon, Greece). Arch für Hydrobiol 147:171–191
15. Dodds WK, Jones JR, Welch EB (1998) Suggested classification of stream trophic state: distributions of temperate stream types by chlorophyll, total nitrogen, and phosphorus. Water Res 32:1455–1462
16. Basu BK, Pick FR (1997) Phytoplankton and zooplankton in a lowland, temperate river. J Plankton Res 19:237–253
17. Jones FH (1984) The dynamics of suspended algal populations in the lower Wye catchment. Water Res 8:25–35
18. Köhler J (1994) Origin and succession of phytoplankton in a river-lake-system (Spree, Germany). Hydrobiologia 289:73–83
19. Reynolds CS, Descy JP (1996) The production, biomass and structure of phytoplankton in large rivers. Arch für Hydrobiol Suppl 113:161–187
20. Neal C, Hilton J, Wade AJ, Neal M, Wickham H (2006) Chlorphyll-*a* in rivers of eastern England. Sci Total Environ 365:84–104
21. Torrecilla NJ, Galve JP, Zaera LG, Retamar JF, Alvarez ANA (2005) Nutrient sources and dynamics in a Mediterranean fluvial regime (Ebro river, NE Spain) and their implications for water management. J Hydrol 304:166–182
22. Picard V, Lair N (2005) Spatio-temporal investigations on the planktonic organisms of the Middle Loire (France), during the low water period: biodiversity and community dynamics. Hydrobiologia 551:69–86
23. Sanchez-Cabeza JA, Pujol L (1999) Study on the hydrodynamics of the Ebro River lower course using tritium as a radiotracer. Water Res 33:2345–2356
24. Navarro E, Bacardit M, Caputo L, Palau A, Armengol J (2006) Limnological characterization and flow patterns of a three-coupled reservoir system and their influence on *Dreissena polymorpha* populations and settlement during the stratification period. Lake Reservoir Manag 22:293–302
25. Pérez MC, Maidana N, Comas A (2009) Phytoplankton composition in the Ebro river estuary, Spain. Acta Bot Croat 68(1):11–27
26. Pérez MC, Comas C, Gonzalez Del Rio J, Sierra JP (2002) Chlorophyceae from the lower Ebro river (Spain). Acta Bot Croat 61(2):99–124
27. Descy JP (1987) Phytoplankton composition and dynamics in the river Meuse (Belgium). Arch für Hydrobiol 78:225–245

28. Sabater S, Artigas J, Durán C, Pardos M, Romaní AM, Tornés E, Ylla I (2008) Longitudinal variation of sestonic chlorophyll and phytoplankton assemblages in the Ebro River. Sci Total Environ 404:196–206
29. Gomà J, Rimet F, Cambra J, Hoffmann L, Ector L (2005) Diatom communities and water quality assessment in mountain rivers of the upper Segre basin (La Cerdanya, Oriental Pyrenees). Hydrobiologia 551:209–225
30. Comín FA (1999) Management of the Ebro river basin; past, present and future. Water Sci Technol 40:161–168
31. de Witt R, Falcon LI, Charpy-Robaud C (2005) Heterotrophic dinitrogen fixation (acetylene reduction) in phosphate-fertilised *Microcoleus chthonoplastes* microbial mat from the hypersaline inland lake "la Salada de Chiprana" (NE Spain). Hydrobiologia 534:245–253
32. Solé A, Gaju N, Esteve I (2003) The biomass dynamics of cyanobacteria in an annual cycle determined by confocal laser scanning microscopy. Scanning 25:1–7
33. Ibáñez C, Prat N, Durán C, Pardos M, Munné A, Andreu A, Caiola N, Cid N, Hampel H, Sánchez R, Trobajo R (2008) Changes in dissolved nutrients in the low Ebro river: causes and consequences. Limnetica 27:131–142
34. Molina Holgado P (2002) Análisis y comparación de la vegetación de las riberas de los ríos Ebro, Tajo y Jarama. Ediciones de la Universidad Autonoma de Madrid, Madrid, p 545
35. Soley S (2010) Interaction between macrophytes and phytoplankton compartments for the uptake and decomposition of phosphorus compounds, the case of the Ebro river (NE Spain). Master thesis. University of Girona, Spain
36. Batalla RJ, Gómez CM, Kondolf GM (2004) Reservoir-induced hydrological changes in the Ebro River basin (NE Spain). J Hydrol 290:117–136
37. Menéndez M, Herrera J, Comín FA (2002) Effect of nitrogen and phosphorus supply on growth, chlorophyll content and tissue composition of the macroalga *Chaetomorpha linum* (O.F. Müll), Kütz, in a Mediterranean Coastal Lagoon. Sci Mar 66:355–364
38. Ollero A (2007) Channel adjustments, floodplain changes and riparian ecosystems of the middle Ebro River: assessment and management. Int J Water Resour Dev 23:73–90
39. Alba-Tercedor J, Sánchez-Ortega A (1988) Un método rápido y simple para evaluar la calidad biológica de las aguas corrientes basado en el de Hellawell (1978). Limnetica 4:51–56
40. Gil-Quilez MJ, Palau A, Fernandez-Manzanal C (2001) Calidad biológica (BMWP) de las aguas del río Cinca (Huesca). Limnetica 20:107–113
41. Oscoz J, Durán C, Pardos M, Gil J, Viamonte A (2008) Evolución histórica de la calidad biológica del agua en la cuenca del Ebro (España) (1990-2005). Limnetica 27:119–130
42. Gallardo B, Garcia M, Cabezas A, Gonzalez E, Ciancarelli C, Gonzalez M, Comin FA (2007) First approach to understanding riparian wetlandsin the middle river Ebro floodplain (NE Spain): structural characteristics and functional dynamics. Limnetica 23:373–386
43. Gallardo B, Garcia M, Cabezas A, Gonzalez E, Gonzalez M, Ciancarelli C, Comin FA (2008) Macroinvertebrate patterns along environmental gradients and hydrological connectivity within a regulated river-floodplain. Aquat Sci 70:248–258
44. Argerich A, Puig MA, Pupilli E (2004) Effect of floods of different magnitude on the macroinvertebrate communities of Matarranya stream (Ebro river basin, NE Spain). Limnetica 23:283–294
45. Cid N, Ibañez C, Prat N (2008) Life history and production of the burrowing mayfly *Ephoron virgo* (Olivier, 1791) (Ephemeroptera: Polymitarcyidae) in the lower Ebro river: a comparison after 18 years. Aquat Insects 30:163–178
46. Muñoz I, Prat N (1994) A comparation between different biological water quality indexes in the Llobregat Basin (NE Spain). Verh Int Ver Theor Angew Limnol 25:1945–1949
47. Araujo R, Ramos MA (2000) Status and conservation of the giant European freshwater pearl mussel *Margaritifera auricularia* (Spengler, 1793) (Bivalvia: Unionoidea). Biol Conserv 96:233–239
48. Gómez I, Araujo A (2008) Channels and ditches as the last shelter for freshwater mussels. The case of *M. auricularia* and other naiads inhabiting the mid Ebro River basin, Spain. Aquat Conserv: Mar Freshw Ecosyst 18:658–670

49. Lopez MA, Altaba CR, Rouault T, Gisbert E (2007) The European sturgeon *Acipenser sturio* is a suitable host for the glochidia of the freshwater pearl mussel *Margaritifera auricularia*. J Molluscan Stud 73:207–209
50. CH Ebro (2005) Economic evaluation of the zebra mussel invasion in the Ebro basin (Pérez L, Chica C). Internal document, pp 48. http://oph.chebro.es/DOCUMENTACION/Calidad/mejillon/en_inicio.htm
51. Elvira B (1995) Native and exotic freshwater fishes in Spanish river basins. Freshw Biol 33:103–108
52. Leunda PM, Miranda R, Oscoz J (2007) Occurrence and conservation of the threatened endemic cobitid, *Cobitis calderoni*, in the Erro River (Ebro Basin, Spain). Cybium 31:13–18
53. Sostoa A, Lobón-Cerviá J (1989) Fish and fisheries of the river Ebro: actual state and recent history. In: Petts GE, Móller H, Roux AL (eds) Historical change of large alluvial rivers: Western Europe. Wiley, New York, pp 233–247
54. Caiola N, Sostoa A (2002) First record of the Asiatic cyprinid *Pseudorasbora parva* in the Iberian Peninsula. J Fish Biol 61:1058–1060
55. Franch N, Clavero M, Garrido M, Gaya N, Lopez V, Pou-Rovira Q, Queral JM (2008) On the establishment and range expansion of oriental weatherfish (*Misgurnus anguillicaudatus*) in NE Iberian Peninsula. Biol Invasions 10:1327–1331
56. Gisbert E, Lopez MA (2007) First record of a population of the exotic mummichog *Fundulus heteroclitus* (L., 1766) in the Mediterranean Sea basin (Ebro River delta). J Fish Biol 71:1220–1224
57. Elvira B, Almodóvar A (2001) Freshwater fish introductions in Spain: facts and figures at the beginning of the 21st century. J Fish Biol 59:323–331
58. Doadrio I (2002) Atlas y Libro Rojo de los Peces Continentales de España. Dirección General de Conservación de la Naturaleza-Museo Nacional de Ciencias Naturales, Madrid, p 374
59. Peribáñez MA, Elrío ML, Gracia MJ, Fernández de Luco D, Castillo JA, Lucientes J, Cia I (2006) *Phyllodistomum folium* (Trematoda: Gorgoderidae) infecting zebra mussels (*Dreissenia polymorpha*) in the Ebro River, Spain. Parasitol Int 55:143–145
60. Montesinos S, Bea M, Durán C, Losada JA (2009) Determinación de macrófitos en el río Ebro entre Flix y Mora D'Ebre. Teledetección. Agua y desarrollo sostenible.. ISBN 978-84-613-4257-0
61. Blanco S, Bécares E (2009) Una nueva alga invasora llega a España: *Didymosphenia geminata*. Quercus 283:30–35
62. Tomás P (2009) Presencia de *Azolla fuliculoides* en los ríos de la cuenca del Ebro. Actas Tercer Congreso Nacional sobre Especies Exótica Invasoras
63. International pannel on climate Change (IPCC) (2007) Climate change 2007: the physical science basis. In: Solomon S, Qin D, Manning M, Chen Z, Marquis M, Averyt KB, Tignor M, Miller HL (eds) Fourth assessment report. Cambridge University Press, Cambridge, p 996

Persistent Organic Pollutants in Water, Sediments, and Biota in the Ebro River Basin

Alícia Navarro-Ortega and Damià Barceló

Abstract The Ebro river basin is one of the most studied basins in Spain. The *Confederación Hidrográfica del Ebro* (CHE), which is the organization in charge of the management of the basin, has different control networks that are operative since 1992. Besides these control networks, there is also a contribution of scientific studies since 1988 to know the distribution of persistent organic pollutants in the basin. Most of these studies are site specific or consider only one family of compounds. Recently, some scientific studies have focused on the basin as a whole, considering several compounds and matrices.

In this chapter, the contribution of the CHE and the research achievements of the EU funded project AQUATERRA (Integrated Modelling of the river–sediment–soil–groundwater system; advanced tools for the management of catchment areas and river basins in the context of global change, Project number 505428 GOCE), together with other findings of the scientific community, to the knowledge about the contamination in the Ebro river basin will be overviewed.

Keywords Industrial and agricultural impacts, Monitoring campaigns, Sediments, Water

A. Navarro-Ortega (✉)
Department of Environmental Chemistry, Institute of Environmental Assessment and Water Research (IDÆA-CSIC), Jordi Girona 18-26, 08034 Barcelona, Spain
e-mail: alicia.navarro@idaea.csic.es

D. Barceló
Department of Environmental Chemistry, Institute of Environmental Assessment and Water Research (IDÆA-CSIC), Jordi Girona 18-26, 08034 Barcelona, Spain
Institut Català de Recerca de l'Aigua (ICRA), Parc Científic i Tecnològic de la Universitat de Girona, Pic de Peguera, 15, 17003 Girona, Spain

D. Barceló and M. Petrovic (eds.), *The Ebro River Basin*,
Hdb Env Chem (2011) 13: 139–166, DOI 10.1007/698_2010_77,

Contents

Abbreviations

APs	Alkylphenols
BPA	Bisphenol A
CHE	*Confederación Hidrográfica del Ebro*
CYP1A	Cytochrome p450 1A
EPA	Environmental Protection Agency
EU	European Union
HCB	Hexachlorobenzene
HCH	Hexachlorocyclohexane
NP	Nonylphenol
OCs	Organochlorine pesticides
OP	Octylphenol
PAHs	Polycyclic aromatic hydrocarbons
PBDEs	Polybrominated diphenyl ethers
PCA	Principal component analysis
PCDDs/Fs	Polychlorinated dibenzo-*p*-dioxins/polychlorinated dibenzofurans
PCBs	Polychlorinated biphenyls
PPs	Polar pesticides
RCSP	*Red de control de sustancias peligrosas*
TBP	Tributylphosphate
WFD	Water Framework Directive

1 Introduction

Throughout the twentieth century, around ten million chemical compounds were synthesized worldwide. Of these, 1,000–2,000 new substances were synthesized each year, usually without a strict control of their effects on various organisms.

The rapid development achieved by our society, especially in the last 50 years, has resulted in a constant increase in demand for consumer goods and technological progress. This development leads to a subsequent incessant industrial activity that generates lots of waste, many of which are discharged into rivers after inadequate treatment [1]. Apart from industrial development, there is an exponential growth of world population (9,000 million foreseen in 2050) [2] that increases the transport and energy needs, thereby increasing the consumption of fossil fuels. Furthermore, feeding a growing population requires the improvement of agriculture, forcing professionals to use techniques that provide a greater production. The preparation of land for cultivation is often made using chemical fertilizers and crop protection is achieved through the use of pesticides [3]. In 1993, the Environmental Protection Agency (EPA) counted, from the total trade of pesticides, 400 without any testing of their carcinogenic effects or toxicity to wildlife [4].

Consequently, Europe has historically been a hotspot of environmental pressures because of the contamination caused by agricultural, municipal, and industrial activities and high population densities [5, 6]. Such contamination has led to poor water quality in many European river basins [7–12]. In addition, this pollution can cause the accumulation in river sediments of toxic compounds such as pesticides [13], surfactants [14], and alkyl polycyclic aromatic hydrocarbons (PAHs) [15]. These can in turn act as a source to biota [16] and as a potential risk for entire ecosystems [17] if the compounds bioaccumulate, and thereby enter the food chain [18].

This general problem of water environment is also comparable when it comes to studying specific geographic units. One of these units is river basins, which are considered an indivisible geographic unit. For this reason, they must be protected entirety from the pollution problem, as the entire river network is interrelated. To assess the status of a river basin and prevent the pollution events, it is necessary to know the anthropogenic pressures to which it is subjected, whether urban, agricultural or industrial, which will determine the types of contaminants that can be found in the environment. Moreover, hidromorphologic and climatic characteristics play an important role in the flow of pollutants, as rain and irrigation of agricultural fields drag the more soluble compounds into groundwater or result in runoff to surface waters like rivers, lakes or wetlands [19].

The Ebro river basin is probably the most studied basin among the Spanish ones [3]. The existing studies can be divided in two groups: the ones realized by the *Confederación Hidrográfica del Ebro* (CHE) which is the organization in charge of the management of the Ebro river basin, and the ones that are reflected in scientific publications.

Until recently, most studies performed within the Ebro River have been site-specific or focused on a single chemical family [20–24]. However, little is known about the concentration and patterns of a wide spectrum of priority contaminants in the whole Ebro aquatic ecosystem. The inclusion of the Ebro river basin into the European Union (EU) project AquaTerra (contract no. 505428) led to a more complete study concerning the entire river basin, various environmental matrices (sediments and two fish species) and five different chemical families (organochlorine pesticides (OCs), PAHs, organotin compounds, alkylphenols (APs) and polybrominated diphenyl ethers (PBDEs) [25]. In this study, the authors assessed the levels of contaminants in the Ebro river basin and evaluated their environmental risks. They concluded that it is necessary to extend this kind of monitoring campaigns for measuring temporal tendencies and avoid possible deleterious effects on freshwater organisms of any river ecosystem. Consequently, a complete monitoring study of the basin that took into consideration the investigation of contamination over the years was carried out. A detailed monitoring programme of water and sediments for compounds such as PAHs, historical and current polar pesticides (PPs), OCs and APs was performed over the period 2004–2006 [26, 27]. Apart from the extensive monitoring campaigns, other studies within the frame of AquaTerra project were performed at the detected hotspots in the Ebro river basin [28–31].

2 Persistent Organic Pollutants

Persistent organic pollutants consist of a wide range of substances used in industrial applications, applied as pesticides, or generated inadvertently as byproducts of industrial processes. They are ubiquitous pollutants, distributed across the planet, even in regions where they have never been produced or used [32]. Among others, the main groups of pollutants that have been studied in the Ebro river basin include PAHs, PPs, OCs and APs, considered as emerging compounds. Recently, the contamination by dioxin-like compounds has been considered.

PAHs are an important group of organic micropollutants (xenobiotics) because of their widespread distribution in the environment (atmosphere, water and soil) [33]. They are considered ubiquitous contaminants as they can also be transported long distances. It is well known that some PAHs exhibit carcinogenic and/or mutagenic properties [34]. Because of their physico-chemical properties, PAHs, especially the higher molecular weight ones, are hardly degradable and tend to accumulate in the different environmental compartments [35]. Pesticides are compounds designed to inhibit the proliferation of or destroy those organisms that can adversely affect human activity. In addition to their acute toxicity, some pesticides may also cause endocrine disruption effects, as well as carcinogenic activity and hepatic injury [36, 37]. The abuse in the use of pesticides has resulted in contamination of soils which in turn have contaminated the groundwater and surface waters. In many cases, these pesticides degrade rapidly in the environment but their degradation products can be both more persistent and toxic than their precursors [38]. In particular, the contamination from OCs is of great concern because of the high distribution of their residues in the aquatic ecosystem and of their toxic and carcinogenic properties [39]. Because of their high persistence, their hydrophobic nature and their low solubility in water, they are adsorbed on the particulate matter and finally accumulated in sediments [40, 41]. Consequently, sediments, apart from being the final acceptors of these pollutants, act as their secondary contamination sources [42]. The contamination from OCs is very relevant in the Ebro river basin as two chloroalkali plants are located in the lower part of the basin (Flix and Monzón). The APs are degradation products of alkylphenol ethoxylates, which are non-ionic surfactants. These compounds are toxic, persistent and bioaccumulative in the aquatic environment [43]. The first evidence that APs could be endocrine disrupters was published in 1938 [44].

All these contaminants enter into the environment through human activities (urban waste, deposits of the waste cleaning plants and agricultural cultivations) [45] and various other pathways (atmospheric and fluvial transport) [46]. Environmental pollution by these groups of organic compounds has received considerable attention as a result of public awareness towards environmental problems and expectations for good quality of life [13].

Dioxin-like compounds are known to be one of the most harmful persistent organic pollutants of the chlorine based compounds. Dioxins are repeatedly synthesized and decomposed by complicated mechanisms and temperature changes at different locations in municipal solid waste incineration processes and the final streams discharged are commonly exhaust gas and incineration ash. Dioxins in different discharge types and compositions of effluents could be affected by incineration conditions such as temperature, feeding and discharging methods and incinerator type [47].

The brominated flame retardants are considered in [101] and consequently are not considered in this chapter.

3 Legislation

With the publication of Rachel Carson's book, Silent Spring [48], and the struggles of fledgling environmental groups, during the 1970s came the first regulations on the production and use of certain hazardous chemicals. Specifically in 1972, the EPA banned DDTs [49]. This action became the starting point of the legislation on chemicals pollution. Subsequently, several conventions on global pollution took place, such as the Rotterdam Convention (1998) on international trade of certain hazardous chemicals [50], or the Stockholm Convention (2001), which regulates the 12 compounds popularly called "dirty dozen" (aldrin, chlordane, dieldrin, endrin, heptachlor, hexachlorobenzene (HCB), mirex, toxaphene, polychlorinated biphenyls (PCBs), DDTs, Polychlorinated dibenzo-*p*-dioxins and polychlorinated dibenzofurans (PCDD/Fs)) [51]. This list of compounds has been updated in May 2009 and nine other chemicals have been added to the dirty dozen (α- and β-hexachlorocyclohexane (HCH), lindane, pentachlorobenzene, penta- and octabromodiphenylether, hexabromobiphenyl, chlordecone and perfluorinated compounds). Despite this new perception of the problem of pollution and the implicit legislation, some banned compounds such as DDTs are still used in various parts of the world, especially in developing countries [51]. Nowadays, such pollutants are even generated as a byproduct of the manufacture of some OCs.

As a result of this new frame, the EU has published directives aimed at the protection of the river basins from a range of substances included in the priority lists [52, 53]. Moreover, before entering into force, the directives must be transposed into each of the member countries. Often, the adoption is done with the full text proposed by the EU, but adding extensions to accommodate the new law on national legal and economic framework.

Some of the European legislations on environmental matrices include directives such as 76/464/EEC [54], concerning pollution caused by certain dangerous substances discharged into the aquatic environment. The protection of the Ebro river basin was initially on the basis of this directive. Currently, the new water directive is in effect, Directive 2000/60/EC (Water Framework Directive: WFD) [55]. The WFD aims at improving the ecological quality of surface waters and includes a list of 33 priority pollutants, many of which have been analyzed in the Ebro river basin. It has to be taken into account that the list of priority pollutants is increasing over the years. In 2010, compounds like hexabromocyclododecanes, chloroalkanes, PCBs or perfluorooctane sulfonic acids were suggested as candidates for the list of priority pollutants. There is also some legislation concerning the prohibition of plant protection products (pesticides) in agriculture, as the case of the ban of atrazine (2004/248/CE) [56], among others.

Regarding the maximum legislated levels of these compounds, the European Council has recently adopted the Directive 2008/105/CE [57] which sets quality standards for the 33 priority pollutants in surface waters. For sediments, however, there are no legislative limits in Europe and to accomplish the legislation it is only necessary to show an improvement in concentration over the previous year's level.

4 Levels of Contamination

4.1 Confederación Hidrográfica del Ebro

The CHE has monitored the contamination in the whole river basin establishing a control network with sampling sites at which different compounds are analyzed. Among industrial and agricultural pollutants, priority compounds according to the Directive 2000/60/CE have been selected to determine their sources, transport and fate within the water sediment interface of the Ebro river basin. This control network (*Red de control de sustancias peligrosas*: RCSP) was introduced in 1992 with only four sampling sites and since then other sampling sites and compounds have been added to improve the water quality control of the Ebro river basin.

The sampling sites are distributed along the entire basin but focusing the areas potentially polluted because of the emission of contaminant substances. Starting from 1992, the sampling sites have been extended, the list of compounds and the studied matrixes included in the network have been enlarged. Nowadays, the RCSP consists of the analysis of different compounds in water, sediments and biota from

18 sampling sites once a year [58]. The number of compounds analyzed varies from 21 to 44 depending on the sampling site. Although there is no constancy among the results from year to year, this control network can be considered the main quality control activity performed in the Ebro river basin [59].

Considering the last report of the RCSP published by CHE, which corresponds to the year 2007, the quality objectives established by the WFD have been accomplished for the water matrix. On the contrary, an increase in the concentration of PAHs has been detected for sediments in some of the sampling sites: around the city of Zaragoza, where no PAHs were detected on 2006; in Puente la Reina, where the concentration found in 2007 was double of that in the previous year and in Gasteiz, where this concentration was tripled. A large increase in the concentration of organochlorine compounds was also found in Tortosa but there was a widespread decrease in Nonylphenol (NP) concentration along the basin. Apart from the evolution of the concentration in sediments, the CHE considers relevant for 2007 the detection of HCB in Miranda de Ebro and Tortosa at high concentrations, thc detection of NP at 1,170 ng g^{-1} in *Zaragoza* and 2,680 ng g^{-1} in Salvatierra (Euskadi), and the detection of PAHs that reached 3,830 ng g^{-1} in Gasteiz. These results show that in some of the sampling sites of the Ebro river basin, the quality objectives established by the legislation are not accomplished.

Using these results, the chemical status of the water bodies of the Ebro river basin can be evaluated. The chemical status is defined as the expression of the accomplishment of the environmental quality legislation considering the surface water bodies. Among the 643 water bodies defined in the Ebro river basin, 35 were diagnosed as bad chemicals status (5.4%). The distribution of these water bodies is shown in Fig. 1. This bad quality was concentrated around the city of Zaragoza,

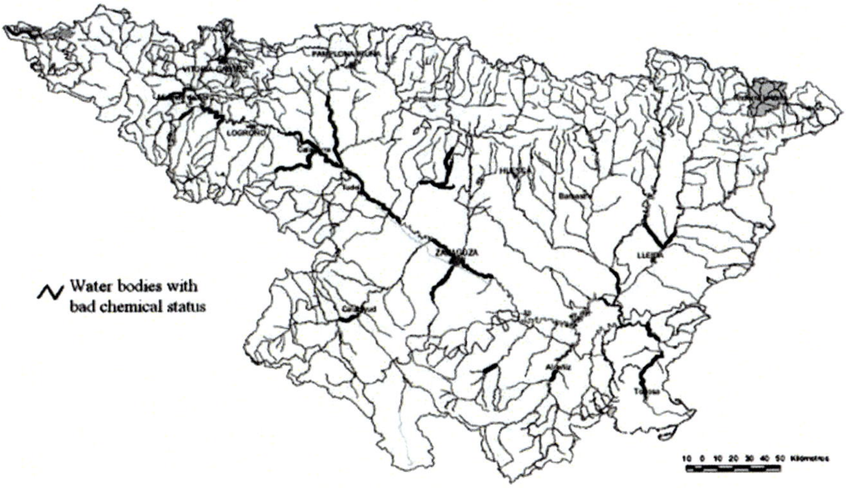

Fig. 1 Chemical status of the Ebro surface water

where many kilometres up and downstream as well as the Rivers Huerva and Gállego showed this status. The other areas with bad quality were around Calatayud and Calahorra. There was consistently bad chemical status around Gasteiz, Miranda de Ebro and Lleida; downstream Monzón and Flix and upstream Tortosa and Reinosa. This last qualification implies that the first 6 km of the Ebro River, between the source in Fontibre and the village of Reinosa, presented bad chemical status. It has to be considered that for the evaluation of the chemical status only 18 sampling sites are controlled and in the others it is only evaluated with respect to the concentration of nitrates that cannot exceed 50 mg L^{-1}. It has also to be considered that many of the defined water bodies have not been diagnosed.

It can be concluded that the Ebro river basin has a significant network of surface water monitoring sites and a considerable programme for water quality analysis is ongoing. Compounds measured include a range of industrial chemicals, pesticides and metabolites, most of which are included as priority pollutants under the WFD. But not all substances are monitored at all sites, nor are they monitored on every occasion a sample is taken. In addition, water is the main sample and very small fractions are sediments, although sediments are an important compartment in a river basin because they can accumulate most of these compounds which can be later redissolved into water [60]. It should be noted that in common with all large-scale sampling regimes, the number and frequency of sampling is relatively low. Therefore, it is possible that higher concentrations of all substances may not be detected [61].

4.1.1 Chemometrical Study of CHE Data

There is a study that compiles the results from the CHE quality network since it was established [62]. The data set used for this study was downloaded from CHE web page. Among all the data sets available from the RCSP, sediment sample data sets covering years from 1996 to 2003 were selected. Before 1996, data sets were rather incomplete. Year 1999 was excluded from the data set under study because most of the compounds analyzed in the control network were not reported for this year. Only the 12 compounds (seven PAHs (naphthalene, fluoranthene, benzo(a)pyrene, benzo(b)fluoranthene, benzo(g,h,i)perylene, benzo(k)fluo-ranthene, indene(1,2,3-cd)pyrene) and five OCs (HCB, hexachlorobutadiene, sum of α-, β-, δ- and γ-HCH isomers, sum of six DDTs (o,p′-DDT, p,p′-DDT o,p′-DDD, p,p′-DDD o,p′-DDE, p,p′-DDE) and sum of 1,3,5-, 1,2,4- and 1,2,3-trichlorobenzene) and nine sampling sites having results for all these years were finally chosen for this study to investigate the temporal trends. Figure 2 shows the location of the sampling sites within the Ebro river basin (chemometrical study).

The multivariate statistical data analysis, using principal component analysis (PCA), of this historical data revealed three main contamination profiles. A first contamination profile was identified as mostly loaded with PAHs. A samples group which includes sampling sites R1 (Ebro river in Miranda de Ebro, La Rioja), T3 (Zadorra river in Villodas, Alava) and T9 (Arga river in Puente la Reina, Navarra), all located in the upper Ebro river basin and close to Pamplona and Vitoria cities,

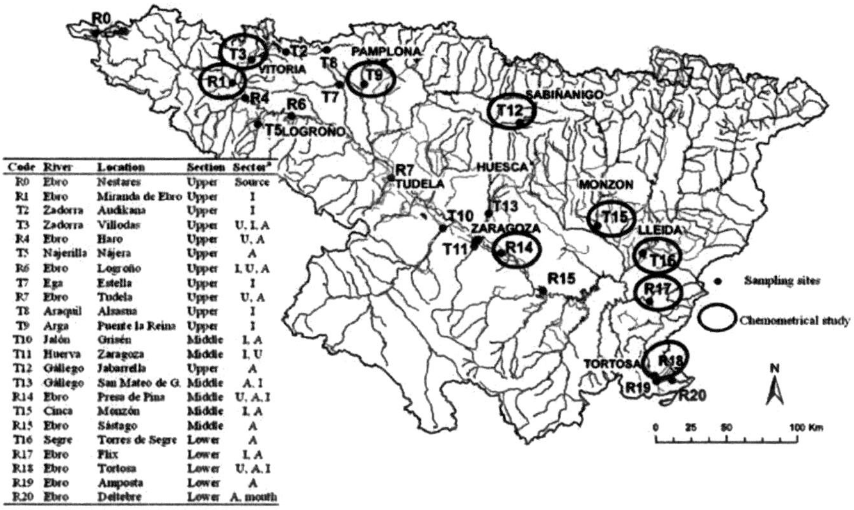

Code	River	Location	Section	Sector[a]
R0	Ebro	Nestares	Upper	Source
R1	Ebro	Miranda de Ebro	Upper	I
T2	Zadorra	Audikana	Upper	I
T3	Zadorra	Villodas	Upper	U, I, A
R4	Ebro	Haro	Upper	U, A
T5	Najerilla	Nájera	Upper	A
R6	Ebro	Logroño	Upper	I, U, A
T7	Ega	Estella	Upper	I
R7	Ebro	Tudela	Upper	U, A
T8	Araquil	Alsasua	Upper	I
T9	Arga	Puente la Reina	Upper	I
T10	Jalón	Grisén	Middle	I, A
T11	Huerva	Zaragoza	Middle	I, U
T12	Gállego	Jabarrella	Upper	A
T13	Gállego	San Mateo de G.	Middle	A, I
R14	Ebro	Presa de Pina	Middle	U, A, I
T15	Cinca	Monzón	Middle	I, A
R15	Ebro	Sástago	Middle	A
T16	Segre	Torres de Segre	Lower	A
R17	Ebro	Flix	Lower	I, A
R18	Ebro	Tortosa	Lower	U, A, I
R19	Ebro	Amposta	Lower	A
R20	Ebro	Deltebre	Lower	A, mouth

Fig. 2 Sampling sites in the Ebro river basin

which is an industrial area, was mostly contaminated by this contamination profile. A second contamination profile was detected loaded mostly with OCs. Samples group including R17 (Ebro river in Flix, Tarragona) and R18 (Ebro river in Tortosa, Tarragona), both located in the lower course of the Ebro river, was highly contaminated by the second contamination profile showing contamination due to agricultural practices. And finally, a third contamination profile more specific for naphthalene was also detected. A group of samples with no clear relation among them (R1-2003, T3-1997, R17-1996, R14-2002 and T12-2001) were highly contaminated by this contamination profile, in which all the samples except R17-1996 were the only ones specifically contaminated by the contamination profile of naphthalene and by no other contamination sources. In the samples group mostly contaminated by the contamination profile of PAHs, the samples also differed among one another depending on the degree of contamination of this third contamination profile. Rest of sampling sites, located in the centre of the river basin, showed lower contributions of the three contamination profiles and therefore they were the less contaminated samples. The PAHs and OCs contamination profiles were diffuse and persistent over the considered years while the naphthalene contamination profile was more specific. For these three resolved sources, no temporal trends were observed in the sampling sites under study.

4.2 *Scientific Studies*

The Ebro river basin has been the subject of numerous studies that have resulted in many scientific publications. These publications are focused on analysing the levels of pollutants that occur in various matrices. The first publications considering the Ebro river basin date from 1988 [22], some years before that the CHE established the RCSP for the analysis of chemical compounds in the basin. Thirty one publications that studied organic pollutants in water, sediments or biota have been considered. The studies collected were divided according to the matrix analyzed to give a better view of the existing data in the area of the Ebro basin. For each of the matrix a map is presented, including the geographical location of the studies and accompanied by graphs with the range of concentrations of each chemical family analyzed (Figs. 3–5). A most extensive study of the polar pesticides present in the Ebro Delta is included in [104].

4.2.1 Water

This is the most studied matrix in the basin and the first to be analyzed, because during the 1980s and 1990s, the analytical methods for other matrices were not well developed. Clearly, the majority of existing studies focus on the Ebro Delta (74%), considered as a vulnerable zone because of industrial and agricultural pressures [21, 22, 63–78]. The Ebro Delta was also the only studied area until 1996 (Fig. 3). After this year, other areas were considered, such as around Flix [20], because of the existence of an industry for organochlorine compounds, or La Rioja [5] and the middle course of the Ebro river [79], because of the agricultural pressures that can provoke contamination by pesticides. Between 2000 and 2006, there was not any publication concerning the analysis of Ebro surface water. Among the 19 existing water studies, there is only one that has considered the total length of the Ebro basin [80], in which

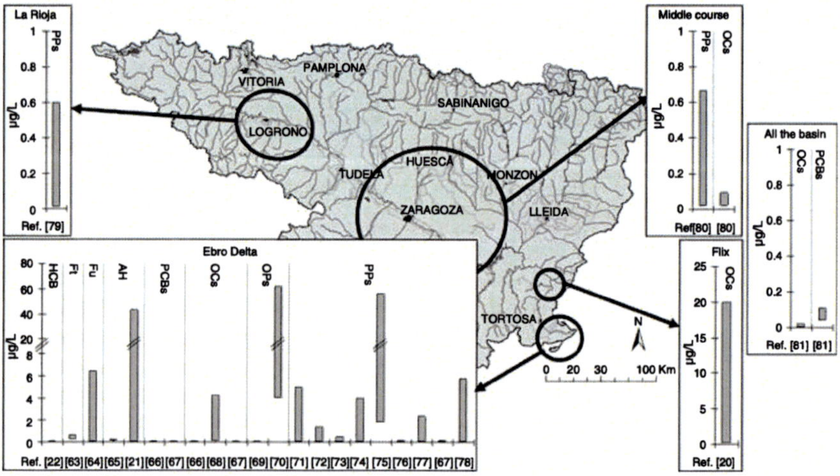

Fig. 3 Scientific studies in water. *Ph* pharmaceuticals; *Ft* phthalates; *Fu* fungicides; *HA* acidic herbicides; *HCB* hexachlorobenzene; *PCBs* polychlorinated biphenyls; *OCs* organochlorine pesticides; *PPs* polar pesticides

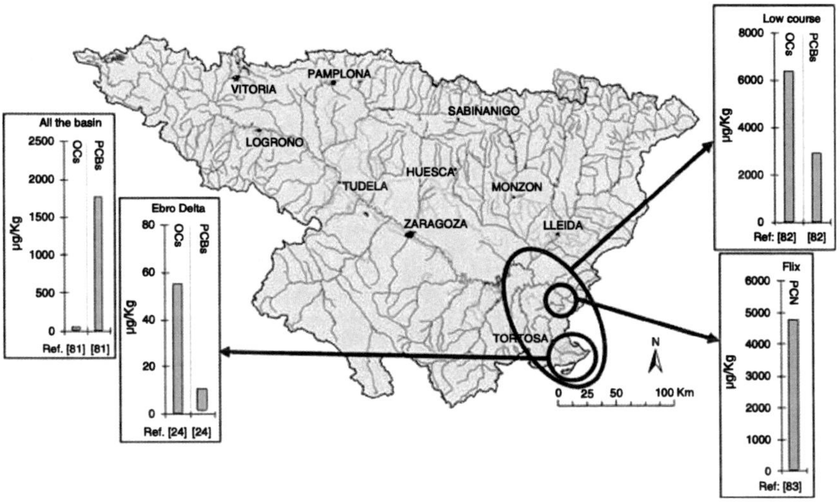

Fig. 4 Scientific studies in sediments. *HBCD* hexabromocyclododecane; *PCBs* polychlorinated biphenyls; *OCs* organochlorine pesticides; *PCN* polychlorinated naphthalene; *DDTs* dichloro diphenyl trichloroethane

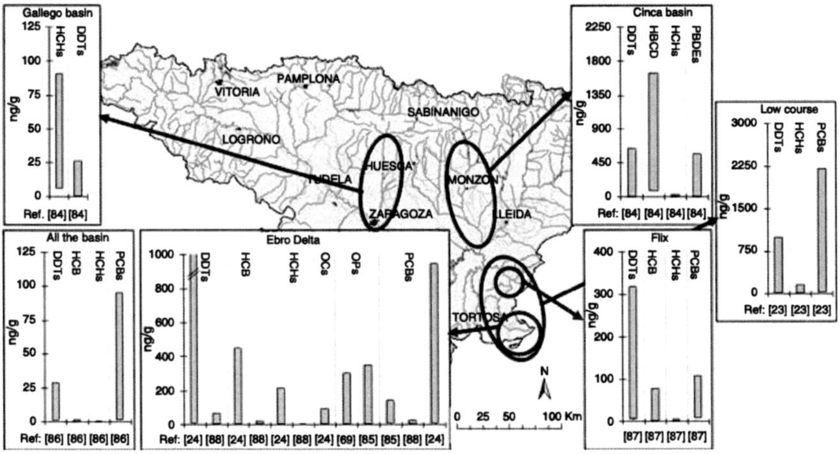

Fig. 5 Scientific studies in biota. *DDTs* dichloro diphenyl trichloroethane; *HCB* hexachlorobenzene; *HCHs* hexachlorocyclohexanes; *OCs* organochlorine pesticides; *OPs* phosphorate pesticides; *PCBs* polychlorinated biphenyls; *HBCD* hexabromocyclododecane

OCs and PCBs were studied. On the contrary, a much larger group of compounds has been studied in the Delta, but in all cases each of the studies was only focused on one group of compounds. The concentrations that have been found for organochlorine compounds (OCs, PCBs, HCB), phthalates and drugs did not exceed 5 μg L^{-1}, with the exception of Flix, where concentrations up to 20 μg L^{-1} for OCs were found. In contrast, the concentrations of the other groups of compounds were higher and more varied, for example, reaching 60 μg L^{-1} of PPs in the Delta, as these compounds have a greater affinity for the aqueous phase.

4.2.2 Sediments

This matrix was not analyzed in the Ebro river basin until 1999. The first study was the largest study to date, because it combined the analyses of OCs and PCBs in water and sediment samples [80]. It was not until 2004 that sediments from the Ebro basin were analyzed again, but in this case the study was focused on the lower course of the Ebro [81], with special attention to the Ebro Delta [24], and the area of Flix [82], downstream of the location of the organochlorine compounds industry in this village, where concentrations of up to 5,000 μg kg^{-1} of polychloronaphthalene were found. The number of studies that consider the sediment matrix is much lower than for water, and the number of chemical families is smaller, focused mainly on organochlorine compounds and PCBs (Fig. 4).

4.2.3 Biota

In this section, studies considering any aquatic species have been considered. Similar to the water matrix, biota has been extensively studied, including species of fish like carp (*Cyprinus carpio*), barbel (*Barbus graellsii*) and mosquitofish (*Gambusia affinis*) among others [23, 83–85], besides zebra mussels (*Dreissena polymorpha*) and crayfish (*Procambarus clarkii*) [86]; viperine snake (*Natrix maura*) [87] and invertebrates (*Physella acuta, Hirudo medicinalis, chironomid larvae, Hydrous pistaceus, Helochares lividus*) and vertebrates (*Rana perezi*) [24]. The first study on the Ebro river basin biota was focused on the freshwater fish species from the Ebro Delta and found concentrations up to 300 ng g^{-1} of organophosphorous pesticides [69]. Half of the existing studies focus on the Ebro Delta, where different families of chlorine compounds have been analyzed. The maximum concentration detected rises to 10,750 ng g^{-1} of DDTs in invertebrates [24]. Other areas have been considered for the analysis of their biota, like the Cinca basin, the Gállego basin or the area surrounding the village of Flix where mainly DDTs, HCB, HCHs have been considered [23, 83, 86]. Among the eight existing biota studies, there is only one that has considered the total length of the Ebro basin [85], in which DDTs, HCB, HCHs and PCBs were studied in two different fish species.

4.3 AquaTerra Achievements

The studies realized within the frame of the AquaTerra project have involved a great number of researchers from different disciplines. In this chapter, the ones concerning organic contamination in water, sediments and aquatic biota have been considered. They have been divided depending on the area considered for the study, the first group includes the ones that consider the entire basin (general monitoring campaigns) while the second group includes the ones that have selected a hotspot of the Ebro river basin.

4.3.1 General Monitoring Campaigns

The first AquaTerra study consisted of a collaboration with the CHE for the investigation of priority organic pollutants in sediments and fish from 18 sampling sites collected along the Ebro river basin [25]. The sampling sites covered industrial, urban and agricultural areas. Four methods were used to detect 20 OCs, eight PAHs, three organotin compounds, two APs and 40 PBDEs in purified extracts.

In this study, the levels of contaminants in the Ebro river basin and their availability to fish were assessed, and the environmental risk was evaluated. The type of contamination and its distribution throughout the Ebro basin reflected the human activities carried out, and it was possible to localize areas with high concentrations of DDTs, PBDEs and CBs in both sediment and fish. PAHs were spread throughout the river at levels lower than the predicted effect levels, whereas organotin contribution was high in some heavily industrialized areas although these contaminants were not accumulated in fish. The contamination pattern was site specific and no downstream increase in concentration of pollutants was observed, but rather a generalized low level diffuse pollution. Target compounds were detected in sediments at 0.01–2,331 mg kg^{-1} dry weight, and only OCs and PBDEs were accumulated in benthopelagic fish. Toxicological assessment was performed according to predicted environmental levels and revealed sites where adverse effects could occur.

The results of this study showed that within freshwater systems, sediments can act as drainage for pollutants and a source to biota. Also, sediment suspension might be the cause of the low-level diffuse contamination found at all sites. This type of contamination is more difficult to tackle, and for that reason, after the first approach included in Lacorte et al. [25] a monitoring programme including the measurement of temporal tendencies during 3 years in an extended collection of sampling sites was performed [26, 27] and it is summarized in the following pages.

Apart from persistent organic pollutants considered in the monitoring programme, others compounds like pharmaceuticals [88] and drugs [89] were also analyzed in the entire Ebro river basin, and the results obtained are presented in [102, 103].

Site Selection and Sampling

The monitoring undertaken in the study of persistent organic pollutants comprised six sampling campaigns during 3 years, between 2004 and 2006. Each year two sampling campaigns were carried out: the first one in June and the second one in

October. Water samples were taken in all the sampling campaigns while sediment samples were taken once a year in the October sampling campaign. The monitoring included 23 sampling sites covering the whole Ebro river basin (11 at the Ebro River and 12 at the main tributaries) from the most vulnerable sites, according to proximity to big cities, agricultural areas or industrial activities and knowledge on historical contamination episodes. Their specific locations are shown in Fig. 2, and are numbered from 1 to 20 following the Ebro river flow, from north-west to south-east. "R" indicates a site on the Ebro River whereas "T" indicates a tributary site. Among all the sampling sites, one 6 km downstream the Ebro source (R0) and another one just before the sea (R20) are included. Table 1 lists the locations of each sampling site, the corresponding river, the section of the river basin and the main economical activities in the area.

Levels Obtained in Water and Sediments

During the 3 years of sampling campaigns, 132 water samples and 65 sediment samples were collected and analyzed. As a result, a total of 9938 concentrations were obtained. To summarize the temporal trends both in water and sediments, the compounds were grouped in four groups: PAHs, APs, OCs and PPs.

In the case of water, PPs were detected in a very high percentage of samples (86%), followed by APs with 75%, but with concentrations lower than 5 $\mu g\ L^{-1}$. It has to be noted that APs were detected at concentrations one order of magnitude higher than PPs. PAHs appeared in a much smaller percentage of samples (29% in 2004) and were detected at very low concentrations and OCs were not detected in this matrix.

However, the situation was opposed for the matrix of sediments, where PAHs and OCs were detected in 100% of the samples at concentrations up to 10,000 and 1,000 ng g^{-1} respectively, whereas PPs were detected only in 29% and were found at concentrations much lower than other groups.

The distribution of these groups between the two matrixes is due to their physical properties: they may have more affinity for sediments (lipophilic compounds such as PAHs and OCs) or for the aqueous phase (hydrophilic compounds such as PPs). In general, the concentrations in the sediment samples of these four groups were much higher than in water, reaching levels above 6,000 ng g^{-1} in the case of APs and PAHs. APs were detected in a high percentage of samples in the two matrixes (75% in water and 100% in sediment) and in addition at high concentrations, reflecting their discharge into the aquatic environment because of the wide use of non-ionic surfactants in many industrial and domestic activities.

Results in Water

Figure 6 summarizes statistical data for all the detected compounds. The frequency of detection varied from 1.5 to 72.0% considering the results from the six monitoring campaigns. The compounds with the lowest frequency of detection were

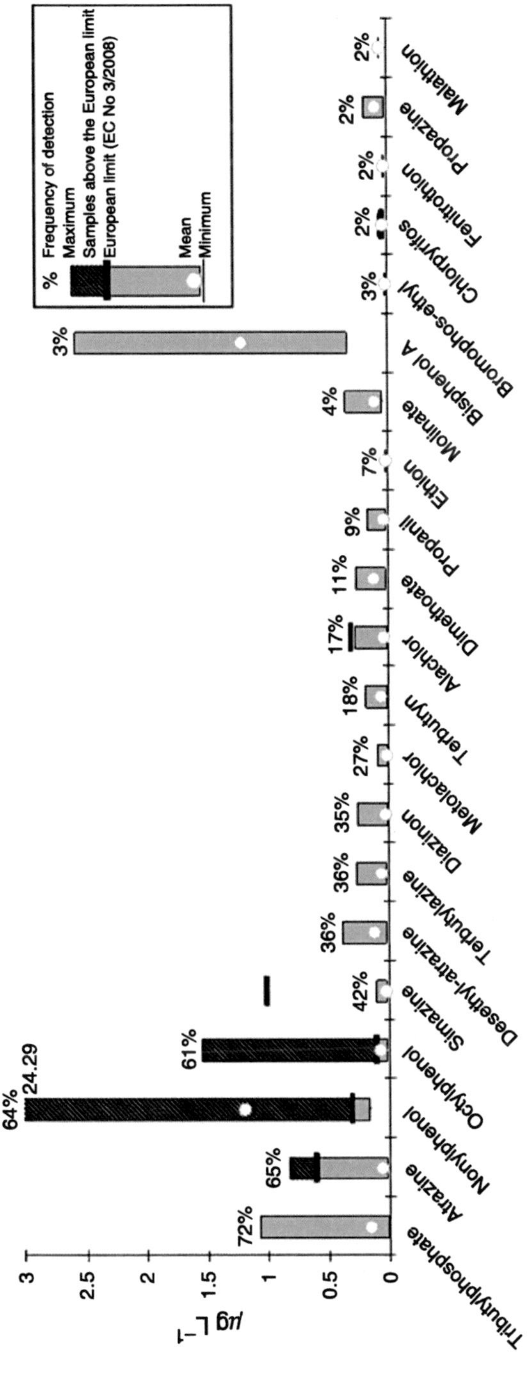

Fig. 6 Descriptive statistics of water data

malathion, propazine, fenitrothion and chlorpyrifos, with detection frequencies between 1.5 and 2.3%. Apart from these compounds, there was another group of five compounds with less than 10% frequency of detection that included propanil, ethion, molinate, bisphenol A (BPA) and bromophos-ethyl. The compounds with highest frequency of detection were tributylphosphate (TBP), atrazine, NP and octylphenol (OP), which were detected in more than 60% of the samples analyzed. Except for atrazine, the other three compounds have an industrial origin. Organophosphorus pesticides are the largest family of analyzed compounds. However, they were detected less frequently than the triazine and chloroacetanilide pesticides.

The compounds that presented highest maximum concentrations were the same as the ones with highest frequencies of detection, with the exception of BPA, which was detected in a few samples but at high concentrations, indicating a point source contamination of this compound. The compounds with highest median concentrations were again BPA and NP, and also desethyl-atrazine and propazine. Comparison of mean and median values of all the variables showed that in most of the cases the mean values were between 1.5 and 2 times higher than the median values. This large difference between mean and median values shows that concentration data were skewed towards low values because of the existence of punctual high values. Compounds that fulfilled this criterion were propanil, chlorpyrifos, alachlor, TBP, OP and NP. They were the ones with more extreme values among the samples. In most cases, standard deviations had the same order of magnitude as mean values. One-third of the compounds showed larger values for the standard deviations than for the mean, especially OP and NP, indicating a large variability of the concentration of these compounds.

Existing legislation has been used to compare the results obtained in the Ebro basin with the recently approved Directive 2008/105/EC [90] that includes water environmental quality standards for the priority substances included in the WFD. The substances that have concentrations in some samples above these limits are also the three compounds with highest frequency of detection (atrazine, NP and OP), with the exception of TBP that does not appear in the considered legislation. In particular, for NP, 75% of the samples were contaminated with concentrations higher than the limit proposed by the WFD (0.3 $\mu g\ L^{-1}$). Most of these samples are around the city of Zaragoza. This indicated that one important contamination problem in the Ebro basin is due to industrial compounds. Another finding is that pesticide contamination in the Ebro River is presently below the legislated values.

The PCA analyses revealed two main sources of contamination in water from the Ebro river basin. The first was mainly associated to pesticides commonly used for agricultural practices, while the second was influenced by industrial compounds. In general, the total concentrations of pesticides found in the June sampling campaigns were higher than in the October campaigns for almost all the sampling sites. This is most likely due to the field application of pesticides, which takes place in May/June, and the short response of these compounds once in the environment. An increasing tendency of the total concentration of the pesticides is also observed while approaching the middle and lower sections of the river. This may also represent a cumulative effect because of an increasing use of the land for agricultural purposes as we go down the river and adding up of influences from the stations above. The second contamination source was dominated by alkylphenols and TBP which were widely distributed all over the river basin, indicating therefore a potential risk for ecosystems.

Results in Sediments

Descriptive statistics data for sediments is summarized in Fig. 7 and compared to legislated levels. The frequency of detection of each compound varied from 2 to 100%, with parathion-methyl, pentachlorobenzene, endosulfan sulphate, isodrin, BPA, propanil, alachlor, molinate, heptachlor, dieldrin and endrin aldehyde being the less frequently detected (between 2 and 3%). This group of compounds is not included in the figure for the purpose of simplification. Pyrene, chrysene and benzo(a)pyrene appeared in all the analyzed samples while the rest of PAHs had more than 55% of appearance, confirming that this family is very widespread in the sediments along the Ebro. The four lighter PAHs, with two and three aromatic rings (acenaphthene, fluorene, naphthalene and acenaphthylene) were the ones less frequently detected because of their higher volatility, solubility and biodegradability [91] and therefore less bioaccumulated in the sediments. Considering the group of PPs, only five out of the 25 analyzed appeared in the sediment matrix and with less than 13% of appearance. These compounds have higher affinity for water [92], and were not supposed to appear in the sediment matrix because of their more polar characteristics. Only diazinon had a 12% of appearance because of its more widespread use as pesticide [93]. As expected, OCs appear more often than PPs, up to 97%, with an important presence of DDTs, especially 4,4-DDT, although this compound is forbidden in Spain for agricultural purposes since 1994 [94], probably due to accumulation and slow degradation of this compound in the sediments.

The compounds that presented higher maximum concentration were the most ubiquitous PAHs, all with more than 1,000 μg kg^{-1}, together with NP with a maximum concentration of almost 6,000 μg kg^{-1}. NP presented also median and mean concentrations of 377 and 876 μg kg^{-1}, respectively, more than five times higher than the rest of compounds, and also showed its presence in more than 87% of the samples. Comparison of mean and median values of all the variables showed that mean values were always higher than the median values, with the only exception of TBP, indicating that data were skewed towards low values with some peaks of contamination. This behaviour produces also large data dispersion for the different sampling campaigns as well as sampling sites, which can be observed in a standard deviation higher than the mean.

To evaluate the potential risk of Ebro river sediments, the concentrations detected were compared to actual legislation. The WFD does not specifically address sediment management [95]. In contrast to this, the Canadian sediment quality guidelines for the protection of aquatic life [96] provide data for different types of sediments and include 33 compounds, PAHs and DDTs among them. All the PAHs included in the Canadian guidelines are present above the limit, ranging from 3% of the samples for naphthalene to 90% for dibenzo(a,h)anthracene. Special attention has to be paid to benzo(a)pyrene and dibenzo(a,h)anthracene, considered

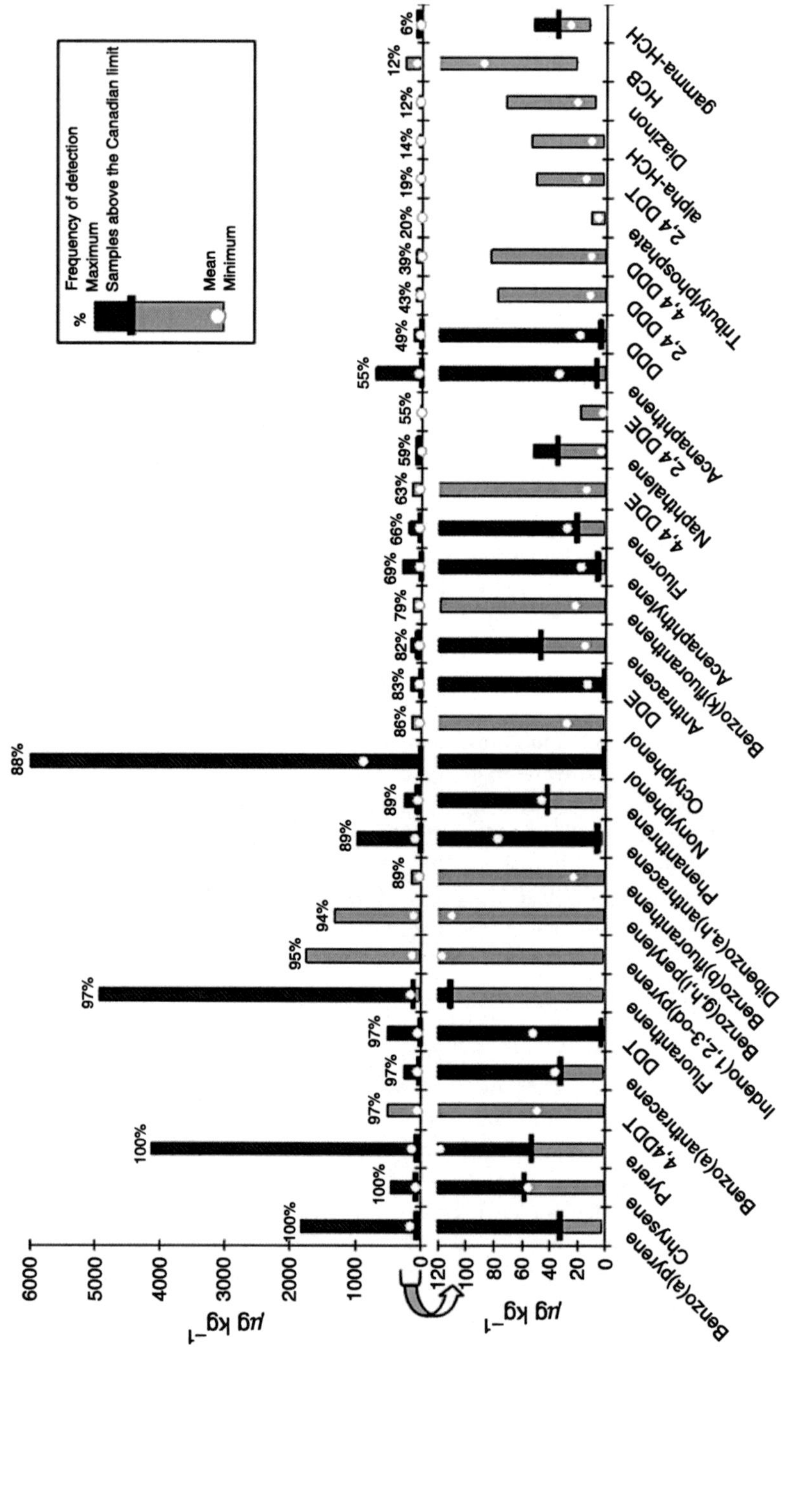

Fig. 7 Descriptive statistics of sediment data

as the most carcinogenic ones among PAHs [97]. These two compounds have 65 and 90% of samples above the Canadian limit, respectively, resulting in a high potential risk for the living organisms in the Ebro river ecosystem.

In the case of DDTs, the three compounds (DDT, DDE and DDD) are considered as the sum of both 2,4 and 4,4 isomers. Half of the samples were above the limits proposed by Canadian guidelines and for DDT more than 95% of the samples were above this limit. The highest concentration levels were ranging from 15.3 to 351 μg kg^{-1}. Two compounds with very small percentage of appearance in the samples, γ-HCH and dieldrin, have concentrations above the Canadian guidelines.

In contrast to the water results [92], in which only atrazine, NP and OP were above the legislated limits, the sediments present more elevated concentrations with 18 out of the 20 compounds legislated having some samples above the limits. This different behaviour shows that the water is a more changeable matrix that reflects the current contamination while the sediments behave as a reservoir for the contaminants as a result of their accumulation potential.

Overall, the PCA analysis revealed two main sources of contamination in sediments from the Ebro river basin. The first one was mostly loaded with PAHs while the second one was loaded with all the DDTs together with HCB. PAHs were widespread all over the sampling area. The two samples more contaminated by PAHs belong to the R0 sampling site, which is a sampling site only 6 km downstream of the Ebro spring (North of Spain). This high PAHs contamination area detected in this a priori unpolluted area can be attributed to important active mining activities during the last two centuries. Among all the pesticides analyzed, only DDTs and HCB appeared at high concentrations, although both compounds are forbidden in Spain. The most OCs contaminated sites correspond to the lower course of the Ebro, downstream the industrial area near the city of Flix (close to the Ebro delta), where a large production of chlorinated pesticides had been carried out in the past. These sampling sites can also receive the impact of the agricultural practices in the Ebro Delta. Contrarily to the results obtained for the water matrix, the sediment contamination did not show any temporal or geographical contamination but a widespread presence with some punctual high concentrations.

Comparison Between Water and Sediments Data

Among all the compounds analyzed, only eight (OP, NP, TBP, BPA, diazinon, propanil, alachlor and molinate) appeared in both matrices. The most noticeable fact is that NP is the compound that contributes more than the other compounds in both water and sediments, as it has medium value for log K_{OW} (4.48) and the concentrations found were high enough to enable its distribution between the two matrices, between 69 and 5,999 μg kg^{-1} in the solid matrix. These high concentrations in the sediments are in the same range as in other rivers of the world. This compound shows a lower importance in the PCA analysis because of its appearance as punctual pollutant, mainly around the industrial area of Zaragoza. The rest of the profile shows some differences between water and sediments. Water moves constantly while the sediments are more bound to one location, and in consequence, the correspondence is not complete as the water indicates the punctual state of the basin and sediments the historical accumulation.

4.3.2 Hotspots Studies

A part from the general monitoring studies, other studies have been performed focusing the hotspots areas of the Ebro river basin, mainly the Ebro delta [30, 31] and the Vero-Cinca tributaries [98–100], and specific groups of compounds like semi-polar pesticides, DDTs and dioxin-like compounds.

The aim of the investigation carried out by Barata et al. [30] in the Ebro Delta was to evaluate toxicity effects of pesticides in aquatic invertebrates using in situ bioassays with the local species, *Daphnia magna*. Investigations were carried out during the main growing season of rice (May–August). The results obtained indicated high levels of pesticides in water, with peak values of 487 μg L^{-1} for bentazone, 8 μg L^{-1} for methyl-4-chlorophenoxyacetic acid, 5 μg L^{-1} for propanil, 0.8 μg L^{-1} for molinate and 0.7 μg L^{-1} for fenitrothion. Measured biological responses denoted severe effects on grazing rates; a strong inhibition of cholinesterases and carboxylesterases, which are specific biomarkers of organophosphorous and carbamate pesticides; and altered patterns of the antioxidant enzyme catalase and the phase II metabolizing enzyme glutathione *S*-transferase. Correlation analysis with pesticide residue levels converted to toxic units relative to its acute 48-h median lethal concentration of *D. magna* indicated significant and negative coefficients between the dominant pesticide residues and the observed biological response, thus denoting a clear cause-and-effect relationship. The results emphasize the importance of considering specific (biomarkers) as well as more generalized and ecologically related (grazing) in situ responses to identify and evaluate biological effects of environmental contaminants in the field.

Kuster et al. [31] developed a method to investigate the occurrence of the target pesticides in a total of 52 water samples collected monthly (from May to August 2005) at 14 selected locations in the rice cultivation area of the Ebro river delta. The study showed high levels, in the μg L^{-1} range, of bentazone, 2-methyl-4-chlorophenoxyacetic acid, propanil, molinate and atrazine, in basically all the samples investigated. The remaining pesticides were present at lower levels (<0.1 μg L^{-1}) or only detected sporadically (e.g. fenitrothion and malathion). The sampling campaign performed in July showed comparatively higher levels than the other three campaigns.

Two studies were performed in the Vero-Cinca tributaries and the Flix area considering the contamination produced by dioxin-like compounds and their biological impact. Quirós et al. [98] developed specific primers for the quantification of cytochrome p450 1A (CYP1A) and metallothionein-1 and -2 gene expression by QRT-PCR in barbel in order to assess their suitability in biological effect monitoring of dioxin-like compounds and applied them to Barbel populations from different sites on the Ebro River basin, that showed a good correlation between the historical records of organochlorine pollution, CYP1A expression levels and EROD activity. These results demonstrate the utility of barbel CYP1A-mRNA expression as a biomarker in field studies. The tools and protocols developed here are likely to apply to other species of the genus *Barbus*, with some 700 species distributed throughout most of the Old World. On the other side, Eljarrat et al. [99] evaluated the environmental impact associated with PCDDs/Fs and dioxin-like PCBs in the Ebro River basin. Sediments and fish from several species were sampled at three sites with different historical pollution records, including the Barbastro area with different industrial activities, and the Flix and Monzón sites, associated with heavy organochlorine compound pollution. Seventeen toxic PCDDs/Fs and 12 dioxin-like PCBs were analyzed by GC–MS. The results obtained indicated significant accumulation of dioxin-like PCBs, but not PCDDs/Fs, in sediments and fish at the Flix site compared to that at the other sites. Concomitantly, CYP1A expression, a known indicator for pollution by dioxins and dioxin-like PCBs, was significantly elevated in barbel (*B. graellsii*) from the Flix site, compared to that in the population from the Barbastro site. CYP1A expression correlated with the concentration of dioxin-like PCBs in the fish fat, whereas no significant correlation was found with PCDDs/Fs concentrations. These data suggest a significant biological impact at the Flix site, closely related to the presence of dioxin-like PCBs, whereas the PCDDs/Fs contribution to this impact appears to be non-significant, at least in the studied sites.

Finally de la Cal et al. [100] evaluated the DDT contamination along the Cinca River which receives input from different activities, including an industry where DDT is used as an intermediate in the production of dicofol. In this study, sampling sites were selected up- and downstream from this industry. Sediments and fishes (59 bleaks (*Alburnus alburnus*) and 23 barbels (*Barbus graellsi*)) were collected in 2002 and analyzed using a new and rapid selective pressurized liquid extraction (SPLE) method. DDT and its metabolites were found in sediments and fishes at levels ranging from 9 to 94 $\mu g\ kg^{-1}$ dry weight and from not detectable level to 2,098 $\mu g\ kg^{-1}$ wet weight, respectively. The highest values corresponded to samples collected just downstream the industry. Thirty kilometres downstream from the factory, levels were clearly lower, showing a weakening of the impact. p,p′-DDE isomer comprised up to 50% and 70% of total DDT measured in sediment and fish, respectively. When compared with values obtained in a previous study in 1999, a generalized drop of the levels in all matrixes (77–97%), was observed. No meaningful differences were found between the two fish species studied or between the two tissues (muscle and liver) analyzed.

5 Future Directions

As a continuation of the extensive monitoring of persistent organic pollutants that was started with the inclusion of the Ebro river basin in the EU Project AquaTerra, a new monitoring programme has been started in 2010. The Ebro river basin has been included as a representative Mediterranean watershed in a new project from the Spanish Ministry of Science and Innovation included in the programme Consolider

2010. This new project, with the title "Assessing and predicting effects on water quantity and quality in Iberian rivers caused by global change (SCARCE-CSD2009-00065)", is a multipurpose project that aims to describe and predict the relevance of global change impacts on water availability, water quality and ecosystem services in Mediterranean river basins of the Iberian Peninsula, as well as their impacts on the human society and economy.

Apart from of investigating the concentration of priority pollutants to assess the temporal and geographic trends as it has been done in the Ebro river basin until nowadays, the SCARCE project pretends to evaluate the consequences of the climate change in the water quality as well as predicting new environmental risks derived from water scarcity. This will be done by determining the presence of new priority (i.e. perfluorinated compounds) and emerging toxicants in Mediterranean river ecosystems in the Iberian Peninsula. In a subsequent step, the effects of chemical and environmental stressors on the biota will be assessed by combining field and experimental studies. The effects of multiple stressors will be addressed from a multi-biomarker perspective [27].

The new monitoring campaigns are already planned and included the sampling of water, sediments, fish, algae, bacteria and macroinvertebrates in 25 sampling sites during high, medium and low flow. A wide range of contaminants (priority and emerging pollutants: pharmaceuticals, drugs of abuse, perfluorinated compounds, flame retardants, endocrine disruptors, polar pesticides, ...) will be included in the analysis. This extensive monitoring will give new information on the Ebro river basins that will help understand the fate and impact of pollutants in the Ebro river basin and bridge chemical contamination with biological effects. In addition, the results from SCARCE will help refine the River Basin Management Plans demanded by the WFD.

Acknowledgements This work was initiated by the European Union under the Global Change and Ecosystems (FP6) Water Cycle and Soil Related Aspects (AquaTerra, Project number 505428 GOCE) and the Ingenio-consolider 2010 project SCARCE (CSD2009-00065) from the Spanish Ministry of Science and Innovation. Alícia Navarro-Ortega provided support by a grant of the Departament d'Universitats, Recerca i Societat de la Informació de la Generalitat de Catalunya (2004FI 00856). This work reflects only authors' views and the European Community is not liable for any use that may be made of the information contained.

References

1. Doménech X (1993) Química Ambiental. In: Ediciones M (ed) El impacto ambiental de los residuos. Madrid, Spain
2. Population Division (1999) The world at six billion. U. Nations
3. Hildebrandt A (2008) Análisis y vigilancia de plaguicidas en aguas subterráneas y suelos. Tesi doctoral, in Química Ambiental, Institut d'Investigacions Químiques i Ambientals, Consell Superior d'Investigacions Científiques, Barcelona, Spain
4. USEPA (1993) Pesticide registration progress report. EPA 738-R-93-001. United States Environmental Protection Agency, Washington, USA

5. Hildebrandt A, Guillamon M, Lacorte S, Tauler R, Barceló D (2008) Impact of pesticides used in agriculture and vineyards to surface and groundwater quality (North Spain). Water Res 42(13):3315–3326
6. Barth JAC, Grathwohl P, Fowler HJ, Bellin A, Gerzabek MH, Lair GJ, Barceló D, Petrovic M, Navarro A, Négrel P, Petelet-Giraud E, Rijnaarts H, Darmendrail D, Langenhoff A, de Weert J, Slob A, van der Zaan BM, Gerritse J, Frank E, Gutierrez A, Kretzschmar R, Gocht T, Steidle D, Garrido F, Jones KC, Meijer S, Moeckel C, Marsman A, Klaver G, Vogel T, Bürger C, Kolditz O, Broers HP, Baran N, Joziasse J, Von Tümpling W, Van Gaans P, Merly C, Chapman A, Brouyère S, Batlle Aguilar J, Orban P, Tas N, Smidt H (2009) Mobility, turnover and storage of pollutants in soils, sediments and waters: achievements and results of the EU project AquaTerra. A review. Agron Sustain Dev 29:161–173
7. Spalding RF, Exner ME (1993) Occurrence of nitrate in groundwater – a review. J Environ Qual 22(3):392–402
8. Novotny V (1999) Diffuse pollution from agriculture – a worldwide outlook. Water Sci Technol 39(3):1–13
9. Gunningham N, Sinclair D (2005) Policy instrument choice and diffuse source pollution. J Environ Law 17(1):51–81
10. Loos R, Gawlik BM, Locoro G, Rimaviciute E, Contini S, Bidoglio G (2009) EU-wide survey of polar organic persistent pollutants in European river waters. Environ Pollut 157 (2):561–568
11. Vryzas Z, Vassiliou G, Alexoudis C, Papadopoulou-Mourkidou E (2009) Spatial and temporal distribution of pesticide residues in surface waters in northeastern Greece. Water Res 43(1):1–10
12. Palma P, Kuster M, Alvarenga P, Palma VL, Fernandes RM, Soares AMVM, López de Alda MJ, Barceló D, Barbosa IR (2009) Risk assessment of representative and priority pesticides, in surface water of the Alqueva reservoir (South of Portugal) using on-line solid phase extraction–liquid chromatography–tandem mass spectrometry. Environ Int 35:545–551
13. Zhou JL, Maskaoui K, Qiu YW, Hong HS, Wang ZD (2001) Polychlorinated biphenyl congeners and organochlorine insecticides in the water column and sediments of Daya Bay, China. Environ Pollut 113(3):373–384
14. Ying GG, Williams B, Kookana R (2002) Environmental fate of alkylphenols and alkylphenol ethoxylates – a review. Environ Int 28(3):215–226
15. Yunker MB, Macdonald RW, Goyette D, Paton DW, Fowler BR, Sullivan D, Boyd J (1999) Natural and anthropogenic inputs of hydrocarbons to the Strait of Georgia. Sci Total Environ 225(3):181–209
16. Crane M (2003) Proposed development of sediment quality guidelines under the European water framework directive: a critique. Toxicol Lett 142(3):195–206
17. Carvalho FP, Villeneuve JP, Cattini C, Tolosa I, Montenegro-Guillen S, Lacayo M, Cruz A (2002) Ecological risk assessment of pesticide residues in coastal lagoons of Nicaragua. J Environ Monit 4(5):778–787
18. Fernandez P, Grimalt JO (2003) On the global distribution of persistent organic pollutants. Chimia 57(9):514–521
19. Doménech X (1999) Química de la contaminació. In: Ediciones M (ed) Madrid, Spain
20. Amaral OC, Otero R, Grimalt JO, Albaiges J (1996) Volatile and semi-volatile organochlorine compounds in tap and riverine waters in the area of influence of a chlorinated organic solvent factory. Water Res 30(8):1876–1884
21. Santos TCR, Rocha JC, Barceló D (2000) Determination of rice herbicides, their transformation products and clofibric acid using on-line solid-phase extraction followed by liquid chromatography with diode array and atmospheric pressure chemical ionization mass spectrometric detection. J Chromatogr A 879(1):3–12
22. Grimalt JO, Gomezbelinchon JI, Llop R, Albaigès J (1988) Water-phase distribution of hexachlorobenzene in a deltaic environment (Ebro delta, Western Mediterranean). Chemosphere 17(10):1893–1903

23. Lopez-Martin J, Ruiz-Olmo J, Borrell A (1995) Levels of organochlorine compounds in freshwater fish from Catalonia, N.E. Spain. Chemosphere 31(6):3523–3535
24. Pastor D, Sanpera C, González-Solís J, Ruiz X, Albaigès J (2004) Factors affecting the organochlorine pollutant load in biota of a rice field ecosystem (Ebro delta, NE Spain). Chemosphere 55(4):567–576
25. Lacorte S, Raldua D, Martinez E, Navarro A, Diez S, Bayona JM, Barceló D (2006) Pilot survey of a broad range of priority pollutants in sediment and fish from the Ebro river basin (NE Spain). Environ Pollut 140(3):471–482
26. Navarro A, Lacorte S, Barceló D (2010) Occurrence and transport of pesticides and alkylphenols in water samples along the Ebro river basin. J Hydrol 383:18–29
27. Navarro-Ortega A, Tauler R, Lacorte S, Barceló D (2010) Occurrence and transport of PAHs, pesticides and alkylphenols in sediment samples along the Ebro river basin. J Hydrol 383:5–17
28. de la Cal A, Eljarrat E, Barceló D (2003) Determination of 39 polybrominated diphenyl ether congeners in sediment samples using fast selective pressurized liquid extraction and purification. J Chromatogr A 1021(1–2):165–173
29. Guerra P, Cal ADL, Marsh G, Eljarrat E, BarcelÃ D (2009) Transfer of hexabromocyclododecane from industrial effluents to sediments and biota: case study in Cinca river (Spain). J Hydrol 369(3–4):360–367
30. Barata C, Damasio J, López MA, Kuster M, De Alda ML, Barceló D, Riva MC, Raldúa D (2007) Combined use of biomarkers and in situ bioassays in *Daphnia magna* to monitor environmental hazards of pesticides in the field. Environ Toxicol Chem 26(2):370–379
31. Kuster M, de Alda MJL, Barata C, Raldúa D, Barceló D (2008) Analysis of 17 polar to semi-polar pesticides in the Ebro river delta during the main growing season of rice by automated on-line solid-phase extraction–liquid chromatography–tandem mass spectrometry. Talanta 75(2):390–401
32. Gómez Gutiérrez A (2008) Contaminants orgànics persistents a la conca mediterrània. El cas del delta de l'Ebre. Tesi doctoral, in Departament de Química Ambiental, Institut d'Investigacions Químiques i Ambientals, Consell Superior d'Investigacions Científiques, Barcelona, Spain
33. Luo X-J, Chen S-J, Mai B-X, Yang Q-S, Sheng G-Y, Fu J-M (2006) Polycyclic aromatic hydrocarbons in suspended particulate matter and sediments from the Pearl River Estuary and adjacent coastal areas, China. Environ Pollut 139(1):9–20
34. Sánchez-Bayo F (2006) Comparative acute toxicity of organic pollutants and reference values for crustaceans. I. Branchiopoda, Copepoda and Ostracoda. Environ Pollut 139 (3):385–420
35. Berset JD, Ejem M, Holzer R, Lischer P (1999) Comparison of different drying, extraction and detection techniques for the determination of priority polycyclic aromatic hydrocarbons in background contaminated soil samples. Anal Chim Acta 383(3):263–275
36. Igbedioh SO (1991) Effects of agricultural pesticides on humans, animals and higher plants in developing countries. Arch Environ Health 46(4):218–224
37. Ritter L, Solomon KR (1995) A review of the persistent organic pollutants: DDT, aldrin, dieldrin, endrin, chlordane, heptachlor, hexachlorobenzene, mirex, toxaphene, polychlorinated biphenyls, dioxins and furans. In: T.I.P.o.C.S. (IPCS) (ed) Geneve, Suisse
38. Pehkonen SO, Zhang Q (2002) The degradation of organophosphorus pesticides in natural waters: a critical review. Crit Rev Environ Sci Technol 32(1):17–72
39. Iwata H, Tanabe S, Sakal N, Tatsukawa R (1993) Distribution of persistent organochlorines in the oceanic air and surface seawater and the role of ocean on their global transport and fate. Environ Sci Technol 27(6):1080–1098
40. Kang YH, Sheng GY, Fu JM, Mai BX, Zhang G, Lin Z, Min YS (2000) Polychlorinated biphenyls in surface sediments from the Pearl River delta and Macau. Mar Pollut Bull 40 (9):794–797

41. Wiberg PL, Harris CK (2002) Desorption of p, p′-DDE from sediment during resuspension events on the Palos Verdes shelf, California: a modeling approach. Continent Shelf Res 22 (6–7):1005–1023
42. Lee KT, Tanabe S, Koh CH (2001) Distribution of organochlorine pesticides in sediments from Kyeonggi Bay and nearby areas, Korea. Environ Pollut 114(2):207–213
43. Díaz A, Ventura F, Galcerán MT (2002) Development of a solid-phase microextraction method for the determination of short-ethoxy-chain nonylphenols and their brominated analogs in raw and treated water. J Chromatogr A 963:159–167
44. Dodds EC, Lawson W (1938) Molecular structure in relation to oestrogenic activity. Compounds without a phenanthrene nucleus. Proc R Soc Lond B Biol Sci 125(839):222–232
45. Galanopoulou S, Vgenopoulos A, Conispoliatis N (2005) DDTs and other chlorinated organic pesticides and polychlorinated biphenyls pollution in the surface sediments of Keratsini harbour, Saronikos gulf, Greece. Mar Pollut Bull 50(5):520–525
46. Pierard C, Budzinski H, Garrigues P (1996) Grain-size distribution of polychlorobiphenyls in coastal sediments. Environ Sci Technol 30(9):2776–2783
47. Kim KH, Seo YC, Nam H, Joung HT, You JC, Kim DJ, Seo YC (2005) Characteristics of major dioxin/furan congeners in melted slag of ash from municipal solid waste incinerators. Microchem J 80(2):171–181
48. Carson R (2001) La primavera silenciosa. In: Drakontos (ed) Editorial Crítica, Barcelona, Spain
49. USEPA (2008) Persistent bioaccumulative and toxic (PBT) Chemical Program. DDT. United States Environmental Protection Agency. Available from: www.epa.gov
50. Rotterdam Convention (2004) Rotterdam convention. Available from: http://www.pic.int
51. Stockholm Convention (2001) Stockholm convention on persistent organic pollutants (POPs). Available from: http://www.pops.int
52. European Commission (1998) 98/83/EC. Official Journal of the European Communities
53. European Commission (2006) 2006/118/EC. Official Journal of the European Communities
54. European Council (1976) Directive 76/464/EEC 4 May 1976 on pollution caused by certain dangerous substances discharged into the aquatic environment of the community. Official Journal of the European Communities
55. European Council (2000) Water Framework Directive: Directive 2000/60/EC of the European Parliament and of the council of 23 October 2000 establishing a framework for community action in the field of water policy. Official Journal of the European Communities
56. European Commission (2004) Decision 2004/248/CE of 10 March 2004 concerning the non-inclusion of atrazine in Annex I to council directive 91/414/EEC and the withdrawal of authorisations for plant protection products containing this active substance. Official Journal of the European Communities
57. European Council (2008) Directive 2008/105/EC of the European Parliament and of the council of 16 December 2008 on environmental quality standards in the field of water policy, amending and subsequently repealing council directives 82/176/EEC, 83/513/EEC, 84/156/EEC, 84/491/EEC, 86/280/EEC and amending directive 2000/60/EC of the European Parliament and of the Council. Official Journal of the European Communities
58. Laboratorio del Área de Calidad de Aguas C.H.d.E. (2005) Red de Control de sustancias peligrosas. Agua, sedimentos y biota. Año 2004. Área de Calidad de Aguas, Confederación Hidrográfica del Ebro, Zaragoza, Spain
59. Pinilla López-Oliva JL (2002) Red de Control de Sustancias Peligrosas. Agua, sedimentos y biota. Área de Calidad de Aguas de la Confederación Hidrográfica del Ebro, Zaragoza, Spain
60. Eggleton J, Thomas KV (2004) A review of factors affecting the release and bioavailability of contaminants during sediment disturbance events. Environ Int 30(7):973–980
61. Chapman A, Bardos P (2005) AquaTerra deliverable Integrator 1.1 F: Ebro river basin characterisation. In: Department of Plant Science (ed) University of Reading, Reading, United Kingdom

62. Navarro A, Tauler R, Lacorte S, Barceló D (2006) Chemometrical investigation of the presence and distribution of organochlorine and polyaromatic compounds in sediments of the Ebro river basin. Anal Bioanal Chem 385(6):1020–1030
63. Peñalver A, Pocurull E, Borrull F, Marcé RM (2000) Determination of phthalate esters in water samples by solid-phase microextraction and gas chromatography with mass spectrometric detection. J Chromatogr A 872(1–2):191–201
64. Readman JW, Albanis TA, Barceló D, Galassi S, Tronczynski J, Gabrielides GP (1997) Fungicide contamination of Mediterranean estuarine waters: results from a MED POL pilot survey. Mar Pollut Bull 34(4):259–263
65. Chirón S, Papilloud S, Haerdi W, Barceló D (1995) Automated online liquid–solid extraction followed by liquid chromatography-high-flow pneumatically assisted electrospray mass–spectrometry for the determination of acidic herbicides in environmental waters. Anal Chem 67(9):1637–1643
66. Cid-Montañés JF, Risebrough RW, Delappe BW, Marino MG, Albaigès J (1990) Estimated inputs of organochlorines from the river Ebro into the northwestern Mediterranean. Mar Pollut Bull 21(11):518–523
67. Gómez-Gutierrez AI, Jover E, Bodineau L, Albaigès J, Bayona JM (2006) Organic contaminant loads into the Western Mediterranean Sea: estimate of Ebro river inputs. Chemosphere 65(2):224–236
68. Aguilar C, Borrull F, Marcé RM (1997) Determination of pesticides in environmental waters by solid-phase extraction and gas chromatography with electron-capture and mass spectrometry detection. J Chromatogr A 771(1–2):221–231
69. Barceló D, Porte C, Cid J, Albaigès J (1990) Determination of organophosphorus compounds in Mediterranean coastal waters and biota samples using gas-chromatography with nitrogen–phosphorus and chemical ionization mass-spectrometric detection. Int J Environ Anal Chem 38(2):199–209
70. Lacorte S, Molina C, Barceló D (1993) Screening of organophosphorus pesticides in environmental matrices by various gas-chromatographic techniques. Anal Chim Acta 281 (1):71–84
71. Durand G, Bouvot V, Barceló D (1992) Determination of trace levels of herbicides in estuarine waters by gas and liquid-chromatographic techniques. J Chromatogr 607(2):319–327
72. Readman JW, Albanis TA, Barceló D, Galassi S, Tronczynski J, Gabrielides GP (1993) Herbicide contamination of Mediterranean estuarine waters – results from a Med Pol Pilot Survey. Mar Pollut Bull 26(11):613–619
73. Chirón S, Alba AF, Barceló D (1993) Comparison of online solid-phase disk extraction to liquid–liquid-extraction for monitoring selected pesticides in environmental waters. Environ Sci Technol 27(12):2352–2359
74. Barceló D, Chiron S, Fernández Alba A, Valverde A, Alpendurada MF (1996) Monitoring pesticides and metabolites in surface water and groundwater in Spain. Herbicide Metabolites in Surface Water and Groundwater, vol 630, pp 237–253
75. Santos TCR, Rocha JC, Alonso RM, Martínez E, Ibáñez C, Barceló D (1998) Rapid degradation of propanil in rice crop fields. Environ Sci Technol 32(22):3479–3484
76. Gascón J, Salau JS, Oubina A, Barceló D (1998) Monitoring of organonitrogen pesticides in the Ebro river. Preliminary loadings estimates. Analyst 123(5):941–945
77. Aguilar C, Ferrer I, Borrull F, Marce RM, Barceló D (1999) Monitoring of pesticides in river water based on samples previously stored in polymeric cartridges followed by on-line solid-phase extraction–liquid chromatography-diode array detection and confirmation by atmospheric pressure chemical ionization mass spectrometry. Anal Chim Acta 386 (3):237–248
78. Köck M, Farré M, Martínez E, Gajda-Schrantz K, Ginebreda A, Navarro A, Alda MLd, Barceló D (2010) Integrated ecotoxicological and chemical approach for the assessment of pesticide pollution in the Ebro river delta (Spain). J Hydrol 383(1–2):73–82

79. Claver A, Ormad P, Rodriguez L, Ovelleiro JL (2006) Study of the presence of pesticides in surface waters in the Ebro river basin (Spain). Chemosphere 64(9):1437–1443
80. Fernández MA, Alonso C, González MJ, Hernández LM (1999) Occurrence of organochlorine insecticides, PCBs and PCB congeners in waters and sediments of the Ebro river (Spain). Chemosphere 38(1):33–43
81. Bosch C, Olivares A, Faria M, Navas JM, del Olmo I, Grimalt JO, Piña B, Barata C (2009) Identification of water soluble and particle bound compounds causing sublethal toxic effects. A field study on sediments affected by a chlor-alkali industry. Aquat Toxicol 94(1):16–27
82. Carrizo D, Grimalt JO (2006) Rapid and simplified method for the analysis of polychloronaphthalene congener distributions in environmental and human samples by gas chromatography coupled to negative ion chemical ionization mass spectrometry. J Chromatogr A 1118(2):271–277
83. Raldúa D, Ferrando P, Durán C, Pedrocchi C (1997) The influence of place of capture, sex, and season on the organochlorine pesticide content in barbel (*Barbus graellsi*) from Northeastern Spain. Chemosphere 35(10):2245–2254
84. Porte C, Barceló D, Albaigès J (1992) Monitoring of organophosphorus and organochlorinated compounds in a rice crop field (Ebro delta, Spain) using the mosquitofish as indicator organism. Chemosphere 24(6):735–743
85. Lavado R, Urena R, Martin-Skilton R, Torreblanca A, del Ramo J, Raldúa D, Porte C (2006) The combined use of chemical and biochemical markers to assess water quality along the Ebro river. Environ Pollut 139(2):330–339
86. Faria M, Huertas D, Soto DX, Grimalt JO, Catalan J, Riva MC, Barata C (2010) Contaminant accumulation and multi-biomarker responses in field collected zebra mussels (*Dreissena polymorpha*) and crayfish (*Procambarus clarkii*), to evaluate toxicological effects of industrial hazardous dumps in the Ebro river (NE Spain). Chemosphere 78(3):232–240
87. Santos X, Pastor D, Llorente GA, Albaiges J (1999) Organochlorine levels in viperine snake Natrix maura carcasses from the Ebro delta (NE Spain): sexual and size-related differences. Chemosphere 39(15):2641–2650
88. Gros M, Petrovic M, Barceló D (2007) Wastewater treatment plants as a pathway for aquatic contamination by pharmaceuticals in the Ebro river basin (northeast Spain). Environ Toxicol Chem 26(8):1553–1562
89. Postigo C, López de Alda MJ, Barceló D (2010) Drugs of abuse and their metabolites in the Ebro river basin: occurrence in sewage and surface water, sewage treatment plants removal efficiency, and collective drug usage estimation. Environ Int 36(1):75–84
90. European Council (2008) Common Position (EC) No 3/2008 of 20 December 2007 adopted by the council, acting in accordance with the procedure referred to in Article 251 of the treaty establishing the European community, with a view to the adoption of a directive of the European Parliament and of the council on environmental quality standards in the field of water policy and amending directives 82/176/EEC, 83/513/EEC, 84/156/EEC, 84/491/EEC, 86/280/EEC and 2000/60/EC. Official Journal of the European Communities, p 1–15
91. Hildebrandt A, Lacorte S, Barceló D (2009) Occurrence and fate of organochlorinated pesticides and PAH in agricultural soils from the Ebro river basin. Arch Environ Contam Toxicol 57:247–255
92. Navarro A, Lacorte S, Barceló D (2010) Occurrence and transport of pesticides and alkylphenols in water samples along the Ebro river basin. J Hydrol 383:18–29
93. Damásio J, Tauler R, Teixidó E, Rieradevall M, Prat N, Riva MC, Soares AMVM, Barata C (2008) Combined use of *Daphnia magna* in situ bioassays, biomarkers and biological indices to diagnose and identify environmental pressures on invertebrate communities in two Mediterranean urbanized and industrialized rivers (NE Spain). Aquat Toxicol 87 (4):310–320
94. BOE (1994) Orden de 4 de febrero de 1994 por la que se prohíbe la comercialización y utilización de plaguicidas de uso ambiental que contienen determinados ingredientes activos peligrosos

95. SedNet (2008) European sediment network. Available from: http://www.sednet.org/content/view/67/106/#legislation
96. Canadian Council of Ministers of the Environment (2002) Canadian sediment quality guidelines for the protection of aquatic life. Canadian Council of Ministers of the Environment, Winnipeg
97. IARC (1991) Monographs on the evaluation of carcinogenic risks to humans. Available from: http://monographs.iarc.fr/ENG/Monographs/PDFs/index.php
98. Quirós L, Piña B, Solé M, Blasco J, López MA, Riva MC, Barceló D, Raldúa D (2007) Environmental monitoring by gene expression biomarkers in *Barbus graellsii*: laboratory and field studies. Chemosphere 67(6):1144–1154
99. Eljarrat E, Martínez MA, Sanz P, Concejero MA, Piña B, Quirós L, Raldúa D, Barceló D (2008) Distribution and biological impact of dioxin-like compounds in risk zones along the Ebro river basin (Spain). Chemosphere 71(6):1156–1161
100. de la Cal A, Eljarrat E, Raldúa D, Duran C, Barceló D (2008) Spatial variation of DDT and its metabolites in fish and sediment from Cinca river, a tributary of Ebro river (Spain). Chemosphere 70(7):1182–1189
101. Eljarrat E, Raldúa D, Barceló D (2010) Origin, occurrence, and behavior of brominated flame retardants in the Ebro River basin. Hdb Env Chem DOI 10.1007/698_2010_70
102. Gros M, Petrovic M, Ginebreda A, Barceló D (2010) Sources, occurrence, and environmental risk assessment of pharmaceuticals in the Ebro river basin. Hdb Env Chem DOI 10.1007/698_2010_72
103. Postigo C, de Alda ML, Barceló D (2010) Illicit Drugs along the Ebro river basin: Occurrence in surface and wastewater and derived consumption estimations. Hdb Env Chem DOI 10.1007/698_2010_79
104. Lopez de Alda M, Köch M, Martínez E, la Farré M, Navarro A, Ginebreda A, Barceló D (2010) Polar pesticides at the Ebro River delta: Occurrence and toxicity in water and biota. Hdb Env Chem (in this volume)

Origin, Occurrence, and Behavior of Brominated Flame Retardants in the Ebro River Basin

E. Eljarrat, D. Raldúa, and D. Barceló

Abstract This work is part of the research included in the European project AQUATERRA, focused on the study of different persistent organic pollutants in different risk zones along the Ebro River basin. This chapter summarizes brominated flame retardant research in the Ebro area. Within a monitoring program at different risk zones in this river, two high contaminated areas were detected. The first one is located along the Cinca River, a tributary of Ebro River, downstream a heavily industrialized town (Monzón) and data showed a high HBCD contamination in this area. The second one is located along the Vero River, a tributary of Cinca River, downstream an industrial park in Barbastro. In this case, high contamination of deca-BDE-209 was found. Our work included the analysis of sediments and biota, with special attention on aspects such as temporal trends, bioavailability, and bioaccumulation of these contaminants. Moreover, an attempt of identification of source contamination was carried out, with the analysis of industrial effluents. In both cases, the industry responsible of the contamination was identified.

Keywords Bioaccumulation, Bioavailability, Brominated flame retardants, Hexabromocyclododecane, Polybrominated diphenyl ether

---

E. Eljarrat (✉) and D. Raldúa
Department of Environmental Chemistry, IDAEA, CSIC, Jordi Girona 18-26, 08034 Barcelona, Spain
e-mail: eeeqam@cid.csic.es

D. Barceló
Department of Environmental Chemistry, IDAEA, CSIC, Jordi Girona 18-26, 08034 Barcelona, Spain
Catalan Institute for Water Research (ICRA), Parc Científic i Tecnològic de la Universitat de Girona, Pic de Peguera 15, 17003 Girona, Spain

D. Barceló and M. Petrovic (eds.), *The Ebro River Basin*, 167
Hdb Env Chem (2011) 13: 167–187, DOI 10.1007/698_2010_70,

Contents

Abbreviations

ABS	Acrylonitrile-butadiene-styrene
BFR	Brominated flame retardant
dw	Dry weight
EPS	Expandable polystyrene
ESI	Electrospray ionization
GC	Gas chromatography
HBCD	Hexabromocyclododecane
LC	Liquid chromatography
LOD	Limit of detection
lw	Lipid weight
MS	Mass spectrometry
MS-MS	Tandem mass spectrometry
NCI	Negative chemical ionization
PBDE	Polybrominated diphenyl ether
POP	Persistent organic pollutant
PVC	Polyvinyl chloride
QqLIT	Quadrupole linear ion trap
QqQ	Triple quadrupole
SIM	Selected ion monitoring
SPLE	Selective pressurized liquid extraction
SRM	Selected reaction monitoring
TBBPA	Tetrabromobisphenol-A
ww	Wet weight

1 Introduction

Brominated flame retardants (BFRs) are comprised of diverse classes or chemical compounds used in a variety of commercial applications. They are used in plastics, textiles, electronic circuitry, and other materials to prevent fires. The estimated global consumption of BFRs shows that their usage is on the rise: during the period from 1990 to 2000, there was more than a doubling of usage from approximately 145–310 kilotons [1].

Currently, the contamination in aquatic environment includes, together with other persistent organic pollutants (POPs), a variety of BFRs. The studies performed in the last 15 years have mainly concerned polybrominated diphenyl ethers (PBDEs), and in particular those congeners which derive from the Penta-mix formulation [2, 3]. PBDEs are typically produced at three different degrees of bromination, i.e., Penta-BDE, Octa-BDE and Deca-BDE. Of the three main technical mixtures in use, the penta-BDE and octa-BDE mixtures are currently being phased out in Europe. Consequently, a shift in production of deca-BDE mixtures for these lower brominated PBDE mixtures has taken place. A recent report of The Bromine Science Environmental Forum estimated the total market demand for the major commercial BFRs in 2001 [4]. This shows the dominance of tetrabromobisphenol-A (TBBPA) (59% of total world usage) and the deca-mix PBDE formulation (27% of total world usage) in volume terms. Recent reports also suggest that usage of hexabromocyclododecane (HBCD) is increasing [5] and that because attention is now switching to this compound which is widely used.

Similar to other POPs, BFRs (except TBBPA) appear to be lipophilic and bioaccumulate in biota and humans [5]. A considerable number of studies of PBDEs in wildlife have been undertaken since the mid-1980s, when Jansson et al. [6] first indicated that these compounds were present in samples collected remote from local sources and so may have become ubiquitous environmental contaminants [7, 8].

There is some discussion of how bioavailable and bioaccumulative BDE-209 is. Moreover, there is still a concern that BDE-209 may debrominate in the environment to form less-brominated BDE congeners which are more bioavailable than BDE-209 itself. A preliminary study on uptake and debromination of BDE-209 in caged rainbow trout following dietary exposure [9] showed a slow but measurable uptake of BDE-209 and the presence of lower brominated PBDEs. BDE-209 has been found generally linked to sediments, but some recent studies also reported low levels of this compound in aquatic biota samples from some locations in Europe [10–12] and from Japan [13].

Technical 1, 2, 5, 6, 9, 10-HBCD is produced industrially by addition of bromine to *cis-trans-trans*-1,5,9-cyclododecatriene. This process leads theoretically to a mixture of 16 stereoisomers (six pairs of enantiomers and four mesoforms) and the product usually is a mixture of the three diastereoisomers α-, β- and γ-isomer [14]. Normally, the γ-isomer is the most dominant in the commercial mixtures (ranging between 75 and 89%), followed by α- and then β-isomer (10–13% and 1–12%, respectively) [15]. The dissimilarities in the structure of α-, β- and γ-isomer might raise differences in polarity, dipole moment and in solubility in water. For example, the solubility of α-, β- and γ-HBCD in water was 48.8, 14.7, and 2.1 μg/L, respectively. Therefore, these different properties may explain the differences observed in their environmental behavior [16]. Covaci et al. [17] and Morris et al. [18] found that in sediments, the distribution of HBCD isomers was the same of commercial mixture: γ-HBCD is the most abundant isomer. However, in biota α- isomer predominates over the γ-isomer, and little or no presence of β-isomer has been observed.

Moreover, α-, β- and γ-HBCD diastereoisomers are chiral. Thus HBCD have three pairs of enantiomers (+)α, (−)α, (+)β, (−)β, (+)γ and (−)γ. The enantiomers have identical physicochemical properties and abiotic degradation rates, but may have different biological and toxicological properties and therefore different biotransformation rates. These transformations may result in nonracemic mixtures of the enantiomers that were industrially synthesized as racemates [16, 19]. The rates of metabolisation process of the enantiomers of chiral environmental pollutants may be significantly different [20].

2 Area of Study

The study area is located in the northeast of Spain, along the Cinca and Vero rivers, in the Ebro River basin (Fig. 1). Six different sampling stations were selected at the Cinca River: site C1 (Puente de las Pilas) and C2 (La Boquera), 20 and 12 km, respectively, upstream from Monzón; site C3, just downstream from Monzón, a heavily industrialized town with a very important chemical industry; site C4 (Alcolea de Cinca), site C5 (Chalamera), and site C6 (Fraga), 27, 30, and 67 km downstream of site C3, respectively. Moreover, three sampling stations were selected at the Vero River: site V1 (Castillazuelo), 11 km upstream from an industrial park; site V2 (Barbastro), just 1 km downstream from the industrial park, and site V3 (La Boquera), 4 km downstream of site V2.

Different sampling campaigns were carried out from 2002 to 2006 (Table 1). Surficial sediments (0–2 cm) were collected at each different selected site. Attempts were made to collect several fishes from the same locations as the sediment samples. Fish specimens were collected by DC electric pulse. Fish were killed, weighted, and the fork length of each fish was measured. Different fish species were sampled: barbels (*Barbus graellsii*), bleaks (*Alburnus alburnus*), southwestern nases (*Chondrostoma toxostoma*) and carps (*Cyprinius carpio*). Whole fish was analyzed for bleaks and southwestern nases, whereas muscle and liver tissues were analyzed for barbels and carps. Muscle samples were collected from below the dorsal fin. The body cavity of each fish was opened and the liver was removed and preserved frozen at −20°C with the muscle sample until analysis.

In order to determine the sources of contamination, some water samples, including wastewaters and effluents from different industries were also sampled. Along the Cinca River and in the industrial area of Monzón, industrial effluents from two different industries were selected: the first one produced EPS (Expandable polystyrene) treated with flame retardants and ABS (Acrylonitrile–butadiene–styrene), and the second one produced PVC (Polyvinyl chloride). As regards the Vero River, three industries were sampled: the first one, a textile industry which produced polyester fibers treated with flame retardants, the second one produced epoxy

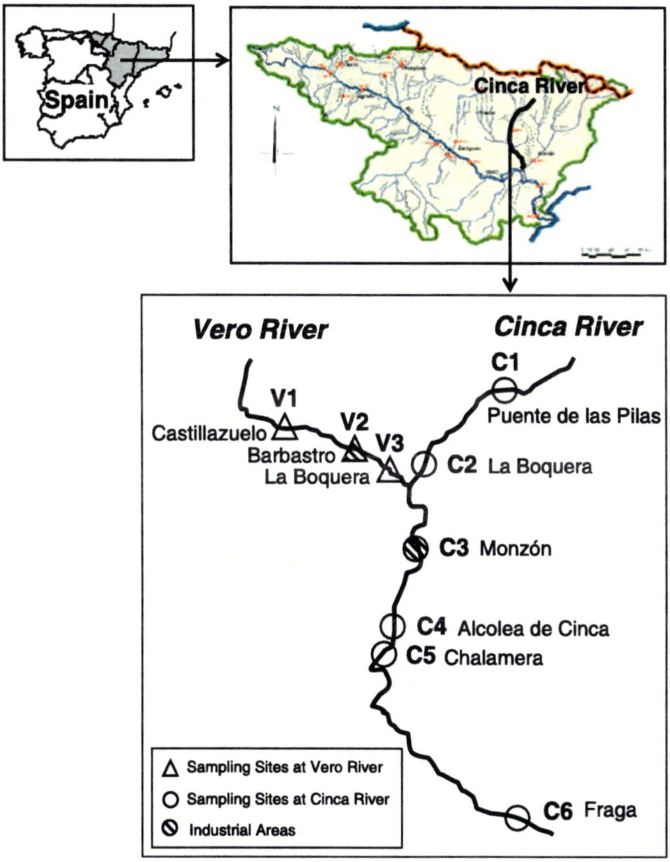

Fig. 1 Geographical location of the area of study and sampling stations along the Cinca and Vero rivers

resins, and the third one is focused on the polyamide polymerization. Moreover, the effluent of the industrial park at the discharge site to the Vero River was also collected.

3 Analytical Methodology

Forty seven PBDE congeners were included in the analytical work: three mono-BDEs (1,2,3), seven diBDEs (7,8,10,11,12,13,15), eight triBDEs (17,25,28,30,32, 33,35,37), six tetraBDEs (47,49,66,71,75,77), seven pentaBDEs (85,99,100,116, 118,119,126), five hexaBDEs (138,153,154,155,166), three heptaBDEs (181,183, 190), four octaBDEs (194,196,197,203), three nonaBDEs (206,207,208), and the decaBDE (209).

Table 1 Sampling campaigns carried out in the area of study between 2002 and 2006

	Cinca river						Vero river		
	Site C1	Site C2	Site C3	Site C4	Site C5	Site C6	Site V1	Site V2	Site V3
October 2002	1 Sediment 4 Barbel[a] (M + L) –	1 Sediment 8 Barbel (M + L) 12 Bleak[b] (WF)	1 Sediment 6 Barbel (M + L) 12 Bleak (WF)	– – –	1 Sediment 5 Barbel (M + L) 8 Bleak (WF)	– – –	–		
November 2004	1 Sediment 6 Southwestern nase[c] (WF) –	1 Sediment 6 Southwestern nase (WF) –	1 Sediment – –	– – –	1 Sediment 6 Southwestern nase (WF) –	– – –	1 Sediment 6 Barbel (M) –	1 Sediment – –	1 Sediment 8 Barbel (M) 2 Carp[d] (M)
November 2005	–						1 Sediment 8 Barbel (M)	1 Sediment –	1 Sediment 5 Barbel (M)
June 2006	1 Sediment	–	1 Sediment	1 Sediment	–	1 Sediment	–		

[a]Barbel *Barbus graellsii*
[b]Bleak *Alburnus alburnus*
[c]Southwestern nase *Chondrostoma toxostoma*
[d]Carp *Cyprinus carpio*
M muscle tissue; *L* liver tissue; *WF* whole fish

A rapid and simple method for PBDE and HBCD determinations in sediment and fish samples was used. The analytical method was based in selective pressurized liquid extraction (SPLE) [21] without further cleanup step and analysis by gas chromatography coupled to mass spectrometry (GC-MS), working with negative ion chemical ionization (NCI) [22, 23].

3.1 Sample Preparation

Sediments and fishes were freeze dried. Lyophilized sediment samples were ground and homogenized by sieving through a stainless steel 0.2-mm sieve, and stored in sealed containers at −20°C until analysis. One gram dry weight (dw) for sediment samples and 0.5 g dw for fish samples were spiked with 3–30 ng of ^{13}C-BDE-209 (150 pg/μL, Wellington Labs., Guelph, Ontario, Canada). Spiked samples were kept overnight to equilibrate. SPLE [21] was carried out using a fully automated ASE 200 system (Dionex, Sunnyvale, CA, USA). A 22 mL extraction cell was loaded by inserting two cellulose filters into the cell outlet, followed by 6 g of alumina (0.063–0.200 mm, from Merck, Darmstadt, Germany). Spiked sediment samples were ground with alumina and cooper (<63 μm, from Merck, Darmstadt, Germany) (1:2:2), whereas fish samples only with alumina (1:2). Mixtures were loaded into the extraction cell on top of alumina. The dead volume was filled with Hydromatrix (Varian Inc., Palo Alto, USA), and the cell was sealed with the top cell cap. Extraction cell was heated to 100°C and filled with hexane:CH_2Cl_2 (1:1) mixture until the pressure reached 1,500 psi. After an oven heat-up time of 5 min under these conditions, two static extractions of 10 min at constant pressure and temperature were developed. After this static period, fresh solvent was introduced to flush the lines and cell, and the extract was collected in the vial. The flush volume amounted to 80–100% of the extraction cell. The extraction was cycled twice. The volume of the resulting extract was about 35 mL.Water samples (2 L) were filtered through 1-μm glass microfibre filters followed by 0.45 μm nylon membrane filters (Whatman). Filtrate water was then mixed with methanol (30%). Extraction of the filtrate water was performed using an automated sample processor Aspec XL (Automated Sample Preparation with Extraction Columns) fitted with a 817 switching valve and an external 306 LC pump, for selection and dispensing of samples, respectively, through the solid phase extraction (SPE) cartridges, from Gilson (Villiers-le-Bel, France). Cartridges (C_{18} (200 mg), from Isolute) were conditioned by passing 6 mL of hexane (1 mL/min) followed by 6 mL of CH_2Cl_2 (1 mL/min), 6 mL of methanol (1 mL/min) and 6 mL of LC-grade water (1 mL/min). 500 mL of the water sample were percolated through the cartridge at 5 mL/min. After sample loading, cartridges were dried by applying vacuum and subsequently eluted with 3 mL of methanol (1 mL/min) followed by 3 mL of CH_2Cl_2 (1 mL/min) and 3 mL of hexane (1 mL/min). The particulate matter retained on filters was extracted using PLE without alumina. An additional cleanup step was performed by SPE with alumina cartridges (5 g, from Isolute). Cartridges were conditioned with

20 mL hexane. The sample volume loaded was 1 mL, and the elution step was performed with 30 mL hexane:CH_2Cl_2 (1:2).

All the extracts (sediment, fish and water) were finally concentrated to incipient dryness and redissolved with 15 μL of PCB-209 at 1 ng/μL (Lab. Dr. Ehrenstorfer, Augsburg, Germany), 15 μL of Cl-BDE-208 at 2 ng/μL [24] and 20 μL of CH_2Cl_2 prior to the analysis by GC-NCI-MS.

3.2 *Instrumental Analysis*

GC-NCI-MS analyses were performed on a gas chromatograph Agilent 6890 connected to a mass spectrometer Agilent 5973 Network (Agilent Technologies España, Madrid, Spain). A HP-5 ms (30 m × 0.25 mm i.d., 0.25 μm film thickness) containing 5% phenyl methyl siloxane (model HP 19091 S-433) capillary column was used for the determination of congeners from mono- to hepta-BDEs. The temperature program was from 110°C (held for 1 min) to 180°C (held for 1 min) at 8°C/min, then from 180 to 240°C (held for 5 min) at 2°C/min, and then from 240 to 265°C (held for 6 min) at 2°C/min, using the splitless injection mode during 1 min and injecting 2 μL. The operating conditions were as follow: ion source temperature = 250°C, ammonia as chemical ionization moderating gas at an ion source pressure of 1.9×10^{-4} torr [22]. For the determination of congeners from octa- to deca-BDE, a DB-5 ms (15 m × 0.25 mm i.d., 0.1 μm film thickness) containing 5% phenyl methyl siloxane capillary column was used with helium as the carrier gas at 10 psi. The temperature program was from 140 (held for 1 min) to 325°C (held for 1 min) at 10°C/min (held for 10 min), using the splitless injection mode during 1 min [23]. Both experiments were carried out monitoring the two most abundant isotope peaks from the mass spectra corresponding to $m/z = 79$ and 81 ($[Br]^-$). For octa- to deca-BDE experiments, $m/z = 487$ and 489 for nona- and deca-BDE, and $m/z = 497$ and 499 for ^{13}C-deca-BDE, were also included. Confirmation criteria for the detection and quantification of PBDEs should include the following: (a) all m/z monitored for a given analyte should maximize simultaneously ±1 s, with signal to noise ratio ≥3 for each; (b) the ratio between the two monitored ions should be within 15% of the theoretical. Quantification of mono- to hepta-BDEs was carried out by internal standard procedure using PCB-209 as internal standard, whereas octa- to deca-BDEs were quantified using ^{13}C-BDE-209 and Cl-BDE-208 as internal standards.

HBCD determinations were carried out simultaneously with mono- to hepta-BDEs. Technical HBCD consists of three diastereomers: α, β and γ-HBCD. HBCD can be determined by GC-MS, but until now, the three diastereomers have not been separated by this technique. However, as apparently the response factors of the three diastereomers do not differ very much [25], HBCD can be quantified as total HBCD by GC-MS. Quantification was carried out by external standard.

Using the described methodology, recoveries ranged from 53 to 84% for sediment, and between 52 and 103% for fish samples. Detection limits (LODs) were in the range of 6–50 pg/g dw for sediments, and in the range of 2–19 pg/g wet weight (ww) for fish samples. Relative standard deviations of the method were in the range of 1–13%.

3.3 HBCD Analysis by LC-MS-MS

3.3.1 Isomer-Specific Determination by LC-QqQ-MS-MS

HBCD stereoisomers can be easily separated using reversed-phase liquid chromatography, and determined by mass spectrometry (LC-MS) or tandem mass spectrometry (LC-MS-MS) [26]. An isomer-specific LC-MS-MS method was developed. Separation of α, β and γ-HBCD stereoisomers was performed on a Waters 2690 HPLC system (Milford, MA, USA) coupled to a Waters Micromass Quattro triple quadrupole (QqQ) mass spectrometer, equipped with a Z-spray electrospray ionization (ESI) interface (Manchester, UK). A 2.1 × 150 mm Symmetry C_{18} 5 μm column and 2.1 × 10 mm guard column supplied by Waters (Massachusetts, USA) was used with a 0.25 mL/min eluent flow. The injection volume was set at 20 μL. Experiments were carried out using water (A), acetonitrile (B), and methanol (C) as mobile phases. The gradient elution started at an initial composition of 60% A/10% B and 30% C and was ramped to 10% A/22.5% B and 67.5% C in 11.30 min and those conditions were maintained for 18.70 min, then initial conditions were reached again in 10 min and were maintained for additional 10 min. A Quattro LC Micromass Spectrometer was operated in ESI negative ion mode using selected reaction monitoring (SRM) for $[M-H]^{-1}$ (*m/z* 640.6) → Br^- (*m/z* 79.0 and 81.0). Argon of 99.999% purity was used as collision gas at a pressure of 3.54×10^{-3} mbar. All data were acquired and processed using MassLinx 4.0 software. The quantification was carried out by the isotopic dilution method, using d_{18}-α- and d_{18}-γ-HBCD solutions as quantification and recovery standards, respectively. Instrumental LOD was between 12 and 75 pg, whereas method LOD including PLE followed by LC-MS-MS was from 7 to 22 ng/g dw for fish and between 9 and 30 ng/g dw for sediment samples (Table 2).

3.3.2 Isomer-Specific Determination by LC-QqLIT-MS

Another methodology for the analysis of HBCD diastereoisomers (α, β and γ), based in LC- quadrupole linear ion trap (QqLIT)-MS method was developed [27]. The LC system used was an Agilent HP 1100 binary pump (Agilent Technologies, Palo Alto, CA, USA) with a Symmetry C_{18} column (2.1 × 150 mm, 5 μm) preceded by a C_{18} guard column (2.1 × 10 mm) supplied by Waters (Massachusetts, USA). Experiments were carried out in negative ionization mode using H_2O: methanol (3:1 v/v) as eluent A and methanol as eluent B, at a flow rate of 0.25 mL/min. The injection volume was set at 4 μL. The elution program started

Table 2 LODs for different analytical methods used for HBCD analysis

	Information	Chromatographic column	MS detection	Quantification	LOD		
					Instrumental	Method (sediment)	Method (fish)
GC-NCI-MS	Total HBCD	HP-5 ms (30 m × 0.25 mm, 0.25 μm)	SIM mode: *m/z* 79 and 81	External standard	–	–	–
LC-QqQ-MS-MS	Isomer-specific	Symmetry C18 (2.1 × 150 mm, 5 μm)	SRM mode: *m/z* 641 to *m/z* 79 *m/z* 641 to *m/z* 81	Isotopic dilution	12–75 pg	9–30 ng/g dw	7–22 ng/g dw
LC-QqLIT-MS	Isomer-specific	Symmetry C18 (2.1 × 150 mm, 5 μm)	SRM mode: *m/z* 641 to *m/z* 79 *m/z* 641 to *m/z* 81	Isotopic dilution	0.5–1.2 pg	–	–
	Enantiomer-specific	Nucleodex β-PM (4.0 × 200 mm, 5 μm)	SRM mode: *m/z* 641 to *m/z* 79 *m/z* 641 to *m/z* 81	Isotopic dilution	0.3–1.5 pg	0.3–1.8 ng/g dw	1.2–3.9 ng/g dw

at an initial composition of 100% A and was ramped to 10% A in 17 min and initial conditions were reached again in 3 min and maintained for an additional 15 min.

Mass spectrometric analysis was performed with a hybrid triple quadrupole/linear ion trap Applied Biosystem MSD Sciex 4000QTRAP (Applied Biosystems, Foster City, USA) instrument equipped with a Turbospray ESI interface. For target quantitative analyses, data acquisition was performed in SRM, recording the transitions between the precursor ion and the two most abundant fragment ions. The developed instrumental method display excellent LODs in SRM mode between 0.5 and 1.2 pg (Table 2).

4 Results and Discussion

4.1 *Cinca River*

4.1.1 Sediment

PBDEs were detected in all the sediment samples at concentrations ranging from 2 to 131 ng/g dw (Table 3). HBCD was detected only in samples collected downstream of Monzón, with levels ranging between 48 and 2,658 ng/g dw. In these samples, HBCD contamination was greater than that observed for PBDEs. Site C3 was found to be the most contaminated zone followed by site C5 > site C1 = site C2. As expected, PBDE and HBCD levels were greater near the site of industrial impact [28].

Our PBDE results were consistent with reported data for river sediments. PBDEs were determined in Swedish river sediments at 8–50 ng/g dw [29]. Similar values were found in Japanese river sediments, with concentration levels between 21 and 59 ng/g dw [30]. Higher levels up to 1,400 ng/g dw were found in a downstream area of a manufacturing plant in United Kingdom [31] and at 120 ng/g dw downstream of an area with textile industries [29]. As regards data for HBCD, Sellström et al. [29] reported concentration levels between nd and 1,600 ng/g dw in river sediments from a Swedish river with numerous textile industries.

Of 47 congeners included in the analytical work, eight different PBDEs were detected, ranging from tetra- to deca-brominated compounds: tetra-BDE-47, penta-BDE-99, penta-BDE-100, penta-BDE-118, hexa-BDE-153, hexa-BDE-154, hepta-BDE-183, and deca-BDE-209. However, BDE-99 could not be quantified due to coelutions with breakdown products of HBCD. This coelution has been previously described by Covaci et al. [32].

Our HBCD results were consistent with reported data for river sediments. Sellström et al. [29] reported concentration levels between nd and 1,600 ng/g dw in river sediments from a Swedish river with numerous textile industries. Morris et al. [18] determined HBCD in river and estuarine sediment samples. HBCD was detected in all sediments analyzed. The maximum value for total HBCD

Table 3 Results (expressed in ng/g dw and ng/g ww for sediment and biota samples, respectively) obtained for samples collected along the Cinca river

				Site C1	Site C2	Site C3	Site C4	Site C5	Site C6
October 2002	Sediment	Total PBDEs		2.4	2.6	42	–	40	–
		HBCD		nd	nd	514	–	90	–
	Barbel[a]	Total PBDEs	Muscle	1.3	4.5	281	–	–	–
			Liver	0.2	0.4	237	–	–	–
		HBCD	Muscle	nd	nd	530	–	–	–
			Liver	nd	nd	554	–	–	–
	Bleak[a]	Total PBDEs		–	4.7	555	–	232	–
		HBCD		–	nd	1,501	–	760	–
November 2004	Sediment	Total PBDEs		14	38	131	–	87	–
		HBCD		nd	nd	1,613	–	866	–
	Southwestern nase[a]	Total PBDEs		2.0	38	–	–	520	–
		HBCD		nd	nd	–	–	4,863	–
June 2006	Sediment	HBCD		nd	–	2,658	77	–	48

[a]Mean values

concentration was 1,700 μg/kg dw in freshwater sediments from the River Skerne (England). In the River Tees the mean concentration was 510 μg/kg dw. HBCD was also detected in sediments from Scheldt (The Netherlands) basin with values between 38 and 950 μg/kg dw.

The isomeric profile of HBCD found in contaminated sediment samples from Monzón was similar to that of commercial HBCD formulations, with γ-HBCD being the most abundant isomer (between 94 and 99% contribution to the total HBCD). The same findings were observed in sediments from lake Winnipeg, where HBCD contamination was dominated by γ-isomer [4].

4.1.2 Biota

Table 3 shows the mean concentrations of BDE congeners, as well as of HBCD, in the selected sites. PBDEs were detected in all the biota samples at concentrations ranging from 0.2 to 555 ng/g ww. HBCD was detected only in samples corresponding to sites downstream Monzón. In these samples, HBCD contamination was similar or greater than that observed for PBDEs. HBCD levels ranged from 90 to 4,863 ng/g ww. Similar to our findings in sediment samples, site C3 was found to be the most contaminated zone followed by site C5 > site C1 = site C2 [28, 33].

The majority of published data in fish samples expressed PBDE levels in ng/g lipid weight (lw), making difficult the comparison with our results expressed in ng/g ww. However, some works also used the same unities of this study. Zennegg et al. [34] analyzed whitefish from different Swiss lakes, and reported PBDE levels between 1.6 and 7.4 ng/g ww. Carps from Detroit river and Des Plaines river [35] presented PBDE concentration levels around 5 ng/g ww and 12 ng/g ww, respectively. All these values were in accordance with our results in sites out of industrial impact (sites C1 and C2).

Of 47 congeners included in the analytical work, 17 different PBDEs were detected, ranging from di- to hepta-brominated compounds. However, BDE-209, the dominant congener of the total PBDE contamination in sediments, was not detected in biota samples. Luross et al. [36] found a BDE congener distribution in aquatic biological matrices dominated by BDE-47 and followed by BDE-99, BDE-100, BDE-153, BDE-154, BDE-28, and BDE-183. However, the distribution found in our study showed a clear predominance of hexa-BDE congeners 153 and 154, as well as of hepta-BDE-183. These data correlate well with distribution observed in sediment samples. The same findings were reported by Rice et al. [35], who analyzed carps from Des Plaines River (USA). The dominant isomers were two hepta-BDE congeners (BDE-181 and BDE-183) as well as two hexabromo congeners (BDE-153 and BDE-154).

Comparison between levels found in muscle tissues and livers showed that similar contamination was detected in both matrices. However, slightly higher BFR levels were found in muscle tissues. Ratios between total PBDE levels in muscle and in liver corresponding to the same individual fish ($R_{M/L}$) were calculated for samples collected at the most contaminated site (site C3). The mean value of these $R_{M/Ls}$ was 1.1. The same $R_{M/Ls}$ were calculated for HBCD contamination, with a similar mean value ($R_{M/L} = 1.3$). In previous study on cod and whiting fishes, PBDE concentrations have been found to be lower in muscle tissue than in liver, whereas higher values were detected in muscle of herring [37].

4.1.3 Bioavailability and Bioaccumulation

When sediment concentrations were compared to those in biota collected at the same sites along Cinca River, high fish to sediment ratios were seen. Specially, levels of BDE-47, BDE-153, and BDE-154 in fish were high compared with levels in sediment (13–21 fold) indicating the high bioavailability of these congeners. Moreover, our calculated fish to sediment ratios correlated well with those reported by Sellström et al. [29]. These large fish to sediment ratios indicate that these contaminants are bioavailable and are taken up readily by the fish. This is also supported by the fact that for some BDE congeners detectable levels were found in fish but not in corresponding sediments: 17 different PBDEs were detected in fish samples, whereas only eight PBDE congeners were found in sediments. Solely BDE-209, the main BDE congener found in sediments from the area, was not found in biota samples. The main reason for their absence in biota seems to be its relatively low bioaccumulation potential. This may be due either to a low uptake rate for this very large molecule or a relatively rapid excretion after biotransformation. The low bioconcentration hazard of BDE-209 was also reported by other authors [38].

The fact that HBCD is found in fish indicates that it is also bioavailable. However, the low fish to sediment ratio (below 1) does not agree with the high ratios reported by Sellström et al. [29] for pikes from a Swedish River. Pike is a predator at the top of the food chain and the higher fish to sediment ratio in this specie could indicate the biomagnification of this compound through the aquatic trophic chain.

Previous studies showed that PBDE levels increased with the age of the fish, indicating bioaccumulation [39]. Fish length is directly related to fish age. Figure 2 shows the measured concentrations of HBCD as a function of fish length at sites C3 and C5. The correlation coefficients indicate acceptable correlations ($R^2 = 0.71$).

Additionally, an in situ experiment was performed to evaluate HBCD bioaccumulation. In site C5, "clean" barbels were exposed to the environment. As a control, the same experiment was performed in site C2. Twenty specimens of barbel were caged into each one of two stainless steel devices that were placed on the river-bed at the two locations. Nine specimens were immediately frozen to be analyzed as blanks. After 15 days of exposure, no mortality was observed in the cages. Table 4 shows levels of HBCD accumulated in the caged barbels exposed in sites C2 and C5. Accumulation at site C2 was negligible, as values measured after the exposure were even lower than those of the fishes not exposed. However, at site C5, after the

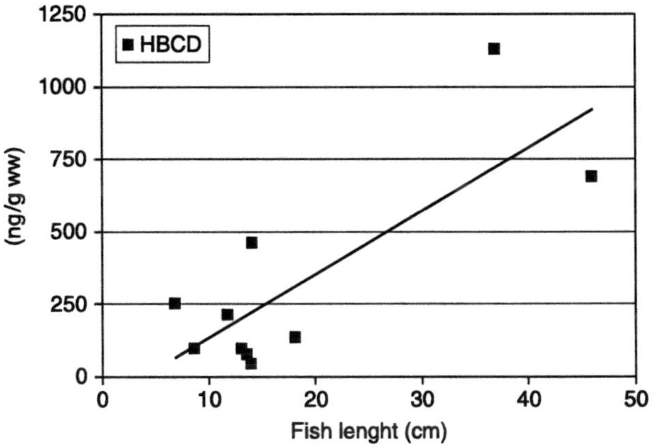

Fig. 2 Length of Cinca river barbel (sites C3 and C5) vs. concentrations of HBCD in the muscle. Line is linear regression on the data

Table 4 Results (mean value (expressed in ng/g lw) ± standard deviation) for caged barbels exposed to an *in situ* bioaccumulation experiment

	α-HBCD	β-HBCD	γ-HBCD	Total HBCD
Not exposed	63 (±9)	30 (±8)	92 (±2)	184 (±15)
Site C2	15 (±2)	2.5 (±1.3)	12 (±3)	28 (±4)
Site C5	333 (±61)	58 (±7)	15,127 (±2,605)	15,517 (±2,673)

15 days of exposure, concentration in caged barbel increased up to 1,337 ng/g ww or 15,518 ng/g lw.

An isomeric analysis was carried out for these barbel samples. For site C5, the proportion of the γ- isomer was between 96 and 98% whereas α-isomer accounted for 2–3% and β-isomer for 0.4%. On the other hand, in barbel not exposed and those from site C2, the predominance of the γ-isomer dilutes. In fact, such profile is unusual in biological samples; as reviewed by Covaci et al. [17], biota shows a higher percentage of α-HBCD than β- or γ-HBCD. The data obtained in our study could be due to a recent technical HBCD exposition.

4.1.4 Source of Contamination

Two industries were suspected to cause the HBCD contamination along the Cinca River: the first one produced EPS treated with flame retardants and ABS, and the second one produced PVC. Analysis of industrial effluents of each industry revealed that industry which produced EPS and ABS is the responsible of the HBCD contamination, with concentration levels around 5 μg/L (Table 5).

Table 5 HBCD and BDE-209 contamination in industrial effluents

	HBCD (ng/L)	BDE-209 (ng/L)
EPS and ABS production	4,980	nd
PVC production	nd	nd
Polyester fibers production	nd	5
Epoxy resins production	nd	45
Polyamide polymerization	nd	2,600
Industrial park effluent	nd	105

4.2 Vero River

4.2.1 Sediment

Table 6 shows the concentrations of BDE congeners from the different selected sites during the two sampling campaigns. PBDEs were detected in all the sediment samples at concentrations ranging from 11 to 14,400 ng/g dw. The site V2 was found to be the most contaminated zone followed by site V3 > site V1. As expected, PBDE levels were greater near the site of industrial impact [40].

Sediments affected by the source contamination (sites V2 and V3) showed a congener pattern clearly dominated by the BDE-209, which was present at concentrations varying from 1,910 to 12,500 ng/g dw. The highest BDE-209 level was found at the sampling site closest to the source of contamination (V2-November 2005, located 5 m downstream from the effluent discharge) and contamination decreased with distance from the industrial area. High levels of BDE-154 (844 ng/g dw) were also detected in this sediment sample. Moreover, some octa- and nona-BDEs were also present at relatively high levels: 37–39 ng/g dw for octa-BDEs, and 169–375 ng/g dw for nona-BDEs. The distribution of octa- and nona-BDE congeners in this sediment sample correlates well with the pattern observed in the technical decaBDE formulation containing 97–98% of BDE-209 and up to 3% nona-BDEs [41].

Some studies have also reported PBDE levels in sediments collected near industry facilities. Sellström et al. [29] reported total PBDE concentration levels between nd and 364 ng/g dw (nd-360 ng/g dw for BDE-209) in river sediments from a Swedish river with numerous textile industries. Higher PBDE levels up to 1,400 ng/g dw (nd-399 ng/g dw for BDE-209) were found downstream from a foam manufacturing plant using PBDEs in the United Kingdom [31]. BDE-209 burdens up to 4,600 ng/g dw have also been reported on suspended particulates from Dutch surface waters, decreasing with distance from a textile facility [42]. Our PBDE results were considerably higher than those reported for sediments collected near industrial areas, probably because of the small dilution factor of the Vero River at this area, which has an average flow of 2.1 m^3/s.

4.2.2 Biota

PBDEs were detected in all the fish muscle samples at concentrations ranging from 28 to 2,092 ng/g lw (Table 6). Similar to our findings in sediment samples, samples

Table 6 Results (expressed in ng/g dw and ng/g lw for sediment and biota samples, respectively) obtained for samples collected along the Vero river

			Site 1	Site 2	Site 3
November 2004	Sediment	BDE-209	7.5	5,395	1,911
		Total PBDEs	11	5,531	1,930
	Barbel[a]	BDE-209	nd	–	67
		Total PBDEs	54	–	791
	Carp[a]	BDE-209	–	–	80
		Total PBDEs	–	–	1,560
November 2005	Sediment	BDE-209	27	12,459	7,454
		Total PBDEs	30	14,395	7,767
	Barbel[a]	BDE-209	nd	–	195
		Total PBDEs	83	–	1,007

[a]Mean values

from the site downstream of the textile industry (V3) were found to be much more contaminated than those collected at the site upstream of the factory (V1). The comparison between the two sampling campaigns showed that levels during 2005 were slightly higher than those found in 2004, probably reflecting the same situation observed for sediment samples with an increase of contamination with time.

Of 23 congeners included in the analytical work, 12 different PBDEs were detected, ranging from tri- to deca-brominated compounds. The most relevant finding is the presence of significant concentrations of BDE-209 in fish samples collected downstream of the factory. In published works related to PBDEs in aquatic biota samples, BDE-209 was not detected or detected only in a minority of the samples and always in concentrations around the limit of detection. The study of BDE-209 levels in aquatic biota samples is of special interest because its bioavailability has been questioned due to its large molecular size resulting in decreased uptake rates [31]. Another possible explanation of low BDE-209 concentrations in aquatic biota may be due to relatively rapid biotransformation. Therefore, and despite high concentrations in abiotic matrices, some authors argue that BDE-209 would not pose an environmental threat. Our work demonstrated that BDE-209 is bioavailable: 14 out of 15 biota samples (barbels and carps) collected downstream the source of contamination, showed BDE-209 levels ranging between 20 and 707 ng/g lw. BDE-209 values in fish collected during the second sampling campaign, with a mean value of 195 ng/g lw and a median value of 86 ng/g lw, were higher than those found in barbels collected during November 2004 (mean value of 67 ng/g lw and median value of 32 ng/g lw). These results are in accordance with the higher values found also in sediment samples collected at the same site (V3).

During the last 5 years, a few studies have been published demonstrating that aquatic biota is able to accumulate BDE-209. Lepom et al. [43] first reported the occurrence of BDE-209 in freshwater fish in Europe, with concentrations up to 37 ng/g lw. Similar results were obtained by Burreau et al. [10], who detected BDE-209 in three different fish species from rural waters of the Lumparn estuary in the

Baltic Sea with concentrations up to 116 ng/g lw. The BDE-209 levels in fish reported here are on average higher than those reported elsewhere.

It is interesting to note the different congener distribution found in fish collected up- and down-stream the source of contamination (Fig. 3). For samples at the control site (V1), the distribution showed a clear predominance of the tetra-BDE-47 (69 and 77% for the first and the second sampling, respectively). The predominance of BDE-47 is consistent with other studies on freshwater fish species [5, 44]. However, in fish from the contaminated area (V3), the BDE-47 contribution to the total PBDE burden decreased to 29%. In contrast, samples from V3 contained a high contribution of hexa-BDE-154, with values between 16% (sampling of 2004) and 20% (sampling of 2005). The contribution of this hexa-BDE congener was only 5 and 7% for the first and the second sampling, respectively, at the site V1. The higher predominance of BDE-154 in the contaminated area could be related to high contamination of this congener in the sediment of this area. But, it could be also attributed to biotransformation of deca-BDE-209. Stapleton et al. [45] studied the debromination of BDE-209 by caged carp following dietary exposure. They concluded that debromination of BDE-209 occurs in carp. Although they did not detect any accumulation of BDE-47 and BDE-99, accumulation of other PBDEs, such as BDE-154, occurs. Finally, it should be pointed out that the deca-BDE-209 contribution in samples collected downstream of the industrial park ranged from 6 to 7% during the first sampling campaign, to 19% in the sampling carried out during 2005. For fish collected during 2005, BDE-209 was the third highest contributing PBDE congener, indicating the relevance of the deca-congener in these samples.

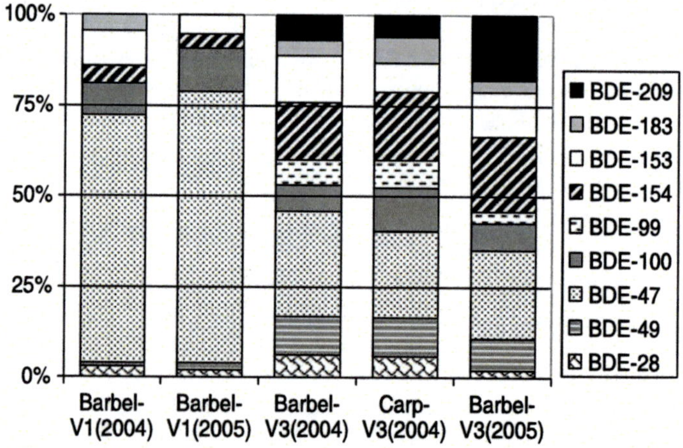

Fig. 3 Percentage contribution of various congeners to the total PBDEs detected in fish samples from the control site (V1) and contaminated area (V3) during the two sampling campaigns (2004 and 2005)

4.2.3 Bioavailability and Bioaccumulation

When sediment concentrations were compared to those in barbel or carp collected at the same site along the Vero River, fish to sediment ratios were obtained. Levels of BDE-47 in fish were high compared with levels in sediment indicating the high bioavailability of this congener, and in accordance with previous ratios obtained in the Cinca river study. The bioavailability decreases with increase in bromination degree, with a fish-to-sediment ratio around 0.1 for the hepta-congener BDE-183. BDE-209 is bioavailable but the bioavailability potential is lower than that observed for other BDE congeners. Much lower are the fish to sediment values: 0.0013 and 0.0011 for the 2004 and 2005 sampling campaigns, respectively. It should be pointed out that an important factor affecting the ratios may be that the system is not in equilibrium because PBDEs are released directly to the river in textile industry effluents, and new inputs are ongoing. The low fish to sediment ratios obtained could be an indication of a recent release of BDE-209 from the industrial park that has contaminated the sediment but not yet been taken up by barbels or carps.

4.2.4 Source of Contamination

Three industries were suspected to cause the BDE-209 contamination along the Vero River: the first one, a textile industry which produced polyester fibers treated with flame retardants, the second one produced epoxy resins, and the third one is focused on the polyamide polymerization. Moreover, the effluent of the industrial park at the discharge site to the Vero River was also analyzed. Analysis of industrial effluents of each industry revealed that industry focused on the polyamide polymerization is the main responsible of the BDE-209 contamination, with concentration levels around 2,600 ng/L (Table 5). Nevertheless, the two other industries also contribute in some way to the total contamination.

5 Conclusions and Perspectives

In this work, two high contaminated areas within the Ebro river basin were detected. The first one showed a high HBCD contamination whereas the second one presented high levels of deca-BDE-209.

The first contaminated area, along the Cinca River and downstream of Monzón, reveals the covariant nature of PBDE and HBCD concentrations in the fish examined here. The amount of each contaminant is highly correlated within each fish. A likely explanation of this correlation is that the contaminants are being acquired from the same source (Cinca River). Moreover, PBDEs and HBCD are bioavailable as indicated by large fish to sediment ratios. Some correlations were found between the concentration levels of BFRs and the length and weight of specimen fishes, showing that these contaminants bioaccumulate in barbels and bleaks. The fact that HBCD is found in fish indicates that it is bioavailable and bioaccumulates. The limited knowledge of their environmental behavior makes it use as a substitute for PBDEs questionable. More studies are needed to determine the distribution, fate and effects of HBCD, to feed into risk assessments for this compound.

The work carried out in the second contaminated area, along the Vero River and downstream of Barbastro, shows that BDE-209 is bioavailable from sediment to fish and that bioaccumulation of highly brominated congeners is possible in highly polluted aquatic environments. Moreover, BDE-209 is probably a source material for lower brominated products, i.e., hexa-BDE-154, that are more readily accumulated and to a greater extend and with a greater degree of persistency. The high BDE-209 concentrations in sediments emphasize the need for a thorough understanding of the fate of this compound. Finally, the BDE-209 releases from some industries, such as textile, epoxy resins and polyamide polymerization, merit special scrutiny. The 2001 US EPA's Toxics Reduction Inventory-estimated emissions by industry type indicates that textile production was by far the largest deca-BDE discharger to US surface waters and publicly-owned wastewater treatment works [46].

Acknowledgments This research study was funded by the European Union under the Global Change and Ecosystems (FP6) Water Cycle and Soil Related Aspects: Integrated modeling of the river-sediment–soil–groundwater system; advanced tools for the management of catchment areas and river basins in the context of global change (AQUATERRA, Project number 505428), and by the Spanish Ministry of Education and Science (Project numbers CTM2005-25168-E and CTM2005-07402-C02-02/TECNO). This work reflects only the author's views and the European Community is not liable for any use that may be made of the information contained therein. Confederación Hidrográfica del Ebro is thanked for its assistance in the industrial effluent sampling.

References

1. Arias PA (2001) Proc 2nd Int Workshop on Brominated Flame Retardants. Stockholm, Sweden, 17
2. Strandberg B, Dodder NG, Basu I, Hites RA (2001) Environ Sci Technol 35:1078
3. Hale RC, Alaee M, Manchester-Neesvig JB, Stapleton HM, Ikonomou MG (2003) Environ Int 29:771
4. Law RJ, Allchin CR, de Boer J, Covaci A, Herzke D, Lepom P, Morris S, Tronczynski J, de Wit CA (2006) Chemosphere 64:187
5. de Wit C (2002) Chemosphere 46:583
6. Jansson B, Asplund L, Olsson M (1987) Chemosphere 16:2343
7. Manchester-Neesvig JB, Valters K, Sonzogni WC (2001) Environ Sci Technol 35:1072
8. Voorspoels S, Covaci A, Schepens P (2003) Environ Sci Technol 37:4348
9. Kierkegaard A, Balk L, Tjärnlund U, de Wit CA, Jansson B (1999) Environ Sci Technol 33:1612
10. Burreau S, Zebühr Y, Broman D, Isaac R (2004) Chemosphere 55:1043

11. Leonards P, Vethaak D, Brandsma S, Kwadijk C, Micic D, Jol J, Schout P, de Boer J (2004) Proceedings of the third international workshop on brominated flame retardants BFR2004, Toronto, Canada, pp 283, 6–9 June 2004
12. Johansson I, Héas-Moisan K, Guiot N, Munschy C, Tronczynski J (2006) Chemosphere 64:296
13. Akutsu K, Obana H, Okihashi M, Kitagawa M, Nakazawa H, Matsuki Y, Makino T, Oda H, Hori S (2001) Chemosphere 44:1325
14. Petersen M, Hamm S, Schäfer A, Esser U (2004) Organohalogen compd 66:226
15. Barontini F, Petarca L, Cozzani V (2001) Ind Eng Chem Res 40:3270
16. Janák K, Covaci A, Voorspoels S, Becher G (2005) Environ Sci Technol 39:1987
17. Covaci A, Gerecke A, Law R, Voorspoels S, Kohler M, Heeb N, Leslie H, Allchin C, de Boer J (2006) Environ Sci Technol 40:3679
18. Morris S, Allchin C, Zegers B, Haftka J, Boon J, Leonards P, van Leeuwen S, de Boer J (2004) Environ Sci Technol 38:5497
19. Marvin C, MacInnis G, Alaee M, Arsenault G, Tomy G (2007) Rapid Commun Mass Spectrom 21:1925
20. Eljarrat E, Guerra P, Barceló D (2008) Trends Analyt Chem 27:847
21. de la Cal A, Eljarrat E, Barceló D (2003) J Chromatogr A 1021:165
22. Eljarrat E, Lacorte S, Barceló D (2002) J Mass Spectrom 37:76
23. Eljarrat E, de la Cal A, Barceló D (2004) Anal Bioanal Chem 378:610
24. Christiansson A, Teclechiel D, Eriksson J, Bergman A, Marsh G (2006) Chemosphere 63:562
25. QUASIMEME. Report workshop analysis brominated flame retardants, Barcelona, Spain, Aberdeen, UK: FRS Marine Laboratory, 21–23 October 2002
26. Budakowsky W, Tomy G (2003) Rapid Commun Mass Spectrom 17:1399
27. Guerra P, Eljarrat E, Barceló D (2008) Rapid Commun Mass Spectrom 22:916
28. Eljarrat E, de la Cal A, Raldúa D, Duran C, Barceló D (2004) Environ Sci Technol 38:2603
29. Sellström U, Kierkegaard A, de Wit C, Jansson B (1998) Environ Toxicol Chem 17:1065
30. Watanabe I, Kawano M, Tatsukawa R (1995) Organohalogen Compd 24:337
31. Allchin CR, Law RJ, Morris S (1999) Environ Pollut 105:197
32. Covaci A, Voorspoels S, de Boer J (2003) Environ Int 29:735
33. Eljarrat E, de la Cal A, Raldúa D, Duran C, Barceló D (2005) Environ Pollut 133:501
34. Zennegg M, Kohler M, Gerecke AC, Schmid P (2003) Chemosphere 51:545
35. Rice CP, Chernyak SM, Begnoche L, Quintal R, Hickey J (2002) Chemosphere 49:731
36. Luross JM, Alaee M, Sergeant DB, Cannon CM, Whittle DM, Solomon KR (2002) Chemosphere 46:665
37. Boon J, Lewis W, Tjoen-A-Choy M, Allchin C, Law R, de Boer J, Ten Hallers-Tjabbes C, Zegers B (2002) Environ Sci Technol 36:4025
38. Andersson O, Blomkvist G (1981) Chemosphere 10:1051
39. Haglund PS, Soc DR, Buser HR, Hu J (1997) Environ Sci Technol 31:3281
40. Eljarrat E, Labandeira A, Marsh G, Raldua D, Barceló D (2007) Chemosphere 69:1278
41. Alaee M, Arias P, Sjödin A, Bergman A (2003) Environ Int 29:683
42. de Boer J, Wester PG, van der Horst A, Leonards PEG (2003) Environ Pollut 122:63
43. Lepom P, Karasyova T, Sawal G (2002) Organohalogen Compd 58:209
44. Labandeira A, Eljarrat E, Barceló D (2007) Environ Pollut 146:188
45. Stapleton HM, Alaee M, Letcher RJ, Baker JE (2004) Environ Sci Technol 38:112
46. Hale RC, La Guardia MJ, Harvey E, Gaylor MO, Mainor TM (2006) Chemosphere 64:181

Illicit Drugs Along the Ebro River Basin: Occurrence in Surface and Wastewater and Derived Consumption Estimations

Cristina Postigo, Miren López de Alda, and Damià Barceló

Abstract The study of illicit drugs and metabolites in the aquatic environment has a double objective: (1) to investigate the environmental concentrations of this group of emerging contaminants in order to assess their potential ecotoxicological risk and (2) to estimate drug abuse by the population at the community level. The present work reports on the occurrence of illicit drugs and metabolites in waste and surface waters collected along the Ebro River basin (NE Spain) and evaluates the contribution of discharged treated wastewaters to the presence of these compounds in river waters. Concentrations of drug residues in influent wastewaters were used to back calculate illicit drug use in the areas served by the investigated wastewater treatment facilities.

Cocaine, benzoylecgonine, ephedrine, and ecstasy were identified as the most ubiquitous and abundant compounds in the area of study. Heroin, 6-acetyl morphine, lysergic compounds, and Δ^9-tetrahydrocannabinol (THC) were the compounds less frequently identified in the investigated samples. Overall, removal of illicit drugs and metabolites in the investigated wastewater treatment plants was satisfactory. However, ecstasy, methamphetamine, nor-LSD, and 11-nor-9-carboxy-THC were occasionally found at higher concentrations in effluent than in influent waters. Dilution of discharged treated waters resulted in total levels of illicit drugs and metabolites in surface waters at the low ng/L range. Estimates of illicit drug use pointed out cocaine as the most consumed drug in the area of study, followed by cannabis, amphetamine, heroin, ecstasy, and methamphetamine.

C. Postigo and M.L. de Alda (✉)
Institute of Environmental Assessment and Water Research, IDAEA-CSIC, Jordi Girona 18-26, 08034 Barcelona, Spain
e-mail: mlaqam@cid.csic.es

D. Barceló
Institute of Environmental Assessment and Water Research, IDAEA-CSIC, Jordi Girona 18-26, 08034 Barcelona, Spain
Catalan Institute for Water Research (ICRA), Emili Grahit 101, Edifici H2O, Parc Científic i Tecnològic de la Universitat de Girona, 17003 Girona, Spain

D. Barceló and M. Petrovic (eds.), *The Ebro River Basin*,
Hdb Env Chem (2011) 13: 189–208, DOI 10.1007/698_2010_79,

Keywords Ebro River, Illicit drugs, Liquid chromatography, Mass spectrometry, Surface water, Wastewater, Water analysis

Contents

1 Introduction

According to the World Drug Report 2009, between 172 and 250 million people consume illicit drugs annually worldwide [1]. Up to 21 million people aged 15–64 were estimated to have used opiates and cocaine globally at least once in 2007, between 16 and 51 million people used amphetamine-group substances, and cannabis users ranged from 143 to 190 million persons [1]. These high figures of consumption are sustained by annual high volumes of production, e.g., up to 774 tons of heroin and morphine, more than 800 tons of cocaine, between 230 and 640 tons of amphetamines and methamphetamines, up to 137 tons of ecstasy-like drugs, and between 13,300 and 66,100 tons of cannabis herb per year [1].

The high amounts in which these substances are consumed and produced have conferred illicit drugs and their human metabolites a pseudo-persistent character in the environment. Like over-the-counter and prescribed pharmaceuticals, illicit drugs are metabolized after consumption and different proportions of the parent compound and metabolic by-products are excreted via urine or feces and flushed into the sewage system toward wastewater treatment facilities, if existing. However, these substances are poorly or incompletely removed by conventional wastewater treatment processes [2, 3]. As a consequence, illicit drugs and metabolites are continuously introduced via wastewater treatment plant (WWTP) effluents into the aquatic media. In fact, this constitutes the main route of entry of this type of compounds into the environment as direct disposal is unlikely.

Illicit drugs and metabolites have been recognized as a group of emerging contaminants of concern [4] with potential, still unknown, ecotoxicological effects.

Several analytical methodologies have been developed and implemented in the past few years to determine the levels of these substances in diverse aqueous matrices, e.g., wastewater and surface water [5–21]. Studies in this line have been performed so far in the United States [5, 10, 15, 22, 23] and in several European countries such as Italy [9, 18, 21, 24, 25], Switzerland [9, 25], the United Kingdom[21, 25–28], Belgium [29, 30], Ireland [8], Germany [14], and Spain [6, 7, 12, 13, 19, 31–34].

The study of the environmental occurrence and fate of illicit drugs and metabolites provides the required information to further investigate their potential toxicological or cumulative effects on the ecosystems. Besides, drug residues in environmental waters reflect illicit drug use in any investigated area [23, 25, 28, 35, 36]. Estimation of illicit drug use at the community level by means of a sewage epidemiology approach has been presented as a powerful tool to complement the official methodologies commonly used to estimate drug use by the population, e.g., statistics, and medical and criminal indices [25, 35, 37]. Contrary to traditional methods, this approach provides near real-time information in an economical and anonymous way, which makes it appropriate to allocate, implement, and evaluate prevention measures to reduce illicit drug use.

This chapter presents the results from a monitoring program carried out in the Ebro River basin to evaluate the occurrence of illicit drugs and metabolites in this area. Among the aims of this study were also to investigate the removal efficiency of some selected WWTPs located along the basin and to assess the contribution of WWTPs effluents to the levels of these emerging contaminants in the receiving river waters. Moreover, the sewage epidemiology approach was applied to estimate collective use of illicit drug in various main cities of the basin and figures obtained were compared with those obtained in similar studies performed in Spain and other European countries and with those provided by official estimates.

2 Sample Collection and Analysis

2.1 Sample Collection

Water sampling took place in October 2007 and July 2008. Location of the sampling sites along the river basin is depicted in the scheme shown in Fig. 1. Selected WWTPs render service to urban areas with populations above 40,000 inhabitants (see Fig. 1) and all together serve about 1.5 million people, which represent half of the total population living in the basin. Influent and effluent waters from seven WWTPs were collected as 24-h composite samples. In the investigated WWTPs, a primary treatment or primary settling usually precedes a secondary treatment based in most cases in conventional activated sludge (CAS) systems. Only WWTP4 is equipped with a different secondary treatment process that consists of biological filters. Hydraulic retention times in the selected WWTPs range from 7 to 10 h (WWTP2, WWTP5, and WWTP6) to more than 30 h (WWTP1,

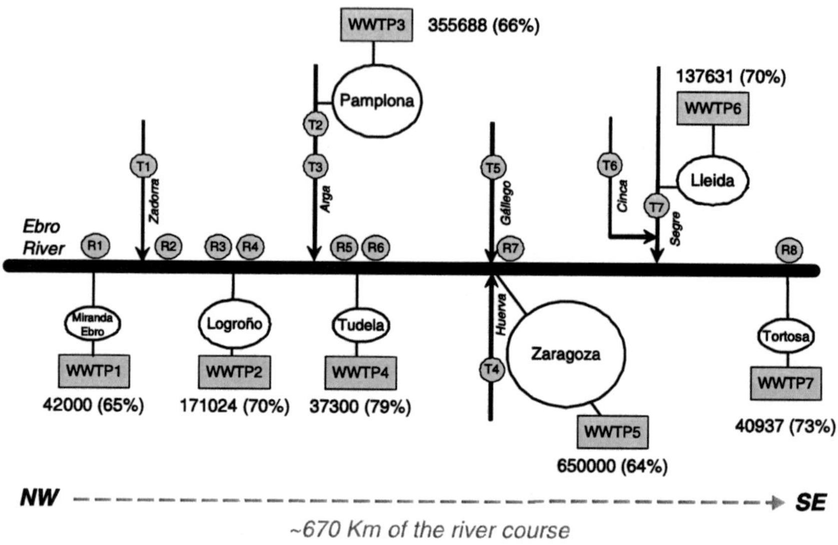

Fig. 1 Scheme of sampling type and location. *Circles* surrounding the name of the investigated urban areas depict the relative size of the population served by the corresponding WWTP. The figures provided in the scheme indicate total inhabitants of the served municipalities according to the INEbase and the percentage of inhabitants aging 15–64 in each case. R: Ebro River, T: tributary

WWTP7). On the other hand, solids retention times are between 3 and 4 days (WWTP5, WWTP6) and 19 days (WWTP7).

River water samples were collected as grab samples from 15 different locations along the basin. As it is shown in Fig. 1, only eight out of the 15 river sampling sites are located in the Ebro River itself (R1–R8, the last one was sampled only in 2008), whereas the rest are on tributaries: Zadorra River (T1, sampled only in 2008), Arga River (T2 and T3), Huerva River (T4), Gállego River (T5), Cinca River (T6) and Segre River (T7). Waters downstream the discharge point of the WWTPs (R2, R4, R6, R7, T2, T7, R8) were collected in order to assess the impact of WWTP effluents in receiving waters. However, due to the inaccessibility of the river edge in some cases, distances between the discharge point and the location sampled downstream are variable.

All water samples were collected in amber-polyethylene terephtalate (PET) bottles and were kept at 4°C during shipment. Upon reception in the laboratory, samples were vacuum filtered through 1 μm glass fiber filters, followed by 0.45 μm nylon membrane filters, and were stored in the dark at −20°C until analysis.

2.2 Analytical Protocol

Up to 16 illicit drugs and metabolites belonging to five different chemical classes (cocainics, cannabinoids, opioids, amphetamine-like, and lysergic compounds) were analyzed in the collected samples. High purity (>97%) standard solutions of the target compounds and deuterated analogs used as surrogate standards (SS) for quantification were obtained from Cerilliant (Round Rock, TX) as solutions in methanol or acetonitrile at concentrations of 1 or 0.1 mg/mL. Individual stock solutions and working standard mixtures were prepared by appropriate dilution of each analyte solution with methanol and stored in the dark at −20°C. The standard mixtures (concentration of surrogate standards 20 ng/mL) were used to prepare the aqueous calibration solutions.

The analyses of illicit drugs and metabolites in the sampled waters were performed following a previously described and validated fully automated method based on on-line solid phase extraction–liquid chromatography–tandem mass spectrometry (on-line SPE–LC–MS/MS) [19].

Sample preconcentration was performed by means of an automated on-line SPE sample processor Prospekt-2 (Spark Holland, Emmen, The Netherlands). Oasis HLB cartridges (Waters, Barcelona, Spain) were used to preconcentrate cannabinoids present in the water samples whereas isolation of the rest of the compounds was done in PLRPs cartridges (Spark Holland). Before extraction, influent samples were diluted with HPLC water (1:9, v/v) to reduce matrix interferences and to fit some analyte concentrations, e.g., cocaine (CO) and benzoylecgonine (BE), within the linear calibration range. A sample volume of 5 mL was spiked with the internal standard mixture (at 20 ng/L) in order to correct for potential losses during the analytical procedure, as well as for matrix effects. Elution of the analytes to the LC system was done with the chromatographic mobile phase.

LC–MS/MS analyses were performed with a system consisting of an Agilent HP 1100 pump (Agilent Technologies, Palo Alto, CA) connected in series with an hybrid triple quadrupole-linear ion trap mass spectrometer 4000QTrap (Applied Sciex, Foster City, CA). Chromatographic separation was achieved with a Purospher Star RP-18-end-capped column (125 × 2 mm, 5 μm) preceded by a guard column of the same packing material (Merck, Darmstadt, Germany) and a mobile phase consisting of gradient acetonitrile/water at a constant flow rate of 300 μl/min. Ionization of the compounds was performed with an electrospray (ESI) source. Cannabinoids were analyzed in the negative ionization (NI) mode whereas the rest of the analytes were ionized in the positive (PI) mode. Data acquisition was performed in the selected reaction monitoring (SRM) mode. Two compound specific SRM transitions (precursor ion → product ion) were monitored to achieve four identification points per target analyte (2002/657/EC) [38]. More details on the experimental conditions of the analytical methodology are described in Postigo et al. [19].

3 Illicit Drugs and Metabolites in Wastewater Treatment Plants

3.1 *Occurrence in Wastewaters*

The frequency of detection of each investigated analyte in the different aqueous matrices and the levels at which the studied classes of illicit drugs and metabolites

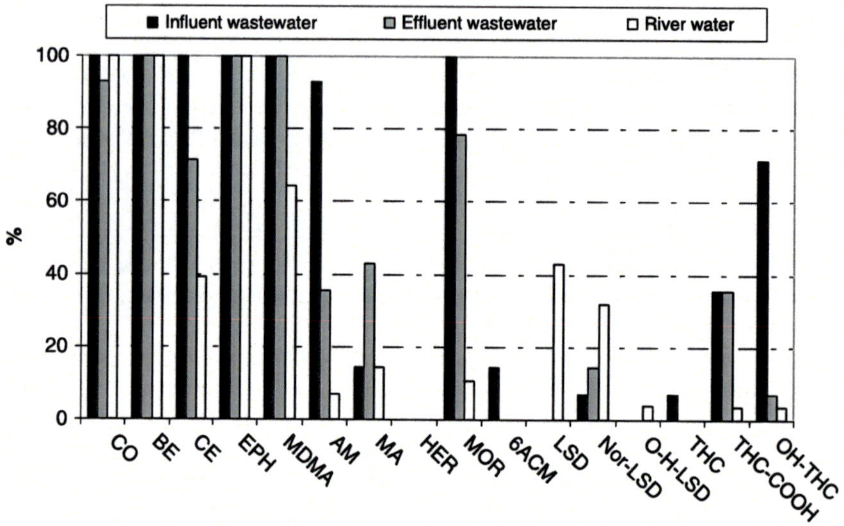

Fig. 2 Frequency of appearance of each investigated illicit drug and metabolite in the various studied aqueous matrices

were present at the inlet and the outlet of the selected WWTPs are shown in Figs. 2 and 3, respectively.

As it can be observed in Fig. 2, three out of the 16 investigated compounds, namely, heroin, lysergic acid diethylamide (LSD), and its metabolite 2-oxo, 3-hydroxy-LSD (O-H-LSD), were not detected in any wastewater sample. Two other target analytes, 6-acetyl morphine (6ACM) and Δ^9-tetrahydrocannabinol (THC), were only present in influent wastewaters and with low detection frequencies. The most ubiquitous compounds, present in all influent and effluent wastewater samples analyzed, were the cocaine metabolite benzoylecgonine, and the amphetamine-like compounds ephedrine (EPH) and 3,4-methylenedioxymethamphetamine (MDMA or ecstasy). Cocaine, cocaethylene (CE, transesterification product of cocaine formed after the joint consumption of cocaine and ethanol), and morphine (MOR) were detected in all influent, but not in all effluent wastewaters (see Fig. 2).

As expected, levels of illicit drugs and metabolites were higher in influent than in effluent waters. Total drug residue levels in influent waters were between 1.5 and 4.7 μg/L, being CO, EPH, and BE the main contributors to the total drug residues amount. Overall, cumulative levels in waters after treatment were one order of magnitude below those determined in the WWTPs inlet and they usually did not surpass 500 ng/L. Only the waters discharged during 2008 by WWTP4 (1,024 ng/L) and WWTP5 (702 ng/L) presented total drug residue values above this level. The main drug classes that remain in the water after treatment were cocainics and amphetamine-like compounds as it can be observed in Fig. 3.

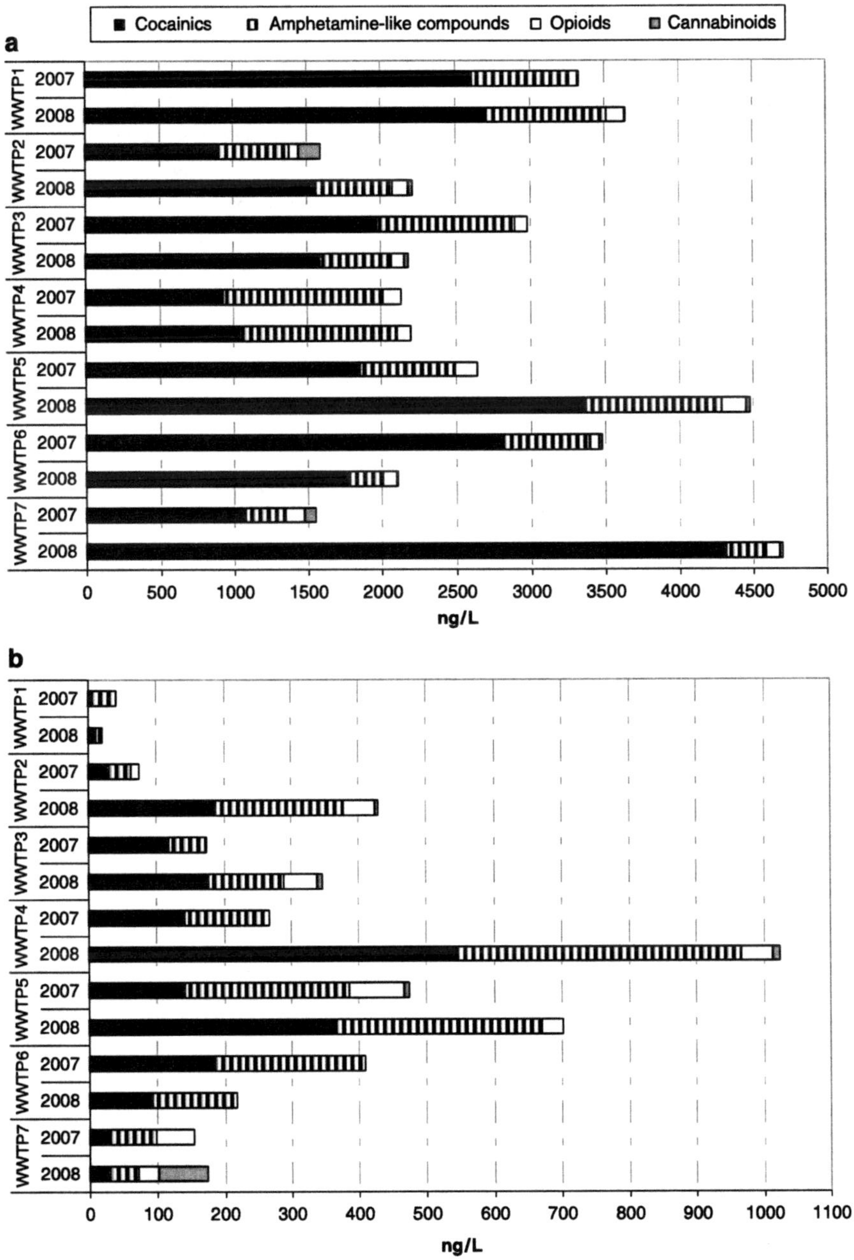

Fig. 3 Total levels of each investigated chemical class of illicit drugs and metabolites in the sampled wastewater: (**a**) Influent wastewaters, (**b**) Effluent wastewaters

Figure 4 shows the concentration range determined for each investigated drug and metabolite in the surface and wastewaters collected in the Ebro River basin, and its comparison with the drug levels found in similar studies carried out in other locations of Spain [6, 7, 12, 13, 19, 31–33] and in various countries of the American [5, 10, 15, 22, 23] and the European continents [8, 9, 11, 14, 16–18, 21, 24–30]. Drug residue levels found in influent and effluent wastewaters from the Ebro River basin were mostly between the minimum and maximum concentrations described for these compounds in the peer-reviewed literature. Only, few compounds, e.g.,

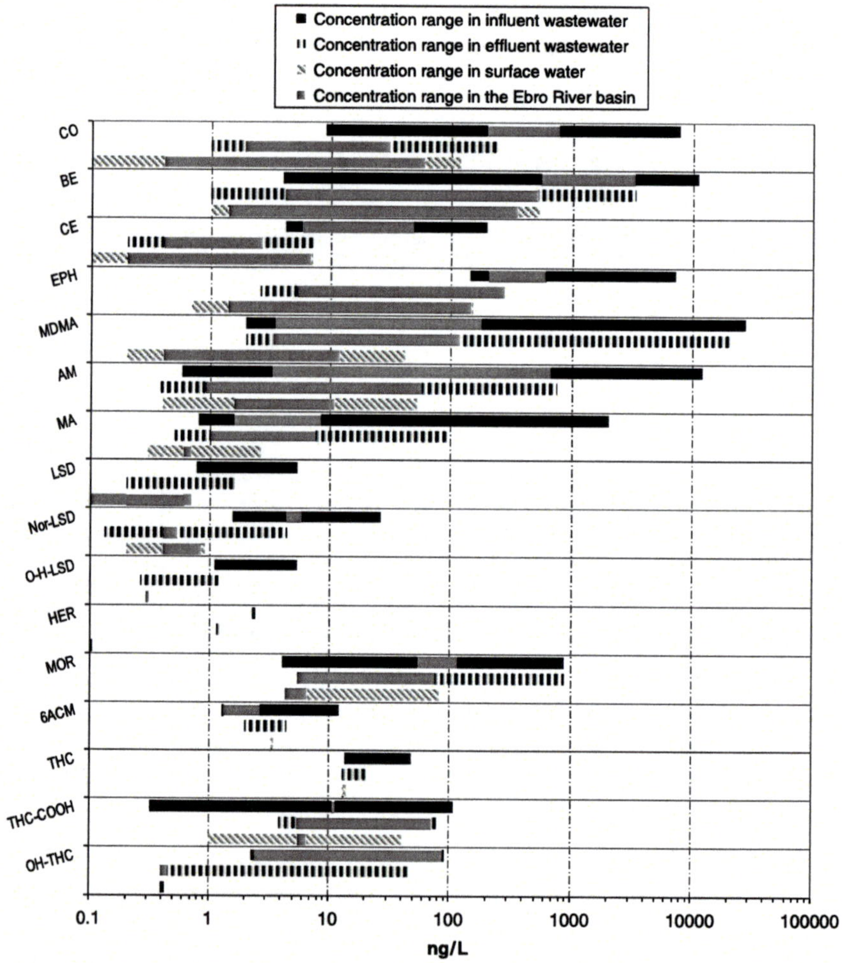

Fig. 4 Range of concentrations determined for each analyte in the different investigated aqueous matrices. For those analytes detected only once in a specific matrix (e.g., nor-LSD and cannabinoids), the concentration range expands from the method limit of determination to the only positive value reported

OH-THC, MOR, and 6ACM presented concentrations below or above those previously reported.

3.2 Removal Efficiency of Illicit Drugs and Metabolites

Levels found in WWTPs effluents were considered to reflect the removal efficiency of the investigated WWTPs. Thus, considering the amount of each compound in the effluent of a WWTP, its removal was expressed as the percentage of analyte lost during the water treatment process, relative to the influent concentration. The removal efficiency of the total spectrum of illicit drugs and metabolites investigated in the different sampled WWTPs in the Ebro River basin ranged from 45 to 94%. The poorest total removals were observed in the WWTPs 4 and 5 that use biological filters and a CAS system with a very low solid retention time (3–4 h) as secondary treatment processes, respectively.

Figure 5 shows the average removal observed for each compound in the WWTPs investigated in the Ebro River basin and its comparison with the removal data reported in the peer-reviewed literature. In agreement with the results of previous works [6–9, 12, 19, 27, 29, 31, 32], most of them performed in WWTPs equipped

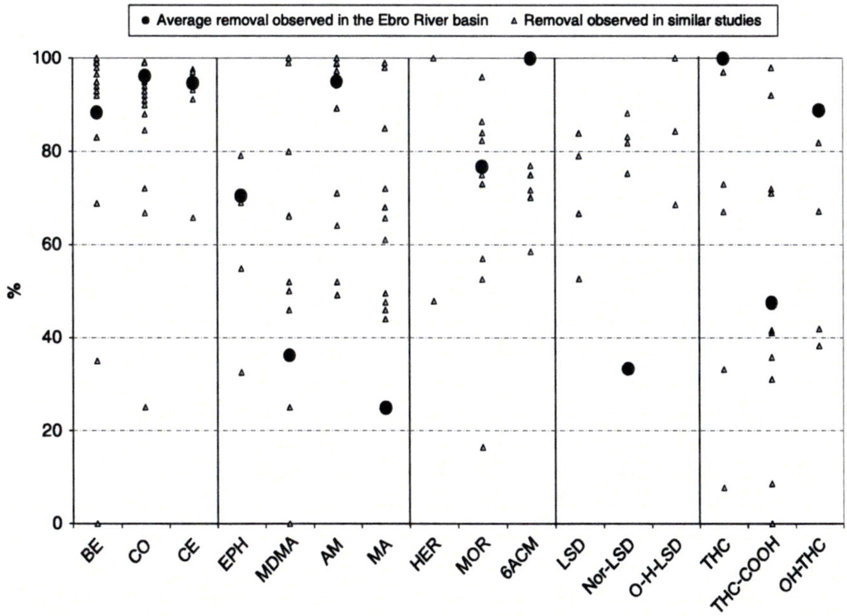

Fig. 5 Average removal of the investigated compounds in the investigated WWTPs of the Ebro River basin

with CAS systems as secondary treatment, the chemical class most efficiently removed in the Ebro River basin was the cocainic (93% on average).

Contrary to other studies, 6ACM and THC presented 100% removal in this study, but due to their low frequency of detection in raw wastewaters and their absence in treated wastewaters (see Figs. 2 and 5). This fact also affects the overall satisfactory removal of opioids (88%) and cannabinoids (79%). However, removal of morphine, the opioid most frequently detected in wastewaters ranged between 46 and 100%, variability also observed by other authors. In the case of cannabinoids, removals reported are very diverse. The average poor removal observed for THC-COOH in the investigated WWTPs from the Ebro River basin (48%) results from occasionally higher concentrations of this analyte in treated wastewaters compared to raw wastewaters. This finding has also been reported in other studies [7, 19].

In agreement with previous works, elimination of amphetamine-like compounds was diverse. Amphetamine and ephedrine are the best removed amphetamine-like compounds in the Ebro River basin. On the other hand, MDMA and methamphetamine were poorly removed compared to other studies as they were found randomly at higher concentrations in effluents than in influents. Similar to these compounds and in contrast to previous studies, the only lysergic analyte determined in wastewaters, nor-LSD, was poorly or not eliminated during wastewater treatment processes (see Fig. 5).

4 Illicit Drugs and Metabolites in Surface Water

The most ubiquitous, and also abundant, compounds in the analyzed surface waters were CO, BE, and EPH. On the other hand, HER, 6ACM, and THC were not positively identified in any of the investigated river water samples. The rest of the target analytes, except lysergic compounds, presented the lowest frequency of appearance in this aqueous matrix (see Fig. 2). The levels at which the investigated classes of illicit drugs and metabolites were detected in surface waters are shown in Fig. 6.

Total levels of illicit drugs and metabolites in surface waters reached up to 578 ng/L (T4), but they were usually lower than 100 ng/L, or even 40 ng/L in most of the sampling locations. Levels of illicit drugs and metabolites in surface waters are directly related to the concentration of these substances in the effluent treated waters, the flow of the discharged treated waters, and that of the receiving surface waters. According to the flows discharged by the WWTPs and those measured in the rivers downstream the WWTPs outlets [39], dilution factors in the investigated area ranged from 3 (T2) to 1,785 (R8). The lowest dilution factors were observed in the tributaries whereas the highest ones took place, as expected, in the last stretch of the Ebro River, close to its mouth. Levels of illicit drugs and metabolites in the last stretch of the Ebro River are, however, not significantly lower than upstream as also the amount of WWTPs discharges increases downstream.

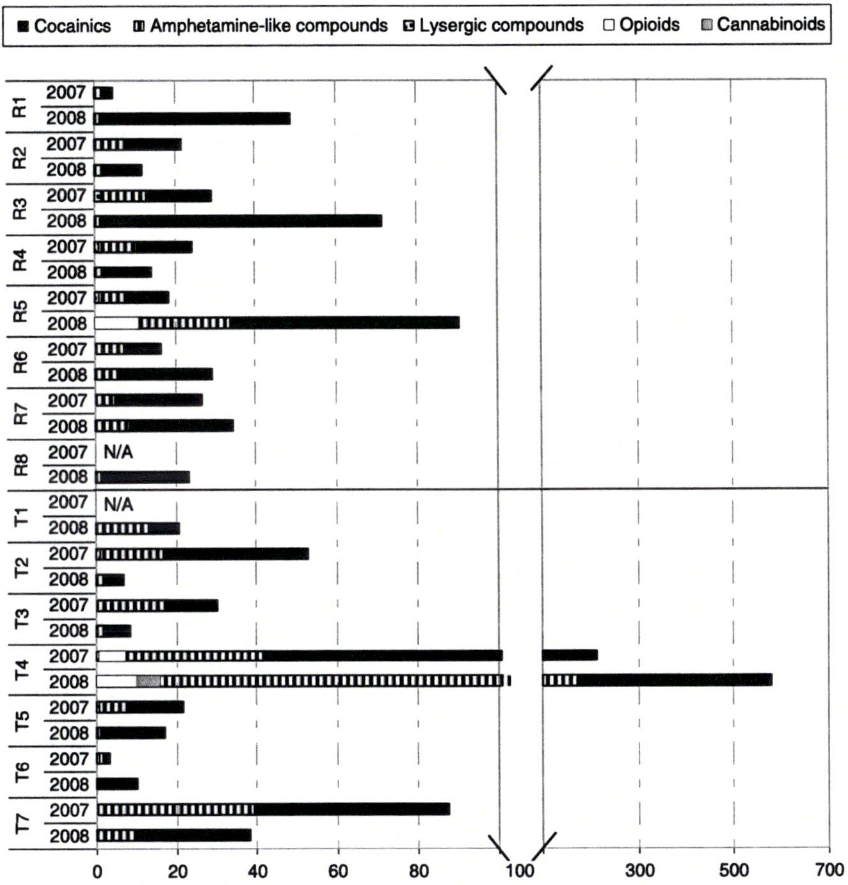

Fig. 6 Total levels of each investigated chemical class of illicit drugs and metabolites in the sampled surface waters

Overall, levels discharged by WWTPs were higher in the sampling campaign carried out in 2008 than that in 2007. However, levels found in the surface waters were higher in 2007 than that in 2008, as a result of higher river flow rates during the sampling campaign of 2008 and, hence, higher dilution factors. In fact, the rainy episodes that took place during the second half of the hydrogeological year 2007–2008 ended a 2-year drought period and helped to recover the water reserve in the basin.

Concentrations of each target analyte in surface waters are in agreement with those reported in previous studies [7, 8, 11, 13, 14, 17, 18, 21, 26, 29–31, 33] (see Fig. 4). In the case of ephedrine and lysergic compounds, the concentrations found are the highest levels reported so far in surface waters. Concerning lysergic compounds, despite the fact that they were found in the pg/L level (below 1 ng/L), these concentrations are too high to be real, taking into account the doses in which

LSD is consumed (μg) and the low excretion rates reported for LSD (1%) and its metabolites (1.2% for nor-LSD and 2–25% for O-H-LSD) in the peer-reviewed literature [40–42]. Moreover, these compounds were rarely present in the investigated wastewaters and O-H-LSD, the LSD metabolite excreted at the highest rate, was absent in the surface waters. All this reasoning leads to suspect the existence of a potential interference in the analyses.

5 Estimation of Illicit Drug Use in the Ebro River Basin

5.1 *Sewage Epidemiology Approach*

Back-calculation of illicit drug use is straightforward from the drug residues found in influent sewage waters. The sewage epidemiology approach was proposed by Daughton in 2001 [43] and put into practice for the first time by Zuccato et al. in 2005 [24]. Briefly, levels of consumption indicators (ng/L) are normalized with the flow entering the WWTP (m^3/day) and the people served by the wastewater treatment facility (inhabitants) [44] and corrected by the corresponding average excretion rate and the molar mass ratio between the parent drug and the metabolite, whenever the consumption indicator used is not the original drug itself. Table 1 summarizes the drug residues selected as consumption indicators of illicit drug use in this study, as well as the average excretion rates, the correction factors, and the average drug doses applied in each case [35].

The consumption indicator is the metabolic byproduct excreted at the highest rate. It may be a metabolite, as it is the case for cocaine (BE) and heroin (MOR), or the drug itself, as it is the case of amphetamine-like compounds. THC, the most psychoactive cannabinoid of the cannabis herb, is highly metabolized before excretion, thus, the consumption indicator selected (THC-COOH) presents an excretion rate of 0.6%. Despite the fact that OH-THC presents a slightly higher excretion rate (2%), this analyte was not selected to back calculate cannabis use due

Table 1 Consumption indicators used to back calculate drug use and average excretion rates, correction factors, and average doses applied in each case

	Consumption indicator	Average excretion rate (%)	Correction factor	Average dose (mg)
Cocaine	Benzoylecgonine	45	2.3	100
Ephedrine	Ephedrine	75	1.3	25
Amphetamine	Amphetamine	30	3.3	50
Methamphetamine	Methamphetamine	43	2.3	40
MDMA	MDMA	65	1.5	100
Heroin	Morphine	42	3.1	30
Cannabis	THC-COOH	0.6	152	125

to its high instability and to achieve comparable results with works already published in this line [25, 31].

In the case of heroin, the indicator used is nonexclusive of heroin consumption, as it is also excreted to a high extent after administration of therapeutic morphine and in minor proportions after use of other opioids, such as codeine, pholcodine, and ethylmorphine. In order to obtain comparable results to previous studies [25, 31], an average medical administration of morphine of 10 mg/capita/year [45] and an excretion rate of 85% of morphine after therapeutic use [25, 31] were considered in the calculations.

Legal and illegal use of amphetamine-like compounds could not be differentiated as the analytical methodology used in the water analyses do not distinguish the enantiomeric forms of these compounds. Usually, the (R)-(−) enantiomer presents a lower stimulant activity and thus, it presents some legal applications. On the other hand, the illegal (S)-(+) enantiomer presents a high stimulating activity and thus, it is the one with a higher presence in the drug black market.

Consumption was normalized to the adult population (aging 15–64) in all cases, but for MA and MDMA that were normalized to the young adult population (aging 15–34), potential consumers of these type of drugs.

5.2 *Drug Usage in the Main Urban Areas of the Ebro River Basin*

Estimates of the average use of each drug in the main urban areas of the Ebro River basin, expressed as g/day/1,000 inhabitants and doses/day/1,000 inhabitants, are shown in Fig. 7.

Cocaine use in the basin was estimated to vary between 1 and 2.4 g/day/1,000 inh. aging 15–64, which is equivalent to 10–24 doses of cocaine/day/1,000 inh. aging 15–64. This yield an average cocaine consumption in the whole studied area of 1.8 g (or 18 doses)/day/1,000 inh. aging 15–64. A similar figure for cocaine consumption, 1.4 g/day/1,000 inh. aging 15–64, was obtained by means of the sewage epidemiology approach in Catalonia (NE Spain) [32]. Despite the fact that most previous studies performed in this line in other European locations do not specify the age segments used to normalize data, and thus, they are not fully comparable, cocaine consumption in the basin was higher than that reported for the cities of Milan (Italy), Lugano (Switzerland), and London (United Kingdom) (below 1 g/day/1,000 people) [25] and that observed throughout Belgium (rarely surpassing 1 g/day/1,000 people), but lower than the maximum reported for South Wales (3.7/day/1,000 people) [28].

The average use of amphetamine was below 0.30 g/day/1,000 inh. aging 15–64 (10 doses/day/1,000 inh. aging 15–64) in all locations but in that served by WWTP4, where amphetamine use was significantly higher, reaching 1.75 g (58 doses)/day/1,000 inh. aging 15–64. The reported average use of amphetamine in the area of Catalonia is significantly lower (0.08 g/day/1,000 inh. aging 15–64) than that

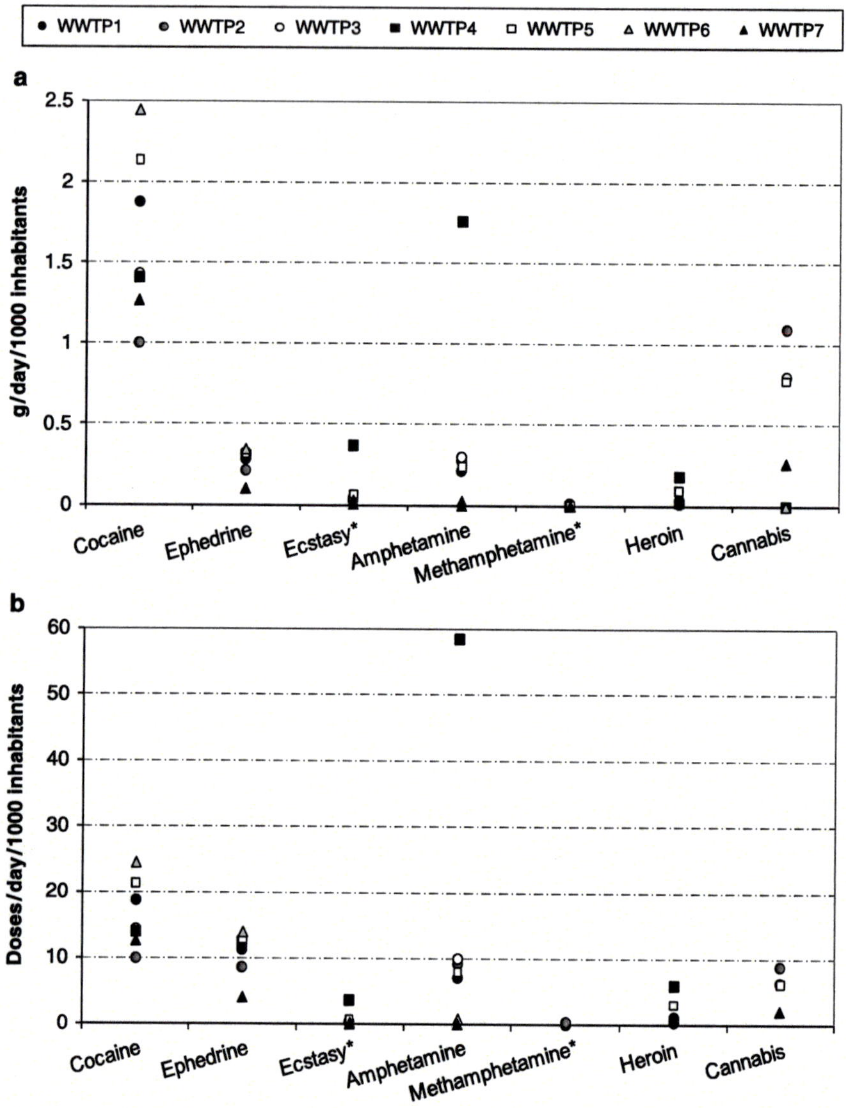

Fig. 7 Average illicit drug use in the areas served by the investigated WWTPs. *Data are normalized by 1,000 inhabitants aged 15–64, but in the case of methamphetamine and ecstasy that were normalized by 1,000 inhabitants aged 15–34

observed in the Ebro River basin [32]. Amphetamine use in the Ebro River basin also surpasses those estimated for the cities of Lugano, where this drug was not present in the influent sewage waters, Milan (below 0.01 g/day/1,000 people), and London (0.08 g/day/1,000 people) [25]. However, it is one order of magnitude below the amphetamine use reported for South Wales (2.50 g/day/1,000 people) [28].

Similar to amphetamine use, consumption of ecstasy was observed to be similar in all the studied locations of the Ebro River basin (0.01–0.07 g/day/1,000 inh. aging 15–34), but in that served by the WWTP4 (0.37 g/day/1,000 inh. aging 15–34). Overall, the average use in the basin is lower than that reported by Huerta–Fontela for Catalonia (0.40 g/day/1,000 inh. aging 15–34) [32] and those observed for Milan and Lugano (below 0.01 g/day/1,000 people) [25].

Methamphetamine residues were found only in two locations (WWTP1 and WWTP2), yielding an average methamphetamine use of around 0.001 g (0.03 doses) and 0.01 g (0.4 doses)/day/1,000 inh. aging 15–34, respectively. These figures are lower than the ones reported for the region of Catalonia (0.02 mg/day/1,000 inh. aging 15–34) [32] and for the cities of Milan (0.01 g/day/1,000 people) and London (0.006 g/day/1,000 people) [25].

The average use of ephedrine, a drug mainly used with therapeutic purposes to treat rhinitis, sinusitis, and depressive states among others, ranged between 0.10 and 0.35 g/day/1,000 inh. aging 15–64, which is equivalent to 4 and 14 doses/day/1,000 inh. aging 15–64, respectively.

Consumption of heroin in the basin was confirmed by the presence of its exclusive minor metabolite 6ACM in influent sewage waters; however, its use was calculated from morphine loads after subtracting the estimated therapeutic use of morphine. The average use of heroin in the basin ranged from 0.01 to 0.18 g/day/1,000 inh. aging 15–64 (0.4–6 doses/day/1,000 inh. aging 15–64). The maximum use of heroin in the Ebro River basin is similar to the average reported for Catalonia (0.14 g/day/1,000 people) [31] and higher than that estimated for Milan (0.07 g/day/1,000 people) and Lugano (0.10 g/day/1,000 people), but lower than that observed in London (0.21 g/day/1,000 people) [25].

Average consumption of cannabis ranged between 0.27 and 1 g/day/1,000 inh. aging 15–64 (2–9 doses/day/1,000 inh. aging 15–64), which is much lower than that reported for the Catalonia area (3.5 g/day/1,000 people) [31] and for the cities of Milan (3 g/day/1,000 people), Lugano (6.5 g/day/1,000 people), and London (7.5 g/day/1,000 people) [25].

Considering that the consumption pattern observed in the area of the Ebro River basin studied, which covers about half of the population living in the basin, could be representative of the whole Spanish country, the estimated average consumption data were used to calculate the annual consumption of each drug in the whole basin and in the Spanish territory. According to the extrapolated figures, which are shown in Table 2, around 21 tons of cocaine, 8 tons of cannabis, 3 tons of amphetamine and ephedrine, 300 kg of ecstasy and heroin, and 7.5 kg of methamphetamine are annually consumed in Spain. These amounts would move in the black market for around 1,100 million Euros.

According to the estimates performed, cocaine is the most consumed drug, followed by cannabis, amphetamine, heroin, MDMA, and methamphetamine. On the contrary, official national annual prevalence data points out cannabis as the most abused illicit drug, followed by cocaine, MDMA, amphetamines, and heroin. This suggests that the drug consumption pattern in the studied areas of the Ebro River

Table 2 Main figures of drug use in the studied area of the Ebro River basin and extrapolation to the Spanish territory

	Average consumption rate in the studied area (doses/day/1,000 inh. 15–64)	Total amount (kg) consumed per year in the studied area	Total amount (tons) consumed per year in Spain (extrapolation)	€/Year[b] (Spain)
Cocaine	17.93	621.14	20.86	696,001,368
Amphetamine	9.19	95.51	3.21	57,003,875
Ecstasy[a]	0.60	8.32	0.29	11,211,406
Methamphetamine[a]	0.05	0.22	7.58×10^{-3}	134,704
Ephedrine	12.01	104.02	3.49	68,470,173
Heroin	0.79	8.26	0.28	10,473,941
Cannabis	5.44	235.52	7.91	263,906,894
Total		1072.99	36.04	1,107,202,361

[a]Referred to the young adult population (aging 15–34): 379,615 and 13,044,727 young adults in the studied area and in the Spanish territory, respectively

[b]Based on the drug prices reported by the Spanish Drug Observatory (Observatorio Español sobre Drogas, OED): 33,365 €/kg of cocaine, 37,756 €/kg of heroin, 1,283 €/kg of cannabis (hachis), 9.8€/250 mg of ecstasy (MDMA), 4.9€/25 mg (1 unit) of amphetaminic pharmaceuticals e.g., ephedrine, 17,771 €/kg of speed (methampehtamine and amphetamine) – [46]

basin is different from that in Spain, but the presence of biases in the official methods, and also in the sewage epidemiology approach, cannot be discarded either.

Concerning the potential biases of the methodology applied in this study, amphetamine use may indeed be overestimated because the enantiomeric forms and thus, legal and illegal uses are not differentiated, and because it is also a metabolite formed after administration of other illegal substances such as fenethylline and fenproporex. Also heroin use may be overestimated if the considered use of therapeutic morphine is lower than real and because also other sources of morphine, e.g., codeine, ethylmorphine, and pholcodine, were not taken into account. On the other hand, cannabis use may be underestimated as the cannabinoids monitored are highly lypophilic substances (log Kow = 5.5–7) and thus, they partition to a lower extent in the aqueous phase and preferably sorb on organic matter and suspended solids. A higher figure for cannabis use would be likely obtained if the concentrations of cannabinoids would have been also measured in sewage suspended solid matter.

6 Conclusions

The present work reports on the presence of illicit drugs and metabolites in waste and surface waters of the Ebro River basin. In agreement to previous works done in this line in other locations of Spain and in diverse European countries, the most abundant and ubiquitous compounds in waters were cocaine and its major metabolite benzoylecgonine, and the amphetamine-like compounds ephedrine and ecstasy. Removal of these compounds during wastewater treatment processes was considered satisfactory for most compounds, but not enough to avoid the presence of these compounds at the ng/L level in the Ebro River and some of its tributaries. The effects that the presence of such concentrations in natural aquatic ecosystems may have on the biota continuously exposed to them are still unknown. Thus, the production of data on the ecotoxicity of these biologically active substances is crucial for the environmental risk assessment of this group of emerging contaminants.

Illicit drug consumption in the areas served by the investigated WWTPs was back-calculated from the drug residues found in the influent wastewaters. In the light of the results obtained, cocaine is the most abused drug, followed by cannabis, amphetamine, heroin, MDMA, and methamphetamine. This profile slightly differs from national official estimates. Deviations may result from a different drug consumption pattern in the studied area compared to the Spanish territory, or from potential biases in the approach used in this study or in official estimates.

Estimations of illicit drug consumption through the applied methodology could be refined by increasing the knowledge on local drug use patterns and on the metabolism and the environmental fate of drugs. The sewage epidemiology approach constitutes a powerful tool to fight drug abuse as it allows obtaining realistic, reproducible, and what is highly important, near real-time data on drug abuse by means of an anonymous, objective, and economical way.

Acknowledgments This work has been supported by the EU Project MODELKEY [GOCE 511237] and by the Spanish Ministry of Science and Innovation through the projects CEMAGUA (CGL2007-64551/HID) and SCARCE (Consolider-Ingenio 2010 CSD2009-00065). It reflects the authors' views only. The EU is not liable for any use that may be made of the information contained in it. Cristina Postigo acknowledges the European Social Fund and AGAUR (Generalitat de Catalunya, Spain) for their economical support through the FI predoctoral grant.

References

1. United Nations Office on Drugs and Crime, UNODC. World drug report 2009. Available at: http://www.unodc.org/documents/wdr/WDR_2009/WDR2009_eng_web.pdf
2. Petrovic M, López De Alda MJ, Díaz-Cruz S, Postigo C, Radjenovic J, Gros M, Barceló D (2009) Fate and removal of pharmaceuticals and illicit drugs in conventional and MBR wastewater treatment plants and by riverbank filtration. Philos Trans R Soc A 367:3979–4003
3. Petrovic M, Radjenovic J, Postigo C, Kuster M, Farre M, López de Alda MJ, Barceló D (2008) Emerging contaminants in waste waters: sources and occurrence. In: Barcelo D, Petrovic M (eds) Handbook of environmental chemistry, vol 5. Water pollution. p 1–35
4. Richardson SD (2008) Environmental mass spectrometry: emerging contaminants and current issues. Anal Chem 80:4373–4402
5. Bartelt-Hunt SL, Snow DD, Damon T, Shockley J, Hoagland K (2009) The occurrence of illicit and therapeutic pharmaceuticals in wastewater effluent and surface waters in Nebraska. Environ Pollut 157:786–791
6. Bijlsma L, Sancho JV, Pitarch E, Ibáñez M, Hernández F (2009) Simultaneous ultra-high-pressure liquid chromatography–tandem mass spectrometry determination of amphetamine and amphetamine-like stimulants, cocaine and its metabolites, and a cannabis metabolite in surface water and urban wastewater. J Chromatogr A 1216:3078–3089
7. Boleda MR, Galcerán MT, Ventura F (2007) Trace determination of cannabinoids and opiates in wastewater and surface waters by ultra-performance liquid chromatography–tandem mass spectrometry. J Chromatogr A 1175:38–48

8. Bones J, Thomas KV, Paull B (2007) Using environmental analytical data to estimate levels of community consumption of illicit drugs and abused pharmaceuticals. J Environ Monit 9:701–707
9. Castiglioni S, Zuccato E, Crisci E, Chiabrando C, Fanelli R, Bagnati R (2006) Identification and measurement of illicit drugs and their metabolites in urban wastewater by liquid chromatography–tandem mass spectrometry. Anal Chem 78:8421–8429
10. Chiaia AC, Banta-Green C, Field J (2008) Eliminating solid phase extraction with large-volume injection LC/MS/MS: analysis of illicit and legal drugs and human urine indicators in US wastewaters. Environ Sci Technol 42:8841–8848
11. Gheorghe A, Van Nuijs A, Pecceu B, Bervoets L, Jorens PG, Blust R, Neels H, Covaci A (2008) Analysis of cocaine and its principal metabolites in waste and surface water using solid-phase extraction and liquid chromatography–ion trap tandem mass spectrometry. Anal Bioanal Chem 391:1309–1319
12. González-Mariño I, Quintana JB, Rodríguez I, Rodil R, González-Peñas J, Cela R (2009) Comparison of molecularly imprinted, mixed-mode and hydrophilic balance sorbents performance in the solid-phase extraction of amphetamine drugs from wastewater samples for liquid chromatography–tandem mass spectrometry determination. J Chromatogr A 1216:8435–8441
13. Huerta-Fontela M, Galcerán MT, Ventura F (2007) Ultraperformance liquid chromatography–tandem mass spectrometry analysis of stimulatory drugs of abuse in wastewater and surface waters. Anal Chem 79:3821–3829
14. Hummel D, Löffler D, Fink G, Ternes TA (2006) Simultaneous determination of psychoactive drugs and their metabolites in aqueous matrices by liquid chromatography mass spectrometry. Environ Sci Technol 40:7321–7328
15. Jones-Lepp TL, Alvarez DA, Petty JD, Huckins JN (2004) Polar organic chemical integrative sampling and liquid chromatography–electrospray/ion-trap mass spectrometry for assessing selected prescription and illicit drugs in treated sewage effluents. Arch Environ Contam Toxicol 47:427–439
16. Kasprzyk-Hordern B, Dinsdale RM, Guwy AJ (2007) Multi-residue method for the determination of basic/neutral pharmaceuticals and illicit drugs in surface water by solid-phase extraction and ultra performance liquid chromatography–positive electrospray ionisation tandem mass spectrometry. J Chromatogr A 1161:132–145
17. Kasprzyk-Hordern B, Dinsdale RM, Guwy AJ (2008) Multiresidue methods for the analysis of pharmaceuticals, personal care products and illicit drugs in surface water and wastewater by solid-phase extraction and ultra performance liquid chromatography–electrospray tandem mass spectrometry. Anal Bioanal Chem 391:1293–1308
18. Mari F, Politi L, Biggeri A, Accetta G, Trignano C, Di Padua M, Bertol E (2009) Cocaine and heroin in waste water plants: a 1-year study in the city of Florence, Italy. Forensic Sci Int 189:88–92
19. Postigo C, López De Alda MJ, Barceló D (2008) Fully automated determination in the low nanogram per liter level of different classes of drugs of abuse in sewage water by on-line solid-phase extraction–liquid chromatography–electrospray-tandem mass spectrometry. Anal Chem 80:3123–3134
20. Van Nuijs ALN, Tarcomnicu I, Bervoets L, Blust R, Jorens PG, Neels H, Covaci A (2009) Analysis of drugs of abuse in wastewater by hydrophilic interaction liquid chromatography–tandem mass spectrometry. Anal Bioanal Chem 395:819–828
21. Zuccato E, Castiglioni S, Bagnati R, Chiabrando C, Grassi P, Fanelli R (2008) Illicit drugs, a novel group of environmental contaminants. Water Res 42:961–968
22. Loganathan B, Phillips M, Mowery H, Jones-Lepp TL (2009) Contamination profiles and mass loadings of macrolide antibiotics and illicit drugs from a small urban wastewater treatment plant. Chemosphere 75:70–77
23. Banta-Green CJ, Field JA, Chiaia AC, Sudakin DL, Power L, De Montigny L (2009) The spatial epidemiology of cocaine, methamphetamine and 3, 4-methylenedioxymethamphetamine (MDMA) use: a demonstration using a population measure of community drug load derived from municipal wastewater. Addiction 104:1874–1880

24. Zuccato E, Chiabrando C, Castiglioni S, Calamari D, Bagnati R, Schiarea S, Fanelli R (2005) Cocaine in surface water: a new evidence-based tool to monitor community drug abuse. Environ Health A Global Access Sci Source 4:1–7
25. Zuccato E, Chiabrando C, Castiglioni S, Bagnati R, Fanelli R (2008) Estimating community drug abuse by wastewater analysis. Environ Health Perspect 116:1027–1032
26. Kasprzyk-Hordern B, Dinsdale RM, Guwy AJ (2008) The occurrence of pharmaceuticals, personal care products, endocrine disruptors and illicit drugs in surface water in South Wales, UK. Water Res 42:3498–3518
27. Kasprzyk-Hordern B, Dinsdale RM, Guwy AJ (2009) The removal of pharmaceuticals, personal care products, endocrine disruptors and illicit drugs during wastewater treatment and its impact on the quality of receiving waters. Water Res 43:363–380
28. Kasprzyk-Hordern B, Dinsdale RM, Guwy AJ (2009) Illicit drugs and pharmaceuticals in the environment – forensic applications of environmental data. Part 1: Estimation of the usage of drugs in local communities. Environ Pollut 157:1773–1777
29. Van Nuijs ALN, Pecceu B, Theunis L, Dubois N, Charlier C, Jorens PG, Bervoets L, Blust R, Neels H, Covaci A (2009) Spatial and temporal variations in the occurrence of cocaine and benzoylecgonine in waste- and surface water from Belgium and removal during wastewater treatment. Water Res 43:1341–1349
30. Van Nuijs ALN, Pecceu B, Theunis L, Dubois N, Charlier C, Jorens PG, Bervoets L, Blust R, Neels H, Covaci A (2009) Cocaine and metabolites in waste and surface water across Belgium. Environ Pollut 157:123–129
31. Boleda MR, Galcerán MT, Ventura F (2009) Monitoring of opiates, cannabinoids and their metabolites in wastewater, surface water and finished water in Catalonia, Spain. Water Res 43:1126–1136
32. Huerta-Fontela M, Galcerán MT, Martin-Alonso J, Ventura F (2008) Occurrence of psychoactive stimulatory drugs in wastewaters in north-eastern Spain. Sci Total Environ 397:31–40
33. Huerta-Fontela M, Galcerán MT, Ventura F (2008) Stimulatory drugs of abuse in surface waters and their removal in a conventional drinking water treatment plant. Environ Sci Technol 42:6809–6816
34. Postigo C, López de Alda MJ, Barceló D (2010) Drugs of abuse and their metabolites in the Ebro River basin: occurrence in sewage and surface water, sewage treatment plants removal efficiency, and collective drug usage estimation. Environ Int 36:75–84
35. Postigo C, López de Alda MJ, Barceló D (2008) Analysis of drugs of abuse and their human metabolic byproducts in water by LC–MS/MS: a non-intrusive tool for drug abuse estimation at the community level. Trends Anal Chem 27:1053–1069
36. Van Nuijs ALN, Pecceu B, Theunis L, Dubois N, Charlier C, Jorens PG, Bervoets L, Blust R, Meulemans H, Neels H, Covaci A (2009) Can cocaine use be evaluated through analysis of wastewater? A nation-wide approach conducted in Belgium. Addiction 104:734–741
37. Van Nuijs ALN, Castiglioni S, Tarcomnicu I, Postigo C, López De Alda MJ, Neels H, Zuccato E, Barceló D, Covaci A (in press) Illicit drug consumption estimations derived from wastewater analysis: a critical review. Sci Total Environ
38. 2002/657/EC, Commission Decision of 12 August 2002 implementing Council Directive 96/23/EC concerning the performance of analytical methods and the interpretation of results
39. SAIHEbro, Automatic hidrologic information system of the Ebro River basin. Available at: http://195.55.247.237/saihebro/index.php?url=/principal
40. Baselt RC (2004) Disposition of toxic drugs and chemicals in man, 7th edn. Biomedical Publications, Foster City, p 1254
41. Canezin J, Cailleux A, Turcant A, Le Bouil A, Harry P, Allain P (2001) Determination of LSD and its metabolites in human biological fluids by high-performance liquid chromatography with electrospray tandem mass spectrometry. J Chromatogr B Biomed Sci Appl 765:15–27
42. Nelson CC, Foltz RL (1992) Chromatographic and mass spectrometric methods for determination of lysergic acid diethylamide (LSD) and metabolites in body fluids. J Chromatogr Biomed Appl 580:97–109

43. Daughton CG (2001) Illicit drugs in municipal sewage in Pharmaceuticals and personal care products in the environment: scientific and regulatory issues. In: Daughton CG, Jones-Lepp TL (eds) ACS symposium series 791. The American Chemical Society, Washington, DC, p 116–139
44. INEbase, National Statistics Institute on-line database. http://www.ine.es/
45. Zuccaro P, Palmi I, Pacifici R, Da Cas R, Raschetti R (2006) L'utilizzo dei farmaci enalgesici nella terapia del dolore. La Rivista Italiana di Cure Palliative 1:41–45
46. Spanish Drug Observatory, OED (2007) Informe 2007 del observartorio español sobre drogas: Situación y tendencias de los problemas de drogas en España. Available at: http://www.pnsd.msc.es/Categoria2/publica/pdf/oed-2007.pdf

Sources, Occurrence, and Environmental Risk Assessment of Pharmaceuticals in the Ebro River Basin

Meritxell Gros, Mira Petrovic, Antoni Ginebreda, and Damià Barceló

Abstract Freshwaters are an essential resource which must be available not only in the required quantity but also in a precise quality. Only less than 1% of the world's freshwater resources are readily available for human use, and even this resource is unevenly distributed among the countries. In developed countries, their contamination is the adverse outcome of incomplete wastewater treatment and improper disposal of sewage into surface and groundwater, which are the major renewable resources for sustainable drinking water production. In a vast array of contaminants of anthropogenic origin reaching our water supplies, pharmaceutically active compounds (PhACs) are among the ones with the biggest input into the environment. In the case of the Ebro river basin, pharmaceuticals can enter the environment through the effluent discharges of 186 WWTPs, and at some points, especially in drought periods, the effluents may represent a significant percentage of the total flow of the river. Besides these WWTP discharges other environmental exposure pathways of PhACs can be land applications (e.g., biosolids and water reuse) or concentrated animal feeding operations (CAFOs). The fact that existing WWTPs only treat wastewaters produced by a 62.8% of the basins total population and therefore,

M. Gros (✉)
Institut Català de Recerca de l'Aigua (ICRA), c/Emili Grahit 101, 17003 Girona, Spain
e-mail: mgros@icra.cat

M. Petrovic
Department of Environmental Chemistry, IDAEA-CSIC, c/Jordi-Girona 18-26, 08034 Barcelona, Spain
Institució Catalana de Recerca I Estudis Avançats (ICREA), Passeig Lluís Companys 23, 80010 Barcelona, Spain

A. Ginebreda
Department of Environmental Chemistry, IDAEA-SIC, c/Jordi-irona 18-8034 Barcelona, Spain

D. Barceló
Institut Català de Recerca de l'Aigua (ICRA), c/Emili Grahit 101, 17003 Girona, Spain
Department of Environmental Chemistry, IDAEA-SIC, c/Jordi-Girona 18-26, 08034 Barcelona, Spain

D. Barceló and M. Petrovic (eds.), *The Ebro River Basin*,
Hdb Env Chem (2011) 13: 209–237, DOI 10.1007/698_2010_72,

that a certain percentage of water remains untreated, may result in lower water quality in some affected areas.

Even though pharmaceutical products are, until now, not included in the list of priority or dangerous substances of the Water Framework Directive (2000/60/EC), and thus, no environmental quality standards are stipulated (Directive 2008/105/EC), substances discharged into a basin should be controlled, as the same directive clearly establishes.

This chapter will review the levels and distribution of pharmaceuticals detected in both waste and river waters from the Ebro river Basin and gives an example about the use of established hazard indexes to estimate the possible risks posed by the pharmaceutical levels detected towards different aquatic organisms (algae, daphnia, and fish). Results presented in this chapter were integrated in the FP6 European Union project AQUATERRA (contract no. 505428).

Keywords Occurrence, Pharmaceuticals, Risk assessment, Wastewater treatment

Contents

Abbreviations

AOPs	Advanced oxidation processes
CAFOs	Confined animal feeding operations
CAS	Conventional activated sludge
CHE	Confederación Hidrográfica del Ebro
EC_{50}	Half maximal effective concentration
ERA	Environmental risk assessment
EMEA	European medicines agency
EPI	Enhanced product ion scan
EU	European Union

FDA	Food and Drug Administration
FP6	6th Framework Programme
HRT	Hydraulic retention time
HQ	Hazard quotient
IDA	Information dependent acquisition
IP	Identification points
IPPC	Integrated Pollution Prevention and Control
K_{biol}	Biodegradation rate
K_d	Sludge-water distribution coefficient
K_H	Henry's constant
LC_{50}	Lethal concentration 50% (dose required to kill half the members of a tested population after a specified test duration)
LC-MS/MS	Liquid chromatography-tandem mass spectrometry
MBR	Membrane bioreactor
MDL	Method detection limits
MEC	Measured environmental Concentrations
NOEC	Nonobserved effect concentration
NSAIDs	Nonesteroideal antiinflammatory drugs
PEC	Predicted environmental concentration
PNEC	Predicted noneffect concentration
PhACs	Pharmaceuticals
QqQ	Triple quadrupole mass analyzer
QqLIT	Triple quadrupole-linear ion trap mass analyzer
RE	Removal efficiency
SRT	Sludge retention time
$t_{1/2}$	Half-lives
WFD	Water Framework Directive
WWTP	Wastewater treatment plant

1 Introduction

Freshwaters are an essential resource which must be available not only in the required quantity but also in a precise quality. Nevertheless, in the last years, this quality has been threatened as a consequence of the increasing industrial activity, population growth and agricultural practices. In fact, only less than 1% of the world's freshwater resources are readily available for human use, and even this resource is unevenly distributed among the countries. A lack of water to meet daily needs is a reality for many people around the world, and water scarcity already affects every continent. In the underdeveloped countries, water scarcity forces people to rely on unsafe sources of drinking water. In the developed countries, this problem is much less critical where contamination of drinking water is the adverse outcome of incomplete wastewater treatment and improper disposal of sewage into surface and groundwater, which are the major renewable resources for sustainable drinking water production. However, research is documenting with increasing frequency that many chemical constituents that have not been considered historically as contaminants are present in natural waters at global scale [1, 2].

In a vast array of contaminants of anthropogenic origin reaching our water supplies, pharmaceutically active compounds (PhACs) are among the ones with the biggest input into the environment. After their prophylactic and therapeutic usage a large fraction of pharmaceuticals is discharged into wastewaters either unchanged or metabolized. Several studies revealed that these substances are partially removed during wastewater treatment processes, being therefore discharged into freshwater ecosystems [3, 4]. Besides their removal during wastewater treatment processes, levels of pharmaceuticals in river waters may also depend on the flow rate of receiving waters. Dilution that occurs between the point of discharge (i.e., WWTP) and downstream usage and/or groundwater wells may not be enough in some cases. Although there are few studies describing the spatial and temporal distribution of pharmaceuticals in rivers receiving wastewater discharges, it is estimated that in the majority of European countries, effluent wastewaters are diluted in a factor between 10 and 100 [5]. In the case of the Ebro, pharmaceuticals can enter the environment not only through the effluent discharges of 186 WWTPs but also from confined animal feeding operations (CAFOs) and through land applications (e.g., biosolids and water reuse). At some points, and especially in drought periods, these effluent discharges may represent a significant percentage of the total flow of the river. The fact that existing WWTPs only treat wastewaters produced by a 62.8% of the basins total population and therefore, that a certain percentage of water remains untreated, may result in lower water quality in some affected areas.

It is difficult to predict what public health and toxic effects would arise from the occurrence of pharmaceuticals in freshwater ecosystems. Although some studies pointed out that many PhACs do not exhibit acute toxicity, they can show chronic effects at concentrations close to the ones detected in wastewaters [6]. Some of the most notorious adverse effects that pharmaceuticals might have in the environment besides acute or chronic toxicity, are resistant development of pathogenic bacteria, due to the occurrence of antibiotics, genotoxicity, and endocrine disruption. Indeed, antibiotic resistant *E. coli* has been demonstrated in surface water [7], while resistance plasmid transfer in aquatic environments has been observed by several experiments [8, 9]. The phenomenon of antibiotic resistance is becoming a subject of concern and awareness within the scientific community. Reservoirs of antibiotic resistance in humans and in animals can interact in different ways, but food and water are the most probable vectors of transmission to the intestinal flora. Furthermore, some PhACs such as antidepressants, β-blockers, or lipid regulators can be prone to bioconcentration/bioaccumulation in aquatic organisms [10]. Nevertheless, there is still a great lack of knowledge whether resistance transference is possible at the very low concentrations found in the environment [11] or to what extent can PhACs reach humans through food-chain biomagnification.

On this context, it is important to set up analytical methods and monitoring programs in order to characterize what are the pharmaceuticals detected in environmental waters and their concentrations levels. Moreover, risk assessment studies, used to estimate the hazards posed by pharmaceuticals and their transformation products, is fundamental and a prerequisite for a comprehensive protection of the environment.

Even though pharmaceutical products are, until now, not included in the list of priority or dangerous substances of the Water Framework Directive (2000/60/EC), and thus, no environmental quality standards are stipulated (Directive 2008/105/EC), substances discharged into a basin should be controlled, as the same directive clearly establishes [12]. Water quality and management along the Ebro river basin is controlled by the Confederación Hidrográfica del Ebro (CHE), which regularly performs monitoring programs to control its state. This monitoring network is mainly based on the control of regulated priority and dangerous pollutants (the ones included in the WFD). Nowadays, the monitoring network consists of the analysis of different compounds in water, sediments and biota from 18 sampling sites once a year.

This chapter will review the levels and distribution of pharmaceuticals detected in both waste and river waters along the Ebro river Basin and gives an example about the use of established hazard indexes to estimate possible risks posed by the pharmaceutical levels detected towards different aquatic organisms (algae, daphnia, and fish). Results presented in this chapter were integrated in the FP6 European Union project AQUATERRA (contract no. 505428), which aimed to provide the scientific basis for an improved river basin management through a better understanding of the river–sediment–soil–groundwater system as a whole, by integrating both natural and socio-economic aspects at different temporal and spatial scales.

2 Sampling Sites

The Ebro river basin is the most extensive basin in the country, representing the 17.3% of the Spanish peninsular territory. The Ebro itself, with 910 km length, receives waters from several tributaries, which altogether represent 12,000 km of waterway network [13].

Regarding the analysis of pharmaceuticals, two different monitoring campaigns were carried out within the framework of AQUATERRA project: (1) a first study which covered the determination of 28 multiple-class pharmaceuticals of major human consumption (including mainly analgesics and antiinflammatories, lipid regulators, psychiatric drugs, antibiotics, antihistamines and β-blockers), taking samples in June 2005 and November 2005 [14] and (2) a broader survey, where an extended list of 73 pharmaceuticals was analyzed over four sampling periods (June and November 2006, October 2007 and July 2008) [15]. Target pharmaceuticals investigated in both campaigns are indicated in Table 1. In both studies, influent and effluent wastewaters from seven WWTPs, located in the main cities along the basin (see Fig. 1) and their subsequent receiving river waters were monitored. Table 2 summarizes the characteristics of the WWTP studied. As indicated, the majority of the plants have a primary and secondary treatment, operating with conventional

Table 1 List of target pharmaceuticals analyzed in each monitoring program

Therapeutic group	Compounds	CAS Number	Sampling campaings	
			First (2005)	Second (2006–2008)
Analgesics and anti-inflammatories	Ketoprofen	22071-15-4	X	X
	Naproxen	22204-53-1	X	X
	Ibuprofen	15687-27-1	X	X
	Indomethacine	53-86-1	X	X
	Diclofenac	15307-86-5	X	X
	Mefenamic acid	61-68-7	X	X
	Acetaminophen	103-90-2	X	X
	Salicylic acid	69-72-7	–	X
	Propyphenazone	479-92-5	X	X
	Phenylbutazone	50-33-9	–	X
	Phenazone	60-80-0	–	X
	Codeine	76-57-3	–	X
Lipid regulators and cholesterol lowering statin drugs	Clofibric acid	882-09-7	X	X
	Bezafibrate	41859-67-0	X	X
	Fenofibrate	49562-28-9	–	X
	Gemfibrozil	25812-30-0	X	X
	Mevastatin	73573-88-3	X	X
	Pravastatin	81093-37-0	X	X
	Atorvastatin	134523-00-5	–	X
Psychiatric drugs	Paroxetine	61869-08-7	X	X
	Fluoxetine	54910-89-3	X	X
	Diazepam	439-14-5	–	X
	Lorazepam	846-49-1	–	X
	Carbamazepine	298-46-4	X	X
Histamine H_2 receptor antagonists	Loratadine	79794-75-5	X	X
	Famotidine	76824-35-6	X	X
	Ranitidine	66357-35-5	X	X
	Cimetidine	51481-61-9	–	X
Tetracycline antibiotics	Tetracycline	60-54-8	–	X
	Doxycycline	564-25-0	–	X
	Oxytetracycline	79-57-2	–	X
	Chlortetracycline	57-62-5	–	X
Macrolide antibiotics	Erythromycin	114-07-8	X	X
	Azithromycin	83905-01-5	X	X
	Roxithromycin	80214-83-1	–	X
	Clarithromycin	81103-11-9	–	X
	Josamycin	16846-24-5	–	X
	Tylosin A	1401-69-0	–	X
	Spiramycin	–	–	X
	Tilmicosin	10850-54-0	–	X
Sulfonamide antibiotics	Sulfamethoxazole	723-46-6	X	X
	Sulfadiazine	68-35-9	–	X
	Sulfamethazine	57-68-1	–	X
Fluoroquinolone antibiotics	Ofloxacin	82419-36-1	X	X
	Ciprofloxacin	85721-33-1	–	X
	Enrofloxacin	93106-60-6	–	X
	Norfloxacin	70458-96-7	–	X
	Danofloxacin	112398-08-0	–	X

(*continued*)

Table 1 (continued)

Therapeutic group	Compounds	CAS Number	Sampling campaings	
			First (2005)	Second (2006–2008)
	Enoxacin	74011-58-8	–	X
Other antibiotics	Trimethoprim	738-70-5	X	X
	Chloramphenicol	56-75-7	–	X
	Metronidazole	443-48-1	–	X
	Nifuroxazide	965-52-6	–	X
B-blockers	Atenolol	29122-68-7	X	X
	Sotalol	3930-20-9	X	X
	Metoprolol	37350-58-6	X	X
	Propranolol	525-66-6	X	X
	Timolol	26839-75-8	–	X
	Betaxolol	63659-18-7	–	X
	Carazolol	57775-29-8	–	X
	Pindolol	13523-86-9	–	X
	Nadolol	42200-33-9	–	X
B-agonists	Salbutamol	18559-94-9	–	X
	Clenbuterol	37148-27-9	–	X
Barbiturates	Butalbital	77-26-9	–	X
	Pentobarbital	76-74-4	–	X
	Phenobarbital	50-06-6	–	X
Antihypertensives	Enalapril	75847-73-3	–	X
	Lisinopril	83915-83-7	–	X
Diuretic	Hydrochlorothiazide	58-93-5	–	X
	Furosemide	54-31-9	–	X
Antidiabetic	Glibenclamide	10238-21-8	–	X
To treat cancer	Tamoxifen	10540-29-1	–	X

Compounds determined in each campaign are marked with an X

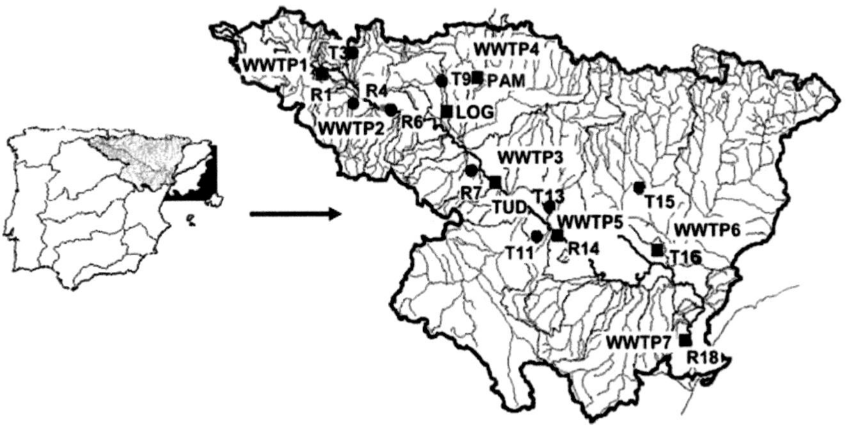

Fig. 1 Map of the sampling sites, indicating all wastewater treatment plants (WWTP), receiving river waters located downstream each plant (indicated with *an square*) and other river waters considered as "hot-spots" (*highlighted as a dot*)

Table 2 Characteristics of the WWTP monitored

WWTP	Location	Receiving river water	Wastewater treated	HRT (h)	Primary treatment	Secondary treatment
WWTP1	Miranda Ebro	Vallas	Urban	32	–	Activated sludge
WWTP2	Logroño	Iregua	Urban and industrial	8	Primary settling	Activated sludge
WWTP3	Tudela	Ebro	Urban	18	Primary settling	Biologic filters
WWTP4	Pamplona	Arga	Urban and industrial	9.5	Primary settling	Activated sludge
WWTP5	Zaragoza	Ebro	Urban	10	Primary settling	Activated sludge
WWTP6	Lleida	Segre	Urban	6–10	Primary settling	Activated sludge
WWTP7	Tortosa	Ebro	Urban	33	–	Activated sludge

activated sludge, except one whose biological treatment is with biologic filters. Nevertheless, the main differences between them lie in their hydraulic retention times. Both time-averaged influent and effluent samples were collected to calculate removal rates of target compounds during treatment processes.

Furthermore, in the second monitoring, besides receiving river waters eight extra samples were included. These sites were selected as possible "hot-spots" for freshwater contamination, due to their location close to the most populated cities and main agricultural areas in the basin, as illustrated in Fig. 1, where points indicated as "R" correspond to the Ebro itself, while tributaries are highlighted as "T." It should be worth mentioning that sampling sites investigated coincided with those selected by the CHE as "hot-spots" of water pollution, except three areas (labeled as TUD, PAM and LOG in the figure) which were added by the authors due to their interest, since they receive the discharge of the wastewater treatment plant.

3 Analysis and Prioritization of the Pharmaceuticals To Be Monitored

The chemical analysis of pharmaceutical residues in the environment has been closely tied to the analytical capabilities, especially in the mass spectrometry field. The development of more sensitive and versatile equipments has enabled their detection at the low environmental levels and has also provided the required tools to ensure a precise identification, according to the criteria fixed by the increasingly stringent European guidelines. Recent trends are focused towards the development and application of generic methods that permit simultaneous analysis of multi-class compounds, including acidic, neutral, and basic pharmaceuticals [16, 17]. A single method for the analysis of various pharmaceuticals belonging to different compound classes has several advantages, such as shorter analysis time, reduced field sampling and overall cost reduction. In all these methodologies, liquid chromatography-tandem mass spectrometry (LC-MS/MS) is the technique of choice, being triple quadrupole (QqQ) equipments the most widely employed. Even though their sensitivity, selectivity, and efficiency is good enough to detect target compounds at the low levels found in the environment, qualitative information needed to support the structural elucidation of compounds is lost. This drawback can be overcome by using novel mass spectrometry analyzers, such as quadrupole-linear ion trap (QqLIT). This hybrid MS instruments are not only a powerful tool for the quantitative analysis of parent compounds but also for their unequivocal identification and confirmation, thanks to the scan combinations that can be achieved when performing Information Dependent Acquisition (IDA).

Prior to the monitoring of the pharmaceutical residues in both waste and river waters, appropriate multiresidue analytical methods, based on off-line SPE followed by LC-ESI-MS/MS were developed in both studies [18, 19]. In all these methods, all compounds are extracted in a single step, using hydrophilic–lipophilic polymeric sorbents, speeding up considerably sample preparation. For the simultaneous analysis of 28 pharmaceuticals, a triple quadrupole (QqQ) instrument was used whereas for the determination of 73 pharmaceuticals, a QqLIT tandem mass spectrometer was employed. With the latter instrument, while quantitative analysis was performed in SRM mode, unequivocal identification and confirmation of target compounds was carried out by using the Information Dependent Acquisition (IDA) function, combining the SRM as the survey scan and an enhanced product ion (EPI) scan of each specific analyte, recorded at three different collision energies, as the dependent scan. An EPI scan corresponds to a daughter ion scan when the third quadrupole operates as a linear ion trap mass analyzer. Spectra achieved were afterward compared with library data, based on EPI spectra at the three collision energies used. With this approach, compound identification and confirmation fulfil the stringent criteria set by the EU regulations (EU Commission Decision 2002/657/EC), obtaining more than enough identification points (IPs) to ensure an accurate identification of target compounds in the samples [18]. In both methodologies, recoveries obtained were generally higher than 50% for both surface and wastewaters, and the overall variability of the method was below 15%, for all compounds and matrices tested. Regarding method detection limits (MDLs), when working with the QqQ instrument they varied between 1 and 160 ng L^{-1}, depending on the compound and matrix. Nevertheless, with the QqLIT instrument, higher sensitivity was achieved, reaching detection limits from 0.1 and 55 ng L^{-1}, again depending on the type of water analyzed.

Selection of target pharmaceuticals (see Table 1) was based on the following criteria: (1) the sales and practices in Spain (according to National Health system), (2) compound pharmacokinetics (the percentage of excretion as nonmetabolized substance), (3) their occurrence in the aquatic media (data taken from other similar studies), and (4) on data provided by environmental risk assessment approaches, which link the calculation of predicted environmental concentrations (PEC) with toxicity data in order to evaluate which compounds are more liable to pose an environmental risk for aquatic organisms [20–22]. In the current European

Medicines Agency (EMEA) regulations, it is set a threshold safety value of 10 ng L^{-1}, and compounds whose PEC exceeds this quantity, have to be subjected to toxicity tests and can therefore, be considered as potential candidates to be included in monitoring programs [23].

4 Occurrence of Pharmaceuticals in Wastewaters

Levels (expressed as range of concentrations between minimum and maximum levels and median values) for the most frequently detected (found in more than 80% of the samples analyzed) and representative active substances of each therapeutic group are shown in Fig. 2a, b for wastewater influent and effluents, respectively. Results presented are a summary of the two aforementioned surveys performed along the Ebro river basin [14, 15]. Each plot includes 42 measures (seven WWTP during six sampling periods), with the exception of the compounds that were only investigated in the second survey (see Table 1). In this case, graphics were built from 28 measures (seven WWTP during four sampling campaigns).

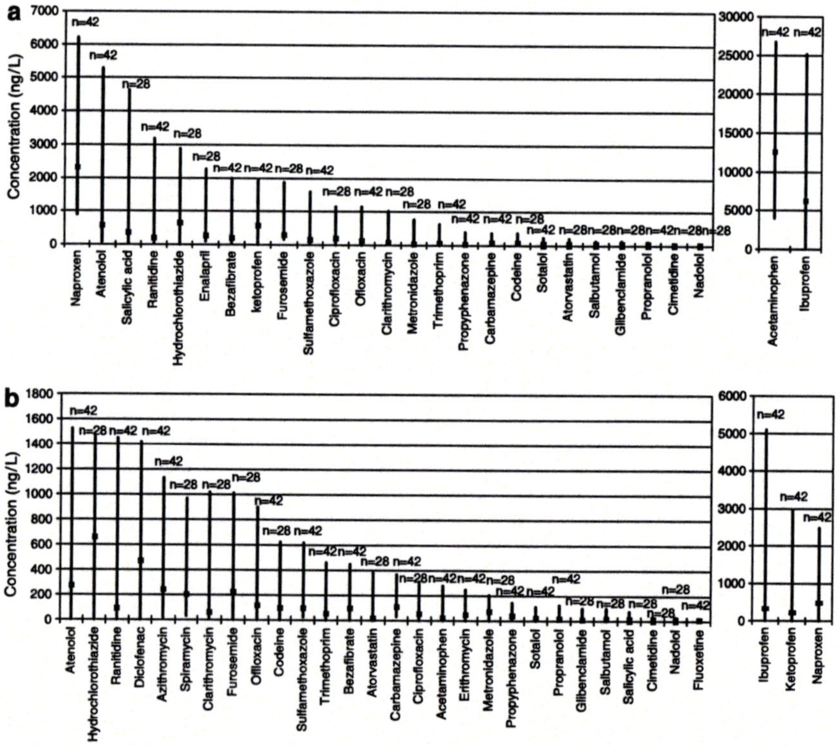

Fig. 2 Levels of pharmaceuticals detected in wastewater influent (**a**) and effluent (**b**), where *n* means the total number of samples analyzed to what concentrations indicated refer to

Comparing influent and effluent concentrations, levels in the former, were higher than in the latter, as expected, which ranged from high nanogram per liter up to microgram per liter. Even though levels in the outlets are generally one order of magnitude lower than in the inlets, concentrations are still significant (high nanogram per liter), reaching sometimes microgram per liter. Compounds more frequently detected in effluent wastewaters are the ones showing average and low removal rates, such as β-blockers, antibiotics (especially macrolides), psychiatric drugs, diuretics, some analgesics (codeine and phenazone type analgesics), lipid regulators, antihistamines, β-agonists, and antidiabetics. Although nonesteroidal antiinflammatory drugs (NSAIDs) are highly removed after wastewater treatment, they are still ubiquitous in effluents, probably due to the fact that their concentrations in influents are so high, that levels remaining in the effluents are still significant.

In fact, the ubiquity of drugs is related to specific sales and practices in each country. For example, analgesics and antiinflammatory drugs, cholesterol lowering statins, antibiotics, antihistamines, antidepressants, and antihypertensives are the families of drugs with major consumption in Spain, among other groups, according to the National Health system. Indeed, this coincides with the results presented in this section, where higher levels corresponded to nonsteroidal antiinflammatory drugs (NSAIDs). Compounds having major significance within this group are acetaminophen, ibuprofen (with individual concentrations from around 0.5 to around 26 $\mu g\ L^{-1}$ in influents) followed by diclofenac, naproxen, salicylic acid and ketoprofen, with median concentrations ranging from 0.3 to 2 $\mu g\ L^{-1}$ (see Fig. 2). Conversely, concentrations in the outlets decreased considerably for almost all substances, showing median concentrations around 0.5 and 0.6 $\mu g\ L^{-1}$ for ibuprofen, ketoprofen, naproxen and diclofenac or even lower for codeine (0.1 $\mu g\ L^{-1}$) and salicylic acid (0.01 $\mu g\ L^{-1}$), which is the main metabolite of aspirin (acetylsalicylic acid).

Besides NSAIDs, other therapeutic groups detected at significant levels in both influent and effluent wastewaters were, β-blockers, antihistamines, antihypertensive agents diuretics, and lipid regulators with the addition of macrolide antibiotics and psychiatric drugs in effluent wastewaters. Most significant compounds within these groups are atenolol, ranitidine, enalapril, furosemide, hydrochlorothiazide, bezafibrate, atorvastatin, pravastatin, sulfamethoxazole, ofloxacin, ciprofloxacin, metronidazole, trimethoprim clarithromycin, azithromycin, spiramycin and fluoxetine, diazepam, and lorazepam for psychiatric drugs (see Fig. 2 for concentrations). Antibiotic losses to the environment can be considered substantial, due to their widespread consumption in human and veterinary medicine. Concerning psychiatric drugs, the antiepileptic carbamazepine is one of the most prominent drugs with a long history of clinical usage. It has proved to be very recalcitrant, as it bypasses sewage treatment. Common WWTP influent concentrations are in the order of magnitude of several hundreds of nanogram per litre (median values of 0.1 $\mu g\ L^{-1}$). Nevertheless, serotonin reuptake inhibitors and benzodiazepines were found at much lower concentrations and with less frequency, with the exception of fluoxetine, which was detected in all samples in effluent wastewaters.

Table 3 Total loads (expressed as gram per day per 1,000 inhabitants) of pharmaceuticals detected in WWTP effluents that are afterwards discharged into receiving river waters. Loads are expressed as the range of the amounts detected in each sampling period altogether with average and mean values

WWTP	Range loads	Average loads	Median loads
WWTP1	0.3–0.9	0.7	0.8
WWTP2	0.2–1.1	0.7	0.8
WWTP3	0.4–1.6	1.2	1.1
WWTP4	0.4–1.5	0.8	0.7
WWTP5	1.2–3.5	2.2	2.2
WWTP6	1.5–5.8	2.8	1.8
WWTP7	0.6–2.0	1.1	0.9

Finally, the β-agonist salbutamol and the antidiabetic glibenclamide were frequently found as well with median values around 0.02 μg L^{-1}.

On the other hand, the analgesic phenylbutazone, the antidepressant paroxetine, tetracycline antibiotics doxycycline and chlorotetracycline, the antibiotics tilmicosin, danofloxacin, enoxacin, nifuroxazide, the β-blockers betaxolol, carazolol, pindolol, the β-agonist clenbuterol, barbiturates, the antihypertensive lisinopril, and the drug to treat cancer tamoxifen were never detected.

When converting total concentration of all pharmaceuticals detected in effluent wastewaters into environmental loads (gram per day per 1,000 inhabitants), it can be concluded that WWTPs are major contributors of their occurrence in freshwaters, since significant quantities are discharged into receiving river waters. This statement is supported by the information given in Table 3, which shows the total loads of pharmaceuticals found in treated wastewaters that are afterwards discharged in receiving river waters. Loads are indicated as a range for the levels calculated in each WWTP along all monitoring programs. In all these situations, loads were calculated by multiplying total concentrations (addition of individual concentrations) with the effluent flow rates and then normalized by the population equivalent of each plant.

5 Removal Under Conventional Activated Sludge Wastewater Treatment

The most important pathways for removal of PhACs during wastewater treatment are biotransformation/biodegradation and abiotic removal by adsorption to the sludge [4]. Considering the low values of Henry coefficients (K_H) of most of the PhACs detected in wastewater streams, stripped fractions removed by volatilization can be neglected [24, 25]. In data presented next, the term "removal" is used here for the conversion of a micropollutant to substances other than the parent compound, without distinguishing between abiotic (adsorption) and biotic degradation (transformation by microorganisms).

5.1 Removal Rates in WWTP from the Ebro River Basin

The range of removal rates (%RE) for the most representative compounds of each therapeutic group, in the whole set of WWTPs under investigation, is given in Fig. 3, altogether with their average %RE. It should be highlighted that, in the figure, only pharmaceuticals showing positive removal rates were considered. Therefore, serotonin reuptake inhibitors, benzodiazepines, carbamazepine and macrolide antibiotics (as described below), are not included. Reported overall removal rates varied strongly between individual pharmaceuticals and, therefore, it is difficult to establish a general trend. Nevertheless, results indicate that

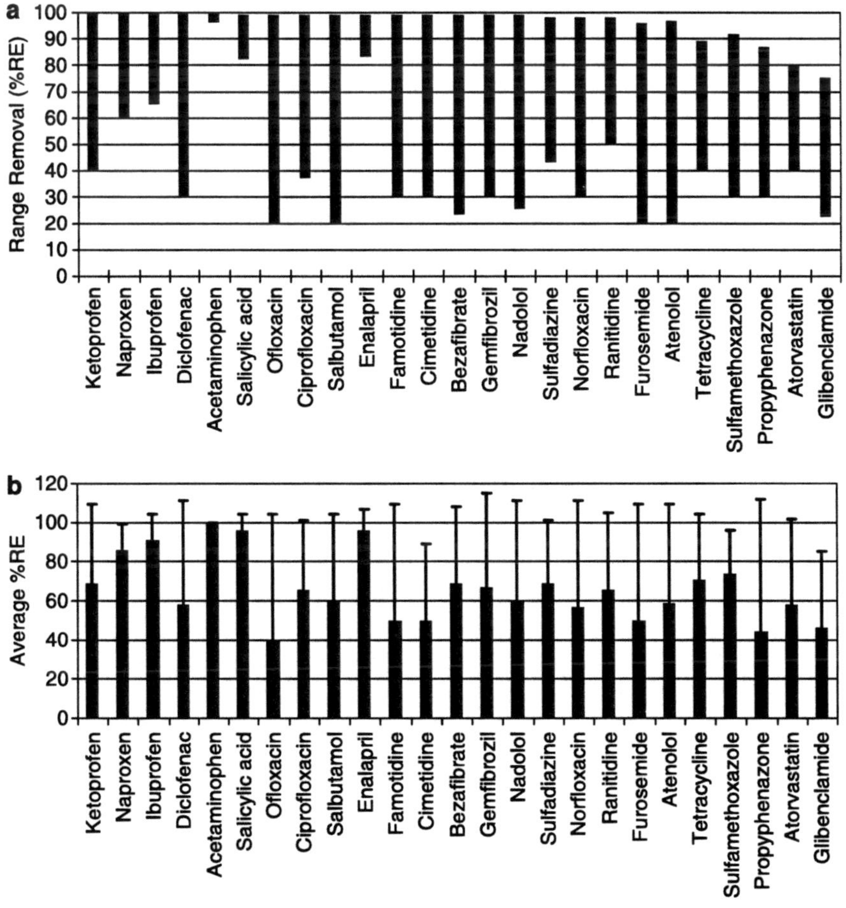

Fig. 3 Range of removal rates (**a**) and average removal efficiencies (**b**) for some of the most ubiquitous pharmaceuticals under conventional activated sludge treatment in the whole set of WWTPs under investigation

elimination of most PhACs is incomplete. In a general extent, three different behaviors were observed (see Fig. 3):

(a) *An increase in concentration along the passage through the WWTPs*: Macrolide antibiotics, the antiepileptic carbamazepine, benzodiazepines, and serotonin reuptake inhibitors showed either poor or no elimination in all WWTP investigated. These results are in good agreement with those reported in the literature. Many other authors reported as well poor elimination of the antiepileptic drug carbamazepine [26–28]. This could be because glucuronide conjugates of carbamazepine can presumably be cleaved in the sewage, and thus increase its environmental concentrations [2].

(b) *No significant to medium removal*: Lipid regulators, fluoroquinolone and tetracycline antibiotics (when detected), cholesterol lowering statin drugs, histamine H_1 and H_2 receptor antagonists, β-blockers, β-agonists, and the antidiabetic glibenclamide were partially degraded, mostly presenting average removal efficiencies between 40 and 60–70%. Concerning diuretics (furosemide and hydrochlorothiazide), their removal range is highly variable (see Fig. 3), with average elimination rates of 50% for furosemide and 32% for hydrochlorothiazide. On the other hand, although sulphonamide antibiotics presented quite high average removal rates (around 70%), in some situations these values were lower (see Fig. 3). Regarding trimethoprim and metronidazole they were only quite efficiently removed in the plants with higher hydraulic retention times, with values ranging from 65 to 80% for both compounds. Removal rates reported in the literature for some of these compounds in conventional activated sludge treatment are similar to the ones reported in the WWTPs monitored. Lindqvist et al. [29] observed unexplained variations of concentration over time for sulphonamide antibiotics, probably because of deconjugation processes that may occur during contact with activated sludge. For example, a significant amount of sulphamethoxazole enters WWTPs in metabolized form as N_4-acetyl-sulphamethoxazole that can be converted back to the original compound [30]. On the other hand, several studies reported that the removal range of the lipid regulators gemfibrozil and bezafibrate are in the range of 46–69% [28, 31] and 36–54% [28]. All these results coincide with the ones reported by Miège et al. [32].

(c) *High removal efficiency:* Only NSAIDs and the antihypertensive enalapril would be fitted in this group, where they are almost fully eliminated in all plants (see Fig. 3). The only exception was diclofenac, whose removal rates varied from no elimination up to 100%.These results are in good agreement with those reported in the literature [25]. While the NSAIDs ibuprofen and acetaminophen are easily degraded in the environment, showing removal rates (RRs) in the range of 80–100%, contradictory results have been reported for the removal of the NSAID diclofenac during conventional activated sludge (CAS) wastewater treatment. No influence of the increased SRT was found on its biodegradation [28]. In some WWTPs, attenuation of 50–70% of diclofenac was reported [2, 24, 33, 34]. In contrast, many studies showed extremely low efficiency of conventional treatment (only 10–30% removal) [29, 35].

5.2 *Influence of Operational Parameters on PhACs Removal*

Although it is not fully elucidated which factors could explain the divergences in removal rates, since many investigations do not report all operational data of the treatment plants, it has been observed that besides compound physico-chemical properties and on the treatment process applied, removal of pharmaceuticals in conventional WWTPs is potentially affected by process operating parameters [25]. These factors are: (1) temperature of operation (higher removal efficiencies have been observed in summer periods in comparison with colder seasons), (2) different kinetic behaviors (degradation rates) of compounds, (3) redox conditions, and (4) sludge retention time (SRT) and hydraulic retention time (HRT). Concerning SRT and HRT, both parameters are considered to have a positive effect on both biodegradation and sorption of PhACs. The prolonged SRT and HRT should provide favorable conditions for the adaptation and diversification of the microbial community of sewage sludge. In fact, higher SRT will allow the enrichment of slowly growing bacteria, and hence a more diverse microbial biocenosis will be established than with lower SRT [4]. The nature of microbial population can have a significant impact on the biodegradation of some persistent PhACs. For example, it was found that longer SRT that provide growth of nitrifying bacteria is favorable for degradation of antibiotic trimethoprim [36]. Moreover, Clara et al. [28] reported a correlation between the observed removal of ibuprofen and bezafibrate and the operating SRT. On the other hand, HRT can more seriously affect compounds with a moderate degradation velocity leading to better elimination of such compounds at higher HRT. The fact that HRT has a significant role on pharmaceutical elimination was demonstrated in the survey conducted along the Ebro river basin, where removal rates were linked with compound half-lives (assuming that compound concentration decreased over time following pseudofirst order kinetics) and with the HRT of each plant [15]. According to the results reported, a minimum HRT is needed to accomplish the complete or high removal of pharmaceuticals. Taking into consideration some representative compounds in each WWTP it could be concluded that: (1) compounds with high removal and degradation rate (low $t_{1/2}$) like all NSAIDs, except diclofenac, and the antihypertensive enalapril and (2) compounds with poor or no elimination and degradation (high $t_{1/2}$), like carbamazepine, HRT does not influence in compound removal. Nevertheless, compounds with medium removal and degradation rate, HRT seems to pay a role, since elimination rates were higher when increasing HRT. Therefore, in a great extent, it could be said that compounds that are biodegradable (high k_{biol} or $t_{1/2}$) and have low K_d values (low sludge–water distribution coefficient, which means that they show low tendency to absorb in sewage sludge) are more influenced by HRT, whereas substances that have high K_d and low K_{biol} (high $t_{1/2}$) are more influenced by SRT. But this is just a general observation, since there are substances, like ibuprofen and other analgesics and antiinflammatories, which show high K_{biol} and $K_{d,}$ that are very well removed independently of SRT and HRT. A similar correlation regarding pharmaceutical removal with SRT could not be performed, since data about this parameter was only available for two plants.

6 Occurrence of Pharmaceuticals in River Waters

6.1 Levels in River Waters

The most frequently detected pharmaceuticals in river waters and their range of concentrations are depicted in Fig. 4. Typical levels range from 10 to100 ng L^{-1} but sometimes they can even reach high nanogram per litre range or low microgram per litre levels. By comparing effluent wastewater and receiving river water profiles, in terms of compounds detected, it can be observed that a similar spectrum of pharmaceuticals are detected in both matrices. Nevertheless, levels in river waters are generally one order of magnitude lower than levels in effluent wastewaters. Therefore, compounds showing average and low removal rates are the ones mostly found in receiving river waters. However, even though analgesics and antiinflammatory drugs are highly removed after wastewater treatment (see previous section), they are also ubiquitous and are present at considerable concentrations in river waters. This could be due to the fact that, although they are efficiently eliminated, concentrations in the inlets are so high, that levels that remain in the effluents are still significant. Nevertheless, for instance, the antihypertensive enalapril, which is also removed over 90% in all WWTP investigated, was never detected in river waters. This could be attributed to either the dilution factor or due to some attenuation due to abiotic processes, such as photo degradation, that might take place.

Levels detected in the Ebro and some of its tributaries are similar to those found in Wales [37], Japan [38], Germany [39], France [40], Switzerland [41] and even in the USA [1], as few examples of the numerous studies available regarding the occurrence of pharmaceuticals in river waters.

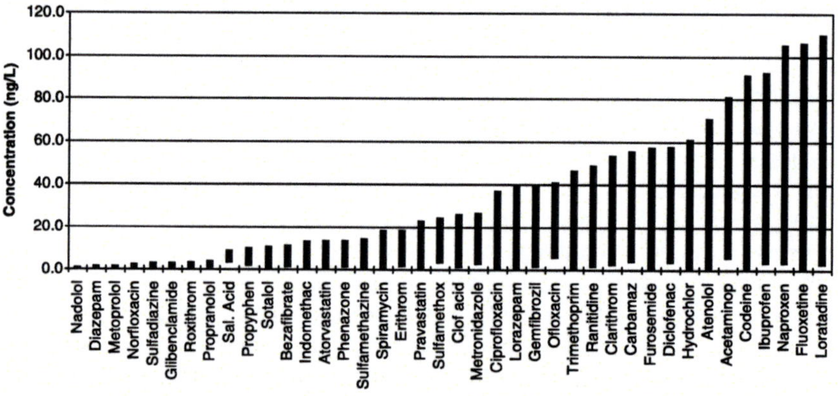

Fig. 4 Range of concentrations, expressed as nanograms per liter, detected for the most representative pharmaceuticals in river waters

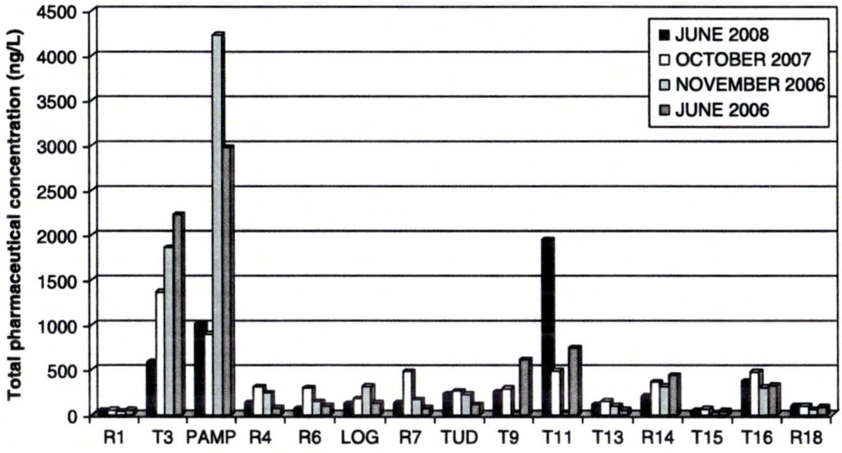

Fig. 5 Total concentrations of pharmaceuticals in the 15 river waters analyzed within the last monitoring program (from June 2006 until July 2008). Sites are ordered from north to south along the basin

On the other hand, in Fig. 5, total concentrations of pharmaceuticals detected in 15 river waters analyzed along the Ebro river basin in four sampling campaigns is illustrated. It can be observed that sites with higher concentrations are the ones labeled as PAM, T3, and T11, which correspond to the Ebro tributaries Arga, Zadorra, and Huerva, respectively, characterized by low river flows (approximately 10 $m^3\ s^{-1}$ for PAM and 0.27 $m^3\ s^{-1}$ for T11). Furthermore, PAM receives the discharge of effluent wastewaters coming from WWTP5. Data presented in Fig. 5 suggests that the discharge of effluent wastewaters and the dilution factor (which is related with the river flow), once pharmaceuticals reach river waters may influence pharmaceutical concentrations detected in these matrices, as explained in more detail in the following section.

6.2 Contribution of Wastewater Effluents and Influence of Hydrological Conditions

In the first study conducted along the Ebro river basin, including the sampling campaigns in 2005, the dilution factor that occurred when pharmaceuticals entered receiving river waters through wastewater effluents was estimated. It was found that the dilution factor was controlled in the areas where effluents were discharged to the Ebro itself, which has a high flow. For instance, the dilution factor averaged 70 in Zaragoza, the area surrounding WWTP5 (1.9 $m^3\ s^{-1}$ of WWTP effluent was mixed with 150 $m^3\ s^{-1}$ of river flow). A similar situation occurred in Lleida (WWTP6), as 0.8 $m^3\ s^{-1}$ of effluent wastewater were mixed with 50 $m^3\ s^{-1}$ of river water, showing a dilution factor of approximately 63. Nevertheless, when wastewater effluents are discharged to secondary rivers, with lower flow, dilution factor is much lower. This is the situation in the tributary Arga, which receives the effluent from WWTP4 where the dilution factor was around a value of 10. Such factors could not be estimated the other WWTP since no data referring to the river flows in these points was found in the CHE database. This is illustrated in Fig. 6 where the most ubiquitous pharmaceuticals of each therapeutic group detected in both effluent waste and river waters in relationship to the dilution factor is shown.

The Ebro river basin is a Mediterranean basin, its flow being characterized by a high variability, which is closely controlled by seasonal rainfall. However, the scenarios foreseen by the IPCC [42] for the Mediterranean climatology seem to point not only to a general reduction of precipitation, but to an increase of seasonality. Therefore, rain will become more irregularly distributed along the year and a more frequent production of extreme hydrologic events, such as floods and droughts, is to be expected to take place in the near to medium future [4]. How these extreme events can affect the presence of pollutants in the water systems is not a straightforward question to answer. A flow decrease may have an obvious direct effect on the dilution factor, giving rise to an increase in the concentration of pollutants and thus to a corresponding increase of risk, towards the aquatic ecosystems [4]. This is illustrated in Fig. 7, which compares mean monthly river flows ($m^3\ s^{-1}$) with the sum of pharmaceutical compounds ($ng\ L^{-1}$) in the Ebro river in the area of Zaragoza and in the Arga river (one of its tributaries). In a great extent, pharmaceutical concentration increased in the periods where the river flow was lower, which is quite consistent (but not conclusive) that there is a kind of mutual relationship. However, for some campaigns, an opposite behavior is observed, which would indicate that dilution flow is not the unique factor governing the concentration levels of pharmaceuticals in receiving river water bodies. Sources of variability are diverse [43] and dilution (flow) is only one among many others. Thus, for instance, analytical error, compound environmental variability due changes of temperature, sediment remobilization, seasonal use of certain drugs, etc., are just a few to be taken into account as contributors to the overall uncertainty.

Nevertheless, a more comprehensive study and more representative data should be necessary to establish sound conclusions about this subject.

7 Risk Assessment and Ecotoxicological Implications

Environmental risk assessment studies to estimate hazards associated to the occurrence of pharmaceuticals are fundamental and a prerequisite for a comprehensive protection of the environment. The current US and European regulatory guidance requires new pharmaceuticals to undergo standard acute toxicity tests (to algae, Daphnia magna, and fish) if the PEC or MEC of the active ingredient is $>1\ \mu g\ L^{-1}$ for the US legislation or $10\ ng\ L^{-1}$, according to the European threshold safety value, set by the European Medicines Agency (EMEA). For the compounds whose

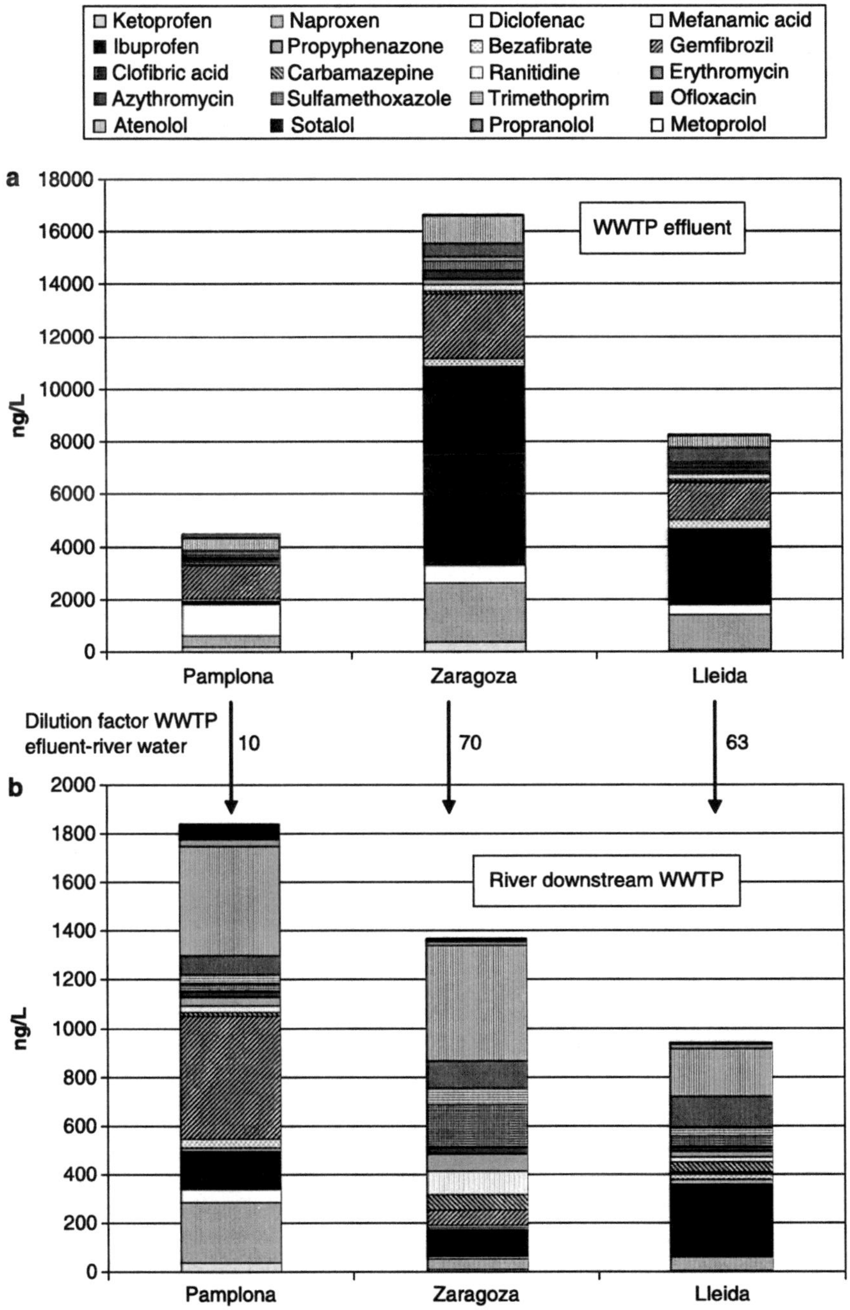

Fig. 6 Concentrations (ng L^{-1}) of some of the most ubiquitous pharmaceuticals detected in (**a**) wastewater effluents and (**b**) river water downstream of three WWTP in the Ebro river basin in relationship to the dilution factor

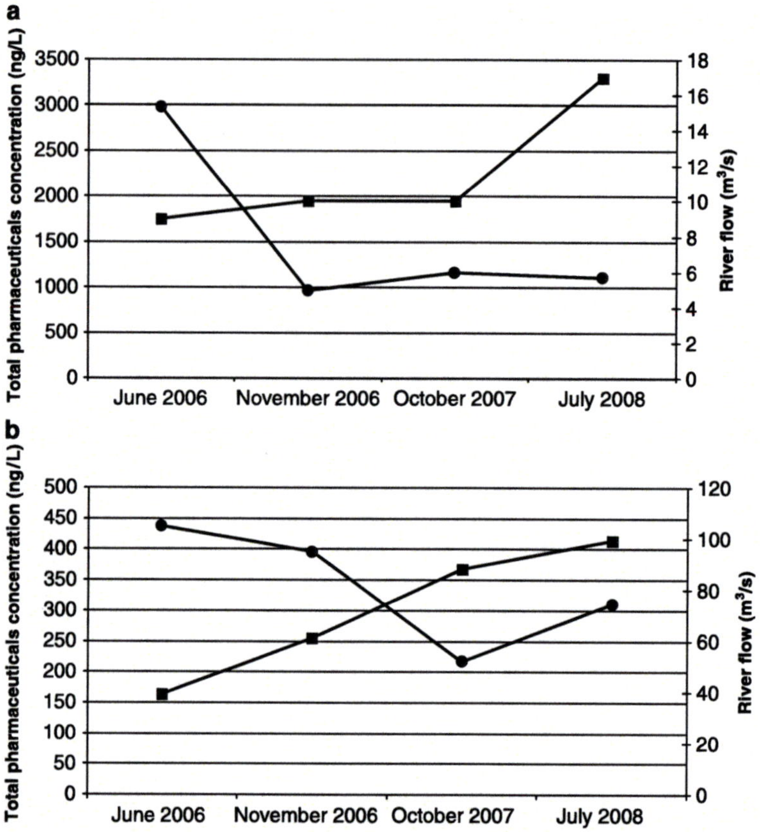

Fig. 7 Comparison between mean monthly flow ($m^3\ s^{-1}$) with the sum of all concentrations levels of the pharmaceuticals detected in (**a**) Arga river the tributary receiving the discharge of WWTP4, and (**b**) Ebro river in Zaragoza, which receives the discharge of WWTP5

PEC exceed these values, as a second tier in the ERA procedure, predicted no-effect concentrations (PNEC) are calculated. PNECs are extrapolated by dividing acute toxicity data, $E(L)_{50}$ values, obtained from standard toxicity tests, or chronic toxicity data, usually expressed as NOEC (nonobserved effect concentrations) by an appropriate assessment factor [22, 44]. In fact, $E(L)_{50}$ values are mainly used when NOEC values are not available. Assessment factors are applied to the data, to take into account the uncertainty in extrapolating laboratory data to environmental considerations. Although it is very difficult to estimate if adverse effects to non target organisms will occur at environmental levels, the hazard quotient could be a useful measure that can be employed to characterize potential ecological risk of a stressor, in this case a pollutant [45]. In most risk assessment approaches, based on EMEA guidelines, this quotient is calculated as the ratio between PEC and PNEC [46, 47]. However, other authors used Measured Environmental Concentrations

(MEC) instead of PEC to evaluate risks posed by pharmaceuticals in a specific site [48]. These quotients are determined for every compound present (or predicted) in the environment, and can be aggregated, usually by simple addition. Moreover, according to the WFD, hazard quotients have to be calculated using taxa of three different representative trophic levels of the ecosystem (namely algae, *Daphnia magna* or fish). Then, if these ratios are lower than 1 (PEC/PNEC<1), no further assessment is necessary [22] whereas if it equals or higher unity, there is a potential environmental hazardous situation and a more extended effect analysis has to be conducted (water sediment effects, specific effects on microorganisms, bioaccumulation study). This approach, which is actually based on short term ecotoxicological lab determinations, is valuable since it provides an a priori identification of risk. However, it does not necessarily reflect the real ecosystem situation, since many additional concurrent factors (stressors) can be equally present.

In this context, in the long term study which covered four sampling periods (between 2006 and 2008), risks towards algae, daphnids and fish were evaluated in both surface and effluent wastewaters [15]. In this case, hazard quotients were estimated for the most representative pharmaceuticals and by dividing measured environmental concentrations (MEC) and PNECs. MECs corresponded to maximum levels detected in both matrices, taking into account all sampling sites, in order to assess risks in the most extreme situations, while PNECs were derived from acute toxicity data (EC_{50}) from the literature divided by an assessment factor of 1,000 [49]. When no experimental values were available, EC_{50} data estimated wit ECOSAR was used [50]. EC_{50} values used to calculate PNECs are presented in Table 4. In fact, the lack of chronic toxicity data was a major hindrance to the effective risk assessment estimation, since pharmaceuticals are most likely to induce chronic rather than acute toxic effects. However, as mentioned before, the use of EC_{50} values, to predict PNEC, is a widely used approach. Moreover, EC_{50} data for all substances was used in order to follow the same criteria for all pharmaceuticals when calculating PNEC values. In this study, the overall relative order of susceptibility was estimated to be algae>daphnia>fish in both matrices. However, in river waters few substances were more sensitive to daphnia rather than algae. Results pointed out that no significant risks could be associated to the presence of pharmaceuticals in river waters, with the exception of the antibiotic erythromycin, the lipid regulator clofibric acid and the antidepressant fluoxetine for daphnia and the antibiotic sulfamethoxazole for algae. Regarding wastewaters, only the cholesterol lowering agent atorvastatin to fish and the antibiotics erythromycin to daphnia and sulfamethoxazole and tetracycline to algae posed an ecotoxicological hazard. Furthermore, some substances presented values close to one, indicating that the margin of safety in these types of waters is narrow.

On the other hand, overall HQs for each individual river site, bioassay and sampling campaign were calculated as described by Ginebreda et al. [12] and using the same EC_{50} values as described above (see Table 4) to calculate PNECs. The fact of considering joint effects (overall hazard quotients) instead of single compound hazards was because these indexes are supposed to follow an additive model for the different compounds involved [12]. Then, overall HQs were

Table 4 EC_{50} (in mg L^{-1}) used to calculate PNEC (by dividing EC_{50} by an assessment factor (AF) of 1,000), for fish, daphnids (*Daphnia magna*) and algae for some of the most ubiquitous pharmaceuticals detected in environmental waters

Compounds	Fish EC_{50} (mg L^{-1})	*Daphnia magna* EC_{50} (mg L^{-1})	Algae EC_{50} (mg L^{-1})	References
Naproxen	34*	15*	22*	[50]
Ibuprofen	5*	9.02	4	[50–52]
Diclofenac	532*	22	14.5	[46]
Acetaminophen	378	9.2	134	[46]
Salycylic acid	1.29*	59*	48*	[46, 50, 52]
Ketoprofen	32*	248*	164	[50]
Indomethacine	3.90*	26*	18*	[50]
Phenazone	3*	6.7*	1.1*	[50]
Propyphenazone	0.8*	3.5*	1.0*	[50]
Codeine	238*	16*	23*	[50]
Bezafibrate	6	30	18	[53]
Gemfibrozil	0.9	10.4	4	[53]
Atorvastatin	0.086			[53]
Pravastatin	1.8			[53]
Clofibric acid	53	0.11	192	[53]
Fluoxetine	1.70*	0.51	0.80*	[54]
Carbamazepine	35.4	76.3	85	[45, 47]
Salbutamol	38*	30*	36*	[50]
Ranitidine	1,076*	63*	66*	[50]
Ofloxacin		31.75		[55]
Ciprofloxacin	2.45×10^5*	991*	938*	[50]
Sulfamethoxazole	562.5	25.20	0.027	[45, 55]
Sulfamethazine	517*	4*	38*	[50]
Clarithromycin	280	25.72	0.09	[46, 47, 55]
Erithromycin*	61.5	7.8*	4.3*	[51]
Roxithromycin*	50.29	7.1	4*	[51, 56]
Tetracycline	220	6*	4*	[46]
Trimethoprim	795*	120.7	16	[45, 46, 50]
Diazepam	28*	14.1	16.5	[52]
Atenolol		200		[53]
Metoprolol	944	63.9	7.9	[46, 47]

Values indicated with * mean that the EC_{50} is estimated with ECOSAR. Data were taken from [50]

calculated, for each organism, as the addition of all individual hazard quotients (hq) for each pharmaceutical. Values achieved for each bioassay are indicated in Fig. 8. Again, and coinciding with what has been mentioned before, the order of susceptibility in river waters was estimated to be algae>daphnia>fish. Comparing each organism response within the different monitoring campaigns, while overall HQ were lower than unity for fish, values higher than one were estimated for some sites (see Fig. 8).

On the other hand, for all bioassays, sites showing highest overall HQs were the ones located close to the most populated areas and, moreover, the ones where the river flow was lower (such as PAM and T3, located nearby the cities of Pamplona and Vitoria, respectively and T9 and T11 close to Zaragoza). All these sampling sites correspond to Ebro tributaries. On the other hand, and as expected, points

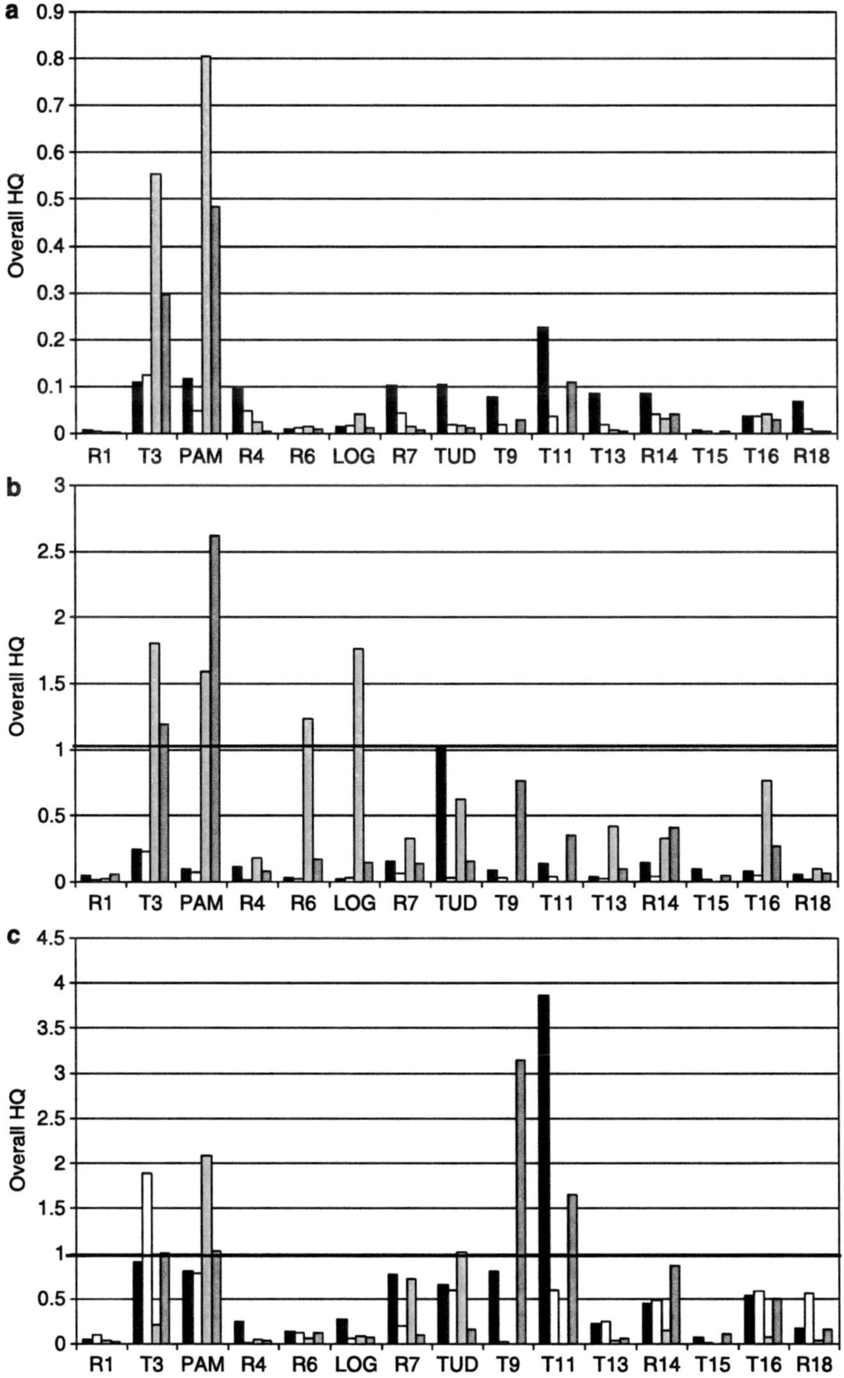

Fig. 8 Overall HQs for each bioassay: fish (**a**) *Daphnia magna* (**b**) and algae (**c**) in every sampling site

corresponding to the Ebro itself presented lower overall HQ, except for a couple of sites during the campaigns carried out in November and June in 2006. By comparing the four different monitoring studies, it was generally observed that HQs were higher for the campaigns performed in June and November 2006. This could be attributed to the fact that, at that period, a 2-year drought was ending and that rains occurring during the hydrological year 2007–2008 may have helped to increase the water reserves in the basin [57].

Moreover, and in order to estimate which compounds could be the ones inducing major hazards to the aquatic organisms tested, the contribution of each compound within every single HQ was evaluated, for the four sampling campaigns and bioassay, by normalizing individual hq to 100, according to the following expression [12]:

$$hq(\%) = \frac{hq}{\mathrm{HQ}}.$$

For brevity, results achieved for each bioassay in only one of the sampling campaigns are indicated in Fig. 9. Even though only one monitoring is illustrated, the most significant pharmaceuticals coincide in all the periods investigated, leading to similar conclusions. Therefore, substances inducing higher risks for fish were the cholesterol regulator atorvastatin, the nonesteroideal antiinflammatory drugs ibuprofen, naproxen and the analgesics salicylic acid and propyphenazone. For daphnia, the most hazardous substances were the lipid regulator clofibric acid, the antibiotic erythromycin, the analgesics acetaminophen (paracetamol) and the NSAIDs ibuprofen and naproxen as well. And finally, for algae, most dangerous pharmaceuticals were found to be the antibiotics sulfamethoxazole and erythromycin, the analgesic propyphenazone and the NSAIDs ibuprofen, diclofenac, and naproxen.

To sum up, it could be concluded, in a general extent, that risks are expected to be higher in areas with lower river flow and that compounds found to induce major hazards coincide with the ones found at the highest concentrations, such as the NSAIDs ibuprofen, diclofenac, naproxen and acetaminophen, lipid regulators, and antibiotics.

The fact that lower risks occur in areas with high river flow indicates that the dilution factor once pharmaceuticals enter river waters can efficiently mitigate possible environmental hazards, while in areas where effluent wastewaters represent a big percentage of the receiving river water flow hazards to aquatic organisms may increase.

8 Strategies to Reduce Pharmaceutical Discharge into Surface Waters

Due to their beneficial health effects and economic importance, the reduction of drug inputs into the environment through restricting or banning their use is not possible. Moreover, the use of pharmaceutical compounds is expected to grow with

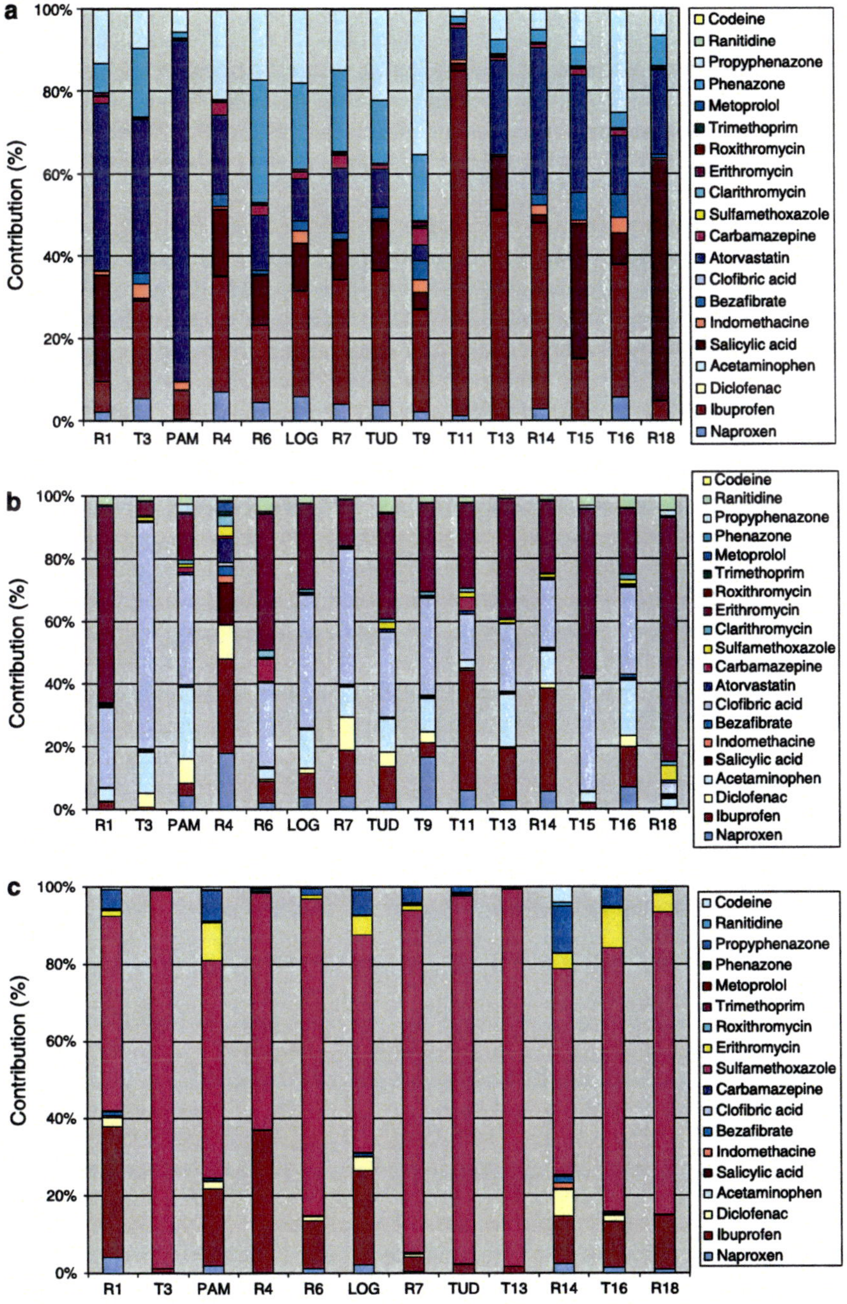

Fig. 9 Percent contribution of different compounds to hazard indexes for (**a**) fish in the monitoring carried out in June 2006, (**b**) *Daphnia magna* in October 2007, and (**c**) algae in November 2006

the increasing age of the population. A wide range of protective actions could be implemented in order to reduce the introduction of pharmaceutical compounds into the environment. Some of these strategies would be based on the regulation of their environmental pathways, perhaps at source through labeling of medicinal products and/or developing disposal and awareness campaigns. Other options are the addition of sewage treatments in hospitals, and to enhance current wastewater treatment techniques in order to eliminate such polar pollutants more efficiently. Some examples of these treatment technologies are Membrane Bioreactors (MBR), ozonization, Advanced Oxidation Processes (AOPs), or other advanced option treatments.

MBR technology is considered the most promising development in microbiological wastewater treatment. It integrates biological degradation of organic matter present in wastewater with membrane filtrations, thus surpassing the limitations of conventional CAS treatment (e.g., limited operational SRT, sludge-settling characteristics). Several studies [24, 30, 58–60] confirmed an advantage of MBR over CAS concerning the reduction of pharmaceutical residues. However, MBR technology is sometimes seen as high risk and prohibitively costly compared with CAS and other more established technologies [4]. Further improvement of the process will increase its cost-effectiveness and MBR technology is expected to play a key role in wastewater treatment in the coming years, where it may ensure enhanced elimination and biodegradation of PhACs, thus reducing the environmental risk posed by those compounds [4].

Regarding ozonization, it is only applied in a limited number of WWTPs after secondary treatment [61]. Several investigations have proven that it is a very effective technique to eliminate pharmaceutical [25, 62, 63]. Oxidation reactions take place due to direct reaction with ozone (O_3), which are very selective or with free $^{\bullet}OH$ radicals, which are generated by ozone decomposition and are very powerful and not selective oxidants. In advanced oxidation processes, O_3 is completely transformed onto $^{\bullet}OH$ radicals and they are recommended when compounds are ozone resistant.

9 Conclusions and Future Research Needs

The analysis of both waste and river waters indicated that pharmaceuticals are widespread pollutants along the Ebro river basin. Removal rates of PhACs in the WWTPs investigated suggested that conventional wastewater treatments are unable to completely remove most of the pharmaceuticals under study. High half-lives observed for the majority of pharmaceuticals in WWTP are an indicator that, in order to enhance compound degradation, higher hydraulic retention times should be required. Nowadays there is a great variety of technologies to upgrade wastewater treatment and to achieve high quality treated effluents. Advanced treatments are a promising tool to eliminate pharmaceuticals more efficiently than conventional treatments. Nevertheless, it is necessary to invest efforts in investigating degradation products generated and their toxicity.

The wide spectrum of substances detected in receiving river waters indicates that WWTP outlets are major contributors of pharmaceuticals in the aquatic environment. However, wastewater treatment must be an obligatory and final treatment step prior to their release into the aquatic media, since load of pharmaceuticals in outlets were considerably reduced after treatment. Dilution factor is controlled in the Ebro river but in other areas where the river flow is low, effluents may represent a significant percentage of the total flow of the river. In fact, higher concentrations of total pharmaceuticals were found in areas with lower river flow, and this could situation could be enhanced in drought periods.

The calculation of hazard quotients (HQs) was a useful tool to estimate the hazards that the occurrence of PhACs may pose to aquatic organisms. It was estimated that the overall relative order of susceptibility was algae>daphnia>fish. Results indicated that the reduction of pharmaceuticals concentration after wastewater treatment, as well as the dilution factor once they are discharged into the receiving river waters, efficiently mitigates possible environmental hazards. Nevertheless, risks are expected to be higher in areas with low river flow.

Results presented in this study clearly suggest that current research regarding the occurrence and fate of pharmaceuticals is required on the following topics:

- There is a need to increase the knowledge about the fate of pharmaceuticals during sewage treatment for implementation of *better removal techniques*.
- Monitoring programs should focus not only on parent compounds but also on metabolites (i.e., conjugates). Moreover, attention should be paid on elucidating the transformation products generated after wastewater treatment and on evaluating their toxicity.
- An important question that should be addressed is whether pharmaceutical residues are *bioavailable* and, if so, what the environmental impact will be.
- Many pharmaceuticals need more investigation about their potential long-term eco-toxicological effects. There is also a general *lack of chronic toxicity data* on pharmaceuticals, in particular in fish. Furthermore, the potential of *combined effects* of pharmaceutical mixtures should be addressed.

References

1. Kolpin D, Furlong E, Meyer M, Thurman E, Zaugg S, Barber L, Buxton H (2002) Environ Sci Technol 36:1202
2. Ternes T (1998) Water Res 32:3245
3. Radjenovic J, Petrovic M, Barceló D (2009) Trends Anal Chem 28:552
4. Petrovic M, López de Alda M, Díaz S, Postigo C, Radjenovic J, Gros M, Barceló D (2009) Philos Trans R Soc A367:3979
5. Alder A, Bruchet A, Carballa M, Clara M, Joss A, Löffler D, McArdell C, Milksh K, Omil F, Tukhanen T, Ternes T (2006). In: Ternes T, Joss A (eds) Consumption and occurrence of human pharmaceuticals, hormones and fragrances, Chap. 2. IWA, London

6. Fent K, Weston A, Caminada D (2006) Aquat Toxicol 76:122
7. Heier H, Tschäpe H (1984) Acta Hydrochim Hydrobiol 12:47
8. Pickup R, Morgan J, Winstanley C (1993) In: Edwards C (ed) Manipulating genetically manipulated microorganisms in the environment. Wiley, Chichester, New York, p 62
9. Dahlberg C, Bergström M, Hermansson M (1998) Appl Environ Microbiol 64:2670
10. Brooks B, Chambliss C, Stanley J, Ramirez A, Banks K, Johnson R, Lewis R (2005) Environ Toxicol Chem 24:464
11. Witte W (2000) J Antimicrob Agents 14:321
12. Ginebreda A, Muñoz I, López de Alda M, Brix R, López-Doval J, Barceló D (2009) Environ Int. doi: 10.1016/j.envint2009.10.003
13. Lacorte S, Raldua D, Martínez E, Navarro A, Díez S, Bayona J, Barceló D (2006) Environ Pollut 140:471
14. Gros M, Petrovic M, Barceló D (2007) Environ Toxicol Chem 26:1558
15. Gros M, Petrovic M, Ginebreda A, Barceló D (2010) Environ Int 36:15
16. Gros M, Petrovic M, Barceló D (2006) Anal Bioanal Chem 386:941
17. Fatta D, Achilleos A, Nikolaou A, Meriç S (2007) Trends Anal Chem 26:515
18. Gros M, Petrovic M, Barceló D (2009) Anal Chem 81:898
19. Gros M, Petrovic M, Gros M (2006) Talanta 70:678
20. Bound J, Voulvoulis N (2006) Water Res 40:2885
21. Castiglioni S, Fanelli R, Calamari D, Bagnati R, Zuccato E (2004) Toxicol Pharmacol 39:25
22. Cooper E, Siewicki T, Phillips K (2008) Sci Total Environ 398:26
23. Straub J (2002) Toxicol Lett 131:137
24. Radjenovic J, Petrovic M, Barceló D (2007) Anal Bioanal Chem 387:1365
25. Suárez S, Carballa M, Omil F, Lema J (2008) Rev Environ Sci Biotechnol 7:126
26. Metcalfe C, Miao X, Koenig B, Struger J (2003) Environ Toxicol Chem 22:2881
27. Clara M, Strenn B, Kreuzinger N (2004) Water Res 38:947
28. Clara M, Strenn B, Gans O, Martínez E, Kreuzinger N, Kroiss H (2005) Water Res 39:4797
29. Lindqvist N, Tuhkanen T, Kronberg L (2005) Water Res 39:2219
30. Gobel A, McArdell C, Joss A, Siegrist H, Giger W (2007) Sci Total Environ 372:361
31. Lishman L, Smyth S, Sarafin K, Kleywegt S, Toito J, Peart T, Lee B, Servos M, Beland M, Seto P (2006) Sci Total Environ 367:544
32. Miège C, Choubert J, Ribeiro L, Eusèbe M, Coquery M (2009) Environ Pollut 157:1721
33. Thomas P, Foster G (2005) Environ Toxicol Chem 24:25
34. Castiglioni S, Bagnati R, Fanelli R, Pomati F, Calamari D, Zuccato E (2006) Environ Sci Technol 40:357
35. Joss A, Zabczynski S, Gobel A, Hoffmann B, Löffler D, McArdell C, Ternes T, Thomnsen A, Siegrist H (2006) Water Res 40:1686
36. Batt A, Kim S, Aga D (2006) Environ Sci Technol 40:7367
37. Thomas K, Hilton M (2004) Mar Pollut Bull 49:436
38. Nakada N, Shinohara H, Murata A, Kiri K, Managaki S, Sato N, Takada H (2007) Water Res 41:4373
39. Wiegel S, Aulinger A, Brockmeyer R, Harms H, Löffler J, H R, Schmidt R, Stachel B, Von Tämpling W, Wanke A (2004) Chemosphere 57:107
40. Togola A, Budzinski H (2008) J Chromatogr A 1117:150
41. Tixier C, Singer H, Oellers S, Müller S (2003) Environ Sci Technol 37:1061
42. Christensen J, Hewitson B, Busuioc A, Chen A, Gao X, Held I, Jones R, Kolli R, Kwon W, Laprise R, Magaña Rueda V, Mearns L, Menéndez C, Raisären J, Rinke A, Sarr A, Whetton P (2007). Regional climate change projections. In: Solomon S, Qin D, Manning M, et al (eds) The physical science basis contribution of working group I to the fourth assessment report of the intergovernamental panel on climate change. Cambridge University Press, Cambridge, United Kingdom and New York, USA
43. Matthies M, Berding V, Beyer A (2004) Environ Toxicol Chem 3:2494
44. Bound J, Voulvoulis N (2004) Chemosphere 56:1143

45. Kim Y, Choi K, Jung J, Park S, Kim P, Park J (2007) Environ Int 33:370
46. Grung M, Kallqvist T, Sakshaug S, Skurtveit S, Thomas K (2008) Ecotoxicol Environ Saf 71:328
47. Huschek G, Hansen P, Maurer H, Krengel D, Kayser A (2004) Environ Toxicol 19:226
48. Santos J, Aparicio I, Alonso E (2007) Environ Int 33:596
49. Sanderson H, Johnson R, Reitsma T, Brain R, Wilson C, Solomon K (2004) Regul Toxicol Pharmacol 39:158
50. Sanderson H, Johnson R, Wilson C, Brain R, Solomon K (2003) Toxicol Lett 144:383
51. Lee Y, Lee S, Lee D, Kim Y (2008) Environ Toxicol Pharmacol 26:216
52. Stauer-Lauridsen F, Birkved M, Hansen L, Holten Lutzhoft H, Halling-Sorensen B (2000) Chemosphere 40:783
53. Hernando M, Agüera A, Fernández-Alba A (2007) Anal Bioanal Chem 387:1269
54. Henry T, Kwon J, Armbrust K, Black M (2004) Environ Toxicol Chem 23:2229
55. Isidori M, Lavorgna M, Nardelli A, Pascarella L, Parrella A (2005) Sci Total Environ 346:87
56. Choi K, Kim Y, Jung J, Kim M, Kim C, Kim N, Park J (2008) Environ Toxicol Chem 27:711
57. Postigo C, López de Alda M, Barceló D (2010) Environ Int 36:75
58. Lesjean B, Gnirrs R, Buisson H, Keller S, Tazi-Pain A, Luck F (2005) Water Sci Technol 52:453
59. Quintana J, Weiss S, Reemtsma T (2005) Water Res 39:2654
60. Reif S, Suárez S, Omil F, Lema J (2008) Desalination 221:511
61. Paraskeva P, Graham N (2002) Water Environ Res 74:569
62. Huber M, Gobel A, Joss A, Hermann N, Loffler D, McArdell C, Ried A, Siegrist H, Ternes T, Von Gunten U (2005) Environ Sci Technol 39:4290
63. Ternes T, Stuber J, Hermann N, McDowell D, Ried A, Kampmann M, Teiser B (2003) Water Res 37:1976

Mercury in Aquatic Organisms of the Ebro River Basin

Luis Carrasco, Josep M. Bayona, and Sergi Díez

Abstract It is well known that chlor-alkali industry has traditionally used the mercury (Hg)-electrolysis process to produce mainly chlorinated solvents. Unfortunately, a fraction of the Hg used during production is released through drainage into the aquatic environment where it can be incorporated to biota. In Europe, Spain and Germany are leading in the number of plants which still use this technology. Moreover, it should be highlighted that three out of the eight chlor-alkali plants in Spain which are still operating with this process are located in the Ebro River basin in the proximities of Sabiñánigo and Monzón cities – along the tributaries Gállego and Cinca Rivers, respectively – and Flix along the Ebro River. Therefore, the mid-low Ebro River watershed might be considered as a hot spot of aquatic mercury pollution in Spain.

This chapter focuses on all the information published up to date about total mercury (THg) and organomercury, with special emphasis on methylmercury (MeHg), in different aquatic organisms sampled along the Ebro River.

First, a brief explanation of the current knowledge regarding the sources and cycling of Hg and its transformation into MeHg is presented. Later, in this chapter, THg and limited data on organomercury levels in aquatic organisms of the Ebro River basin are detailed. The aquatic organisms most commonly studied in the Ebro River basin are zebra mussel, red swamp crayfish, and different fish species, namely European catfish, northern pike, common carp, rudd, roach, barbell, and bleak.

According to the different sentinel species analyzed, THg levels in specimens collected downstream from the impacted areas are 10–20 times greater than upstream levels. It clearly points out the relevance of chlor-alkali plants in terms of mercury river pollution.

L. Carrasco, J.M. Bayona, and S. Díez (✉)
Department of Environmental Chemistry, Institute of Environmental Assessment and Water Research, IDAEA-CSIC, Jordi Girona, 18-26, 08034 Barcelona, Spain
e-mail: sdsqam@cid.csic.es

D. Barceló and M. Petrovic (eds.), *The Ebro River Basin*,
Hdb Env Chem (2011) 13: 239–258, DOI 10.1007/698_2010_71,

Keywords Aquatic organisms, Chlor-alkali plants, Crayfish, Fish, Mercury, Methylmercury, Zebra mussel

Contents

1 Introduction

1.1 Mercury in Aquatic Ecosystems

Mercury (Hg) can occur in a large number of physical and chemical forms with a variety of properties, thus determining complex distribution, bioavailability, and toxicity patterns [1]. The most important chemical forms are elemental Hg (Hg^0), ionic Hg (Hg^{2+} and $Hg_2{}^{2+}$), and alkylmercury compounds. Because of their capability to permeate through biological membranes and to bioaccumulate and to biomagnificate through the trophic chain, alkylmercury compounds are the most toxic mercury species found in the aquatic environment [2].

Due to its chemical inertness, vaporizable nature (enthalpy of vaporization = 59.15 kJ/mol), and low water solubility (at 20°C, 2×10^{-6} g/g), elemental mercury vapor has over one year of residence time, long-range transport, and global distribution in the atmosphere [3–8].

Atmospheric deposition is an important source of mercury for surface waters and terrestrial environments that can be categorized into two different types, wet and dry depositions. Wet deposition during rainfall is the primary mechanism by which mercury is transported from the atmosphere to surface waters and land. Whereas the predominant form of Hg in the atmosphere is Hg^0 (>95%), is oxidized in the upper atmosphere to water-soluble ionic mercury, which is returned to the earth's surface in rainwater. In addition to wet deposition of Hg in precipitation, there can also be dry deposition of Hg^0, particulate (HgP), and reactive gaseous mercury (RGM) to watersheds [9–11]. In fact, about 90% of the total Hg input to the aquatic environment is recycled to the atmosphere and less than 10% reaches the sediments [12]. By current consensus, it is generally accepted that sulfate-reducing bacteria (SRB) are often involved in the transformation of inorganic into organomercury compounds [13]. Hence, methylmercury (MeHg) can be bioaccumulated and biomagnificated throughout the aquatic food chain from plankton to top fish predators [12, 14]. These bioaccumulation and biomagnification processes may result in mercury levels in top fish predators, which could exceed a million times the levels found in water. Fate and transport within an aquatic system, and burial in sediment, depends on such factors as particle loading, organic matter content (OM), stratification, redox chemistry, pH, and the presence/absence of inorganic sulfide phases. Factors favoring the methylation process are higher temperature, lower pH, higher organic matter content, anoxic conditions, and appropriate sulfate (200–500 μM) concentrations [15–17].

The best-documented outbreaks of Hg poisoning are the contamination of the Japanese Minamata Bay in 1956 [18] by an acetaldehyde plant and the poisoning of bread in Iraq in 1972 after wheat had been treated with MeHg fungicide [19]. These cases demonstrate that the intake of MeHg-contaminated food has extreme damages on human beings. Accordingly, aiming to safeguard human health, the Joint FAO/World Health Organization (WHO) Expert Committee on Food Additives (JECFA) [20] set a Provisional Tolerable Weekly Intake (PTWI) of 1.6 μg/kg body weight (bw) or 0.23 μg/kg/day for sensitive groups of the population, such as pregnant women. In the same sense, the U.S. Environmental Protection Agency (EPA) adopted a revised reference dose (RfD) for MeHg of 0.1 μg mercury per kg body weight per day [21]. Indeed, the European Commission Regulation (EC) No 466/2001 sets the maximum level of mercury in fish for human consumption at 0.5 mg/kg wet weight; whereas the maximum limit for top fish predators species, such as tuna fish (*Thunnus* spp.), anglerfish (*Lophius* spp.), or swordfish (*Xiphias gladius*), was set at 1 mg/kg wet weight [22].

The sources from which mercury is released to the environment can be grouped into four categories: (1) natural sources; (2) current anthropogenic releases from mobilization of mercury impurities in raw materials; (3) current anthropogenic releases resulting from mercury used intentionally in products and processes; and (4) re-mobilization of historically-deposited anthropogenic mercury releases worldwide. Important sources from the second group include the following: mercury mining; small-scale gold and silver mining; breakage of thermometers, manometers, thermostats, fluorescent lamps, auto headlamps; dental amalgam fillings; manufacturing, waste treatment, and incineration of products containing mercury; landfills; cremation; and chlor-alkali production. Regarding the last one, the "Chlorine Industry Review 2008–2009" published by Euro Chlor – the association of chlor-alkali process plant operators in Europe – , reported that overall European emissions in 2008 amounted to 0.92 g Hg/tonne chlorine capacity compared to 0.97 g Hg/t in 2007. The average mercury emissions for Western European countries decreased also to 0.91 g/t capacity (http://www.eurochlor.org/upload/documents/document352.pdf).

In Europe, Spain is the third country – after Germany (18) and France (10) – as regards the number of chlor-alkali plants, with a total of nine. Nevertheless, Spain is ahead along with Germany, in respect to the number of plants (8) which are still using Hg-electrolysis technology. The membrane technology represents

Fig. 1 Geographical area of the Ebro River basin and situation of the most relevant areas studied. Map shows the main tributaries

nowadays almost half (49%) of the installed production capacity in Europe, whereas the mercury process accounts for 34% at the beginning of 2009, continuing the progressive phase out of this technology. The diaphragm process still accounts for about 14% of the total capacity.

The Hg-electrolysis technology is one of the major point sources of Hg contamination, and its impact on the environment has been studied worldwide [23–26]. Although mercury cell chlor-alkali industry is obsolete in most of the European Union countries [27], in Spain it will be allowed until the end of 2010.

Finally, it should be mentioned that three out of the eight Spanish chlor-alkali plants operating with the mercury process are located in the Ebro River basin; in the cities of Sabiñánigo and Monzón – along the tributaries Gállego and Cinca Rivers, respectively – and Flix along the Ebro River (Fig. 1). Indeed, mercury emissions from the Flix and Monzón have already been reported [28]. Therefore, the mid-low Ebro River watershed might be considered as a hot spot of aquatic pollution of mercury in Spain.

1.2 Mercury Pollution in the Ebro River Basin

As a consequence of human activities, large quantities of inorganic pollutants coming from different sources (e.g., domestic, industrial, etc.) are released into the Ebro River, where a fraction of them can be accumulated in sediments. In this regard, the occurrence of inorganic Hg and other heavy metals in sediments from different stations along the river basin has been studied [29]. Despite a hot spot in the upper Ebro River close to industrial activities (paper mill, electroplating works, and smelters) described in this work, higher Hg levels have been detected in the lower course of the Ebro River [30]. This fact is attributable to the industrial waste discharge of a chlor-alkali plant, built in 1897 on one bank of the Flix reservoir (Fig. 1), which uses mercury electrodes to the electrolytic preparation of chlorine gas and caustic soda. The chlorine production of the plant is nowadays the highest in Spain, with a capacity of 15×10^4 Mg per year. At present, the major concern in Flix is a 35×10^4 Mg deposit of hazardous industrial solid waste, occupying an area of 4.2 Ha, containing high concentrations of Hg (170 μg/g) at surface sediments and up to 440 μg/g at 100 cm depth.

On the other hand, the industrial activity of the other two mercury cell chlor-alkali plants have caused important Hg pollution in Sabiñánigo (capacity of 25×10^3 Mg chlorine/year) and Monzón (31×10^3 Mg chlorine/year) (see Fig. 1), two small and industrial cities located in the middle course of the Gállego and Cinca Rivers, respectively, two tributaries of the Ebro River [31].

The impact of Hg pollution in both local populations of molluscs and fish has been evaluated [31–34]. High concentrations of mercury were found in all the tested species, thus indicating that surrounding areas of chlor-alkali plants are probably the major point sources of Hg pollution in the Ebro River basin. Furthermore, morphological anomalies and reproductive deficiencies were observed in several fish species at Flix reservoir [35]. Nevertheless, the middle and lower courses of the Cinca and Ebro Rivers, respectively, have a significant ecological, fishing, and agricultural value. In addition, it is suspected that non-negligible amounts of Hg might be transported downstream from the Flix reservoir toward the Ebro Delta area where wildlife reserve and Ramsar sites are located. Therefore, due to its multiple water demands (irrigation, human use, industrial activities, and other basin requirements) and the supposed human consumption of fish, the study of the mercury pollution in the Ebro River watershed is currently under enormous attention.

In the following sections, all the information published up to date about total mercury and organomercury in different aquatic organisms sampled along the Ebro River basin is reviewed.

2 Total Mercury in Aquatic Organisms of the Ebro River Basin

2.1 *Zebra Mussel*

Zebra mussel (*Dreissena polymorpha*) is a freshwater bivalve belonging to the Dreissenidae family. The common name, zebra, refers to the zebra stripes pattern on the shell and the scientific name, *polymorpha*, is derived from the many morphs or forms which occur in the shell color pattern, including albino and solid black or brown. The shell is triangular-shaped and medium-sized, reaching up to 3 cm maximum length [36].

Zebra mussel is native to Eastern Europe; however, it has spread all over the continent during the last 200 years [37]. In the mid 1980s, the species reached North America, where it colonized the Great Lakes and extended by the Mississippi River to the Gulf of Mexico [38]. In the Flix area, local residents found some zebra mussel in 2001. The occurrence of these species in the Ebro River is accidental. In fact, the larvae probably arrived brought in the water where living baits are transported. Other pathways could have been boats or anchors [39].

Initial monitoring of the zebra mussel's occurrence in the Ebro revealed that the species had invaded a 40 km area in the region, comprising the Mequinenza, Riba-roja, and Flix dams, lengthening the downriver [40]. The species causes severe impacts related to two distinctive capabilities, namely: attaching to solid surfaces in very high densities and removing planktonic organisms and particulates by pumping water [41].

In addition to its broad occurrence under contrasting conditions, easy collection, relatively long life span, and resistance to the accumulation of toxic chemicals, zebra mussels provide a link between the pelagic and benthic food webs, possibly creating a rapid pathway for contaminant transfer from sediments to predators such as common carp (*Cyprinus carpio*) and, occasionally, humans [32, 42–47]. Consequently, zebra mussel has been considered a potential sentinel organism for assessing Hg redistribution from point sources in the Ebro River and the degree of bioavailability to river and wetland food webs.

In fact, an extensive work on mercury and organomercury levels in zebra mussels collected in several sampling sites in the Flix reservoir area has been reported [32]. The specific sampling points were located in: (a) hot spot, (b) wildlife reserve located on the river bank opposite the factory, (c) the meander located immediately downstream from the dam, and (d) a reference site located 1–2 km upstream from the reservoir (Fig. 2).

2.1.1 THg Concentrations

The average THg concentration of all tissue samples ($n = 285$) was 0.2482 ± 0.2546 μg/g, wet weight, ww, ranging from 0.0148 to 0.8052 μg/g. The highest levels were found at the factory site, close to the waste dumping point, followed by the meander, wildlife reserve, and reference sites (Table 1). This fact indicates an important mercury uptake by zebra mussels at the hot spot and redistribution to areas located immediately downstream and near to the river bank opposite the dumping site.

The range of THg concentrations found in zebra mussels from the Flix dam area (0.16–6.81 μg/g, dry weight, dw) was the highest ever reported worldwide (Table 2).

In comparison with other studies, which reflect either diffusive pollution or sampling sites located relatively far from the pollution point source, the sampling

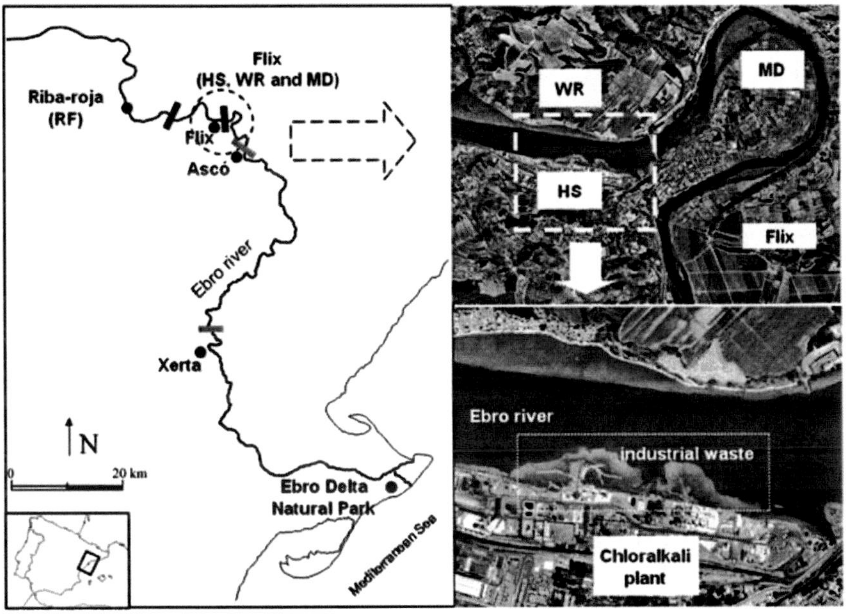

Fig. 2 Location of the lower Ebro River catchment, zebra mussel, crayfish, and fish species sampling sites in the Flix dam area. Samples were taken at several stations in the four areas (RF = control site upstream; WR = wildlife reserve; HS = hot spot or chlor-alkali plant, and MD = meander). Positions of dams (*black bars*) and overflow dams (*grey bars*) are indicated to illustrate the mutual isolation of aquatic organisms sampled

Table 1 Mean and range THg concentrations (μg/g ww) in zebra mussels at the different stations investigated in the Flix area. Adapted from [32]

Sampling site	*n*	Mean THg	Range
Factory (HS)	110	0.5248	0.3534–0.8052
Wildlife reserve (WR)	86	0.0660	0.0294–0.1372
Upstream (RF)	77	0.0276	0.0148–0.0491
Meander (MD)	12	0.1068	0.0846–0.1414

sites of this study were extremely close to the hot spot. This comprehensive study in the lower Ebro River watershed also demonstrated that concentrations decrease when increasing mussel size at all sites.

2.2 *Red Swamp Crayfish*

Red swamp crayfish (*Procambarus clarkii*) is a crustacean native to the South-central United States and Northern Mexico. It has been introduced in all continents except Australia and Antarctica, thus becoming the most widely distributed crayfish all over the world [52]. *Procambarus clarkii* is an omnivorous species, feeding mainly on

Table 2 Comparison of mean concentrations of THg (μg/g dw) in zebra mussels and crayfish from the Flix area collected at diverse worldwide locations. Adapted from [32]

Location	Year/ # sites	THg	Range	References
Zebra mussel (*Dreissena polymorpha*)				
New York State rivers, USA	1991/11	0.1028	0.050–0.380	[74]
Genesee	1991/2	0.2466	0.160–0.380	[74]
Lower Po River, Italy	1992–1993	0.08		[75]
Upper Mississippi River, USA	1995/14		0.02–0.05	[42]
Kleines Haff, Germany	1996/3	0.0837	0.032–0.218	[77]
Northern Italian lakes, Italy	1996/5	0.0704	0.049–0.158	[76]
St. Lawrence River, Canada	1996/14	0.1355	0.103–0.220	[78]
Niagara River, NY, USA	1995/3	0.0931	0.050–0.120	[79]
Lower Ebro River, Spain	2006/4	1.57	0.16–6.81	[32][b]
Flix reservoir, Spain	2006/1	4.37	3.70–6.81	[32][c, d]
Lower Ebro River, Spain	2006/4	1.06		[48][b]
Flix reservoir, Spain	2006/1	3.01		[48][c]
Crayfish (*Procambarus clarkii*)				
Atchafalaya River Basin, USA	1986/2	0.2435	0.062–0.494	[49]
Lower Ebro River, Spain	2006/3	1.49[a]	0.056–3.11	[48]
Flix reservoir, Spain	2006/1	2.67		[48][c]
Cache Creek Settling Basin, USA	1998	0.670		[50]
El Entredicho, Almadén, Spain	2004		2.38–9.06	[51]

[a]No meander station included
[b]All stations (*UP*, *WR*, *FT*, and *MD*)
[c]Only FT station
[d]Value in dry weight basis assuming a moisture content of 88%

algae and other plant material, aquatic insects, snails, and detritus, whereas it forms part of the diets of certain fish, mammals, birds, and even humans [50].

The species dig holes at the bottom and slopes of lakes and canals, causing severe impacts on invaded ecosystems. In the Ebro River, the crayfish is considered responsible for the disappearance of native fauna such as newts, frogs, and the autochthonous crayfish. Furthermore, it affects economically the rice crops in the Delta [53].

As zebra mussel, red swamp crayfish can be used to monitor the aquatic environment for Hg pollution owing to its capacity to accumulate metals in its tissues [54]. Other features, namely abundant populations, long life cycle, widespread distribution, and relatively sedentary lifestyle, also contribute to use this crayfish as bio-indicator of monitoring environmental quality [55].

2.2.1 THg Concentrations

In the Flix area, the determination of THg levels in muscle of red swamp crayfish individuals sampled has been reported [48]. Several specimens were collected in the same stations as zebra mussels.

Mean concentration of THg in tissue pooled samples ($n = 34$) was 1.49 ± 0.22 μg/g dw, ranging from 0.056 to 3.11 μg/g. The highest level was found at the factory site, located close to the waste dumping site. This result indicates an important mercury uptake by the specimens located at the hot spot and a limited redistribution to other areas close to the opposite river bank.

In comparison, higher THg levels (up to 9 μg/g dw) have been reported in the Almadén mercury mining district (Ciudad Real, Spain) [51], which can be regarded as the largest geochemical anomaly of mercury on Earth.

2.3 Fish

2.3.1 Description of the Fish Species Studied

Mercury levels have been determined in seven fish species (European catfish, northern pike, common carp, rudd, roach, barbell, and bleak) from the three mercury pollution hotspots of the Ebro River basin previously described, namely Sabiñánigo, Monzón, and Flix areas and downstream sites (Fig. 1). These species display different diet and habitat and belong to different families: siluridae, esocidae, and cyprinidae.

European catfish (*Silurus glanis*), a siluridae species also known as wels or sheatfish, is the largest European freshwater fish, reaching up to 2.5 m length and more than 100 kg weight. European catfish is considered an opportunistic and omnivorous depredator (with nocturnal predatory activity), consisting its diet on red swamp crayfish and fish, namely: mosquitofish (*Gambusia holbrooki*), northern pike (*Esox lucius*), pumpkinseed (*Lepomis gibbosus*), mullet (*Liza* sp.), and mainly cyprinid species such as common carp (*Cyprinus carpio*), Ebro barbel (*Luciobarbus graellsii*), bleak (*Alburnus alburnus*), roach (*Rutilus rutilus*), and rudd (*Scardinius erythrophthalmus*). Interestingly, birds have also been detected in catfish stomachs [56].

Northern pike (*Esox lucius*), an esocidae species, is considered one of the most widely distributed freshwater fish [57]. It is a piscivorous species, consisting its diet on common carp (*Cyprinus carpio*) and roach (*Rutilus rutilus*). Nevertheless, in the absence of prey fish, invertebrate feeding could be important for this species [58, 59].

Common carp (*Cyprinus carpio*) is a cyprinid fish considered omnivorous, feeding on detritus, plant material (including debris, diatoms, and seeds), molluscs, amphipods (*Echinogammarus* sp.), and phantom midge larvae, mainly by grubbing in sediments [60]. The fact that carp is extremely tolerant to pollution suggests that carp may be a good indicator of environmental degradation [61].

Rudd (*Scardinius erythrophthalmus*), a medium-sized cyprinid species, is an herbivorous; the bulk of its diet consists of plants and algae. However, small invertebrates, such as nematodes, larvae of a freshwater shrimp (*Atyaephyra desmaresti*), larvae of small, littoral chironomids (*Paratanytarsus* sp. and *Cricotopus flavocinctus*), and particularly microcrustaceans such as ostracods, are included.

Roach (*rutilus rutilus*), a medium-sized cyprinid fish, is a planktonic and benthic species, feeding mainly on cladocerans (*D. longispina*), detritus, plant debris, amphipods (*Echinogammarus* sp.), filamentous algae, and ostracods. Roach can thrive on poor quality, even polluted water and displays more capacity of adaptation to different kinds of food than rudd [62].

Barbel (*Barbus graellsii*), a cyprinid species endemic to the Ebro River [63], is a bottom-feeding fish. Finally, the cyprinid fish, *Alburnus alburnus*, also known as bleak, feed mainly on plankton, including crustaceans and insects (http://www.fishbase.org).

2.3.2 Mercury Concentrations in the Ebro River Tributaries

The first well-documented case study of mercury pollution in fishes and its distribution within the fish community of the Ebro River is focused on the Gállego and Cinca tributaries. Mercury levels in barbel, common carp, and northern pike sampled immediately upstream from Gurrea de Gállego and Ballobar, two small towns in the lower course of both tributaries, have been described [31]. Both sampling points are located downstream from Sabiñánigo and Monzon chlor-alkali plants, respectively. Mean THg concentrations in muscle tissue of barbel ($n = 27$), common carp ($n = 5$), and northern pike ($n = 7$) from the lower Gállego River were 2.00 μg/g ww, ranging from 0.25 to 4.41 μg/g; 0.74 μg/g ww (0.30–1.36); and 1.73 μg/g ww (1.20–2.61), respectively. On the other hand, mean THg concentrations in the muscle tissue of the specimens from the lower Cinca tributary were 1.96 μg/g ww (1.02–3.74); 1.44 μg/g ww (1.05–2.59); and 2.80 μg/g ww (2.74–2.87) for barbel ($n = 23$), common carp ($n = 16$), and northern pike ($n = 2$), respectively.

Northern pike (1.93 μg/g) exhibited a significantly higher mean mercury concentration than common carp (1.23 μg/g). This result indicates a biomagnification of mercury from common carp, a larger secondary consumer, to northern pike, a tertiary consumer. Nevertheless, barbel (1.98 μg/g), also a secondary consumer, showed a mean THg level similar to that of northern pike. Different behavioral habits may explain the high mercury levels found in barbel species.

Another well-documented case of feral fish chronically exposed to mercury through the effluent of the chlor-alkali plant is the case of the Cinca tributary [64, 65]. Barbel and bleak were collected upstream (S1) and downstream (S2) a chlor-alkali plant located at Monzon (Fig. 3). It is important to point out that there is no physical barrier between S1 and S2 to prevent fish upstream migration of suspicious contaminated fishes from S2.

In barbel, THg concentrations were determined in both muscle and liver, whereas in bleak, a whole fish homogenate was used for THg analysis. Mean THg concentrations in muscle tissue and liver of upstream barbels ($n = 12$) were 0.137 ± 0.065 μg/g ww, ranging from 0.079 to 0.327 μg/g ww and 0.054 ± 0.036 μg/g ww (0.015–0.148), respectively; whereas mean THg values of downstream barbels ($n = 11$) were 1.484 ± 0.554 μg/g ww (0.570–2.433) and 1.780 ± 0.761 μg/g ww (0.747–3.641), for muscle and liver, respectively. The differences found

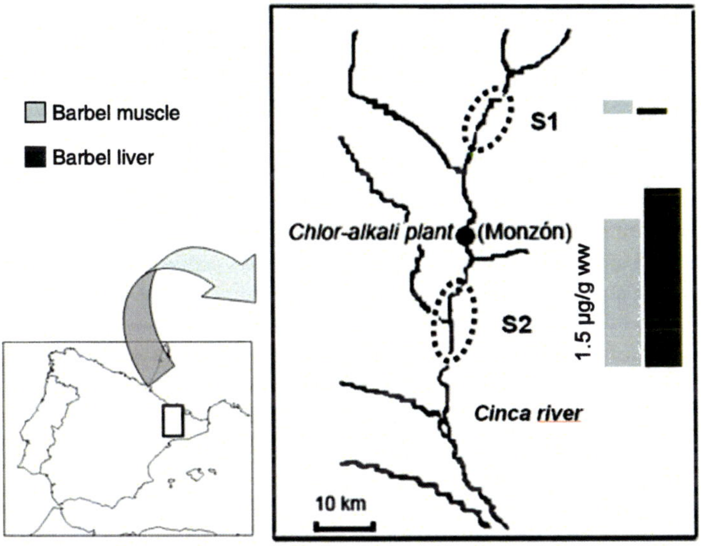

Fig. 3 Map with the sampling sites in the Cinca River, upstream (S1) and downstream (S2) of the chlor-alkali plant of Monzón showing THg levels. Adapted from [65]

in both muscle and liver of upstream and downstream barbels show clearly the impact of the mercury released into the Cinca tributary. In contrast, mercury pollution had no impact on the specimens upstream from Monzón.

In fish, muscle is the main target for organic mercury and liver for inorganic and metallic mercury [66, 67]. Liver/muscle ratios of THg were significantly higher in downstream barbels from the factory, indicating that the effluents from the chlor-akali plant in Monzón discharge large quantities of inorganic Hg.

In this regard, high liver/muscle ratios of mercury concentration were reported in fish exposed to the effluents from a chlor-alkali plant in a contaminated area [68, 69]. Then, liver of fish downstream from the plant also had very high inorganic mercury to THg ratios [69].

On the other hand, mean THg in bleak ($n = 10$) collected upstream from the chlor-alkali plant was much lower (0.083 ± 0.061 µg/g ww; ranging from 0.041 to 0.244 µg/g ww) than that reported in downstream specimens ($n = 20$) (0.927 ± 0.619 µg/g ww; ranging from 0.321 to 2.362 µg/g ww).

Hence, the low THg levels reported in both barbel and bleak specimens sampled upstream the chlor-alkali plant in Monzón, suggest that neither of these fish species migrate upstream the factory, and therefore the fishing area upstream from Monzón does not carry a very high risk to public health, at least for these two fish species studied.

In contrast, in a previous study [33] in which a less number of barbel were captured ($n = 6$–11), specimens shows much higher THg levels (9.90 µg/g ww (range: 2.20–17.86 µg/g ww)) in Monzon and much less in Zaragoza (0.17 µg/g ww (range: 0.08–1.79 µg/g ww)).

2.3.3 Mercury Concentrations in Flix Area and Downstream

The case study of Flix is another well-documented concern about fish that are chronically impacted by Hg throughout a chlor-alkali plant. Here, we review all the data available about the reported levels of Hg. A description of the biological effects is not the aim of this chapter and will not be covered, but we will mention whether or not these effects exist.

The determination of THg levels in liver, kidney, and muscle of common carp from the Flix area and sampling points of interest located downstream has been extensively reported [33, 34]. In this regard, the specimens were collected in five stations, namely Ribaroja reservoir (located 10 Km upstream from Flix); the Flix reservoir (chlor-alkali plant and a waste dumping site); Ascó, Xerta, and the irrigation channels of the Ebro Delta Natural Park, three sites situated 5.6 Km, 37 Km, and 70 Km downstream from Flix, respectively (Fig. 2).

In this study, sampling sites were selected according to accessibility, availability of fish, and mutual isolation regarding fish populations. The dam at Riba-roja forms a large water reservoir upstream Flix, whereas the Flix dam itself forms a much smaller reservoir. On the other hand, an overflow dam separates Ascó and Xerta sites, while between Xerta and Delta there are no physical barriers (see map in Fig. 2). To our knowledge, these four fish populations are physically isolated with each other. Nevertheless, some fishes may be carried downstream in occasionally high-flow episodes, especially throughout the overflow dam, although it seems unlikely for the 60-m high Riba-roja dam.

Mean THg concentrations in liver, kidney, and muscle of common carp from the different sampling points are listed in Table 3.

As can be seen, the highest levels were found in kidney, followed by liver and muscle, thus indicating that both kidney and liver accumulate mercury to higher concentrations than muscle. The same trends than those found in the Gallego and Cinca tributaries (Sect. 2.3.2) were observed for liver/muscle ratios (e.g., organic Hg vs. THg) in Flix and downstream sites, indicating that the effluents from the plant seem to contain large quantities of inorganic Hg.

Average THg concentrations increased progressively at Flix, Ascó, and Xerta sites and decreased at the irrigation channels of the Delta. This result evidences a mercury pollution transport downstream from the Flix plant until Xerta. Differences found when comparing water conductivity at Flix, Ascó, and Xerta sites [48] with

Table 3 THg concentrations (μg/g ww) in carp (n = 6–11) in the lower Ebro River basin

Location	Liver	Kidney	Muscle	References
Zaragoza	0.07 (0.01–0.20)[a]			[33]
Riba-roja	0.03 (0.03–0.05)[b]	0.07 (0.06–0.31)	0.17 (0.07–0.24)	[34]
Flix	0.10 (0.03–0.52)[b]	0.37 (0.11–1.80)	0.36 (0.17–0.57)	[34]
Flix	0.63 (0.27–1.12)[a]			[33]
Ascó	0.34 (0.08–4.01)[b]	2.43 (0.19–12.82)	0.77 (0.19–1.03)	[34]
Xerta	1.27 (0.43–2.20)[b]	3.84 (0.65–6.92)	0.96 (0.52–1.09)	[34]
Ebro Delta NP	0.20 (0.09–0.32)[a]			[33]

[a]Mean values
[b]Median values

Delta irrigation channels (Faria, personal communication) indicate a saline water intrusion in the irrigation channels of the Delta. This saline intrusion may cause a dilution of mercury pollution originated at the chlor-alkali plant, hence resulting in lower average THg concentrations in Delta specimens.

On the other hand, Hg content in sediments might be a key information to understand why the levels of Hg in the biota downstream are increasing. It is well recognized that the rate of remobilization of Hg from sediments (e.g., by desorption and/or methylation) depends on many factors including its chemical form and particle size. Variations of mercury contents in sediments are related to their granulometry and, consequently, to their capacity to retain mercury. Unfortunately, there is a lack of information about the Ebro River sediment characteristics (e.g., particle size, organic matter content, sulfur, and sulfide contents), which could be useful to understand the fate of Hg. However, as it has been previously discussed [34], the distribution of Hg levels in fish muscle is very consistent with the (poor) existing data on Hg pollution in the Ebro River sediments from Riba-roja to Delta. Current analysis of surface sediments shows that maximal values were found at Flix, on the factory residues at the discharge site, with values ranging from 15 to 170 μg/g of Hg [30, 70]. Upstream the Flix factory, mercury values in sediments were typically below 0.5 μg/g [70], whereas for sediments sampled downstream the Flix dam, in Asco, Xerta and up to the Delta, mercury concentrations range from 0.5 to 2 μg/g [30, 70].

According to these results, Hg values in sediments in the low Ebro River Basin seem to change little over the last decade [29]. In addition, it is important to point out that as described in Sect. 1.1 the fate and transport of Hg within an aquatic system, and burial in sediment, depends on several factors such as redox chemistry, pH, or local activity of methyl mercury producing bacteria, among others. Hence, it is perfectly likely that there may be areas downstream from the hot spot with a greater production of MeHg, or merely mercury may become more bioavailable, entering the trophic chain and affecting much more the biota living in these waters.

In summary, physiological differences between four populations of carps from the low Ebro River and their effects correlated well with the amount of Hg found in fishes, with the maximal levels of mercury in tissue and the highest biological impact relocated several kilometers downstream from the discharge site. This probably reveals that the observed effects are more related to long-term and continuous exposure to Hg rather than to current occasional short-term exposures.

Additionally, the determination of THg levels in muscle of some fish species within the Flix reservoir has been studied (Diez, unpublished results). As expected, THg concentrations increased with trophic level. A comparison between mercury levels in muscle tissue of the individual fish species ($p < 0.05$) produced the following order: European catfish > common carp > rudd > roach. Average THg concentrations (in ww) in muscle of European catfish ($n = 9$), common carp ($n = 30$), rudd ($n = 20$), and roach ($n = 5$) were 0.848 ± 0.476 μg/g; 0.333 ± 0.155 μg/g; 0.283 ± 0.163 μg/g; and 0.217 ± 0.199 μg/g, respectively. The relative trophic level of the analyzed species is based on a generalized knowledge of the species and stomach contents of the studied adult fish populations [56, 60, 62, 71].

3 Organomercury in Aquatic Organisms of the Ebro River Basin

Unlike THg, a limited data set about the occurrence of organomercury compounds in biota of the Ebro River has been noticed. MeHg concentrations were only determined in zebra mussel specimens in the Flix hot spot [32] as well as some fish species living in Flix dam (Diez, unpublished results).

3.1 MeHg Concentrations and MeHg/THg Ratios in Zebra Mussel and Fish

In the Flix area, organomercury was determined in the zebra mussel samples ($n = 110$) from the hot spot, and the mean value found was 0.308 ± 0.212 μg/g ww (range: 0.220–0.589 μg/g). MeHg comprised about 60% of the mean THg in zebra mussels, ranging from 50 to 80%. The ratio values reported in the Flix area are similar to those found by Cope and co-workers, 50%, range: 30–70% [42], and slightly higher than those measured in Dreissena polymorpha and Dreissena bugensis in Lake Erie, 30% [46].

On the other hand, mean MeHg levels in the European catfish muscle ($n = 9$), common carp ($n = 8$), rudd ($n = 7$), and roach ($n = 4$) within the Flix reservoir were 0.624 ± 0.379 μg/g ww; 0.290 ± 0.248 μg/g ww; 0.231 ± 0.125 μg/g ww; and 0.155 ± 0.095 μg/g ww, respectively. As a result, MeHg comprised about 80% (range: 50–98%) of the mean THg in all the fish species analyzed within the Flix reservoir (Diez, unpublished results).

Therefore, we can conclude that bivalves have MeHg/THg ratios of 50–80%, which are slightly smaller than the ratio found in fish (usually 80–100%) [72, 73]. The difference in this ratio between bivalves and fish may be due to the trophic level occupied by each organism, since fish occupy a higher trophic level, which facilitates food transfer, bioaccumulation, and biomagnification of organic forms of Hg (i.e., MeHg). Differences in the MeHg/THg ratio among size classes in mussel may be related to either changes in the diet or the detoxification capacity.

4 Concluding Remarks

A comprehensive overview of total mercury and MeHg levels in different aquatic organisms sampled along the Ebro River basin has been presented in this chapter. In this context, a few points were noticed:

(a) Lack of studies, sampling points, and/or monitoring programs in the upper Ebro River basin, e.g., upstream the city of Zaragoza. This fact is probably due to that

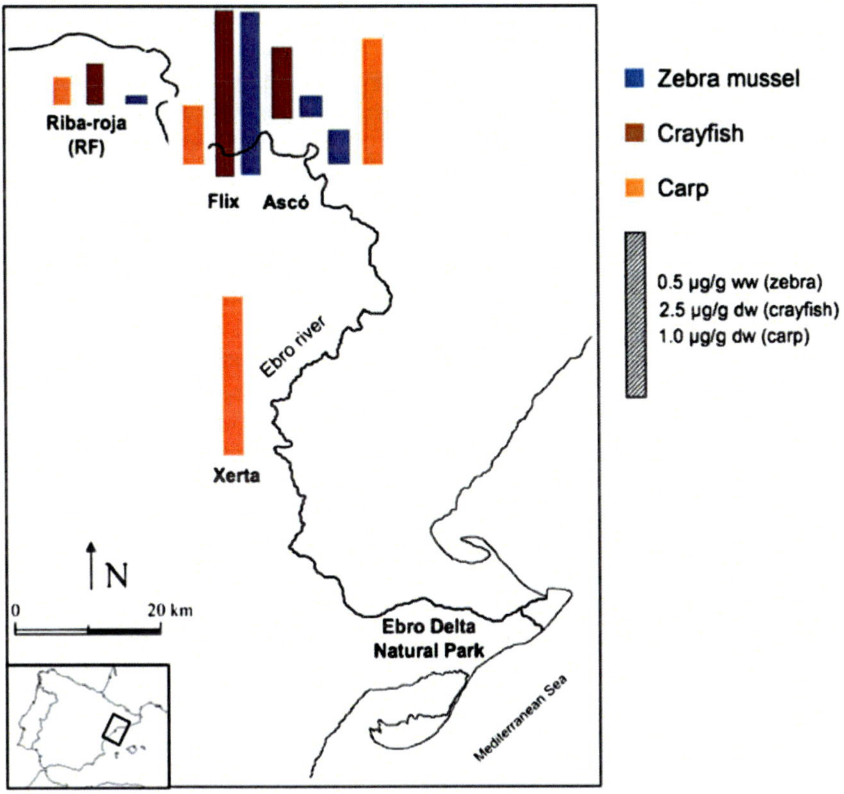

Fig. 4 THg concentrations (μg/g) in aquatic organisms at the different stations investigated in the Flix area and downstream sites

most of the research up to date has been focused around the three chlor-alkali plants, which are located in the middle-lower part of the basin.

(b) As compared to the THg values, there is a noticeable limited data on organomercury levels from biota and sediments of the Ebro River. In this respect, recent publications have shown the importance of knowing the amount of MeHg in aquatic organisms (e.g., ratio of THg/MeHg in fish tissue samples) and sediments. Then, further studies about this topic in biota are needed to better understand the bioaccumulation and biomagnification through the trophic chain and to find possible hot spots where key methylation processes in sediments may take place. With regard to sediments, neither data nor studies on Hg methylation processes are available.

(c) Some studies demonstrated that bivalves such as zebra mussel could be considered as potential sentinel organism for assessing Hg redistribution from point sources and the degree of bioavailability to the river. There is still a lack of studies on mercury and organomercury in target organisms (e.g., certain sentinel fish species, zebra mussel, etc.) of the Ebro River basin.

(d) In order to compare mercury pollution in different areas, a comprehensive monitoring program throughout the Ebro River watershed would be required. Currently, there are many different fish species analyzed and, unfortunately, we are not able to compare the data generated under various studies because organisms, species, and/or areas studied are different. Due to its widespread distribution and capacity of bioaccumulation, red crayfish and zebra mussel are proposed as reliable bioindicators of mercury pollution.

(e) The Hg contamination in resident zebra mussel also showed that the mercury impact at the Flix reservoir was essentially limited to the area surrounding the factory itself [32]. On the other hand, fish data prove the capacity of Hg pollution transport over long distances from the emission sites. Indeed, in the study in which the same fish species was collected upstream and downstream from a chlor-alkali plant [34], results show physiological differences between populations of carps with maximal levels of mercury in fish tissue. It is also noticed that the highest biological impact occurs in sites downstream (e.g., Ascó and Xerta) rather than at the hot spot located at Flix. Finally, as many of the specimens exceeded the maximal legal amount of mercury for human consumption (0.5 mg/Kg), these results point out the need for a strict control, which is expected to lead to a reduction of mercury emissions at the mid-low Ebro River (Fig. 4).

References

1. Leermakers M, Baeyens W, Quevauviller P, Horvat M (2005) Mercury in environmental samples: speciation, artifacts and validation. Trends Analyt Chem 24:383–393
2. Mason RP, Benoit JM (2003) In: Graig PJ (ed) Organometallic compounds in the environment. Willey, Chichester, p 57
3. Biester H, Kilian R, Franzen C, Woda C, Mangini A, Scholer HF (2002) Elevated mercury accumulation in a peat bog of the Magellanic Moorlands, Chile (53 degrees S) – an anthropogenic signal from the Southern Hemisphere. Earth Planet Sci Lett 201:609–620
4. Fitzgerald WF, Engstrom DR, Mason RP, Nater EA (1998) The case for atmospheric mercury contamination in remote areas. Environ Sci Technol 32:1–7
5. Mason RP, Fitzgerald WF, Morel FMM (1994) The Biogeochemical Cycling of Elemental Mercury – Anthropogenic Influences. Geochim Cosmochim Acta 58:3191–3198
6. Roos-Barraclough F, Martinez-Cortizas A, Garcia-Rodeja E, Shotyk W (2002) A 14 500 year record of the accumulation of atmospheric mercury in peat: volcanic signals, anthropogenic influences and a correlation to bromine accumulation. Earth Planet Sci Lett 202:435–451
7. Ross-Barraclough F, Shotyk W (2003) Millennial-scale records of atmospheric mercury deposition obtained from ombrotrophic and minerotrophic peatlands in the Swiss Jura Mountains. Environ Sci Technol 37:235–244
8. Szefer P, Czarnowski W, Pempkowiak J, Holm E (1993) Mercury and major essential elements in seals, penguins, and other representative fauna of the Antarctic. Arch Environ Contam Toxicol 25:422–427
9. Diez S (2009) Human health effects of methylmercury exposure. Rev Environ Contam Toxicol 198:111–132

10. Graydon JA, Louis VLS, Hintelmann H, Lindberg SE, Sandilands KA, Rudd JWM, Kelly CA, Hall BD, Mowat LD (2008) Long-term wet and dry deposition of total and methyl mercury in the remote boreal ecoregion of Canada. Environ Sci Technol 42:8345–8351
11. Lindberg S, Bullock R, Ebinghaus R, Engstrom D, Feng XB, Fitzgerald W, Pirrone N, Prestbo E, Seigneur C (2007) A synthesis of progress and uncertainties in attributing the sources of mercury in deposition. Ambio 36:19–32
12. Clarkson TW, Magos L, Myers GJ (2003) Human exposure to mercury: the three modern dilemmas. J Trace Elem Exp Med 16:321–343
13. Siciliano SD, O'Driscoll NJ, Lean DRS (2002) Microbial reduction and oxidation of mercury in freshwater lakes. Environ Sci Technol 36:3064–3068
14. Grandjean P, White RF, Nielsen A, Cleary D, Santos ECD (1999) Methylmercury neurotoxicity in Amazonian children downstream from gold mining. Environ Health Perspect 107:587–591
15. Benoit JM, Gilmour CC, Mason RP, Heyes A (1999) Sulfide controls on mercury speciation and bioavailability to methylating bacteria in sediment pore waters. Environ Sci Technol 33:951–957
16. Hammerschmidt CR, Fitzgerald WF (2004) Geochemical controls on the production and distribution of methylmercury in near-shore marine sediments. Environ Sci Technol 38:1487–1495
17. Kelly CA, Rudd JWM, Holoka MH (2003) Effect of pH on mercury uptake by an aquatic bacterium: implications for Hg cycling. Environ Sci Technol 37:2941–2946
18. Harada M (1995) Minamata disease – methylmercury poisoning in Japan caused by environmental-pollution. Crit Rev Toxicol 25:1–24
19. Greenwood MR (1985) Methylmercury poisoning in Iraq – an epidemiological-study of the 1971–1972 outbreak. J Appl Toxicol 5:148–159
20. FAO/WHO (2006) Summary and conclusions of the sixty-seventh meeting of the Joint FAO/WHO Expert Committee on Food Additives (JECFA), 20–29 June 2006. JECFA, Rome
21. USEPA (2005) Water quality criterion for the protection of human health. Methylmercury. U.S. Environmental Protection Agency, Office of Science and Technology, Office of Water, Washington, DC
22. EU (2001) Official Journal of the European Communities. pp 1–13
23. Biester H, Muller G, Scholer HF (2002) Estimating distribution and retention of mercury in three different soils contaminated by emissions from chlor-alkali plants: part I. Sci Total Environ 284:177–189
24. Degetto S, Schintu M, Contu A, Sbrignadello G (1997) Santa Gilla lagoon (Italy): a mercury sediment pollution case study. Contamination assessment and restoration of the site. Sci Total Environ 204:49–56
25. Landis MS, Keeler GJ, Al-Wali KI, Stevens RK (2004) Divalent inorganic reactive gaseous mercury emissions from a mercury cell chlor-alkali plant and its impact on near-field atmospheric dry deposition. Atmos Environ 38:613–622
26. Montuori P, Jover E, Diez S, Ribas-Fito N, Sunyer J, Triassi M, Bayona JM (2006) Mercury speciation in the hair of pre-school children living near a chlor-alkali plant. Sci Total Environ 369:51–58
27. OSPAR (2008) Implementation of PARCOM Decision 90/3 on reducing atmospheric emissions from existing chlor-alkali plants. ISBN 978-1-905859-88-7
28. OSPAR (2007) Mercury losses from the chlor-alkali industry (1982–2005). ISBN 978-1-905859-56-6
29. Ramos L, Fernandez MA, Gonzalez MJ, Hernandez LM (1999) Heavy metal pollution in water, sediments, and earthworms from the Ebro River, Spain. Bull Environ Contam Toxicol 63:305–311
30. Grimalt JO, Sánchez-Cabeza JA, Palanques A, Catalán J (2003) ACA/CIRIT final report
31. Raldua D, Pedrocchi C (1996) Mercury concentrations in three species of freshwater fishes from the lower Gallego and Cinca Rivers, Spain. Bull Environ Contam Toxicol 57:597–602

32. Carrasco L, Diez S, Soto DX, Catalan J, Bayona JM (2008) Assessment of mercury and methylmercury pollution with zebra mussel (Dreissena polymorpha) in the Ebro River (NE Spain) impacted by industrial hazardous dumps. Sci Total Environ 407:178–184
33. Lavado R, Urena R, Martin-Skilton R, Torreblanca A, del Ramo J, Raldua D, Porte C (2006) The combined use of chemical and biochemical markers to assess water quality along the Ebro River. Environ Pollut 139:330–339
34. Navarro A, Quiros L, Casado M, Faria M, Carrasco L, Benejam L, Benito J, Diez S, Raldua D, Barata C, Bayona JM, Pina B (2009) Physiological responses to mercury in feral carp populations inhabiting the low Ebro River (NE Spain), a historically contaminated site. Aquat Toxicol 93:150–157
35. Benejam L, Benito J, García-Berthou E (2009) Decreases in condition and fecundity of freshwater fishes in a highly polluted reservoir. Water Air Soil Pollut. doi:10.1007/s11270-009-0245-z
36. Mackie GL (1991) Biology of the exotic zebra mussel, Dreissena-polymorpha, in relation to native bivalves and its potential impact in Lake St-Clair. Hydrobiologia 219:251–268
37. Karatayev AY, Burlakova LE, Padilla DK (1997) The effects of Dreissena polymorpha (Pallas) invasion on aquatic communities in eastern Europe. J Shellfish Res 16:187–203
38. Minchin D, Lucy F, Sullivan M (2002) Zebra mussel: impacts and spread. In: Leppäkoski E, Gollasch S, Olenin S (eds) Invasive aquatic species of Europe: distribution, impacts and management. Kluwer Academic Publishers, Dordreicht, pp 135–146
39. Binimelis R, Monterroso I, Rodriguez-Labajos B (2007) A social analysis of the bioinvasions of Dreissena polymorpha in Spain and Hydrilla verticillata in Guatemala. Environ Manage 40:555–566
40. Ruiz C, Jimenez PJ, Lopez MA, Rofes J, Vinado J (2003) Localización y evaluación de una nueva invasión biológica: el mejillón cebra (*Dreissena polymorpha*) en el Ebro. Jornada técnica sobre el mejillón cebra, Zaragoza
41. Johnson LE, Padilla DK (1996) Geographic spread of exotic species: ecological lessons and opportunities from the invasion of the zebra mussel Dreissena polymorpha. Biol Conserv 78:23–33
42. Cope WG, Bartsch MR, Rada RG, Balogh SJ, Rupprecht JE, Young RD, Johnson DK (1999) Bioassessment of mercury, cadmium, polychlorinated biphenyls, and pesticides in the upper Mississippi river with zebra mussels (Dreissena polymorpha). Environ Sci Technol 33:4385–4390
43. Custer CM, Custer TW (1996) Food habits of diving ducks in the Great Lakes after the zebra mussel invasion. J Field Ornithol 67:86–99
44. Elder JF, Collins JJ (1991) Fresh-water mollusks as indicators of bioavailability and toxicity of metals in surface-water systems. Rev Environ Contam Toxicol 122:37–79
45. French JRP (1993) How well can fishes prey on zebra mussels in eastern North-America. Fisheries 18:13–19
46. Hogan LS, Marschall E, Folt C, Stein RA (2007) How non-native species in Lake Erie influence trophic transfer of mercury and lead to top predators. J Great Lakes Res 33:46–61
47. Zhou QF, Zhang JB, Fu JJ, Shi JB, Jiang GB (2008) Biomonitoring: an appealing tool for assessment of metal pollution in the aquatic ecosystem. Anal Chim Acta 606:135–150
48. Faria M, Huertas D, Soto DX, Grimalt JO, Catalan J, Riva MC, Barata C (2010) Contaminant accumulation and multi-biomarker responses in field collected zebra mussels (Dreissena polymorpha) and crayfish (Procambarus clarkii), to evaluate toxicological effects of industrial hazardous dumps in the Ebro river (NE Spain). Chemosphere 78:232–240
49. Finerty MW, Madden JD, Feagley SE, Grodner RM (1990) Effect of environs and seasonality on metal residues in tissues of wild and pond-raised crayfish in southern Louisiana. Arch Environ Contam Toxicol 19:94–100
50. Hothem RL, Bergen DR, Bauer ML, Crayon JJ, Meckstroth AM (2007) Mercury and trace elements in crayfish from northern California. Bull Environ Contam Toxicol 79:628–632

51. Higueras P, Oyarzun R, Lillo J, Sanchez-Hernandez JC, Molina JA, Esbri JM, Lorenzo S (2006) The Almaden district (Spain): anatomy of one of the world's largest Hg-contaminated sites. Sci Total Environ 356:112–124
52. Huner JV (2002) Procambarus. In: Holdich DM (ed) Biology of freshwater crayfish. Iowa State University Press – A Blackwell Science Company, Ames (IA), pp 541–584
53. Rodriguez-Labajos B (2006) Interlinked biological invasions in the Ebro River. A multi-scale scenario approach. Master thesis. Universitat Autonoma de Barcelona, Cerdanyola del Vallès (Barcelona), pp 22–23
54. Alikhan MA, Bagatto G, Zia S (1990) The crayfish as a biological indicator of aquatic contamination by heavy-metals. Water Res 24:1069–1076
55. Sanchez Lopez FJ, Garcia MDG, Martinez Vidal JL, Aguilera PA, Frenich AG (2004) Assessment of metal contamination in Donana National Park (Spain) using crayfish (Procamburus [Procambarus] clarkii). Environ Monit Assess 93:17–29
56. Carol J, Benejam L, Benito J, Garcia-Berthou E (2009) Growth and diet of European catfish (Silurus glanis) in early and late invasion stages. Fundam Appl Limnol 174:317–328
57. Kerr SJ, Grant RE (1999) Ecological impacts of fish introductions: evaluating the risk. Fish and Wildlife Branch. Ministry of Natural Resources, Peterborough, ON, K.P.R MNR 15 BN, Ontario, p 480
58. Elvira B, Nicola GG, Almodovar A (1996) Pike and red swamp crayfish: a new case on predator-prey relationship between aliens in central Spain. J Fish Biol 48:437–446
59. Paradis Y, Bertolo A, Magnan P (2008) What do the empty stomachs of northern pike (Esox lucius) reveal? Insights from carbon (delta C-13) and nitrogen (delta N-15) stable isotopes. Environ Biol Fishes 83:441–448
60. Garcia-Berthou E (2001) Size- and depth-dependent variation in habitat and diet of the common carp (Cyprinus carpio). Aquat Sci 63:466–476
61. Carol J, Benejam L, Alcaraz C, Vila-Gispert A, Zamora L, Navarro E, Armengol J, Garcia-Berthou E (2006) The effects of limnological features on fish assemblages of 14 Spanish reservoirs. Ecol Freshw Fish 15:66–77
62. Garcia-Berthou E, Moreno-Amich R (2000) Introduction of exotic fish into a Mediterranean lake over a 90-year period. Arch Hydrobiol 149:271–284
63. Doadrio I, Elvira B, Bernat Y (1991) Peces Continentales Españoles. Coleccion Técnica. ICONA-CSIC, Madrid
64. Díez S, Barata C, Raldúa D (2008) Exposure to mercury: a critical assessment of adverse ecological and human health effects. In: Prasad MNV (ed) Trace elements as contaminants and nutrients: consequences in ecosystems and human health. Wiley, New York, pp 345–373
65. Raldua D, Diez S, Bayona JM, Barcelo D (2007) Mercury levels and liver pathology in feral fish living in the vicinity of a mercury cell chlor-alkali factory. Chemosphere 66:1217–1225
66. Burrows WD, Krenkel PA (1973) Studies on uptake and loss of methylmercury-203 by bluegills (Lepomis-Macrochirus Raf). Environ Sci Technol 7:1127–1130
67. Hakanson L (1984) Metals in fish and sediments from the River Kolbacksan water-system, Sweden. Arch Hydrobiol 101:373–400
68. Abreu SN, Pereira E, Vale C, Duarte AC (2000) Accumulation of mercury in sea bass from a contaminated lagoon (Ria de Aveiro, Portugal). Mar Pollut Bull 40:293–297
69. Arribere MA, Guevara SR, Sanchez RS, Gil MI, Ross GR, Daurade LE, Fajon V, Horvat M, Alcalde R, Kestelman AJ (2003) Heavy metals in the vicinity of a chlor-alkali factory in the upper Negro River ecosystem, Northern Patagonia, Argentina. Sci Total Environ 301: 187–203
70. Bosch C, Olivares A, Faria M, Navas JM, del Olmo I, Grimalt JO, Pina B, Barata C (2009) Identification of water soluble and particle bound compounds causing sublethal toxic effects. a field study on sediments affected by a chlor-alkali industry. Aquat Toxicol 94:16–27
71. Garcia-Berthou E, Moreno-Amich R (2000) Rudd (Scardinius erythrophthalmus) introduced to the Iberian peninsula: feeding ecology in Lake Banyoles. Hydrobiologia 436:159–164

72. Bloom NS (1992) On the chemical form of mercury in edible fish and marine invertebrate tissue. Can J Fish Aquat Sci 49:1010–1017
73. Claisse D, Cossa D, Bretaudeau-Sanjuan J, Touchard G, Bombled B (2001) Methylmercury in molluscs along the French coast. Mar Pollut Bull 42:329–332
74. Secor CL, Mills EL, Harshbarger J, Kuntz HT, Gutenmann WH, Lisk DJ (1993) Bioaccumulation of toxicants, element and nutrient composition, and soft tissue histology of zebra mussels (*Dreissena polymorpha*) from New York State waters. Chemosphere 26:1559–1579
75. Camusso M, Balestrini R, Muriano F, Mariani M (1994) Use of fresh-water mussel *Dreissena polymorpha* to assess trace-metal pollution in the lower River Po (Italy). Chemosphere 29:729–745
76. Camusso M, Balestrini R, Binelli A (2001) Use of zebra mussel (*Dreissena polymorpha*) to assess trace metal contamination in the largest Italian subalpine lakes. Chemosphere 44:263–270
77. Wiesner L, Günther B, Fenske C (2001) Temporal and spatial variability in the heavy-metal content of *Dreissena polymorpha* (Pallas) (Mollusca: Bivalvia) from the Kleines Haff (northeastern Germany). Hydrobiologia 443:137–145
78. Kwan KHM, Chan HM, de Lafontaine Y (2003) Metal contamination in zebra mussels (*Dreissena polymorpha*) along the St. Lawrence River. Environ Monit Assess 88:193–219
79. Richman L, Somers K (2005) Can we use zebra and quagga mussels for biomonitoring contaminants in the Niagara River? Water Air Soil Pollut 167:155–178

Pesticides at The Ebro River Delta: Occurrence and Toxicity in Water and Biota

Marianne Köck-Schulmeyer, Miren Lopez de Alda, Elena Martínez, Marinella Farré, Asunción Navarro, Antoni Ginebreda, and Damià Barceló

Abstract Pesticide use has increased worldwide to protect the food supply of the swelling global population. Although it is undisputed that pesticides are essential in modern agriculture, there is a growing concern about environmental contamination from agrochemicals. For example, application of pesticides in the Ebro River delta (NE, Spain) during the rice-growing season is suspected to be one of the major causes behind the shellfish mortality episodes that occur yearly in this area at springtime. In an attempt to disclose the causes of these seafood mortality episodes, this chapter presents the results obtained from a monitoring study carried out in April–June 2008 in the Ebro River delta (NE, Spain), where surface water and seafood samples were analyzed for both toxicity and pesticides. The main conclusion of this study was that pesticides are likely responsible, together with other not investigated parameters, such as metals, for the observed mortality. The results obtained are discussed also in relation to others previously published for the same or other similar areas.

Keywords Ebro River delta · Ecotoxicity · Pesticides · Shellfish · Water analysis

Contents

M. Köck-Schulmeyer, M. Lopez de Alda (✉), E. Martínez, M. Farré, A. Navarro, A. Ginebreda, and D. Barceló
IDAEA-CSIC, c/Jordi Girona 18-26, 08034 Barcelona, Spain
e-mail: mlaqam@cid.csic.es

D. Barceló
IDAEA-SIC, c/Jordi Girona 18-26, 08034 Barcelona, Spain
ICRA Scientific and Technological Park of the University of Girona, Emili Grahit, 101, 17003 Girona, Spain

D. Barceló and M. Petrovic (eds.), *The Ebro River Basin*,
Hdb Env Chem (2011) 13: 259–274, DOI 10.1007/698_2010_89,

1 Introduction

Almost half of the plants and animals listed in the Habitats Directive [1] are present in the Mediterranean region. This large number reflects not only the variety of threats in the region, but also its remarkable abundance of species. There are more plants here than in other European biogeographic regions together [2].

Among these regions is the Ebro River, the Spain's most voluminous and important river in the country. With 910 km, the river accounts for approximately 6% of all riverine waters entering the Mediterranean and drains approximately one-sixth of the Iberian Peninsula, including important urban and industrial areas [3]. But because it is highly exploited by industrial and agricultural activities, the Ebro River basin is the focus of several scientific studies [4–10].

For example, 95 km upstream from the river mouth there is the Flix reservoir, where there are around 200,000–360,000 tons of industrial waste with a high concentration of heavy metals and organochlorines due to the activity of an chloro-alkali industry for more than half a century [11].

Downstream, in the mouth, there is the Ebro River delta, considered one of the most important natural wetland areas in the Mediterranean [12], and one of the river areas most affected by the presence of human activity.

1.1 Ebro River Delta

In this delta, we can contemplate orchards, vegetables, fruit-bearing trees, and mainly rice fields. The delta, which occupies 35,000 ha and goes some 20 km into the sea, is a dynamic and complex structure originated from the alluvial sediments transported by the river, which have given rise, among other physical features, to two coastal shallow bays that extend along both sides of the river mouth [13] (Fig. 1). The Ebro delta is considered one of the most important natural wetland areas in the western Mediterranean also for birdlife preservation. Just to have an idea, this delta also hosts numerous beaches, marshes, and salt pans that provide habitat for more than 300 species of birds. The main economic activity of this area is agriculture, which is mostly dominated by rice (about 80% of the land is dedicated to its production) [14]. A network of channels

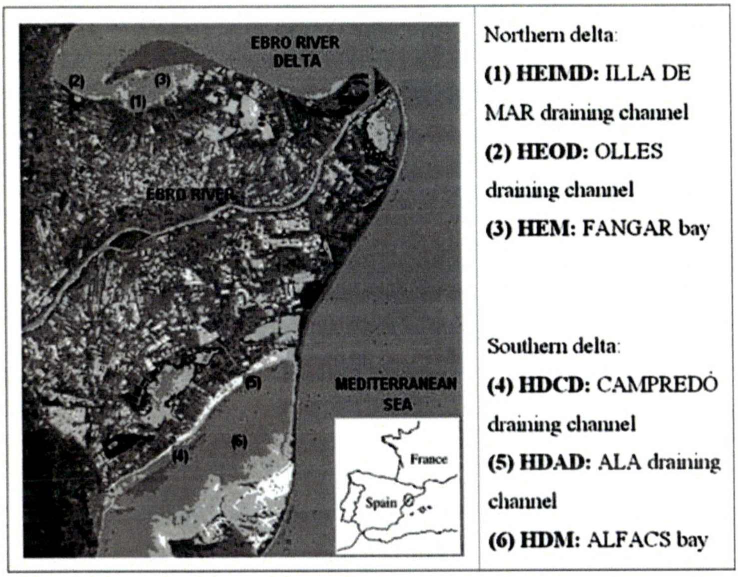

Fig. 1 Map of sampling sites at the Ebro River delta (NE, Spain)

and irrigation ditches constructed by both agricultural and conservation groups help to maintain the ecologic and economic resources of the Ebro delta.

1.2 Shellfish Culture

However, the fertile land is not the only star of the delta. The bays also play an important role in the economy of this delta. Shellfish culture has been well developed in the shallow waters of the two aforementioned bays, and has become the second economic activity of the area after agriculture. In the Ebro delta, there are currently around 164 platforms for the production of mussels and oysters, 74 in Fangar bay (northern), and 90 in Alfacs bay (southern). Annual production for both bays is 3,000 tons of mussels and less than 1,000 tons of oysters [15].

The relative importance of both activities is well reflected on the population labor distribution per activity: 70% is dedicated to agriculture and 15% to shellfish production [14]; however, these two activities are not very compatible. Aquatic organisms are currently being exposed to multiple chemical and environmental stressors with different mechanisms of toxicity, each contributing to a final overall adverse effect [16]. Recently, the shellfish farmers in the Ebro River delta have complained about a loss of production in the periods of rice cultivation that they attribute to the heavy pesticide loads discharged after rice field treatment, and this has raised also public concern about the quality of the water in this area.

1.3 Pesticide Contamination

This high agriculture activity compromises the chemical and ecological status of the Ebro delta. Pesticides used in the area migrate into the delta building up, together with the pollution produced by some industries, a contamination that can be dangerous to its both fauna and flora, and this contamination increases when water from the rice fields is drained into the delta [17].

Many studies have been conducted along the Ebro River to assess the impact of pesticide use on its fauna. For instance, between 1989 and 1991 Solé et al. [3] conducted a study to investigate the bioaccumulation and ecotoxicity of various organophosphorous (OP) pesticides (and other pollutants) in bivalves cultured in the delta. A seasonal fluctuation of pollutants that was related to the biological cycle of the organisms and to the management of the waters in the rice crop fields of the delta was observed. However, levels were in general low and did not reach toxicity thresholds for bivalves, thus concluding that OP pesticides did not pose a threat to bivalves. The authors stressed though that pesticides might be harmful to other more sensitive species such as crustaceans, which in fact are not successfully farmed in the area, and that precautions should be taken during and shortly after field treatment with pesticides [3]. More recently, Damásio et al. [16] have identified endosulfan, propanil, and phenylureas as major pesticides affecting bivalve species exposed to agricultural pollution in the Ebro delta.

Other studies have focused on fish and reptiles. In 2004, Lavado et al. [18] provided the first evidence of endocrine disruption in fish from the Ebro river. Important alterations were detected in carps from Flix, a heavily industrialized area, and in Aragón, an agricultural area. In 2009, Santos and Llorente [19] reported a decline of the viperine snake *Natrix maura* in the delta. This reptile that in 1995 was common in the rice fields (0.93 animals/ha) was no longer found in 2008. In the authors' opinion, this decline would have been caused by the combined effect of the following factors (1) the transformation and degradation of the habitat; (2) the increase in population densities of natural predators, such as herons; (3) the decrease in prey availability (frogs and fish were also observed to decline); (4) the massive use of pollutants in the rice fields; and (5) snake death on local roads and directly by human persecution; and also in the authors' opinion, recovery of the *N. maura* population would require an integral change in agricultural management, including the reduced use of pollutants, among other measures [19].

Because of this massive use of agrochemicals and the ecological effects observed, the study of the occurrence of pesticides (herbicides, insecticides, plant growth regulators, fungicides, etc.) in this area has increased sharply in the last years [13, 20–26], thanks to the support of the water agencies in charge. The two most recently published works [27, 28] investigated temporal and geographical variations for a set of 30 pesticides and industrial compounds in surface waters and sediments along the Ebro river from source to mouth during the period 2004–2006, and found that, first, pesticides have a point source origin in the Ebro delta area (where concentrations reached up to 2,575 ng/L), and, second, the concentration of pesticides follows a seasonal trend, which is characterized by higher levels over the spring–summer period following agriculture application.

1.4 Legislation on Pesticides

The presence of pesticides in water has been subjected to regulation in the European Union (EU) for many decades. In 2008, the Directive 2008/105/EC [29], offspring of the Water Framework Directive (WFD) [30], established environmental quality standards (EQSs), both annual average (AA) and maximum allowable concentrations (MACs), for a series of pesticides and other priority substances (33 in total) in inland and other surface waters, as a first measure to protect surface water against pollution and deterioration. However, it is clear that in the aquatic environment there are many other pollutants and pressures, and that these pollutants and pressures vary geographically. For this reason, the same directive leaves it to the Member States to lay down, where necessary, EQSs for other site-specific pollutants, as well as EQSs for sediment and/or biota, at national level. To this end and for an integrated assessment of the status (chemical and ecological) of the water bodies, the combination of data on chemical concentrations and toxicological evaluation is essential.

The study described in the following section, which has been recently performed by our group in the Ebro delta, is a good example of the integration of chemical and biological tools to obtain a more comprehensive picture of the existing toxic pressures, as supported by the WFD, and of the surveillance of compounds beyond those specifically targeted in the legislation that can, nonetheless, be a problem in a specific area.

2 Case Study: Investigation of the Occurrence and Toxicity of Pesticides in the Ebro River Delta

2.1 Introduction

In response to the concern expressed by the shellfish farmers operating in the Ebro River delta about the potential positive role of pesticides on the oyster and mussel mortalities observed in the area, our group, commissioned by and with the collaboration of the Catalan Water Agency (ACA), carried out a comprehensive study in which chemical and toxicity data were combined to assess potential toxic presures present in the delta. To this end, a combined approach scheme integrating the measurement of various general physicochemical parameters in water, quantitative chemical analysis of pesticides in water and biota, and ecotoxicity assays in water was applied to a series of samples collected at springtime (between mid-April and mid-June 2008) from six selected sites of the delta: the two (northern and southern) bays (Fangar and Alfacs, respectively) and four main draining channels discharging the output water from the rice fields into the bays (Fig. 1). In total, 104 water samples, and 7 oyster and 3 mussel samples (each corresponding to a pool of 8–12 individuals collected alive from the two bays), were investigated.

Analysis of pesticides in water was performed by fully automated online solid-phase extraction-liquid chromatography–tandem mass spectrometry (SPE-LC–MS/MS) [25, 31]. These pesticides (a total of 22 belonging to the classes of triazines, OP, chloroacetanilides, phenylureas, thiocarbamates, acid herbicides, and anilides) were selected on the basis of previously published studies [20, 25], information gathered from the water authorities, and known use in rice crops.

Analysis of pesticides (eight in total, namely, molinate, propanil, fenitrothion, malathion, bentazone, cypermetrine, maloxon, and fenitrothion oxon) in biota was accomplished with a method based on pressurized liquid extraction (ASE), followed by SPE clean-up, and analysis by gas chromatography-mass spectrometry with electron impact ionization (GC/MS-EI).

Ecotoxicity assessment of water samples was carried out, in parallel to chemical analysis, using three standardized bioassays based on the micro-crustacean *Daphnia magna*, the algae *Pseudokirchneriella subcapitata*, and the bioluminescent bacteria *Vibrio fischeri*.

Finally, the pesticide concentrations determined in water and biota, together with the toxicity values of each individual compound, the toxicity data measured in the water samples, and the general physicochemical values were combined and analyzed together to establish potential cause–effect relationships and identify major toxicants or environmental pressures in the area of study. More details can be found in Köck et al. [12].

2.2 Results and Discussion

2.2.1 Levels of Pesticides in Water

Of the 22 pesticides analyzed in water, 21 were found to be present in some or all of the samples analyzed; cyanazine was the only undetected compound. Figure 2 shows the concentration of individual and total pesticides and their frequency of detection in the water samples collected from each of the six sampling sites monitored.

In what compliance with the stipulated legislation is concerned, only alachlor was found to exceed the EQS (MAC of 700 ng/L) in two of the samples investigated (HDAD, May 2; HEIMD, April 21) [29].

Bentazone and MCPA were the most ubiquitous compounds (detected in 100% of the samples), whereas malathion followed by MCPA and molinate were the compounds found at highest concentrations (5,825, 4,197 and 3,590 ng/L, respectively). Pesticide profiles similar to this have been observed previously in the studied area, as

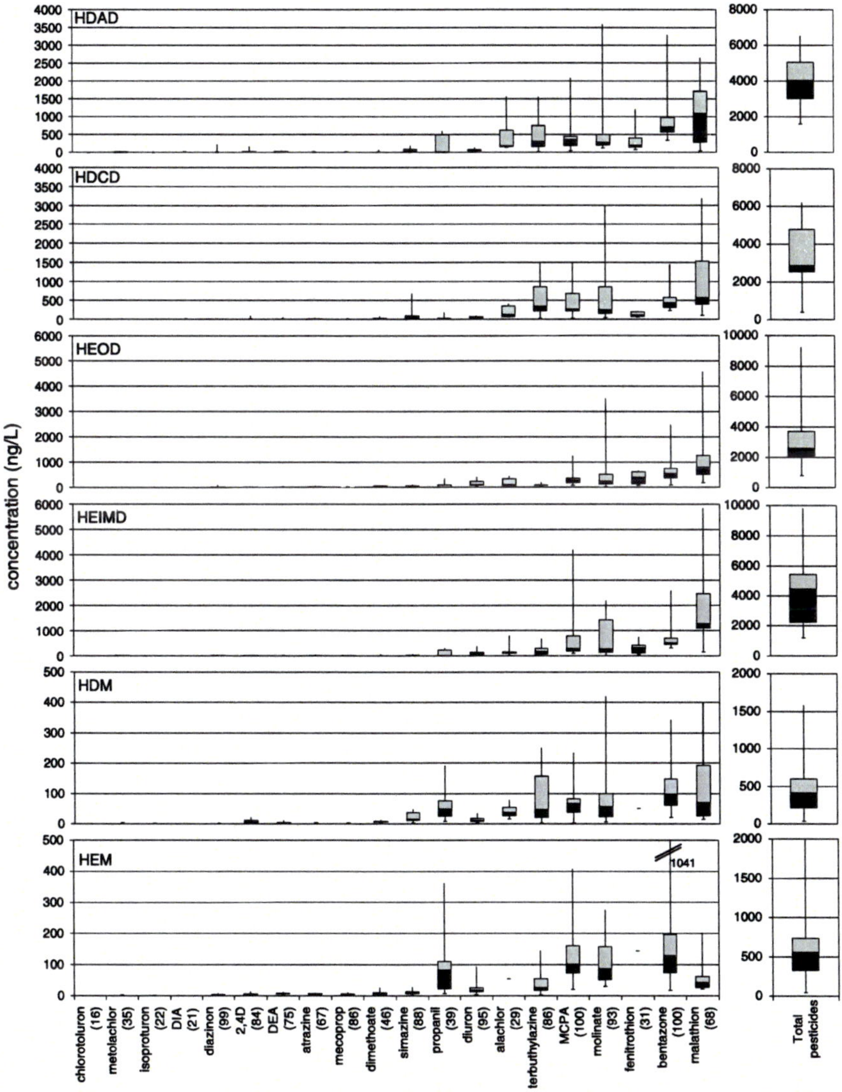

Fig. 2 *Boxplots* of the concentrations of individual and total pesticides and frequency of detection (expressed in % in parentheses) in the water samples collected from each of the six sampling sites. HDAD, HDCD, HEOD, and HEIMD: draining channels; HDM and HEM: bays; *DIA* Deisopropylatrazine; *DEA* desethylatrazine

well as in other Mediterranean estuaries [32]. In 1990–1991, Readman et al. [32] conducted an extensive pilot survey in various selected Mediterranean locations to generate data about herbicides levels in estuarine zones, and found molinate and bentazone to be the most, or among the most abundant herbicides in the Ebro River delta (maximum concentrations 1,400 and 5,500 ng/L, respectively) and in the Po/Italy region (1,750 and 311 ng/L, respectively), which is in line with the rice cultivation activities performed in both areas. Nevertheless, in the study as a whole the most commonly encountered herbicides were atrazine, simazine, alachlor, metolachlor, and molinate. This study provided the first extensive evidence that significant concentrations of some herbicides persist through freshwater and estuarine environments and contaminate marine systems [32]. Later, other works investigating also the presence of pesticides in Mediterranean estuaries (see Table 1) have served to confirm this finding [17, 33–39].

Comparison of the pesticide concentrations (ng/L) found in this study in sites HDCD and HDAD with those measured in a previous study performed in 2005 in the same sampling sites [16, 20] showed a general good agreement for all pesticides except for bentazone, MCPA, propanil, and atrazine, which presented now comparatively lower concentrations, and alachlor, malathion, diuron, and molinate, whose concentrations have increased considerably (Fig. 3).

Comparing the six sampling points monitored in this study (Fig. 4), HEIMD followed by HDAD were the most polluted ones, showing concentration levels of total pesticides above 5 μg/L in 39% and 22% of the samples analyzed, respectively. As expected, the sampling points located in the two bays (HEM and HDM) were the less polluted, due to the dilution effect of the marine water. This dilution effect may explain also the higher contamination level observed in the northern bay as compared to the southern bay, which is comparatively larger in size and depth.

2.2.2 Water Toxicity

Figure 5 illustrates the results of the water toxicity evaluation performed with each of the three assays (24–48 h immobilization of *D. magna*, growth inhibition of *Pseudokirchneriella subcapitata*, and bioluminescence inhibition of *V. fischeri*), and the period of observed shellfish mortality. In 8 of the 14 monitored days, high toxicity values (above 50% inhibition) were recorded for at least one organism in at least one of the six sampling sites. As it can be seen in the figure, peak toxicity values occurred mainly in mid-May coinciding with, or a few days before, the shellfish mortality episodes.

These experimental toxicity values were compared with the theoretical toxicity units calculated for each sample and organism from the pesticide concentrations measured in the samples and their toxicity (50% effective concentration, EC_{50}) towards the corresponding test organisms [12]. This comparison, which is illustrated in Fig. 6, gives an estimation of the extent to which the pesticides measured contribute to the observed experimental toxicity as well as about their relative contribution to it.

According to the results obtained, and assuming that the simple additive approach applied is valid, malathion appears to be the main contributor to *Daphnia* toxicity (98% toxic contribution on average for all samples); molinate, fenitrothion, and malathion would be the most relevant pesticides in the case of *V. fischeri*

Table 1 Main features of Mediterranean estuaries studied for pesticides and corresponding references

Mediterranean estuaries	Area (Ha)	Agricultural activities[a]	References
Ebro delta, Spain	35,000 total 15,000 agriculture 10,000 rice	The Ebro delta is an alluvial plain. Lagoons and marshes cover 20% of the area, the remainder is principally agricultural (mainly rice cultivation)	[12, 16, 20, 32, 33]
Albufera de Valencia, Spain	21,000 total 18,000 rice	The rice fields and pine groves surrounding it form an ecological unit of high tourist, economic, and scientific interest. It is one of the humic zones included in the Convention on Wetlands and it was declared a Nature Reserve in 1986	[17, 34]
Rhône delta, France	75,000 total ¼ of the area agriculture	The Rhone drains 15% of the cultivated surfaces in France. Within its highly populated and industrialized basin, vineyards, horticulture, and corn production dominate agriculture in the upper regions whereas the lower reaches are devoted principally to rice	[32, 35]
Po, Italy	270,000 total 10,000 just rice	It flows through an industrialized and highly cultivated plain which supports diverse crops	[32]
Evros, Greece	20,000 total 9,000 agriculture	The Evros River forms a vast delta of which 9,000 ha is dedicated to agriculture, cereals, sunflowers, and cotton cultivation mainly	[32, 36]
Danube, Ukrania	417,800 total 61,000 agriculture (rice more important)	The cultures were a source of fertilizers and pesticides that have accumulated in the food chain causing physiological changes in animals and plants, as well as the disappearance of some species	[37, 38]
Nile, Egypt	622,000 rice	It is one of the world's largest river deltas and it is a rich agricultural region. Within the delta, land use includes agricultural (primarily cotton) and industrialized activities	[32, 39]

[a]Some information was obtained from DELTA-MED Association (http://www.deltamed.org)

(69, 21, and 8% toxic contribution, respectively); and the alga *P. subcapitata* would be more affected by the presence of some herbicides, such as diuron, simazine, and terbuthylazine (36, 26, and 17% toxic contribution, respectively).

On the other hand, as it can be seen in Fig. 6, experimental and theoretical toxicity values followed fairly similar trends. The differences between them might

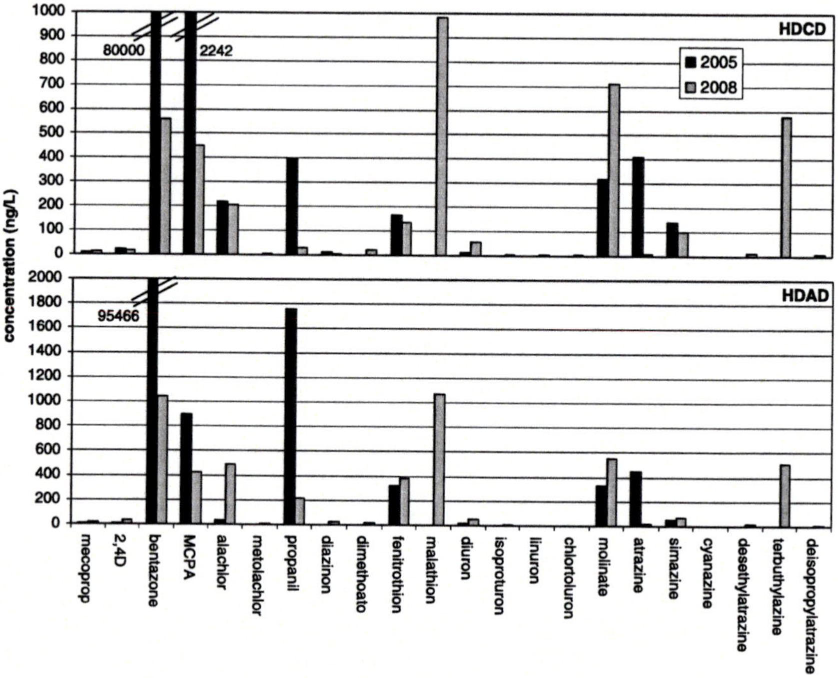

Fig. 3 Comparison of pesticides concentrations (ng/L) reported for sites HDCD and HDAD in this study (average of 18 measurements performed between April and June 2008) and in a previous one (average of four measurements performed monthly between May and August 2005) [16, 20]

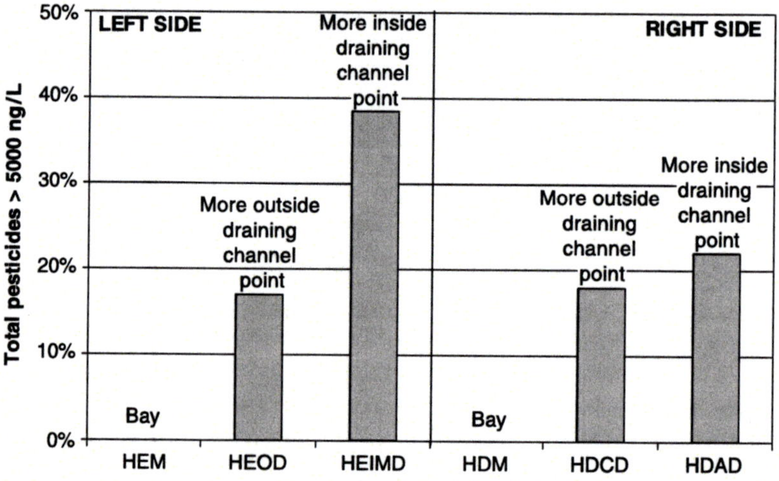

Fig. 4 Percentage of water samples with total pesticides levels >5 μg/L in the six sampling sites investigated

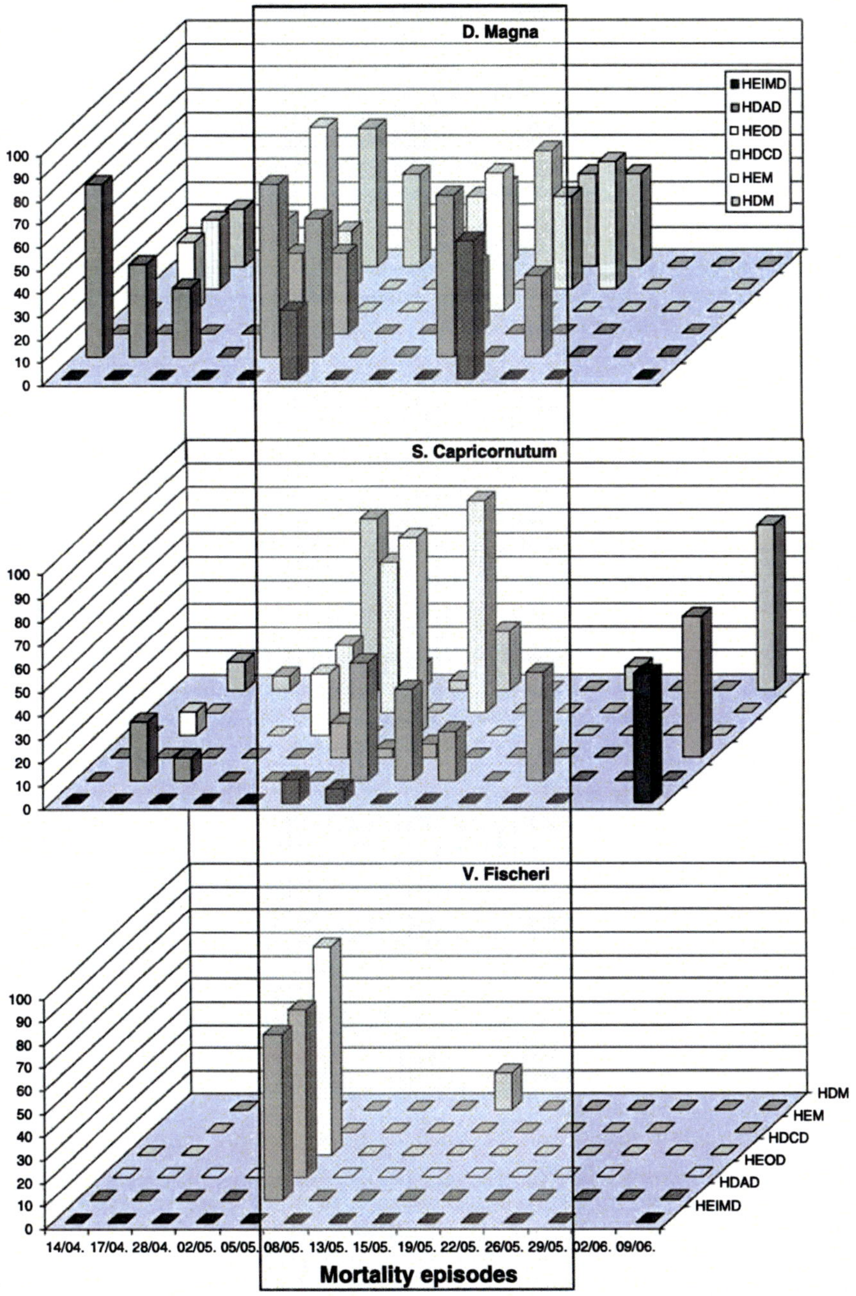

Fig. 5 Inhibition rates against the tested organisms of each water sample investigated

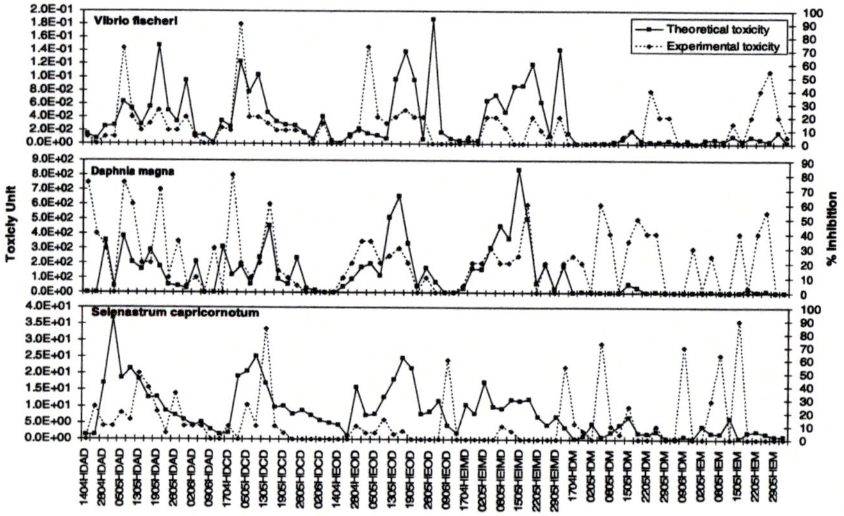

Fig. 6 Comparison between theoretical and experimental toxicity values against each of the three tested organisms for all water samples

be due to synergism or antagonism effects, matrix effects, and/or the presence of other not investigated chemicals (such as different pesticides, heavy metals, etc).

2.2.3 Levels of Pesticides in Shellfish

In shellfish only three of the eight compounds monitored were detected: fenitrothion, which was found in a mussel sample at 3.46 mg/kg, malathion, found in an oyster sample at 53.12 mg/kg, and the malathion degradation product malaoxon, which was found in one oyster and two mussel samples at concentrations between 2.53 and 4.59 mg/kg. As it can be seen in Fig. 7, positive samples (five of ten analyzed) were found in both bays and in scattered days along the period of study, with the sample showing the highest pesticide concentration (53.12 mg/kg of malathion in oysters collected from the northern bay on May 27) coinciding with the period of shellfish mortality.

In all cases, the concentrations of malathion and fenitrothion measured in water (up to 5.8 and 1.2 μg/L, respectively) were below the LC_{50} (lethal concentration 50%) values reported for these compounds in oysters and mussels, which range between 2.7 and 278 mg/L in the case of malathion, and between 10.3 μg/L and 123 mg/L in the case of fenitrothion (http://www.pesticideinfo.org). However, it has to be stressed that these LC_{50} values express acute toxicity, that both malathion and fenitrothion might be bioaccumulated by molluscs (as their detection in biota suggests), and that aquatic organisms are exposed to a variety of contaminants, some of which could show synergetic or additive effects [40]. Further matters of

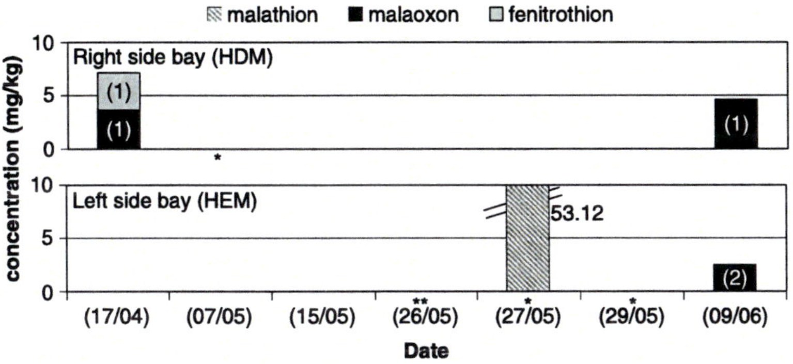

Fig. 7 Levels of pesticides (mg/kg) in shellfish; (1) mussel; (2) oyster; *single asterisk* mortality episodes; *double asterisks* big mortality episodes (around 60%)

concern are (1) the detection itself of malathion in water and biota, provided that since June 2007 its use in plant protection products is forbidden in the European Union (Decision 2007/389/EC) [41] and the period of grace granted for the use of existing stocks expired in December 2008, and (2) the fact that malathion is transformed in the aquatic environment to malaoxon, which is more toxic than the parent compound [42].

2.2.4 Other Environmental Parameters

Agricultural practices are the main responsible source of surface water pollution in the Ebro River basin and this is particularly true in the lower course of the river [43, 44]. However, there are other factors that could contribute to the water conditions and possibly to the shellfish mortality episodes, such as the presence of other contaminants such as metals (copper is often used as algicide), hazardous algal blooms, and/or high temperatures. The elevated temperatures reached in these bays in mid-summer, especially in the Alfacs bay where values can exceed 28°C for several weeks, have been previously pointed out as a cause of mussel mortality [45, 46]. Thus, to investigate some of these possibilities, the water authorities in charge recorded continuously during the period of study in the sampling sites HEM and HDM (i.e., the northern and southern bays, respectively) some physicochemical parameters, including temperature, dissolved oxygen, pH, redox potential and conductivity. The first four parameters showed a normal behavior in both bays (thus excluding, for instance, anoxia episodes), but the conductivity varied notably especially in the northern bay. Such observed variations that are attributed to inputs of fresh water from the draining channels were coincident on time with peaks of pesticides, as well as with the registered mortality episodes in the northern bay, which constitutes a further support to the hypothesis of pesticides being a likely cause of shellfish impairment. Temperature was disregarded as a potential factor since the maximum values reached during the period of study were 21.8°C and 23.1°C in the northern and southern bays, respectively.

3 Conclusions

The great number and variety of environmental conditions that can be behind an adverse ecotoxicological observation makes virtually impossible to establish direct cause–effect relationships. However, the consistency of results between pesticides concentrations (in water and seafood), toxicity, and shellfish (oysters and mussels) mortality episodes in this study suggest that the pesticides used in the delta during the rice-growing season are likely responsible for the observed bivalves mortality, although the contribution of other not investigated factors, such as contamination by metals (e.g., copper) or by other agrochemicals, can obviously not be discarded. Of the various pesticides investigated, malathion and to a lesser extent diazinon and molinate appear as the most relevant compounds. Temperature in this particular case study is not believed to play an important role because it did not exceed 24°C during the monitoring period and the most remarkable mortality episodes took place in the northern bay of the delta, which was identified as the most polluted area, and where temperature was comparatively lower than in the southern bay.

Moving mussel cultivation outside the bays of the Ebro River delta has been suggested by scientists at the Institute for Agricultural Research and Technology (IRTA) and the Spanish Institute of Oceanography (IEO) as a potential solution to prevent shellfish mortality caused by pollution or high temperatures in the summer.

Acknowledgments This work has been supported by the Catalan Water Agency and by the Spanish Ministry of Science and Innovation through the projects SCARCE (Consolider-Ingenio 2010 CSD2009-00065) and CEMAGUA (CGL2007-64551/HID). Merck is acknowledged for the gift of LC columns and Biotage for the gift of SPE cartridges. Marianne Köck gratefully acknowledges the European Social Fund and AGAUR (Generalitat de Catalunya, Spain) for their economical support through a FI predoctoral grant.

References

1. Council of European Economic Community (1992) Directive 92/43/EEC on the conservation of natural habitats and of wild fauna and flora. Off J Eur Community L 206:7
2. Sundseth K (2010) Natura 2000 en la región mediterránea. Unión Europea, Bruselas
3. Sole M, Porte C, Barcelo D et al (2000) Bivalves residue analysis for the assessment of coastal pollution in the Ebro Delta (NW Mediterranean). Mar Pollut Bull 40:746–753
4. Barth JAC, Grathwohl P, Fowler HJ et al (2009) Mobility, turnover and storage of pollutants in soils, sediments and waters: achievements and results of the EU project AquaTerra. A review. Agron Sustain Dev 29:161–173
5. Falco S, Niencheski LF, Rodilla M et al (2010) Nutrient flux and budget in the Ebro estuary. Estuar Coast Shelf Sci 87:92–102

6. Faria M, Huertas D, Soto DX et al (2010) Contaminant accumulation and multi-biomarker responses in field collected zebra mussels (Dreissena polymorpha) and crayfish (Procambarus clarkii), to evaluate toxicological effects of industrial hazardous dumps in the Ebro river (NE Spain). Chemosphere 78:232–240
7. Isidoro D, Aragó R (2007) River water quality and irrigated agriculture in the ebro basin: an overview. Int J Water Resour Dev 23:91–106
8. Martí-Cid R, Huertas D, Nadal M et al (2010) Dietary exposure to organochlorine compounds in Tarragona Province (Catalonia, Spain): health risks. Hum Ecol Risk Assess 16:588–602
9. Moreno-Mateos D, Pedrocchi C, Comín FA (2010) Effects of wetland construction on water quality in a semi-arid catchment degraded by intensive agricultural use. Ecol Eng 36:631–639
10. Postigo C, López de Alda MJ, Barceló D (2010) Drugs of abuse and their metabolites in the Ebro River basin: occurrence in sewage and surface water, sewage treatment plants removal efficiency, and collective drug usage estimation. Environ Int 36:75–84
11. Benejam L, Benito J, García-Berthou E (2010) Decreases in condition and fecundity of freshwater fishes in a highly polluted reservoir. Water Air Soil Pollut 210:231–242
12. Kock M, Farre M, Martinez E et al (2010) Integrated ecotoxicological and chemical approach for the assessment of pesticide pollution in the Ebro River delta (Spain). J Hydrol 383:73–82
13. Manosa S, Mateo R, Guitart R (2001) A review of the effects of agricultural and industrial contamination on the Ebro delta biota and wildlife. Environ Monitor Assess 71:187–205
14. Camp J (1994) Aproximaciones a la dinámica ecológica de una bahía estuárica mediterránea. Universitat de Barcelona, Barcelona
15. Acuicultura I (2007) Los productores del Delta promocionan la ostra. http://www.ipacuicultura.com/ipac/noticia.php?idNoticia=2426. Accessed 2010
16. Damasio J, Navarro-Ortega A, Tauler R et al (2010) Identifying major pesticides affecting bivalve species exposed to agricultural pollution using multi-biomarker and multivariate methods. Ecotoxicology 19:1084–1094
17. Peris E, Requena S, de la Guardia M et al (2005) Organochlorinated pesticides in sediments from the Lake Albufera of Valencia (Spain). Chemosphere 60:1542–1549
18. Lavado R, Thibaut R, Raldua D et al (2004) First evidence of endocrine disruption in feral carp from the Ebro River. Toxicol Appl Pharmacol 196:247–257
19. Santos X, Llorente GA (2009) Decline of a common reptile: case study of the viperine snake Natrix maura in a Mediterranean wetland. Acta Herpetol 4:161–169
20. Barata C, Damasio J, Lopez MA et al (2007) Combined use of biomarkers and in situ bioassays in Daphnia magna to monitor environmental hazards of pesticides in the field. Environ Toxicol Chem 26:370–379
21. Claver A, Ormad P, Rodriguez L et al (2006) Study of the presence of pesticides in surface waters in the Ebro river basin (Spain). Chemosphere 64:1437–1443
22. Gomez-Gutierrez AI, Jover E, Bodineau L et al (2006) Organic contaminant loads into the Western Mediterranean Sea: estimate of Ebro River inputs. Chemosphere 65:224–236
23. Hildebrandt A, Guillamon M, Lacorte S et al (2008) Impact of pesticides used in agriculture and vineyards to surface and groundwater quality (North Spain). Water Res 42:3315–3326
24. Hildebrandt A, Lacorte S, Barcelo D (2009) Occurrence and fate of organochlorinated pesticides and PAH in agricultural soils from the Ebro River basin. Arch Environ Contam Toxicol 57:247–255
25. Kuster M, de Alda MJL, Barata C et al (2008) Analysis of 17 polar to semi-polar pesticides in the Ebro river delta during the main growing season of rice by automated on-line solid-phase extraction-liquid chromatography-tandem mass spectrometry. Talanta 75:390–401
26. Santos TCR, Rocha JC, Barcelo D (2000) Determination of rice herbicides, their transformation products and clofibric acid using on-line solid-phase extraction followed by liquid chromatography with diode array and atmospheric pressure chemical ionization mass spectrometric detection. J Chromatogr A 879:3–12
27. Navarro A, Tauler R, Lacorte S et al (2010) Occurrence and transport of pesticides and alkylphenols in water samples along the Ebro River Basin. J Hydrol 383:18–29

28. Navarro-Ortega A, Tauler R, Lacorte S et al (2010) Occurrence and transport of PAHs, pesticides and alkylphenols in sediment samples along the Ebro River Basin. J Hydrol 383:5–17
29. Council of the European Communities (2008) Decision 2008/105/EC on environmental quality standards in the field of water policy. Off J Eur Community L 348:84
30. Council of the European Communities (2000) Directive 2000/60/EC establishing a framework for Community action in the field of water policy. Off J Eur Community L 327:1
31. Kampioti AA, da Cunha ACB, de Alda ML et al (2005) Fully automated multianalyte determination of different classes of pesticides, at picogram per litre levels in water, by on-line solid-phase extraction-liquid chromatography-electrospray-tandem mass spectrometry. Anal Bioanal Chem 382:1815–1825
32. Readman JW, Albanis TA, Barcelo D et al (1993) Herbicide contamination of mediterranean estuarine waters – results from a Med Pol pilot survey. Mar Pollut Bull 26:613–619
33. Feo ML, Eljarrat E, Ginebreda A et al (2010) Presence of pyrethroid pesticides in water and sediments of Ebro River Delta. J Hydrol doi: 10.1016/j.jhydrol.2010.08.012
34. Gamon M, Saez E, Gil J et al (2003) Direct and indirect exogenous contamination by pesticides of rice-farming soils in a Mediterranean wetland. Arch Environ Contam Toxicol 44:141–151
35. Comoretto L, Arfib B, Talva R et al (2008) Runoff of pesticides from rice fields in the Ile de Camargue (Rhone river delta, France): field study and modeling. Environ Pollut 151:486–493
36. Skoulikidis NT (2009) The environmental state of rivers in the Balkans-A review within the DPSIR framework. Sci Total Environ 407:2501–2516
37. Burgess RM, Terletskaya AV, Milyukin MV et al (2009) Concentration and distribution of hydrophobic organic contaminants and metals in the estuaries of Ukraine. Mar Pollut Bull 58:1103–1115
38. Vosniakos F, Petre J, Pascu L et al (2010) Aquatic ecosystem quality assessment of the Danube Delta in the periods April-October 2007 and 2008. Fresenius Environ Bull 19:20–29
39. Said TO, El Moselhy KM, Rashad AAM et al (2008) Organochlorine contaminants in water, sediment and fish of Lake Burullus, Egyptian Mediterranean Sea. Bull Environ Contam Toxicol 81:136–146
40. KayLynn Newhart (2006) Environmental fate of Malathion. California Environmental Protection Agency Department of Pesticide Regulation, CA, USA
41. Council of the European Communities (2007) Decision 2007/389/EC concerning the non-inclusion of malathion in Annex I to Council Directive 91/414/EEC. Off J Eur Community L 146:19
42. Aker WG, Hu XK, Wang P et al (2008) Comparing the relative toxicity of malathion and malaoxon in blue catfish Ictalurus furcatus. Environ Toxicol Chem 23:548–554
43. Terrado M, Barcelo D, Tauler R (2009) Quality assessment of the multivariate curve resolution alternating least squares method for the investigation of environmental pollution patterns in surface water. Environ Sci Technol 43:5321–5326
44. Terrado M, Barcelo D, Tauler R (2010) Multivariate curve resolution of organic pollution patterns in the Ebro River surface water-groundwater-sediment-soil system. Anal Chim Acta 657:19–27
45. Ramón M, Cano J, Peña BJ et al (2005) Current status and perspectives of mollusc (bivalves and gastropods) culture in the Spanish Mediterranean. Boletín Instituto Español de Oceanografía 21:361–373
46. Ramon M, Fernandez M, Galimany E (2007) Development of mussel (Mytilus galloprovincialis) seed from two different origins in a semi-enclosed Mediterranean Bay (NE Spain). Aquaculture 264:148–159

Biological Effects of Chemical Pollution in Feral Fish and Shellfish Populations from Ebro River: From Molecular to Individual Level Responses

Benjamin Piña, Demetrio Raldúa, Carlos Barata, Melissa Faria, Anna Navarro, Joana Damasio, Alba Olivares, Laia Quirós, Sergi Pelayo, and Marta Casado

Abstract A multilevel approach, from whole animal to molecular level, was applied to the study of the biological impact of chemical pollution in fish and shellfish populations from the rivers Vero, Cinca, and from the Flix reservoir in the Ebro River. The analysis provided a general picture of the health status of the rivers and quantified the physiological effects of different pollutants originating in existing chemical plants discharging in the area. The data show that fish acclimated to very high concentrations of some toxicants, like mercury, whereas organochlorinated compounds (OCs) and poly bromo diphenyl ethers apparently induce permanent negative effects, including oxidative stress, poor condition and fertility, DNA damage, and liver and kidney histological anomalies. Toxic determinants appeared different for vertebrates and invertebrates and suggest that a key difference between both animal groups may be the presence of activable aryl hydrocarbon receptor (AhR), which only occurs in deuterostomata (Chordates, Echynoderma and alikes). The adverse biological effects were recorded up to 30–35 km downstream the different sources, and their distribution differed for OCs and for Hg. Intensive local agricultural practices, rather than pollution from the Ebro's chemical plants, seem to account for adverse biological effects observed in the Ebro Delta.

Keywords *Barbus graellsii*, Biomarkers, Choline esterases, *Cyprinus carpio*, *Daphnia magna*, *Dreissena polymorpha*, EROD, Metallothionein, qRT-PCR, Xenobiotics

B. Piña (✉)
Institute for Environmental Diagnostics and Water Research (IDAEA-CSIC), Jordi Girona 18, 08034 Barcelona, Spain
e-mail: bpcbmc@cid.csic.es
D. Raldúa, C. Barata, M. Faria, A. Navarro, J. Damasio, A. Olivares, L. Quirós, S. Pelayo, and M. Casado
Institute for Environmental Diagnostics and Water Research (IDAEA-CSIC), Jordi Girona 18, 08034, Barcelona, Spain

D. Barceló and M. Petrovic (eds.), *The Ebro River Basin*,
Hdb Env Chem (2011) 13: 275–293, DOI 10.1007/698_2010_78,

Contents

1 Introduction

The Ebro catchment is the largest river basin in Spain. It covers an area of 85,362 km^2 (http://www.chebro.es) and receives the potential influence of three million people. It suffers a considerable ecological impact from different industrial activities located predominantly in the last third of its course. These activities result on the release to the Ebro River and to some of its tributaries of Tm quantities of chlorinated organic compounds, PBDE and other brominated flame retardants, mercury, and other metals. In addition, intensive agricultural practices, mainly concentrated in the last 30 km of the river course and in its Delta, imply the use of large quantity of pesticides and fertilizers. In this paper, we review physiological effects of these pollutants in resident fish and shellfish populations and provide a general picture of the health status of the low course of the Ebro River.

2 Analyzed Sites

The studies revised in this chapter correspond to the last 125 km of the Ebro River, from the Riba-roja dam to the river mouth, plus about 100 km of one of the main Ebro tributaries the Cinca/Vero system (see map in Fig. 1 for location of the sites). This area is heavily dammed to regulate the river flow as well as to provide electric power and facilitate water captation for irrigation channels. Dams and overflow dams relevant for the studies revised here are shown in Fig. 1.

Selection of sites was directed to the analysis of the influence of different facilities and/or activities that a priori could affect natural animal populations from the river. There are at least three industrialized areas that discharge different kind of pollutants into the river system (see Fig. 1 for the location of different cities, towns, and geographical features). The most upstream one is the industrial park

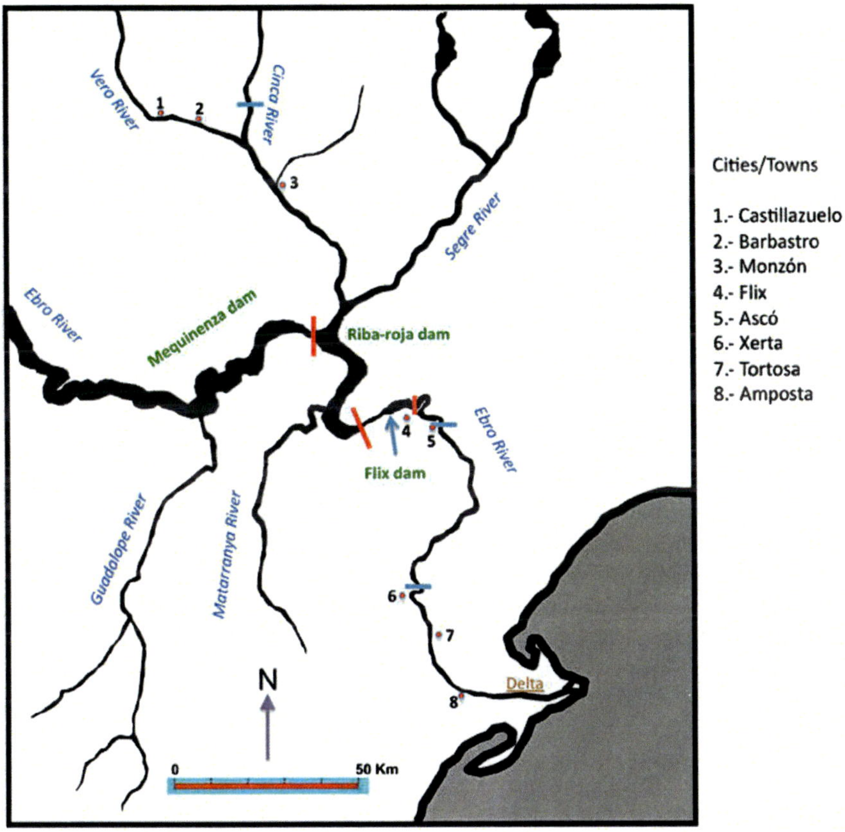

Fig. 1 Map of the lower Ebro River and of its tributaries. River and dam names are indicated in *blue* and *green*, respectively. Positions of relevant dams are indicated by *red bars* whereas overflow dams are indicated by *blue bars*. Cities and towns mentioned in the text are numbered in the map, according to the following key: 1. – Castillazuelo; 2. – Barbastro; 3. – Monzón; 4. – Flix; 5. – Ascó; 6. – Xerta; 7. – Tortosa; 8. – Amposta. The position of the Ebro Delta is also marked. Scale and river courses are only approximate

from Barbastro (14,500 inhabitants), in the Vero River, with different industrial activities including tannery, textile, epoxy resins, and polyamide polymerization. In fact, levels of PBDE-209 in sediments from Vero River downstream of Barbastro loads in PBDE-209 are the highest ever recorded in the world [1].

A second risk zone corresponds to Monzón (17,042 inhabitants), a highly industrialized city in the middle Cinca River. Its industrial activity has caused the historical release of organic and inorganic compounds to the river coming from chlor-alkali industry, from production and utilization of solvents and organochlorine pesticides, and from the use of brominated flame retardants in the production processes. Very high concentrations of mercury have been recorded for sediment and fish samples in the downstream of Monzón [2–4].

The third risk area is located along the Ebro River, in Flix (4,098 inhabitants), where a mercury cell chlor-alkali plant operates since the beginning of the twentieth century. This long operational period, along with the construction of a dam next to the factory around 1960, resulted in the accumulation of high amounts of heavy polluted sediments in the adjacent riverbed. These sediments include Tm quantities of organochlorine compounds as well as mercury and other mining residues [2, 5]. In addition, the plant is still releasing mercury to the river, an industrial procedure to be phased out by 2012. Due to the ecological and economical importance of the low Ebro River, a recovery program has been implemented for the removal of the residues to a controlling disposal site (more information in http://iagua.es).

The fourth and last study area corresponds to the Ebro's Delta. It is a triangular-shaped geological structure that enters about 20 km into the Mediterranean Sea (Fig. 1). It holds 21,600 ha of rice fields, producing 113,500 tons of rice per year, plus a profusion of vegetable crops and orchards, a prosperous aquaculture industry, producing mussels and oysters, and it is surrounded by a very important fishing area. The two coastal splits, one North and the other South of the Delta, enclose two shallow bays of capital importance for fish and mussel production. The geological and hydrological dynamics of the Delta del Ebro have been altered for rice farming by means of upstream barrages and canalization through the Delta. Two main channels, one on each side of the river, bring the water from Xerta-Tivenys to the Delta (some 30 km distance). The water is then distributed to and collected from the rice fields by a network of irrigation and drainage ditches. Rice farming involves the use of more than 25 agrochemicals to treat plagues, with triazines, phenylureas, anilines, and organophosphates being the compounds of broader use [6]. Three sampling sites were located at the ends of the main drainage channels (see Fig. 6a) that collect and transport the water from the southern hemidelta rice fields to one of the sea lagoons. By selecting the main drainage channels instead of individual rice fields, it was possible not only to detect but also to evaluate the effects of most pesticide residues that are applied in the rice fields.

The choice of reference sites is arguably more difficult that the picking of impacted ones, as human impact is overwhelmingly present all on the Ebro River. For Vero River, the selected reference site was located in a pristine area upstream of Barbastro (see Fig. 1). For Cinca River, the reference site was located upstream of Monzon, avoiding also the influence of the Vero River (see Fig. 1). At the Riba-roja dam (210 Hm^3, www.embalses.net), upstream of Flix, Ebro River receives water from several large tributaries. This area was considered as reference for the low Ebro sites. Chemical analyses for different pollutants (see below) of sediments, fish, and shellfish collected in these areas confirmed their suitability as reference sites.

3 Risk Substances

Industrial activities along the Vero, Cinca and low Ebro Rivers, added to the agricultural practices at the Delta, determine the presence of at least four types of pollutants. Organochlorine compounds (OCs) are by large the most important ones as they were produced by thousands of Tm during the 100-plus operational period of the Flix factory. Although they are no longer synthesized in Flix, many Tm of these compounds buried in the residues of the Flix dam are still present in the river, and there are indications that they are leaking from the sediment. Significant concentrations of different OCs, including PCBs, HCHs, HCB, and DDT, and its degradation products, have been detected in sediments and fauna downstream the Flix reservoir [2, 5, 7].

A substance of main environmental and health concern is mercury. Tons of mercury are released annually form Flix and Monzón mercury cell chlor-alkali plants, and there is evidence that it reaches sediments and animals downstream these plants. In fact, a substantial fraction of fish from the low Ebro River (arriving as far downstream as Xerta) contains too much Hg in their bodies to be considered apt for human consumption [8, 9].

Brominated flame retardants, and especially PBDE-209, are particularly relevant in the section of the Vero River downstream Barbastro. Although the industrial activity on this area can be considered moderate, the usually low flow of the river and the absence of strict regulations for these compounds allow concentrations of μg/g of PBDEs in the fat of fish captured in this area [1].

Pesticides arrive in the Ebro River at least from two main sources. Chlorinated pesticides (DDT, HCHs, etc.), currently limited or banned in Europe, are or have been produced by different factories in Monzón and Flix, along with its subproduct HCB. These compounds are present not only in the Flix residues, but also in sediments and fauna all along the low Ebro River. In addition, modern, more sophisticated pesticides, including triazines, phenylureas, anilines, and organophosphates, are currently used in the intensive agricultural practices currently common in the Ebro Delta [6, 7, 10].

4 Exposure and Effect Markers

The putative impact of these contaminants in the exposed fauna can be analyzed at multiple levels. The true ecological impact can be only assessed by a detailed study of the resident populations. Such studies have been indeed performed in the low Ebro River, but they require large efforts in terms of amount of work and time. A more readily approach is to use specific markers, changes in the physiology (in its broad term) of the exposed animals that can be related to exposure to environmental pollutants. These physiological changes can implicate the whole animal (average size, sex ratio, condition factor), cells and tissues (cytological aberrations, histological damage), biochemical parameters (enzymes, hormone levels, DNA, protein and lipid damage, specific protein synthesis) or regulatory mechanisms (changes in gene expression, among others) [11]. Figure 2 shows a diagram of the different levels at which to evaluate biological impacts on biota and the corresponding methodologies to analyze them. The higher the level of the effect, the more

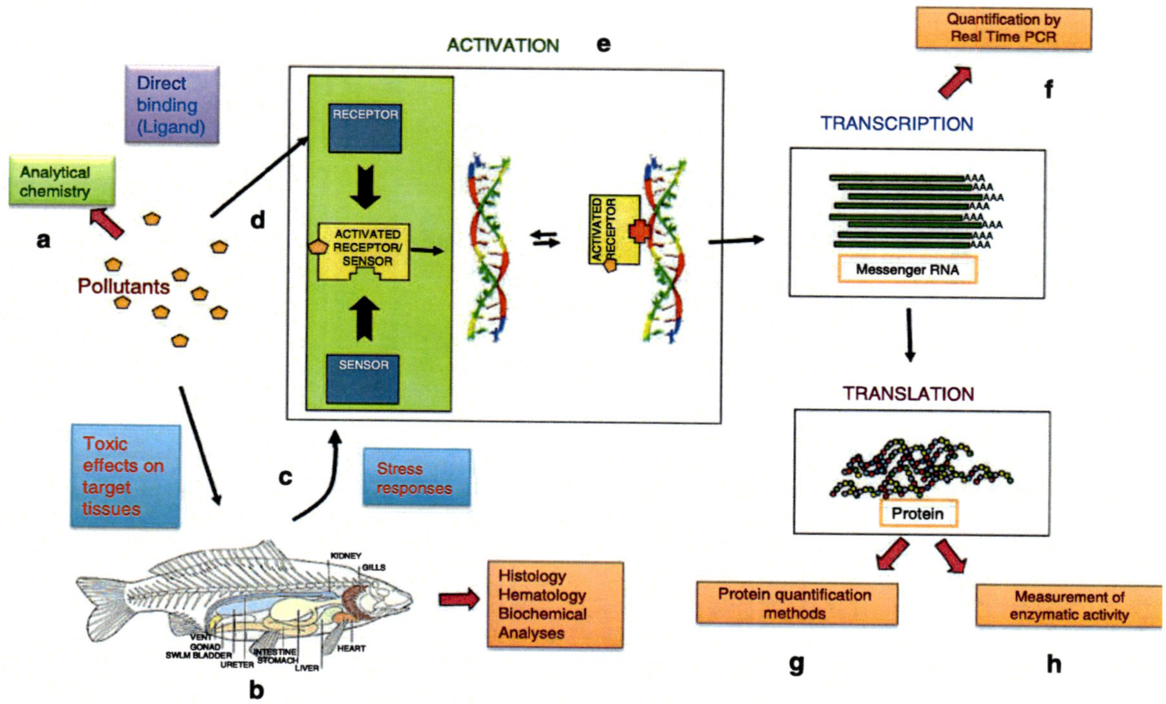

Fig. 2 Different levels of action of toxicants and the corresponding methodologies of monitoring. Environmental pollutants can be quantified by analytical procedures (**a**) and their direct toxic effects analyzed in target tissues by physiological, histological, hematological, or biochemical methods (**b**). A consequence of these toxic effects is the activation of a cellular stress response, by which cells reshape their metabolism to compensate the toxic action (**c**). In addition, many pollutants are able to interact with endogenous receptors and thereby disrupting cellular regulatory mechanisms (**d**). Both stress responses and ectopic receptor activation result in changes on the expression of specific genes (**e**); these changes may be monitored by either measuring the corresponding mRNA (**f**), the concentration of the encoded protein (**g**) or, in the case of enzymes, changes in their enzymatic activity (**h**). It is important to note that stress-regulated genes may become activated or inhibited by toxicants, and that changes in enzymatic activities may occur in the absence of any modification of protein or mRNA levels through the direct interaction of the toxicant with the enzyme

significant and less specific the effect will be. For instance, the impairment of fish growth is a significant, but very unspecific effect. However, induction in the levels of metallothionein in kidney is not such a significant effect, but highly specific of exposure to heavy metals.

The analysis of pollution effects at very different levels is considered an essential feature in ecotoxicology, and several examples of this approach applied to the impact of pollutants in the Ebro basin are reviewed here. At the whole animal level, biometric data (total weight, fork length, liver weight) were used in fish to probe the animal fitness, using standard indices as the condition factor (CF) or the hepatosomatic index (HSI) [12]. Survival and different feeding inhibition analyses were used in *Daphnia magna* in transplantation experiments [13] (Table 1).

Histopathological analyses were performed essentially in fish. The presence of marcophagic aggregates and other histological alterations were analyzed in the liver of barbel (*Barbus graellsii*) and bleak (*Alburnus alburnus*). The proportion of micronucleated erythrocytes in fish blood (MNI) was obtained by flow cytometry techniques and used as diagnostic for cytological and DNA damage [14, 15].

A battery of different biochemical quantitative assays was applied to many different tissues and species. DNA damage and lipid peroxidation assays measure the direct impact of genotoxics and oxidant pollutants [16, 17] whereas alteration of GSH levels in liver is a marker for oxidative stress [18]. Mercury and other heavy metals are known to induce metallothionein levels in different tissues although this effect is variable in different species and organs [19–22].

One of the most common targets of environmental pollutants on the exposed organisms is to impair some key enzymatic systems. Enzymes participating in three main metabolic pathways have been extensively used to study specific kinds of biological impact. Xenobiotic transformation enzymes are able to modify, mainly by oxidation, and eventually help to excrete, mainly by conjugation, different organic pollutants. Enzymes participating in these two steps, called phase I and phase II, increase their activities as a response of the presence of xenobiotics [11]. In the works reviewed here, these enzymes were represented by EROD and different reductases (Phase I) [23, 24] and by UDPTG and GST (Phase II) [25]. A second metabolic pathway relevant to the effect of pollution in fish and shellfish is the oxidative stress response. Many enzymes, like SOD, CAT, and GPX, are activated in the presence of strong oxidants and neutralize them [11]. Finally, different choline esterases, like AChE, ChE, PChE, and CbE, are targets, intended or accidental, of many pesticides, and inhibition of their activities is used as a marker of these pollutants in the environment [26].

Gene expression markers are becoming a very useful tool to monitor exposure to different pollutants [27–29]. Some of them, like CYP1A and MT1 and MT2, are the genetic counterparts of EROD and MT protein quantitative assay, respectively. The results have shown a close relationship between gene expression and biochemical markers in most (but not all) cases [4, 30, 31]. High expression of CYP1A in liver and scales is diagnostic for exposure to dioxin-like compounds [8, 32]. The use of this methodology in scales is one of the first approaches to monitorization of these pollutants (and, partially, of metals) without killing the animals. Vitellogenine, as

Table 1 Methodological approaches used in the papers included in this review

Assay level	Assay	Species	Tissue	Stress monitored
Whole animal	Condition factor	Cc	N/A	General fitness
	Hepatosomatic index	Cc	N/A	General fitness, liver damage
	Survival	Dm	N/A	General toxicity
	Feeding inhibition	Dm	N/A	General toxicity, neurotoxicity
Tissue	Micronuclei index	Cc, Rr, Sg	Blood	DNA damage
	Hepatic histological analysis	Bg, Aa	Liver	Liver damage
Biochemical assays (nonenzymatic)	MT (protein) quantitation	Bg, Dp, Pc	Liver (Bg), digestive gland (Dp, Pc)	Metal poisoning
	Lipid peroxidation	Bg, Cc,Cf, Dp, Pc		Oxidative stress
	DNA damage quantitation	Cc, Dp, Pc	Liver (Cc), digestive gland (Dp, Pc)	DNA damage
	GSH, GSSG quantitation	Cc, Dp, Pc	Liver (Cc), digestive gland (Dp, Pc)	Oxidative stress, mercury poisoning
Enzymatic assays	EROD	Bg, Cc	Liver	Oxidative stress, dioxin-like compounds
	UDPTG	Bg	Liver	Oxidative stress, xenobiotics
	CAT	Bg, Cc, Cf, Dp, Pc, Dm	Liver (Bg, Cc), digestive gland(Cf, Dp, Pc), whole (Dm)	Oxidative stress
	SOD	Cf, Dp, Pc	Digestive gland	Oxidative stress
	NAD/NADP reductases	Bg	Liver	Oxidative stress
	GST	Bg, Cc, Cf, Dp, Pc, Dm	Liver (Bg, Cc), digestive gland (Cf, Dp, Pc), whole (Dm)	Oxidative stress
	GPx	Bg, Cf, Dp, Pc	Liver (Bg), digestive gland(Cf, Dp, Pc)	Oxidative stress
	GR	Bg, Dp, Pc	Liver (Bg), digestive gland(Dp, Pc)	Oxidative stress
	ChE	Bg, Dp, Pc	Brain (Bg), gills (Dp), muscle (Pc)	Neurotoxicity
	AChE	Cf, Dm	Gills (Cf), whole (Dm)	Neurotoxicity
	PChE	Cf	Gills (Cf)	Neurotoxicity
	CbE	Cf, Dp, Pc, Dm	Gills (Cf, Dp), muscle (Pc), whole (Dm)	Neurotoxicity
Gene expression	CYP1A	Bg, Cc	Liver	Oxidative stress, dioxin-like compounds
	MT1,2	Bg, Cc	Liver, kidney (Cc), scales(Cc)	Metal poisoning
	Vg	Cc, Rr	Liver	Estrogenic compounds

Species names: Bg, *Barbus graelsii*; Cc, *Cyprinus carpio*; Cf, *Corbicula fluminea*; Dm, *Daphnia magna*; Dp, *Dreissena polymorpha*; Pc, *Procambarus clarkii* well as egg-specific protein synthetized in female liver, is also detectable as mRNA in males exposed to estrogen-like compounds [29, 33–35]. There are historical reports of feminization of fish in the low Ebro River [36], but a very recent campaign have shown no evidence of estrogenic pollution at present.

5 Sentinel Species

The studies on natural populations rely heavily on the selection of adequate species that would be used as a representative of the impact of the pollution on the whole ecosystem. Selected species have to be abundant at both impacted and reference sites, easy to collect, and, preferably, of ecological or economical impact. In general, predators or scavengers (or filter-feeders, in the case of mollusks) tend to accumulate more pollutants in their bodies than grazers and, therefore, their burden in pollutants is accordingly higher. Size does matter as very small animals are often difficult to collect in enough quantities for chemical or biochemical analyses, whereas very large ones are difficult to capture, handle, and dissect. A second level of choice is the part of the organism to be sampled. Some small organisms, like arthropods or very small fish, may be sampled whole, but in many cases the analysis has to be restricted to a defined organ or structure. A marker may respond to a particular stimulus (hormones, oxidation, metals, etc.) in liver, let's say, but not in brain or muscle. Therefore, limiting the sampling to the tissues known to respond to a particular response increases the sensitivity and reliability of the results. Liver is, without doubt, the preferred organ for fish whereas gills and digestive glands are preferred in bivalves. However, some markers require sampling other tissues, like kidney, brain or even scales in fish, and gonads or mantle in mollusks.

RNA quantitation-based markers have the advantage of using small amounts of sample (from 10 to 100 mg), allowing their application to very small animals or body parts. The downside is that their application requires the knowledge of the exact sequence (or at least a portion of it) of the gene(s) to be monitored. Genomes of very different species are getting sequenced at amazing speed, but they are (and will ever be) restricted to a limited number of species if we compare with the amazing biodiversity present in many environmentally relevant ecosystems. If these techniques are to be applied, only known species, or those with close, better known relatives, are suitable as sentinels. The currently dwarfing costs of DNA sequencing will facilitate in the next future the extension of these techniques in the until-then "orphan" species.

There are several species of fish and other aquatic organisms susceptible of being used as pollution sentinels in the Ebro River (Table Species). However, none of them is present in all sites shown in Fig. 1, at least in enough abundance. From Barbastro to Flix (including the Riba-roja dam), barbel and bleak were preferred for chemical and pollution marker analyses whereas carp (*Cyprinus carpio*), European catfish (*Silurus glanis*), and roach (*Rutilus rutilus*) were the main species sampled in the area from Riba-Roja to Xerta. Zebra mussel (*Dreissena polymorpha*) and the Louisiana crayfish (*Procambarus clarkii*) were selected as invertebrate species in Flix and Ascó whereas *Corbicula fluminea* and *D. magna* were the species used to probe pollution in the Delta, in transplantation experiments.

6 Analysis of Biological Effects

6.1 The Cinca/Vero System

This area was the most profusely studied in the AQUATERRA project in terms of biological effects in fish populations. Barbel and bleak were the sentinel species selected in this area and an array of histological and biochemical tests were used to monitor the impact due to three major sources of pollution: mercury and OCs at Monzón (with a comparison in one of the papers with Flix) and PBDEs in Barbastro [1–4, 37]. Mercury pollution was directly correlated to an increase of MT protein in the liver of barbel captured downstream Monzón when compared to samples captured upstream (Fig. 3a). However, mRNA quantitative analyses failed to show any differences between downstream and upstream Monzón, neither correlated with MT protein levels. Further studies showed that MT mRNA in liver is a rather weak marker for chronic metal pollution in liver (see below) [4]. The presence of degenerative hepatocytes in barbels and bleaks was also linked to mercury poisoning although it can also reflect the impact by other pollutants, like OCs or PBDEs (Fig. 3e).

The effect of OC pollutants, and more specifically, dioxin-like OCs, was tested in barbels both biochemically (EROD) and by CYP1A mRNA quantitation methods (Fig. 3b, [4]). Both markers showed a clear increase on impacted areas downstream Barbastro and Monzón relatively to reference, upstream sites. In this case, a clear correlation was observed between biochemical and gene expression markers, indicating that the increase of EROD activity was, at least in part, due to de novo synthesis of the enzyme CYP1A (Fig. 3b). The observed increase in CYP1A expression in barbel samples from the Cinca River was significantly lower than in those from the more heavily contaminated Flix site. Correlation analysis of biological effect with chemical data suggested that coplanar PCBs, rather than actual dioxins, were responsible for the dioxin-like response observed in Cinca, Vero, and Flix barbels [2, 4].

There are still very few evidences of biological impact attributable to PBDEs. Comparison of barbel populations upstream and downstream Barbastro showed a significant increase of hepatic phase I and II metabolism (reductases, glutathione S transferase, UDPG, and GPX) whereas brain cholinesterase and EROD activity were depleted ([1], Fig. 3c). Histological analyses showed signs of lesions in liver and kidney, also associated to high PBDE burden (Fig. 3d). This was one of the first evidences for a causal relationship between PBDE levels and biochemical and histopathological responses in feral fish [1].

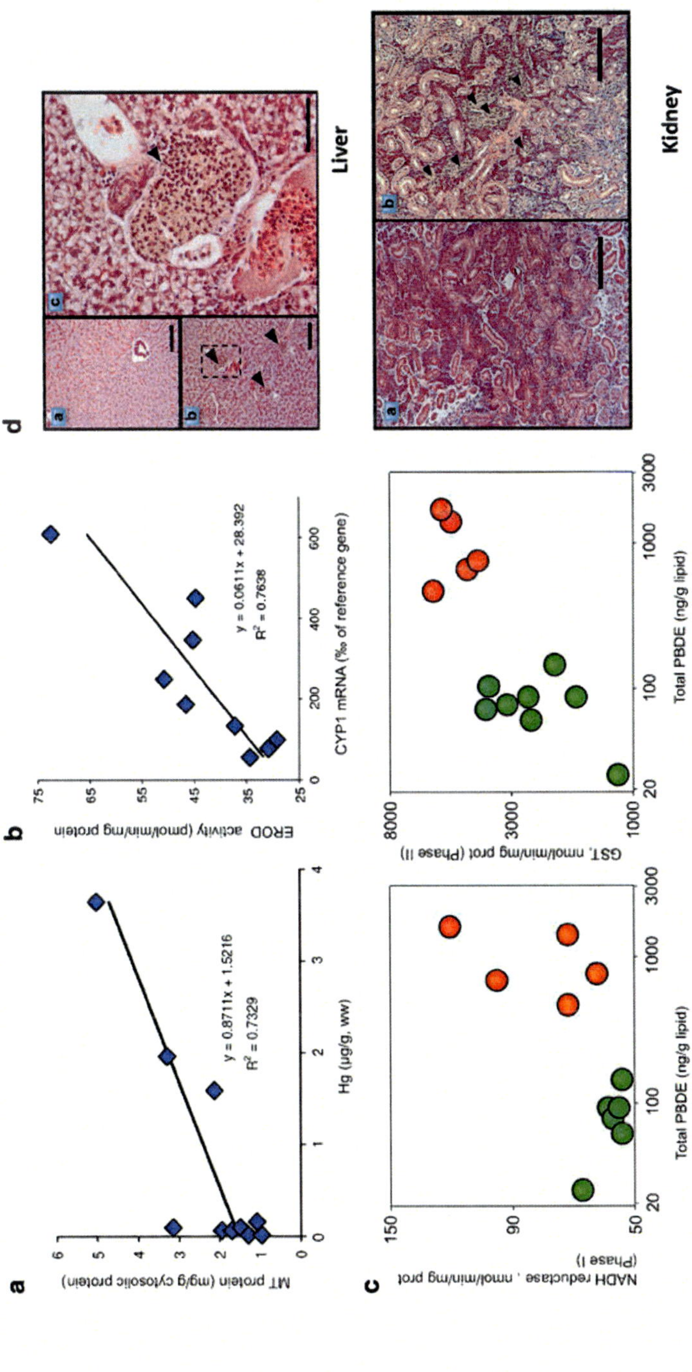

Fig. 3 Summary of marker results from the Vero/Cinca system. **(a)** Correlation between metallothionein protein and total mercury content in barbels from the Cinca River, upstream and downstream of Monzón. **(b)** Correlation between CYP1A mRNA and EROD activity in barbels from the Cinca River. **(c)** Correlation between PDBE burden and NADH reductase (*left*) and GST (*right*) activity in barbels from upstream (*green triangles*) and downstream (*red triangles*) Barbastro industrial site. **(d)** Histological analyses of barbel liver (*upper*) and kidney (*lower panel*). Sub-panels a show normal structures from a specimen captured upstream Barbastro whereas sub-panels **b** and **c** show section corresponding to and impacted specimen captured downstream Barbastro. *Arrowheads* indicate the presence of macrophage aggregates, characterized by the brown lipofuscin–ceroid pigment (*scale bar* = 50 µm). Data from [4] and [1]

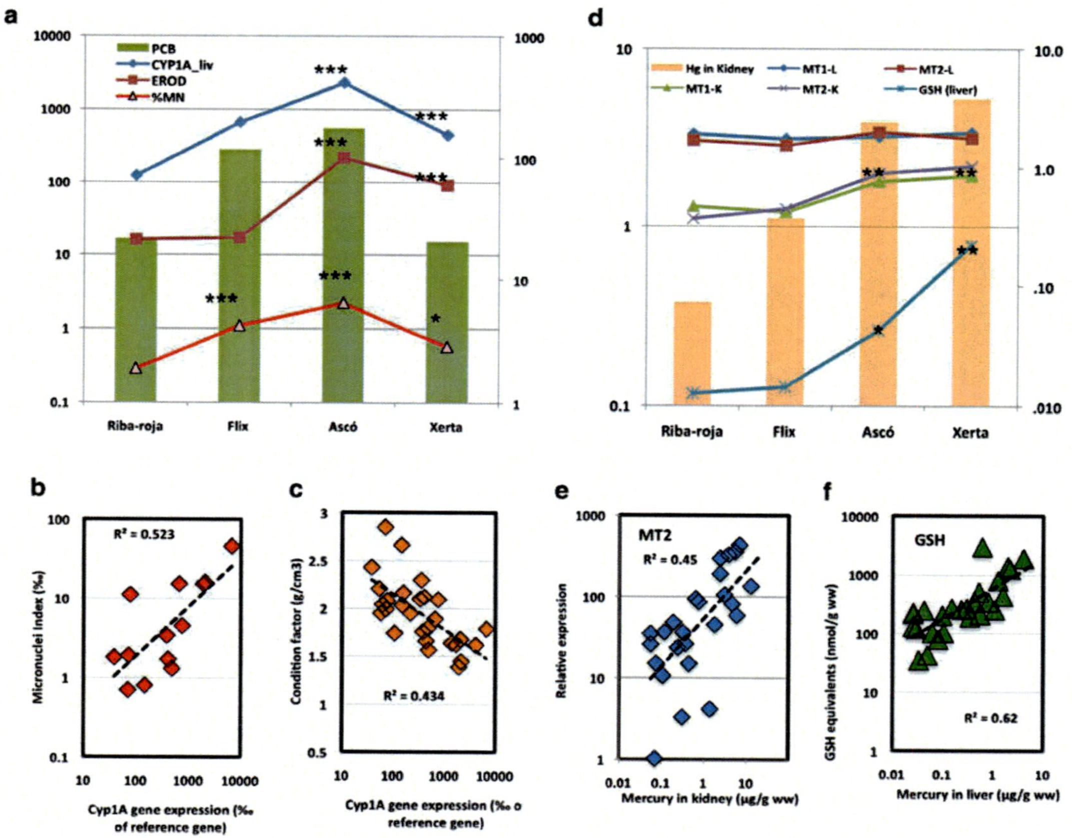

Fig. 4 Data on pollutant burden and its effects on fish in the low Ebro River, from Riba-roja to Xerta. (**a**) Total PCB content in sediments from Riba-roja (reference site), Flix, Ascó, and Xerta in ng/g (*green bars*). Superimposed to this profile, *line graphs* indicate median values for three markers related to

6.2 Low Ebro River (from Riba-Roja to Xerta)

This section of the Ebro River is under the direct impact of the Flix chlor-alkali plant and of its residue sediments. Contamination of Hg and OCs is maximal in these sediments, and there is evidence that it leaks downstream as far as at least to Xerta, 37 km away Flix ([8, 9], Fig. 4a, d). Fish populations in Flix show poorer condition and fecundity than their counterparts sampled in Riba-roja or the Delta [38]. In addition, HSI and CF in carps showed good correlation (positive and negative, respectively) with OC burden (Fig. 4c). These data indicate that the pollution by Flix residues and, specifically, by OCs, affects the growth and health status of fish from the low Ebro River.

Consequently with the effects on fish populations and individuals, carp from Flix or downstream showed higher EROD, CYP1A expression, and MNI than the Riba-roja population, effects correlated with their burden in OCs. These biological effects were higher in fish captured in Flix or Ascó than in fish from Xerta, showing an attenuation of the biological response with the distance to the source (Fig. 4a–c). On the contrary, biological responses associated to mercury poisoning, like higher MT expression in kidney or scales, or GSH content in liver, reached their maximal values in Xerta, in a pattern essentially identical to that of the Hg burden of fish (Fig. 4d–f, [8]). Therefore, mercury and OC pollution showed rather different patterns of distribution along the low Ebro River. The exact mechanism by which Hg burden increase with the distance of the point of discharge remains to be elucidated. It is noticeable that mercury pollution, albeit exceeding in some animals the maximal amount allowed by the UE (0.5 mg/kg in muscle), did not correlate with low HSI or CF content, indicating a good adaptation of the animals to the presence of the pollutant. Only the high MT expression in kidney (some tenfold over the nonimpacted samples) and of GSH in liver (over 100-fold) indicate an acclimatation of the impacted populations to the chronic mercury pollution (Fig. 4e, f, [8]).

The effects of Flix sediment pollution on invertebrates were examined at two levels. Local populations of zebra mussels and crayfish (*P. clarkii*) were sampled right over the Flix residue sediment, across the reservoir, in a meander immediately downstream the dam and in Ascó, in addition to the Riba-roja reference site (Fig. 5a, b). Results from a battery of biochemical biomarkers (Fig. 5c) were similar

Fig. 4 (continued) dioxin-like pollution, CYP1A mRNA abundance (*blue*), EROD enzymatic activity (*brown*, both in liver), and micronuclei index in erythrocytes (*red*) in carp populations from the same areas. *Asterisks* indicate significant differences from the Riba-roja population: *, $p < 0.05$; **, $p < 0.01$; ***, $p < 0.001$, Student's *T*-test. (**b**) Double log correlation between micronuclei index in blood erythrocytes and CYP1A gene expression in liver from the same animals. (**c**) Inverse correlation between condition factor and CYP1A gene expression in liver. (**d**) Profile of Hg content in carp kidney (*salmon bars*) and of expression of two Metallothionein gene in liver (*blue* and *red*) and in kidney (*green* and *purple*). The *pale blue line graphs* correspond to GSH levels in liver. *Asterisks* indicate significant differences from the Riba-roja population as in panel **a**. Note that kidney, but not liver MT expression correlate with mercury burden, as the hepatic GSH levels – panels (**e**) and (**f**). Data from [8] and Olivares et al. unpublished results

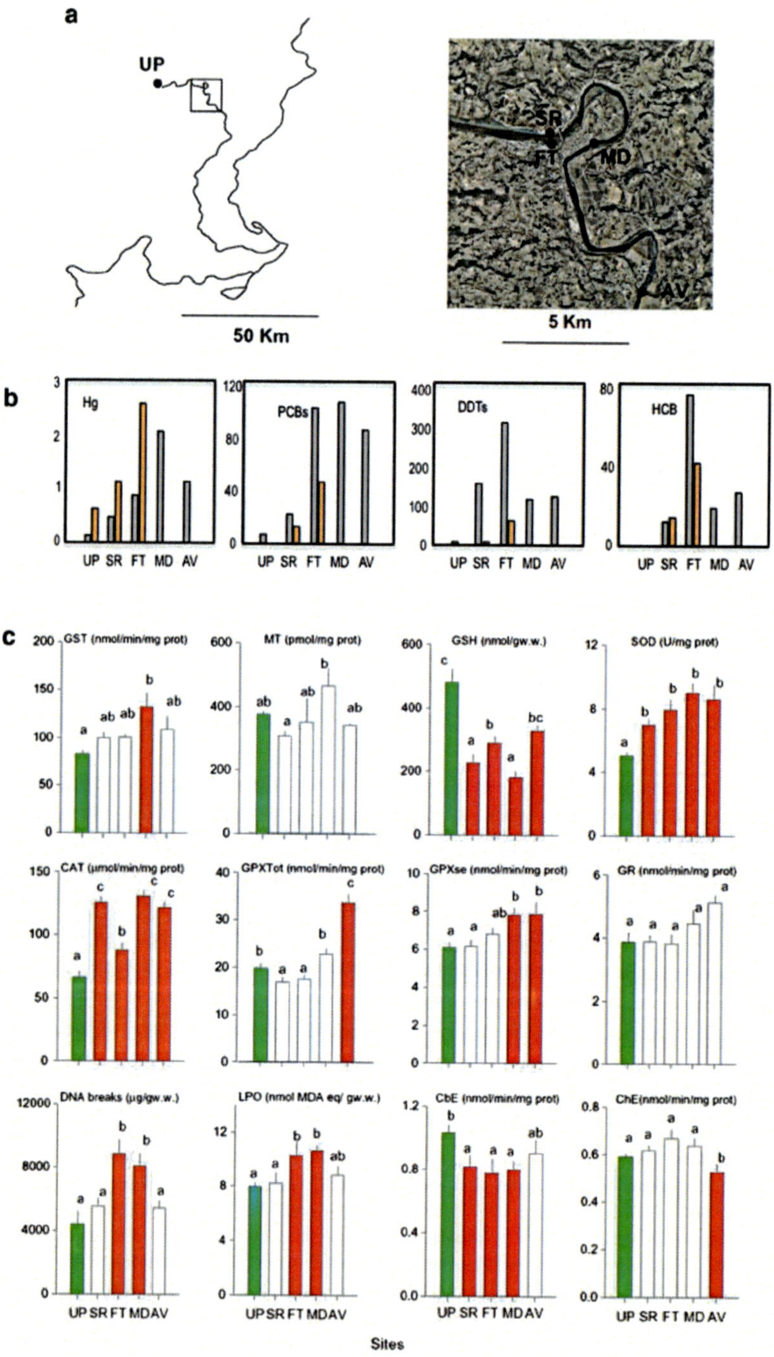
a
UP
SR
FT
MD
AV
50 Km
5 Km
b
Hg
PCBs
DDTs
HCB
UP SR FT MD AV
c
GST (nmol/min/mg prot)
MT (pmol/mg prot)
GSH (nmol/gw.w.)
SOD (U/mg prot)
CAT (µmol/min/mg prot)
GPXTot (nmol/min/mg prot)
GPXse (nmol/min/mg prot)
GR (nmol/min/mg prot)
DNA breaks (µg/gw.w.)
LPO (nmol MDA eq/ gw.w.)
CbE (nmol/min/mg prot)
ChE(nmol/min/mg prot)
UP SR FT MD AV
Sites

for mussels and crayfish, and indicated an oxidative stress in the populations from Flix sediments, the meander and, partially, Ascó, compared to the samples in the Flix reservoir, but not in contact with the contaminated sediments, or Riba-roja. Antioxidant enzymes, like CAT, GPX, and GST, showed elevated activity levels in the impacted populations, which also showed increased DNA damage and lipid peroxidation. MT levels, indicator also in invertebrates of metal pollution, were only slightly elevated in the meander, but not in the Flix, mussel population [39] (Fig. 5c).

The nature of toxic effects of Flix residues on invertebrates was dissected using sublethal *D. magna* feeding tests to analyze the effect of different subfractions of sediment elutriates in the ability of *Daphnia* to feed. The results suggest that lye, on one hand, and heavy metals, on the other, were the most active toxicants for *Daphnia* whereas OCs showed toxic effects only in vertebrate-based assays [40]. This is relevant to the interpretation of the effects of the combination of pollutants present in the Flix residues: Activation of the AhR by exogenous ligands is the key step determining toxicity by OCs in vertebrates, but this process does not occur in protostomata, in which the homologous counterpart of the AhR apparently is devoid of any ligand-binding domain [41, 42].

6.3 *The Ebro Delta*

The analysis of biological effects of pollution on the Ebro Delta focused on the impact of pesticides in invertebrates, both in *D. magna* and in the Asian clam *C. fluminea*. In both cases, control animals were exposed to the water from drainage channels (Fig. 6) and the putative environmental hazards examined by multibiomarker and multivariate methods [43]. In the case of the Asian clams, antioxidant and esterase enzyme responses were reduced and lipid peroxidation levels increased steadily from May in upstream stations to August in drainage channels [44]. Endosulfan, propanil, and phenylureas were identified as the most likely responsible for the observed effects. Results from *D. magna* also showed inhibition of cholinesterases and carboxylesterases, specific markers of organophosphorous and carbamate pesticides, as well as altered patterns of antioxidant enzymes and GST. These toxic effects correlated with the concentrations of propanil, molinate,

---

Fig. 5 Invertebrate population analysis at the low Ebro River. (**a**) Sites of zebra mussel and crayfish sampling. (**b**) Total body burden in Hg (*left*), PCBs (*center left*), DDT + DDE (*center right*), and HCB (*right*) for zebra mussel (*gray bars*) and crayfish (*orange bars*). (**c**) Results from biochemical markers in the four populations of zebra mussel. The corresponding value for the reference site is marked in *green*, values significantly different from them (ANOVA) are marked in *red*. Note that toxicants elevate values of many oxidative stress as well as DNA damage and metallothionein expression whereas they reduce ChE and CbE activity. Maximal toxic effects were recorded in Flix factory (FT) and in the MD (meander) sites. Data from [39]

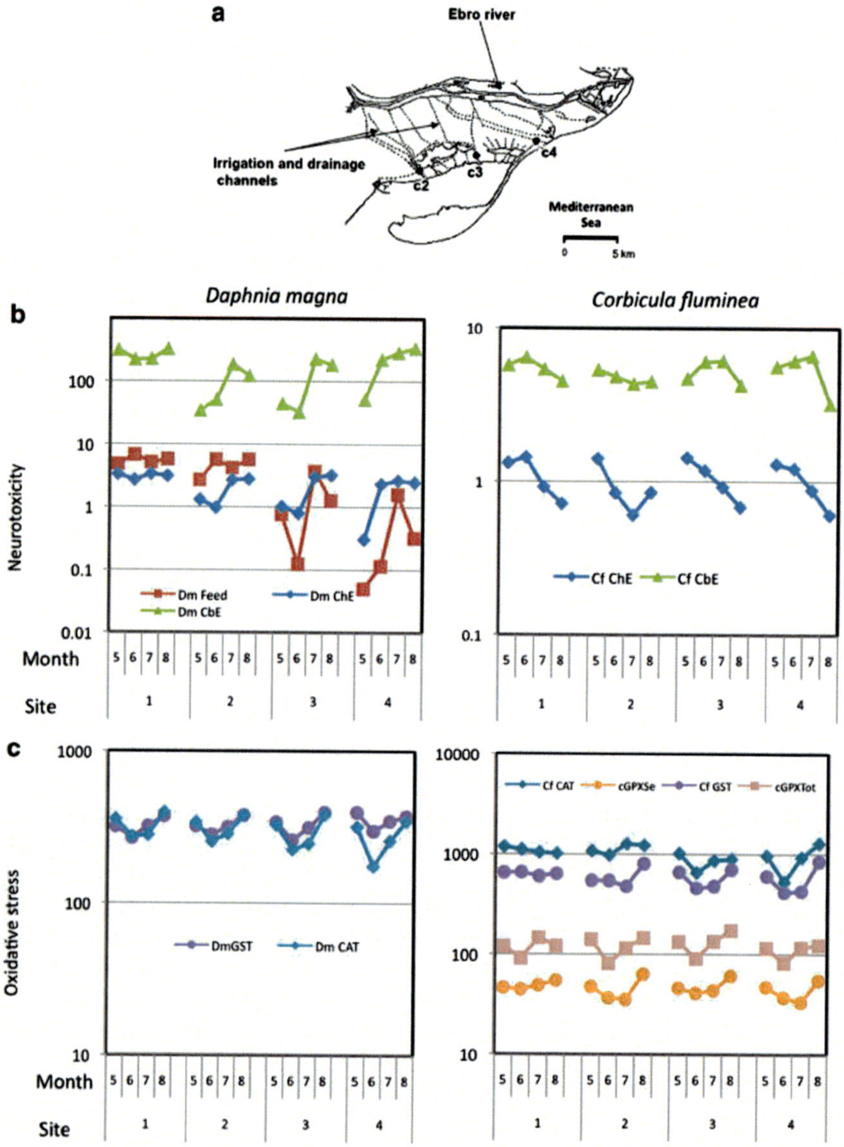

Fig. 6 Biomonitoring of pollution in the Ebro Delta with *Daphnia magna* and *Corbicula fluminea*. (**a**) Map of sampling sites. Site 1 is out of the figure limits, close to Amposta (see Fig. 1). (**b**) Assays for neurotoxic activity (ChE and CbE) and *D. magna* feeding). Note the different pattern for the microcrustacean (sensitive to insecticides), which show a maximal toxic (inhibitory) effect in May and June, and the mollusk (relatively resistant). (**c**) Oxidative stress markers. These markers showed a similar response for both species with maximal effects (activation) in May (month 5) and August (month 8). Data from [43] and [44]. "Dm" and "Cf" identify markers from *D. magna* and *C. fluminea*, respectively

and fenitrothion in water [43, 44]. All these results indicate that agricultural practices, rather than pollution form upstream Ebro, were the main stressors of the invertebrate population in the Delta. This is relevant for the important mussel and oyster-growing aquaculture facilities operating in this area.

7 Concluding Remarks

Survey of biological effects of pollution in the low Ebro River showed a clear impact of the existing chemical plants in the rivers Vero and Cinca, as well as in the Flix reservoir, on the local populations of fish and invertebrates. It also showed that fish can acclimate to very high concentrations of some toxicants, like mercury whereas OCs and PBDEs apparently induce a permanent oxidative stress and other negative effects, including poor condition and fertility, DNA damage, and liver and kidney histological anomalies. It also showed that the toxic determinants are different for vertebrates and invertebrates and suggest that one of the main differences may be the presence of activable AhR, which only occurs in deuterostomata (Chordates, Echynoderma and alikes). The adverse biological effects were recorded up to 30–35 km downstream the source and their distribution differed for OCs and for Hg. Finally, the adverse biological effects observed in the Ebro Delta, at least in invertebrates, were likely related to the intensive local agricultural practices rather than to pollution from the Ebro River.

Acknowledgments This work has been partially supported by the European Union and reflects only the author's views, and the European Community is not liable for any use that may be made of the information contained in it (AQUATERRA, GOCE-CT-2004-505428). Additional support from the Spanish Ministry of Science and Innovation AQUATOXIGEN (CGL2008-01898), ECOTOXIMED (CGL2004-03514), the Spanish Ministry of Environment (MOVITROF, ZEBRAPOP, MMARyM 042/RN08/03.4), and the Catalan Water Authority (ACA, MOVITROF) is also acknowledged.

References

1. Raldua D, Padros F, Sole M, Eljarrat E, Barcelo D, Riva MC, Barata C (2008) First evidence of polybrominated diphenyl ether (flame retardants) effects in feral barbel from the Ebro River basin (NE, Spain). Chemosphere 73(1):56–64
2. Eljarrat E, Martínez MA, Sanz P, Concejero MA, Piña B, Quirós L, Raldúa D, Barceló D (2008) Distribution and biological impact of dioxin-like compounds in risk zones along the Ebro River basin (Spain). Chemosphere 71(6):1156–1161
3. Raldua D, Diez S, Bayona JM, Barcelo D (2007) Mercury levels and liver pathology in feral fish living in the vicinity of a mercury cell chlor-alkali factory. Chemosphere 66:1217–1225
4. Quirós L, Piña B, Solé M, Blasco J, López MA, Riva MC, Barceló D, Raldúa D (2007) Environmental monitoring by gene expression biomarkers in *Barbus graellsii*: laboratory and field studies. Chemosphere 67:1144–1154
5. Grimalt J (1993) http://mediambient.gencat.cat/cat/ciutadans/informacio_ambiental/Flix/estudi.jsp?ComponentID=42291&SourcePageID=42851#1. Last accession, September 2009

6. Terrado M, Kuster M, Raldua D, Lopez de Alda M, Barcelo D, Tauler R (2007) Use of chemometric and geostatistical methods to evaluate pesticide pollution in the irrigation and drainage channels of the Ebro river delta during the rice-growing season. Anal Bioanal Chem 387(4):1479–1488
7. Mañosa S, Mateo R, Guitart R (2001) A review of the effects of agricultural and industrial contamination on the Ebro delta biota and wildlife. Environ Monit Assess 71:187–205
8. Navarro A, Quiros L, Casado M, Faria M, Carrasco L, Benejam L, Benito J, Diez S, Raldua D, Barata C, Bayona JM, Pina B (2009) Physiological responses to mercury in feral carp populations inhabiting the low Ebro River (NE Spain), a historically contaminated site. Aquat Toxicol 93:150–157
9. Carrasco L, Diez S, Soto DX, Catalan J, Bayona JM (2008) Assessment of mercury and methylmercury pollution with zebra mussel (*Dreissena polymorpha*) in the Ebro River (NE Spain) impacted by industrial hazardous dumps. Sci Total Environ 407(1):178–184
10. Lavado R, Urena R, Martin-Skilton R, Torreblanca A, Del Ramo J, Raldúa D, Porte C (2006) The combined use of chemical and biochemical markers to assess water quality along the Ebro River. Environ Pollut 139:330–339
11. van der Oost R, Beyer J, Vermeulen N (2003) Fish bioaccumulation and biomarkers in environmental risk assessment: a review. Environ Toxicol Pharmacol 13:57–149
12. Bervoets L, Van Campenhout K, Reynders H, Knapen D, Covaci A, Blust R (2009) Bioaccumulation of micropollutants and biomarker responses in caged carp (*Cyprinus carpio*). Ecotoxicol Environ Saf 72(3):720–728
13. Damasio J, Tauler R, Teixido E, Rieradevall M, Prat N, Riva MC, Soares AMVM, Barata C (2008) Combined use of *Daphnia magna* in situ bioassays, biomarkers and biological indices to diagnose and identify environmental pressures on invertebrate communities in two Mediterranean urbanized and industrialized rivers (NE Spain). Aquat Toxicol 87(4):310–320
14. al-Sabti K, Metcalfe CD (1995) Fish micronuclei for assessing genotoxicity in water. Mutat Res 343(2–3):121–135
15. Quiros L, Ruiz X, Sanpera C, Jover L, Pina B (2008) Analysis of micronucleated erythrocytes in heron nestlings from reference and impacted sites in the Ebro basin (N.E. Spain). Environ Pollut 155:81–87
16. Olive PL (1988) DNA precipitation assay: a rapid and simple method for detecting DNA damage in mammalian cells. Environ Mol Mutagen 11(4):487–495
17. Huang YL, Cheng SL, Lin TH (1996) Lipid peroxidation in rats administrated with mercuric chloride. Biol Trace Elem Res 52(2):193–206
18. Bainy ACD, Saito E, Carvalho PSM, Junqueira VBC (1996) Oxidative stress in gill, erythrocytes, liver and kidney of Nile tilapia (*Oreochromis niloticus*) from a polluted site. Aquat Toxicol 34:151–162
19. Amiard JC, Amiard-Triquet C, Barka S, Pellerin J, Rainbow PS (2006) Metallothioneins in aquatic invertebrates: their role in metal detoxification and their use as biomarkers. Aquat Toxicol 76(2):160–202
20. Auslander M, Yudkovski Y, Chalifa-Caspi V, Herut B, Ophir R, Reinhardt R, Neumann PM, Tom M (2008) Pollution-affected fish hepatic transcriptome and its expression patterns on exposure to cadmium. Mar Biotechnol (NY) 10(3):250–261
21. Lecoeur S, Videmann B, Berny P (2004) Evaluation of metallothionein as a biomarker of single and combined Cd/Cu exposure in *Dreissena polymorpha*. Environ Res 94(2): 184–191
22. Schlenk D, Zhang YS, Nix J (1995) Expression of hepatic metallothionein messenger RNA in feral and caged fish species correlates with muscle mercury levels. Ecotoxicol Environ Saf 31 (3):282–286
23. Ortiz-Delgado J, Behrens A, Segner H, Sarasquete C (2008) Tissue-specific induction of EROD activity and CYP1A protein in *Sparus aurata* exposed to B(a)P and TCDD. Ecotoxicol Environ Saf 69:80–88
24. Douglas KT (1987) Mechanism of action of glutathione-dependent enzymes. Adv Enzymol Relat Areas Mol Biol 59:103–167

25. Taysse L, Chambras C, Marionnet D, Bosgiraud C, Deschaux P (1998) Basal level and induction of cytochrome P450, EROD, UDPGT, and GST activities in carp (*Cyprinus carpio*) immune organs (spleen and head kidney). Bull Environ Contam Toxicol 60:300–305
26. Escartin E, Porte C (1997) The use of cholinesterase and carboxylesterase activities from *Mytilus galloprovincialis* in pollution monitoring. Environ Toxicol Chem 16(10):2090–2095
27. Piña B, Casado M, Quirós L (2007) Analysis of gene expression as a new tool in ecotoxicology and environmental monitoring. Trends Analyt Chem 26(11):1145–1154
28. Tom M, Auslander M (2005) Transcript and protein environmental biomarkers in fish – a review. Chemosphere 59:155–162
29. Rotchell JM, Ostrander GK (2003) Molecular markers of endocrine disruption in aquatic organisms. J Toxicol Environ Health B Crit Rev 6(5):453–496
30. George S, Gubbins M, MacIntosh A, Reynolds W, Sabine V, Scott A, Thain J (2004) A comparison of pollutant biomarker responses with transcriptional responses in European flounders (*Platichthys flesus*) subjected to estuarine pollution. Mar Environ Res 58:571–575
31. McClain JS, Oris JT, Burton GA Jr, Lattier D (2003) Laboratory and field validation of multiple molecular biomarkers of contaminant exposure in rainbow trout (*Oncorhynchus mykiss*). Environ Toxicol Chem 22:361–370
32. Quiros L, Raldua D, Navarro A, Casado M, Barcelo D, Pina B (2007) A noninvasive test of exposition to toxicants: quantitative analysis of cytochrome P4501A expression in fish scales. Environ Toxicol Chem 26(10):2179–2186
33. Lattier D, Gordon D, Burks D, Toth G (2001) Vitellogenin gene transcription: a relative quantitative exposure indicator of environmental estrogens. Environ Toxicol Chem 20:1979–1985
34. Garcia-Reyero N, Raldúa D, Quirós L, Llaveria G, Cerda J, Barceló D, Grimalt JO, Piña B (2004) Use of vitellogenin mRNA as a biomarker for endocrine disruption in feral and cultured fish. Anal Bioanal Chem 378:670–675
35. Porte C, Janer G, Lorusso LC, Ortiz-Zarragoitia M, Cajaraville MP, Fossi MC, Canesi L (2006) Endocrine disruptors in marine organisms: approaches and perspectives. Comp Biochem Physiol C 143:303–315
36. Lavado R, Thibaut R, Raldua D, Martin R, Porte C (2004) First evidence of endocrine disruption in feral carp from the Ebro River. Toxicol Appl Pharmacol 196(2):247–257
37. Raldúa D, Pedrocchi C (1996) Mercury concentrations in three species of freshwater fishes from the lower Gallego and Cinca Rivers, Spain. Bull Environ Contam Toxicol 57:597–602
38. Benejam L, Benito E, García-Berthou E (2010) Decreases in condition and fecundity of freshwater fishes in a highly polluted reservoir. Water Air Soil Pollut 210(4):231–242
39. Faria M, Carrasco L, Diez S, Riva MC, Bayona JM, Barata C (2009) Multi-biomarker responses in the freshwater mussel *Dreissena polymorpha* exposed to polychlorobiphenyls and metals. Comp Biochem Physiol C Toxicol Pharmacol 149(3):281–288
40. Bosch C, Olivares A, Faria M, Navas JM, del Olmo I, Grimalt JO, Pina B, Barata C (2009) Identification of water soluble and particle bound compounds causing sublethal toxic effects. A field study on sediments affected by a chlor-alkali industry. Aquat Toxicol 94(1):16–27
41. Hahn ME, Karchner SI, Shapiro MA, Perera SA (1997) Molecular evolution of two vertebrate aryl hydrocarbon (dioxin) receptors (AHR1 and AHR2) and the PAS family. Proc Natl Acad Sci USA 94:13743–13748
42. Nebert DW, Puga A, Vasiliou V (1993) Role of the Ah receptor and the dioxin-inducible [Ah] gene battery in toxicity, cancer, and signal transduction. Ann N Y Acad Sci 685:624–640
43. Barata C, Alanon P, Gutierrez-Alonso S, Riva MC, Fernandez C, Tarazona JV (2008) A *Daphnia magna* feeding bioassay as a cost effective and ecological relevant sublethal toxicity test for environmental risk assessment of toxic effluents. Sci Total Environ 405(1–3):78–86
44. Damasio J, Navarro A, Tauler R, Lacorte S, Barceló D, Soares A, López MA, Riva MC, Barata C (2010) Identifying major pesticides affecting bivalve species exposed to agricultural pollution using multi-biomarker and multivariate methods. Ecotoxicology DOI:10.1007/s10646-010-0490-3

Climate Change, Water Resources and Pollution in the Ebro Basin: Towards an Integrated Approach

C.I. Bovolo, S. Blenkinsop, B. Majone, M. Zambrano-Bigiarini, H.J. Fowler, A. Bellin, A. Burton, D. Barceló, P. Grathwohl, and J.A.C. Barth

Abstract The European Water Framework Directive encourages the management of rivers at the basin scale requiring an understanding of the interdependencies between the physical environment and the local hydrology, ecology and climate. The current and future climate of the Spanish Ebro river basin, including its tributary, the Gállego, has been studied extensively within the EU AquaTerra (AT) project along with the hydrology and behaviour of environmental pollutants with respect to climate change. This chapter provides an overview of current research carried out for the Ebro and the Gállego and provides a summary and review of the recent and new AT studies examining climate, water resources and environmental pollutants. AT climate studies suggest that the Ebro will become significantly hotter and drier in the future, especially in summer. For example, regional climate models (RCMs) project a decrease in daily mean precipitation of up to 50.5% in the summer by the 2080s (Bürger et al. Environ Pollut 148:842–854, 2007). Future climate scenarios from an ensemble of RCMs have been used to provide input data to hydrological models of the Ebro and Gállego catchments to simulate the impacts of climate change on the basins. These results suggest that changes in climate will reduce water availability in the area, especially during the

C.I. Bovolo (✉), S. Blenkinsop, H.J. Fowler, and A. Burton
School of Civil Engineering and Geosciences, University of Newcastle upon Tyne, Newcastle upon Tyne, NE1 7RU, UK
e-mail: Isabella.bovolo@ncl.ac.uk

B. Majone, M. Zambrano-Bigiarini, and A. Bellin
Department of Civil and Environmental Engineering, Università di Trento, Trento, Italy

D. Barceló
Consejo Superior de Investigaciones Científicas, Instituto de Investigaciones Químicas y Ambientales de Barcelona, Barcelona, Spain

P. Grathwohl
Institute of Geology, University of Tübingen, Sigwartstr. 10, 72076, Tübingen, Germany

J.A.C. Barth
Lehrstuhl fur Angewandte Geologie, GeoZentrum Nordbayern, Erlangen, Germany

D. Barceló and M. Petrovic (eds.), *The Ebro River Basin*,
Hdb Env Chem (2011) 13: 295–329, DOI 10.1007/698_2010_86,

summer months when irrigation demands are highest. In addition, AT pollutant studies have revealed that persistent organic pollutants such as DDT are found in high concentrations near industrial sites, and heavy metal, pesticide and nutrient pollution are widespread. Problems of poor water quality in the basin may be exacerbated in periods of low water flow as pollutants will be more concentrated due to a lack of dilution with respect to high flow periods. AT studies have shown that this leads to a bioaccumulation of bio-available brominated flame retardants in fish, with potentially serious health effects. Changing climate patterns may influence degradation, turnover, sorption and transport behaviour of pollutant contamination with unknown effects. The effect of climate change on pollutants has not yet been quantitatively assessed with the same rigour as applied to hydrological assessments, and there remains considerable potential for further integration of climate impact assessment within the pollutant monitoring and modelling communities to improve future projections. These studies have important implications for future integrated basin management strategies in the Ebro.

Keywords Basin management, Climate change impacts, Downscaling, Hydrological modelling, Pollutants

Contents

1 Introduction

Water resource management strategies need to balance water supply and demand, particularly in the context of droughts or floods, and contend with problems such as diffuse and point source pollution, poor water quality and ecosystem degradation, all within local political and socio-economic frameworks. As water use in one location may influence water quality and availability in another, it is becoming increasingly important to manage water resources at the river basin scale, especially in view of increasing population and water quality demands paired with increasing competition for water across regions and sectors as well as for issues of governance and regional security [1]. The European Water Framework Directive (WFD) (2000/60/EC), implemented in 2000, addresses the sustainable management of rivers and their associated inland surface waters, transitional waters, coastal waters and groundwater at a basin scale and encourages cooperation across member states containing sections of the same international river basin district [2].

Managing a river at a basin scale in a sustainable manner requires an understanding of the interdependencies between the physical environment and local hydrology, ecology and climate. In addition, to address the long-term management of a river basin, it is important to understand the future impacts of climate change. Climate change can affect multiple aspects of a river basin: the frequency and intensity of disturbance events, e.g. wildfire, floods, droughts and shallow landslides; the hydrological cycle, e.g. precipitation characteristics, runoff, soil water dynamics and sea level rise; soil characteristics through changes in organic matter and nutrient cycling processes; and biological diversity [3]. Climate also affects pollutant fate and transport processes directly; however, long-term land use change driven by climate may have a more significant effect on pollutants such as pesticides in the environment than the change in climate directly [4]. There is a need, therefore, to quantify regional and local changes in future climate (particularly with regards to precipitation and temperature) and to determine the associated impacts on a river basin to develop integrated river basin management strategies for the future.

Hydrological models are useful tools for understanding catchment characteristics and processes. They can also be used with different input scenarios to understand how the catchment may respond to changes in land use or climate. Integrating hydrological models with climate change scenarios is therefore particularly useful for planning and managing water resources and consequently such methodologies are in widespread use. However, there has been comparatively little research on the impacts of climate change on pollutant fate and transport processes. This area of research needs to be further developed and integrated with hydrological and climate studies.

Here, we discuss these issues within the context of the Ebro river basin and one of its sub-catchments, the Gállego. We present climate change scenarios for the basin and discuss the integration of climate change scenarios in general with several aspects of the Ebro's hydrology and also discuss potential impacts of climate change on pollution.

2 The Ebro River Basin in a Hydrological Context

The Ebro river catchment covers 85,534 km^2 and flows from an altitude of 3,404 m in the Pyrenees in Cantabria (NE Spain) southeast to the Mediterranean Sea (Fig. 1), where it has an ecologically and agriculturally important delta that covers more than 150 km^2 [5].

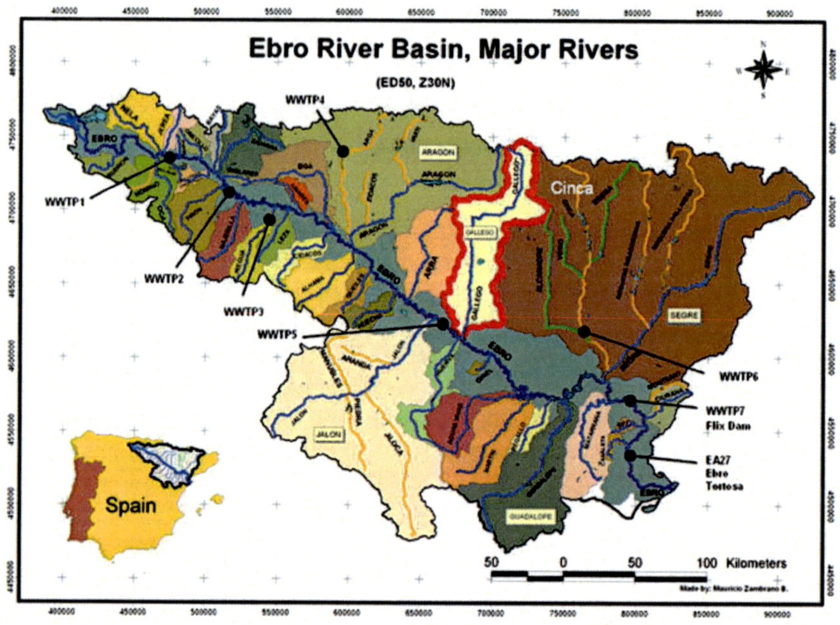

Fig. 1 Major catchments of the Ebro Basin highlighting the Gállego catchment and the approximate location of waste water treatment plants (WWTPs)

Across the catchment, there are at least 349 precipitation gauges and 146 temperature stations that were active in the period 1961–1990 with less than 30% and less than 35% of missing daily data, respectively (see Fig. 2 for the monthly means) [6]. It is important to have as complete records as possible for this period (designated as a "control" in climate modelling experiments), as this provides the basis for comparing observed climates worldwide and for the validation of climate models. Annual and seasonal precipitation in the Ebro is spatially highly variable. For instance, the Ebro basin as a whole receives an annual average precipitation of 620 mm, but in the semi-arid central Ebro basin, precipitation can be as low as 320 mm/year making it one of the driest regions in Europe [7]. In contrast, in the Pyrenees and Cantabrian Mountains annual precipitation can total more than 2,000 mm. The basin suffers from long dry spells in summer and frequent storms in autumn and spring. Annual potential evapotranspiration for the basin is about 700 mm [5].

There are 318 river gauging stations within the Ebro basin [80]. Around 60 of these monitor natural flow regimes and are typically located around the edges of the basin in the medium to higher reaches of the tributary rivers [8]. Others are located on rivers whose streamflow has been altered by reservoirs (see [9] for a review of historical water policy in Spain). In total, 187 reservoirs impound 57% of the mean annual runoff [10]. As an example, annual discharge measured

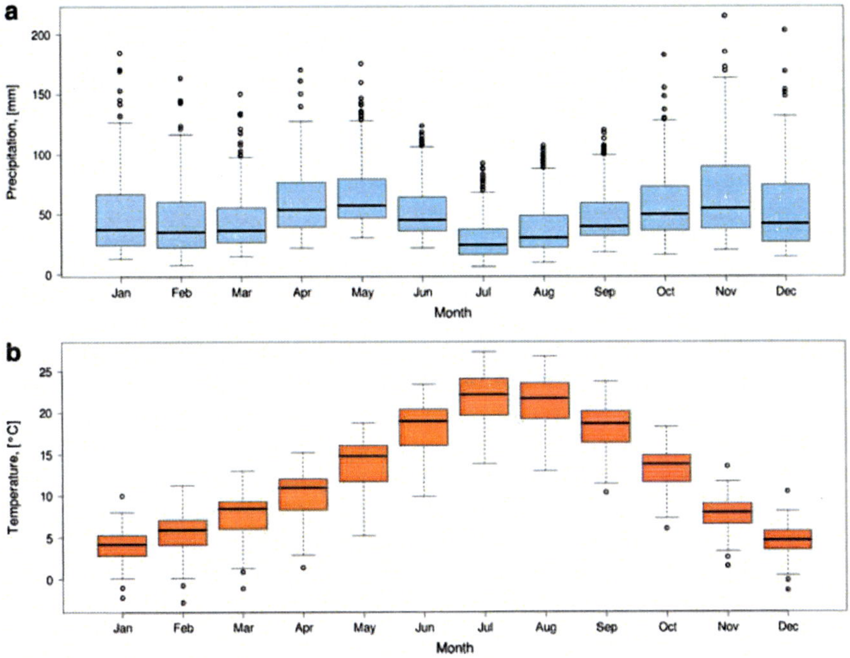

Fig. 2 Monthly average precipitation and temperature during the control period 1961–1990 for stations with over 70% and 65% of data respectively [80]. *Central boxes* show the middle 50% of data; the *thick horizontal line* indicates the mean; the *whiskers* show the range of data (10th to 90th percentiles) and the *dots* show outliers

at the EA27 Ebro Tortosa gauging station, 47.8 km from the outlet (Fig. 1), is about 436 m^3/s (13,768 × 10^6 m^3/year) with a maximum of 715 m^3/s (22,556 × 10^6 m^3/year) for 1960–1961 and a minimum of 136 m^3/s (4,284 × 10^6 m^3/year) for 1989–1990 based on annual water-year (October to September) data from 1960 to 1990 [6]. Peak flow occurs in winter (rainy season in the Cantabrian Mountains) and early summer (due to snow melt in the Pyrenees), while minimum flow occurs in August (Fig. 3).

The Gállego River, a tributary to the Ebro, is a mountainous catchment with an area of 3,969 km^2. It runs north-south from the central Pyrenees at an altitude of 3,056 m and meets the Ebro at Zaragoza at an altitude of 189 m (Fig. 1). There are more than 36 precipitation and 25 temperature gauges in the catchment with daily data spanning 30 years. Streamflows are low in summer and large in winter. According to CHE [5], under a natural regime the annual discharge would be about 34.2 m^3/s (1,087 × 10^6 m^3/year); however, its regime has been altered by the presence of 18 hydroelectric stations and associated reservoirs [11]. Between October and March, water is stored in a number of reservoirs and released for domestic, industrial and hydropower use in line with the minimum flow

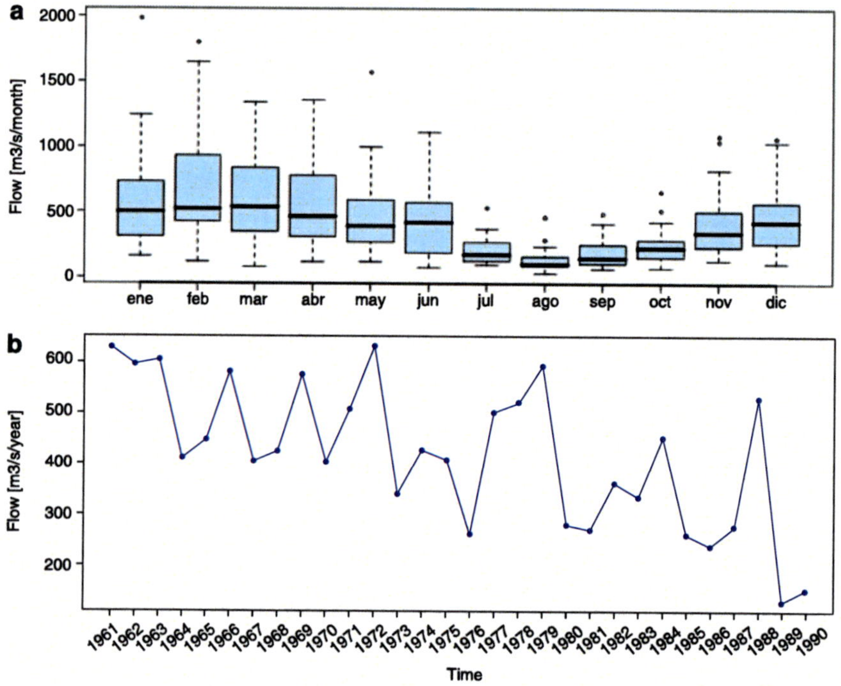

Fig. 3 Measured streamflows at gauging station 027 "Ebro en Tortosa" during the control period 1961–1990. The *upper figure* shows averaged monthly streamflow where the *central boxes* indicate the middle 50% of data; the *thick horizontal line* indicates the mean; the *whiskers* show the range of data (10th to 90th percentiles) and the *dots* show outliers. The *lower figure* shows annual flow at the gauging station. Daily data provided by CHE

requirements of the WFD. Flood risk is managed by allocating extra storage capacity in the reservoirs. Between March and October, irrigation demands are high and priority is usually given to agricultural rather than to hydropower usage.

The Gállego provides water to three main irrigated agricultural districts [12]: the upper Gállego (1,512 ha), the lower Gállego (17,573 ha) and the Riegos de Alto Aragon (33,464 ha) with a total irrigated area of 52,549 ha. Only 16,000 ha of the 52,549 ha are within the Gállego catchment, the remaining 70% are located in the surrounding catchments. The water demand from the lower Gállego irrigation system is about 214×10^6 m^3/year (annual mean discharge of $Q_a = 6.8$ m^3/s), mostly provided by the Ardisa reservoir (Fig. 8d). The demand from the Riegos de Alto Aragon system is about 305×10^6 m^3/year ($Q_a = 9.7$ m^3/s), which is provided by La Sotonera reservoir (Fig. 8d) through the Monegros channel, which exports water to the Eastern Cinca basin (Fig. 1). The irrigated area located within the Gállego river basin and belonging to the Riegos de Alto Aragon system is called La Violada irrigation district and covers an area of approximately 4,000 ha.

2.1 *Current Environmental and Ecological Challenges in the Ebro*

Water and river basin management in Spain is complex with many conflicting forces characterised by regional climatic contrasts, uneven water availability and high water demands from different sources including agriculture (the largest water-using sector), industry, energy and household use [13]. Numerous hydroelectric plants and dams impose a significant regulatory requirement on river flows. In addition, significant water losses due to evaporation from open-water sources and evapotranspiration, combined with intensive use of irrigation for growing more water-intensive but profitable crops such as peaches [14], have resulted in decreased river discharge. The intensification of agriculture throughout the Ebro has resulted in a dramatic increase in productivity but at the cost of diffuse pollution from the historic and ongoing use of pesticides. Combined with pollution from industry and other sources, this has led to deterioration in the quality of river water, and high nutrient levels from agricultural areas have caused eutrophication problems [15]. Salinisation is also an increasing problem due to the increased abstraction of groundwater which causes invasion of aquifers by marine water, and due to the application of brackish irrigation water which concentrates salts in the soils. The historic use of levelling land for irrigation has also exposed large areas of underlying tertiary deposits and sub-soils which are high in saline substances, thereby exacerbating the problem. The use of modern drip irrigation systems, however, is currently being encouraged to help alleviate the problem [13].

The Ebro delta is of ecological significance as it is an important Mediterranean wetland site and bird and fish habitat. The delta was relatively natural until the mid-nineteenth century; then large areas of wetlands and lagoons were converted to paddy fields, market gardening and fruit orchards [16]. Deposition of suspended sediments via irrigation waters resulted in an accretion rate of 0.5 cm/year until 1960. However, there is now almost no sediment accretion, and this has lead to coastal retreat and lowering of the delta plain below sea level [17]. Currently 50% of the delta is below mean high water level and is protected by dikes. This suggests that in the future, relative sea level rise will become the most important climate-induced potential hazard for the Ebro Delta [18].

Management plans for the Ebro basin therefore need to consider several problems including water scarcity, water quality, saltwater intrusion, loss of wetlands and other habitats, eutrophication and pollution. To counteract some of these problems, the current Actuaciones para la Gestión y la Utilización del Agua (AGUA)[1] plan, created in line with the WFD, emphasises water reuse, modernisation of irrigated lands to improve efficiency and desalinisation and indicates the beginning of new management policies [9]. Current and future strategies will need to consider these problems in the context of a changing climate.

[1]AGUA: http://www.mma.es/secciones/agua/entrada.htm

3 Climate Modelling

The main tools used to provide global projections of future climate are general circulation models (GCMs). These are mathematical models based on fundamental physical laws and thus constitute dynamical representations of the climate system. Computational constraints impose a limitation on the resolution that it is possible to realise with such models, and so some unresolved processes are parameterised within the models. This includes many key processes that control climate sensitivity such as clouds, vegetation and oceanic convection [19] of which scientific understanding is still incomplete.

Over recent years, increased computational power and improved efficiency have allowed significant developments and improvements to be applied to climate models [19], including the improved representation of dynamical processes such as advection [20] and an increase in the horizontal and vertical resolution of models. It has also enabled additional processes to be incorporated in models, particularly the coupling of the atmospheric and ocean components of models, the modelling of aerosols and of land surface and sea ice processes. The parameterisations of physical processes have also been improved.

However, for the development of integrated river basin management strategies, it is important to understand and quantify the changes in climate taking place at a regional and local level. GCM horizontal resolutions (~200–300 km) remain unable to provide data at sufficient spatial resolution to capture changes in climate on the river basin scale, particularly in areas of complex topography. Regional detail can, however, be obtained by applying downscaling processes which are classified into two types. Dynamical downscaling methods involve the nesting of high-resolution (~25–50 km) regional climate models (RCMs) into GCMs. Alternatively, statistical downscaling requires the application of statistical methods based on the relationships between large- and local-scale climate variables. There are a wide range of techniques that may be used including bias-correction approaches, simple interpolation, empirical regression relationships, weather typing, neural networks and stochastic weather generator approaches. Dynamical downscaling has the advantage that it produces simulations based on physically consistent processes at a finer resolution than is available from GCMs; however, it is computationally intensive and a relatively limited number of model runs are available. On the other hand, statistical downscaling is cheaper and computationally efficient and can be used to derive variables not available from RCMs. Methods are also largely transferable to other regions but require long, homogeneously observed historical series for calibration. They are also sensitive to the choice of predictor variables and make the assumption of stationarity in the predictor–predictand relationship. A review of downscaling methods within the context of hydrological studies is provided by Fowler et al. [21], while several studies have compared different methods [22, 23].

A further key consideration is that projections of future climate derived from models are associated with significant uncertainties, especially on regional and local scales [20]. These arise in part due to uncertainty in the trajectory of future greenhouse gas emissions but also in the models themselves [24]. Model uncertainty is a consequence of systematic model differences and is also due to different parameterisations of fine-scale physical processes as noted above. One way of addressing these uncertainties is through the use of multi-model ensembles, which has only been made possible with recent advances in computing capabilities [19]. Within AquaTerra [25], projections of future changes in climate for the Ebro basin were therefore derived using RCM output from the European Union Fifth Framework Programme (FP5) PRUDENCE project [26]. This project used RCMs to provide a series of high-resolution (~50 km) simulations of European climate through "time-slice" experiments, generating stationary climate simulations for control (CTRL: 1961–1990) and future scenario (SCEN: 2071–2100) time periods. The experiments used to examine projections for the Ebro basin in AquaTerra (Table 1) were chosen, so that uncertainties in the models could be evaluated for different RCMs with the same bounding GCM and the same RCM in combination with different bounding GCMs, allowing some comparison of the influence of the choice of GCM. However, the full range of uncertainty generated by the choice of GCM boundary conditions is necessarily constrained by the experimental structure provided by the PRUDENCE project [28]. Boundary conditions in this ensemble are derived primarily from HadAM3H [29] and ECHAM4/OPYC [30]. The HadRM3P and Arpège RCM simulations derive boundary conditions from HadAM3P and HadCM3, respectively. Both HadAM3H and HadAM3P are dynamically downscaled to an intermediate resolution from HadCM3 and are thus closely related and may be considered as the same GCM. Reflecting the experiments made available by PRUDENCE, all of the future projections used here assume greenhouse gas and aerosol emissions described by the SRES [31] A2

Table 1 The PRUDENCE regional climate models used in the various AquaTerra studies of the Ebro basin

	RCM	Driving GCM	PRUDENCE acronym (CTRL/SCEN)	AquaTerra acronym (CTRL/SCEN)
1	HIRHAM	HadAM3H A2	HC1/HS1	HIRHAM_H/HIRHAM_H_A2
2	HIRHAM	ECHAM4/OPYC A2	ecctrl/ecscA2	HIRHAM_E/HIRHAM_E_A2
3	RCAO	HadAM3H A2	HCCTL/HCA2	RCAO_H/RCAO_H_A2
4	RCAO	ECHAM4/OPYC A2	MPICTL/MPIA2	RCAO_E/RCAO_E_A2
5	HadRM3P	HadAM3P A2	adeha/adhfa	HAD_H/HAD_H_A2
6	Arpège	HadCM3 A2	DA9/DE6	ARPEGE_H/ARPEGE_H_A2
7	RACMO	HadAM3H A2	Control/Scenario	RACMO_H/RACMO_H_A2
8	CLM	HadAM3H A2	CTL/SA2	CLM_H/CLM_H_A2
9	CHRM	HadAM3H A2	HC_CTL/HC_A2	CHRM_H/CHRM_H_A2
10	REMO	HadAM3H A2	Control/Scenario	REMO_H/REMO_H_A2
11	PROMES	HadAM3H A2	Ref/A2	PROMES_H/PROMES_H_A2

The AquaTerra acronyms are adopted here to provide an easier understanding of the format of each experiment. The first part of each acronym refers to the RCM and the second to the GCM data used to provide the boundary conditions. Scenario simulations have the further suffix A2. For further details on RCM formulations, see [27]. Some AquaTerra studies used a smaller subset of this ensemble (medium–high) scenario (rapid economic growth, low population growth and rapid introduction of new, efficient technology).

Given the characteristics of the PRUDENCE ensemble, it should be noted that the RCM experiments used in AquaTerra do not fully sample uncertainty in potential future climates. As already indicated, the ensemble only uses a limited number of GCMs and further does not fully examine the available combinations of the RCM–GCM matrix. Second, ensembles like PRUDENCE do not sample structural model uncertainty in either a systematic or a random way. Many models possess common parameterisations and algorithms and thus do not constitute totally independent samples of the uncertainty space [24]. The size and composition of the ensemble are a pragmatic use of available resources constituting an "ensemble of opportunity" [32]. Finally, the future projections in this ensemble are limited to the SRES A2 emissions scenario, again due to the limited use of other scenarios within PRUDENCE. Recent developments and the issues surrounding the use of multi-model ensembles are reviewed in detail by Tebaldi and Knutti [32], while advances in applying ensembles and probabilistic methods to hydrological impact studies are reviewed in Fowler et al. [21].

Climate change impact studies in particular have commonly relied on the results of just one climate model; however, New et al. [33] have demonstrated that the output from large model ensembles have the potential to add much useful information to decision-making processes in climate change impact assessments and is a step towards the application of probabilistic scenarios in impacts studies which would allow a more rigorous assessment of uncertainty. Therefore, notwithstanding the limitations discussed here, the work undertaken on the Ebro basin provides a more comprehensive assessment of the magnitude of climate model uncertainty than has previously been incorporated in hydrological studies in this region, and the same framework could readily be applied to a larger ensemble.

Observational data with which to validate the model CTRL experiments were obtained from the CRU TS 2.0 (CRU) global series of observed monthly climate means [34]. This is a gridded global series of monthly climate means for the land surface for the period 1901–2000 and was constructed by the interpolation of station data onto a 0.5° grid. Data for the period 1961–1990 were extracted from the CRU data set for direct comparison with the PRUDENCE output which was re-gridded onto a common 0.5° grid for this purpose.

3.1 AT Climate Studies

Climate studies conducted within the AT project have compared the skill of RCMs in reproducing observed means and extremes, examined how climate change will affect mean precipitation and mean temperature and examined the estimated future occurrence of extreme precipitation and varied aspects of hydrology for Europe, Spain and the Ebro [35, 36].

3.1.1 Control Climate

For the Ebro Basin, despite some biases, the studies have found that the 11 RCMs (Table 1) are able to reproduce the observed annual bimodal distribution of precipitation and the annual temperature cycle (Fig. 4a and d), although they have variable skill in reproducing the spatial distribution and magnitude of monthly means. The RCAO_H experiment produces the largest temperature overestimates

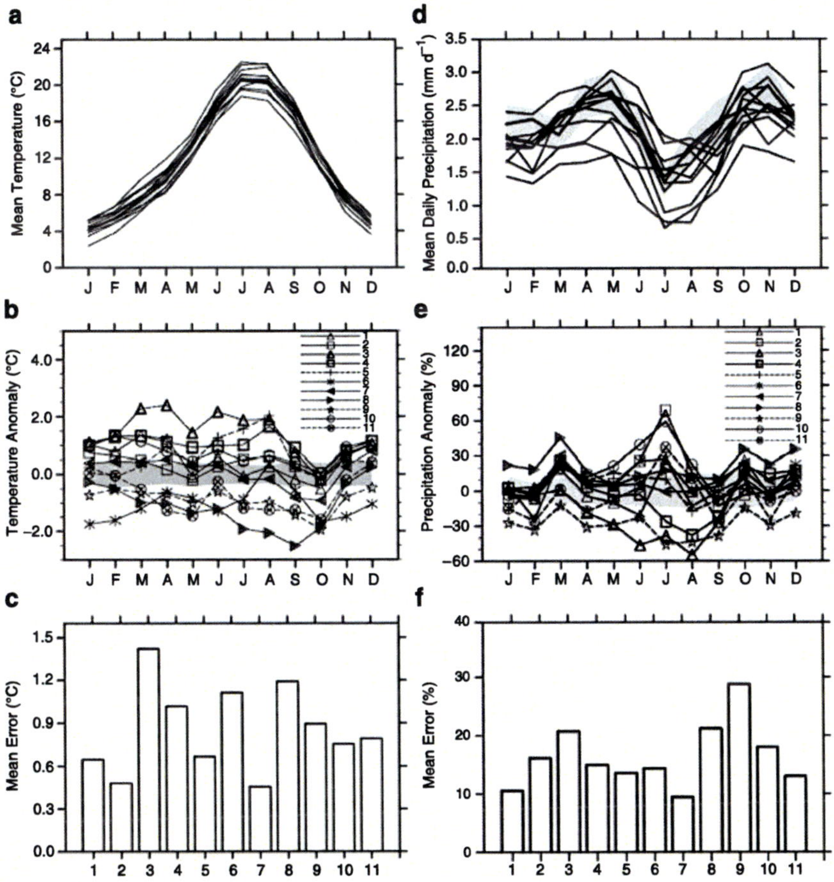

Fig. 4 *Left*: the mean 1961–1990 monthly temperature for the Ebro catchment. Part (**a**) shows the annual cycle, each line representing a different RCM simulation and the *bold line* representing the CRU observed series. The *shading* represents the 95% confidence interval for the estimate of the observed 30-year sample mean. Part (**b**) represents the individual monthly model means as an anomaly from the CRU mean with 95% confidence interval superimposed. Part (**c**) represents the mean absolute annual error for each of the RCMs. *Right*: as for left column but for mean precipitation (**d**) for the Gállego catchment. Model anomalies in parts (**e**) and (**f**) are expressed as a percentage relative to the CRU monthly mean. Model numbers correspond to experiments shown in Table 1. Figure from [35]

for most months (Fig. 4b) and the largest error overall (Fig. 4c), while RCAO_E also significantly overestimates temperature. However, it is evident that model skill is not consistent throughout the year; for example, HAD_H performs reasonably well for most months but overestimates temperature during summer suggesting that RCM performance is dependent on being able to represent specific physical processes. The CLM_H and ARPEGE_H experiments most notably underestimate mean temperature for the Ebro over most of the year, but in the case of the former this is also greatest during summer and autumn. The overall mean model error (Fig. 4c) indicates that RACMO_H and HIRHAM_E perform with most skill, while RCAO_E and CLM_H demonstrate the least skill. However, these coarse averages conceal important detail as the models vary in their ability to reproduce the spatial distribution of temperature. In particular, during the cooler November to March period, some models have less skill in reproducing observed temperatures over the mountainous northern coast of Spain (this is discussed in more detail in the subsequent section examining the Gállego catchment using bias-correction).

As precipitation is more spatially variable, we here examine the model skill for the Gállego sub-catchment. The RCMs reproduce the bimodal distribution of annual precipitation over the Gállego but simulate a large range of values for mean monthly precipitation, particularly from May to August (Fig. 4d), indicating that some RCMs are unable to reliably reproduce summer precipitation processes. This may result from regional climate decoupling from zonal circulation during summer and early autumn when meso- to local-scale processes become more important; for example, Bolle [37] indicates that land–sea temperature gradients and topographical characteristics exert a greater influence at this time. Several models which perform with reasonable skill during other periods fail to capture precipitation processes during these months (Fig. 4e). In particular, HIRHAM_H and REMO_H both overestimate mean precipitation by ~60% in July, while overall summer precipitation is underestimated by ~50% by CHRM_H. Overall, RACMO_H performs with greatest skill (Fig. 4f), while CHRM_H is the least skilful. Again, validation of climate models over an aggregated area can hide important detail and over the Gállego there is little inter-model consistency in the simulation of spatial precipitation patterns (not shown). It should also be noted that this validation considered only mean temperature and precipitation, but this exercise should assess statistics that are appropriate to the impact under consideration. For example, an assessment of flood risk should consider model simulation of extremes, while drought is dependent on longer term variability.

3.1.2 Future Climate

For the future scenarios (SCEN), all of the RCM experiments for the Ebro indicate that temperatures are projected to increase throughout the whole year with the largest increases in summer months [35]. There is considerable uncertainty as to the magnitude of the change in the ensemble with the driving GCM responsible for a substantial part of the large range in projected temperature increase (Fig. 5a).

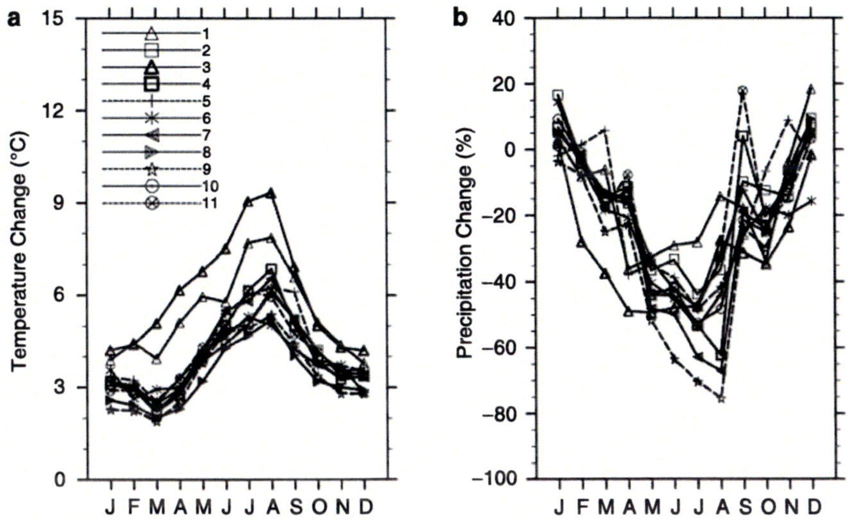

Fig. 5 Projected RCM change in (**a**) mean temperature and (**b**) mean precipitation for the Ebro catchment. Change is for 2071–2100 from the 1961–1990 control period and for precipitation is expressed as a percentage of the control mean. Model numbers correspond to experiments shown in Table 1. Figure taken from [35]

The experiments driven by the ECHAM GCM (HIRHAM_E_A2, RCAO_E_A2) project the largest increases throughout the year, up to ~9°C in summer. Change is not uniform across the basin, however; between March and October, all RCMs suggest greater warming in the interior of the Iberian Peninsula. For precipitation, the RCMs project relatively large decreases in areal average mean summer precipitation for the Ebro for the period 2071–2100, aggravating the already low summer precipitation levels. During August, for example, precipitation is projected to decrease by −14% (HIRHAM_E_A2) and −62% (RCAO_H_A2), and this is compensated by only small increases in precipitation (less than +10%) during winter [35] (Fig. 5b), but again there is considerable uncertainty in the range of projections though for precipitation the driving GCM is less important than the RCM itself. Nevertheless, these results suggest that drought will become an increasing problem in the Ebro Basin region, as relatively little change in the recharge of water during winter months will not compensate for the decreases at other times. In fact, all RCMs project an increase in droughts lasting at least 10 months where the accumulated water deficit exceed 30% of the annual mean precipitation [38].

The studies highlighted above have assessed model skill on a regional scale by averaging RCM results over an entire region and comparing them to areally averaged monthly observations (CRU). However, RCMs show considerable spatial variability in skill within individual experiments with adjacent grid-cells at times exhibiting biases in excess of 40%. Averaging grid-cells by area on a regional basis therefore results in a loss of local detail and may even give a misleading picture of model skill [39]. Furthermore, simulated climate data from an individual model grid-cell do not match the observed climatic means taken at a point, partially due to the inability of RCMs to resolve sub grid-scale processes and topography and also due to the fact that observational point data are not always representative of conditions over a larger area. The standard practice for dealing with this mismatch of scales is to apply an additional statistical downscaling technique to the RCM data [21].

3.2 Bias-Correction

A relatively simple statistical downscaling technique which may be applied quickly to a large number of models is the use of correction factors based on monthly relationships between observed data collected at a particular weather station and the relevant RCM control data set for the appropriate grid-cell [40]. These monthly differences (for temperature) and ratios (for precipitation) between the control and the point observations (i.e. not the gridded interpolated CRU data set) can then be used to correct the daily RCM control and scenario data. This gives bias-corrected scenarios of temperature and precipitation, which can then be used as input to hydrological models for the exploration of various management and policy formulations.

The "bias-correction" is necessary to correct both the absolute magnitude and the seasonal cycle to that of the observations. This approach assumes that the same model biases persist in the future climate and thus GCMs more accurately simulate relative change than absolute values. It provides a correction of monthly mean climate only and does not correct biases in higher order statistics including the simulation of extreme events and persistence.

The bias-correction method has been applied to six of the PRUDENCE RCM control experiments and corresponding future scenarios (from the selection in Table 1) using observations for 1961–1990 for 349 and 146 precipitation and temperature stations, respectively, within the Ebro basin [80] and to 31 suitable gauging stations within the Gállego sub-basin. Individual RCM skill in matching grid-cell scale control data to point observations was assessed and local climate change scenarios were developed. To produce continuous time-series with no missing data, data were interpolated using an inverse distance weighted algorithm [80].

Figures showing the spatial distribution of the annual mean precipitation and temperature for the Ebro for the bias-corrected control and future periods for two of the most extreme RCMs (RCAO_E and HIRHAM_H) are given in [80]. To avoid repetition, the following sections discuss the bias-corrected results for the Gállego. As discussed in [38], these results are similar to those derived for the Ebro.

3.2.1 Bias-Corrected Control Climate

Mean daily precipitation and temperature from the six RCM control experiments are compared with observations for the same period for the Gállego in Fig. 6. Precipitation observations show a bimodal distribution over the catchment, with maxima in May and November and minima in March and June. Seasonal variations are captured relatively well by the RCMs for all cells, although there is a larger range in the mean daily control precipitation compared with mean daily observations at higher elevations. However, inter-model differences decrease and the agreement between RCMs and observations improves for lower elevations [compare the results for the upper and lower Gállego in (Fig. 6)]. Similarly, RCMs show greater variability in reproducing monthly mean daily temperature at higher elevations, suggesting that the models differ in their ability to represent climate processes at altitude.

The driving GCM in particular affects the skill of the RCM experiment in reproducing the monthly observed climatic means which corresponds with similar results noted in other studies (21, 28, 41). The ECHAM-driven RCM experiments generally underestimate precipitation and overestimate temperature compared to observations, while Hadley-driven experiments either span observations or overestimate precipitation and underestimate temperature.

3.2.2 Bias-Corrected Future Climate

Projected changes in bias-corrected RCM output for the period 2071–2100 are shown in Fig. 7. Some RCMs project slightly wetter winters for the period 2071–2100 for the Gállego catchment. All RCMs project much drier spring and summer months. An unweighted average across all models for two sites 9454A (upper Gállego) and 9495U (lower Gállego) suggests a decrease of −27.2% precipitation in April and −48.3% in July, while daily precipitation is projected to increase by only +5.8% to +9.2% in January. Overall, an annual decrease of −21.9% to −17.6% is projected for the lower and upper Gállego, respectively.

RCM projections suggest that mean temperatures in the Gállego will increase substantially throughout the year by an average (across all models and the two sites 9454A and 9495U) of +3.6°C in January to +7.2°C in August, with an average increase of +4.8 to +4.7°C per year for sites 9454A and 9495U, respectively. These projections are consistent with observations for Spain that show significant increases in mean annual and seasonal temperature over the twentieth century [42].

These results suggest that winter precipitation in the Gállego, although projected to increase by 4.1% and 2.2% in December and January, respectively [35], will not be enough to recharge aquifers and rivers which are likely to be depleted due to lower precipitation during the rest of the year (by as much as 52% in July). Water scarcity is therefore likely to become an increasing problem in the future, especially as temperature, and hence evapotranspiration, is projected to increase.

Projected seasonal changes in precipitation suggest that increased storage for winter precipitation may be needed because dams which are designed to store water

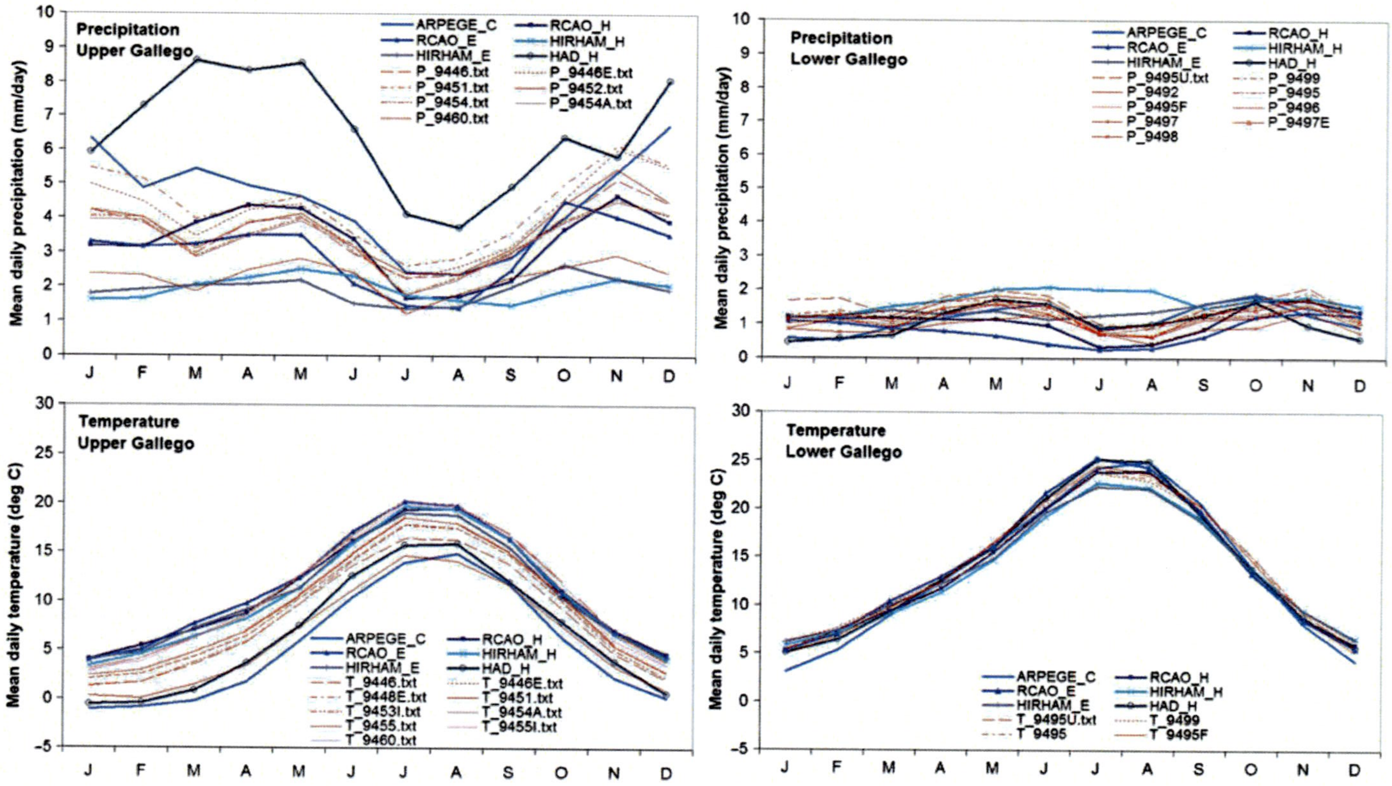

Fig. 6 Mean daily precipitation (mm/day) (*upper*) and temperature (°C) (*lower*) derived from six RCMs for the 1961–1990 CTRL compared with interpolated observations (labelled P_ and T_ respectively) for the same period. Results are shown for RCM cells in the upper (left) and lower (right) Gállego catchment

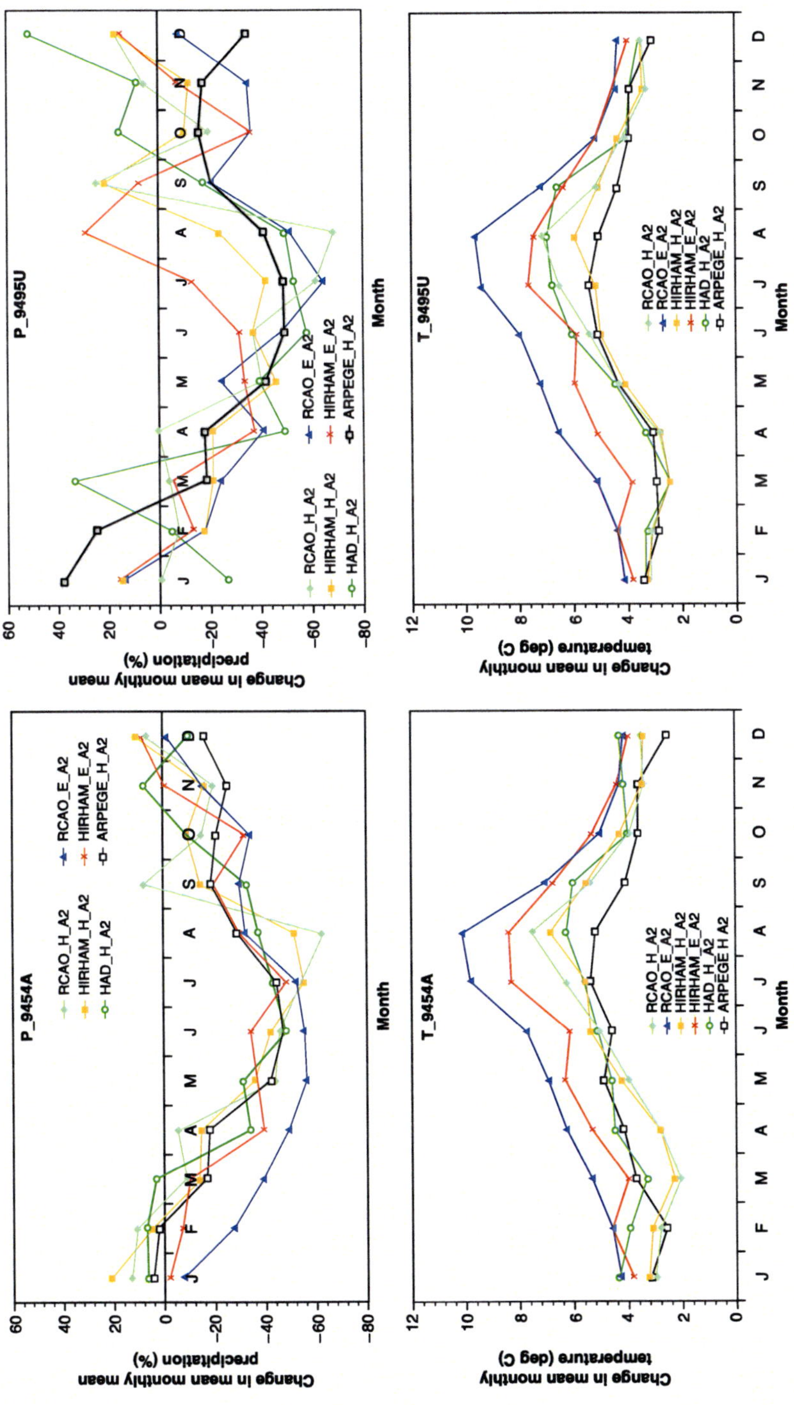

Fig. 7 Projected change in mean precipitation and temperature for the period 2071–2100. Changes in precipitation are expressed as a percentage of the 1961–1990 means for six RCMs in the upper (*left*) and lower (*right*) Gállego

under current winter conditions may not be suitable if winter precipitation increases in the future (see Sect. 4), increasing the likelihood of winter overflow, exacerbating flooding and erosion and reducing the effective amount of water stored for summer use. Summer water shortages may also be exacerbated by higher water demands from the agricultural sector due to current trends to intensify and cultivate more water-intensive (but profitable) crops. There will therefore be a pressing need to regulate and manage flows. Furthermore, water resources management plans will need to ensure that there are still sufficient flows available for the generation of hydroelectricity [10]. For the Gállego, the magnitude of the projected changes in temperature and precipitation and the nature of recent observed changes indicate that river basin managers need to start planning measures to prevent droughts and water resources shortages now.

4 Integration of Climate and Hydrological Models

Regional and local authorities are increasingly relying on computer models of hydrology for quantitative predictions of the impacts of climate change on water supplies, drought or flood frequencies and magnitudes. Hydrological models can integrate information across spatial and temporal scales and help assess the sensitivities of different human and natural systems to perturbations by multiple forcing factors. Within AT, the bias-corrected RCM control and future precipitation and temperature time-series discussed above have been used to drive hydrological models of the Ebro and Gállego catchments to investigate the effect of climate change on water availability.

Zambrano-Bigiarini et al. [80] discusses using bias-corrected daily time-series of precipitation and temperature from two RCMs to drive the SWAT hydrological model for four catchments in the Ebro with different hydrological regimes. The authors find that projected streamflows in the target basins are likely to decrease in the future with the largest decreases found in the more arid regimes.

As the application of bias-corrected RCM data for the Ebro is discussed further in [80], here we introduce the use of bias-corrected RCM data used as input to the GEOTRANSF hydrological model to model the impacts of future climate change on the Gállego basin and demonstrate the integration of high-resolution climate change projections with hydrological modelling.

4.1 Hydrological Modelling of the Gállego Catchment

GEOTRANSF is a semi-distributed modelling system which is built around a conceptual model that limits the model's complexity while maintaining an accurate description of the main mechanisms controlling runoff production and transfer at the basin scale [43]. With this structure GEOTRANSF can be integrated with decision support and socio-economic models. It was used here to model the impacts of future climate change on the Gállego river basin. The predictive capabilities of the model have been previously tested through long-term simulations in the Brenta catchment (NE Italy) where the model was found to be able to reproduce the streamflow at two gauges within the catchment [44]. Here, results of a preliminary study are summarised, whereby bias-corrected precipitation and temperature data for 2071–2100 from the RCAO_E model (one of the driest and hottest RCMs) were used as input to GEOTRANSF to simulate future water availability for irrigation.

In the model, the Gállego catchment has been subdivided into 349 channels with associated sub-catchments aggregated in eight macro-areas. The mean contributing area to the channels is 11 km^2, and streamflow is computed at eight control sections located at the outlet of each macro-area. Calibration has been performed by matching simulated and observed daily streamflow for the period 2001–2005 at four reservoirs (Bubal, Ardisa, La Sotonera and Lanuza) and at four streamgauges (Fig. 8a). Spatially interpolated observed precipitation and temperature data were used to simulate the current hydrology of the Gállego, while bias-corrected RCM future scenario data (as described previously) were used to simulate the projected hydrological conditions in 2071–2100.

Care was taken to integrate the actual operational rules of the four large reservoirs into the model. However, the catchment contains several small, additional non-monitored reservoirs, which are used for hydropower production (Fig. 8c, d). It is likely that these smaller reservoirs are operated in such a way as to store water at night releasing it for hydropower production during the day, when the price of the energy is higher. For this reason, it has been assumed that their influence on daily streamflow is small and can be neglected. The management of the four larger reservoirs is performed under the following constraints: the volume of water stored in the Ardisa reservoir, which feeds La Sotonera reservoir, cannot fall below 1.2×10^6 m^3 to guarantee fish survival, and when La Sotonera exceeds

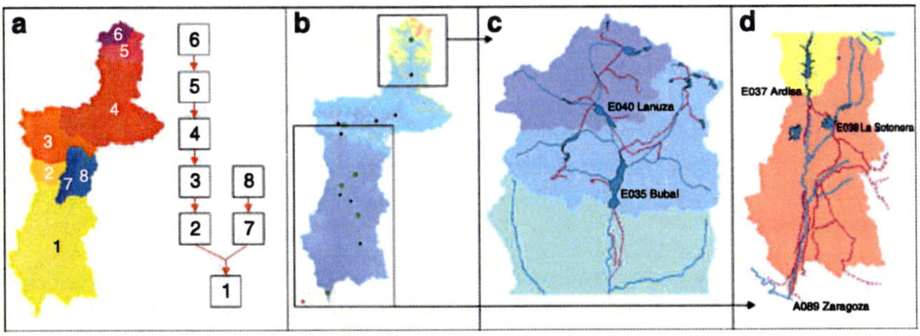

Fig. 8 (**a**) The GEOTRANSF modelling scheme for the Gállego showing macro-areas with nodes at the outlet. Numbers 2 (Ardisa), 5 (Bubal), 6 (Lanuza) and 7 (La Sotonera) refer to reservoir gauging stations, while numbers 1, 3, 4 and 8 refer to stream gauges. (**b**) Raingauges in the Gállego used for the calibration. Reservoirs, rivers (*blue*) and connecting channels (*pink*) in the *upper* (**c**) and *lower* (**d**) Gállego

maximum storage, excess water is released into the Gállego River. This system is completed with several additional irrigation channels, some of them diverting water to the neighbouring Segre basin. The "Canal de Gállego" also transfers water from the Ardisa to La Sotonera reservoir (Fig. 8d). These channels, which export unknown volumes of water, introduce complexity and uncertainty into the model, particularly in the lower Gállego catchment.

The model was run at a daily time-step using the parameters obtained during calibration. Releases of water from La Sotonera and Ardisa reservoirs were calculated using reservoir operational rules inferred from agricultural demand [45].

Results of the simulations derived using the RCAO_E climate projection for the period 2071–2100 (Fig. 9) show that the annual inflow to La Sotonera reservoir (Fig. 8d) is projected to reduce significantly (blue line in Fig. 9) and may fall below the maximum water demand of 421.64×10^6 m^3. Furthermore, the volume transferred from La Sotonera reservoir to the Monegros channel is often lower than the maximum water demand (pink line in Fig. 9), even when the incoming fluxes are above the threshold. This is in part due to a shift in the seasonal distribution of precipitation which is projected to increase slightly in winter and decrease in summer, when agricultural demand reaches its maximum. Furthermore, simulations show that La Sotonera (which has a working capacity of 181.58×10^6 m^3) reaches maximum storage capacity in winter when part of the incoming flow then spills into the river. Due to these losses, the water available for irrigation is smaller than that potentially available.

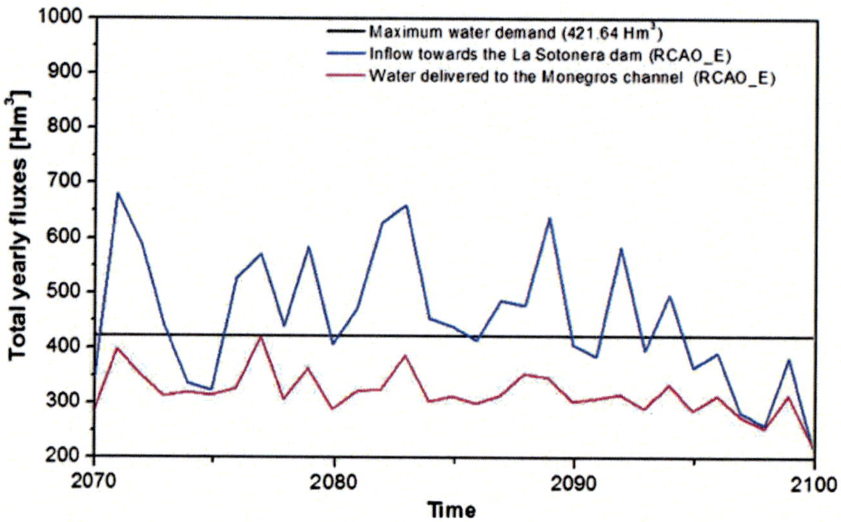

Fig. 9 Annual inflow to La Sotonera dam compared with water delivered to the Monegros channel simulated for 2071–2100 using RCM RCAO_E_A2. Also shown is the maximum value of water required to feed the irrigation system connected to La Sotonera dam (421.64×10^6 m^3) also simulated using RCM RCAO_E_A2

These results suggest that, under the RCAO_E climate change scenario, future water resources in the Gállego may not be sufficient to supply the foreseen maximum irrigation demand, although this is based on just one, extreme member of the PRUDENCE ensemble and additional simulations are needed in order to quantify the possible range of variations. With limited irrigation water, farmers will be limited in their choice of crops, crops will suffer water stress and agricultural yield could decrease considerably, and with it, the income of farmers. Currently, irrigation in the Gállego is predominantly based on inefficient flooding methods, with only a small percentage based on more efficient sprinkler, sprinkler with pump or drip irrigation systems. Significant volumes of water could therefore be conserved by increasing the on-farm irrigation efficiency [46]. To cope with this challenge, a National Irrigation Plan ("Plan National de Regadios," 329/2002) was launched in 2002. The objective of this plan is to restructure, modernise and consolidate agriculture in Spain to reduce the environmental impacts of irrigation. Although this plan encourages the adoption of more efficient irrigation systems, the shift to new technologies may be slow. It may be possible to create additional storage volumes, but building new reservoirs, or increasing the volume of existing ones, is a controversial issue and may create conflicts [e.g. the proposal to build a dam above the Ardisa reservoir (Biscarrues)].

5 Climate and Pollutants

Several AT studies have explored the issue of pollutants in the Ebro through a detailed monitoring campaign. These studies have concentrated on monitoring the current pollution levels in the Ebro and have not taken explicit account of changes that may take place in the future under climate change. Here, these studies are reviewed and important results from these studies are summarised. In addition, an attempt is made to interpret the results in light of potential climate change impacts although further research and integration with the climate change community is needed to understand how pollutant fate and transport processes are affected by a changing climate.

Sites near industrial areas in the Ebro have been found to have the highest concentration of priority contaminants [15, 47–49], while dispersion of agricultural products by drift, runoff and drainage has resulted in residues being found in groundwaters, rivers, coastal waters and lakes far from point sources [50]. Priority contaminants in aquatic environments include persistent organic pollutants (POPs) such as dichlorodiphenylethylenes (DDT) and polybrominated diphenyl ethers (PBDEs).

5.1 *Dichlorodiphenylethylenes*

DDT was historically used in Spain as a pesticide from the mid-1950s to mid-1960s [51] but, as it forms hazardous metabolites as it breaks down, its use was restricted in Spain in 1977 [52] and in most developed countries by the 1980s. However, Spain is one of the few countries in the world (together with China, Brazil and India) which allows its use as an intermediate in the production of dicofol (an organochloride pesticide) [53].

DDT and some of its breakdown products are relatively resistant to breakdown by enzymes and higher organisms found in the soil and are therefore highly persistent in soil, sediment and biota [54]. Despite significant reduction in the use of DDT in the Ebro since the 1960s, DDT still continues to cycle in the ecosystem, with high levels widely detected in both sediments and fish [47, 51, 52].

High sediment toxicity was found near a chloralkali plant in the lower Ebro [48]. Rather than being transported downstream, however, these pollutants are trapped in local sediments by the Flix dam (Fig. 1) which forms a natural barrier impeding sediment transport and the mixing of fish populations [47, 48]. Due to the high accumulation of pollutants behind this and other dams, there is now a pressing need to dredge these sediments to clear the pollutants which have accumulated over many years.

Physical processes such as dredging and fish trawling can, however, cause a net release of sequestered contaminants from the sediment to the overlying waters. Resuspension by storm events and bioturbation can also have similar effects [55]. Changes in the hydrological cycle, such as extreme rainfall events or flooding, can mobilise DDT and other pollutants stored in soil and sediments through erosion and transport processes [55]. As DDT and its metabolites are highly toxic to fish and affect the development and behaviour of biochemical processes, DDT pollution is of particular concern for fish productivity and distribution, and through bioaccumulation, to humans [54].

5.2 Polybrominated Diphenyl Ethers

Brominated flame retardants derive from several commercial applications such as plastics, textiles, electronic circuitry and other materials to prevent fires [56, 57, 81]. This group of pollutants includes stable and toxic PBDEs which have a high liposolubility resulting in its bioaccumulation and biomagnification along the food chain rendering them dangerous pollutants [58]. PBDEs consist of penta-, octa- and deca-BDEs. Penta-BDEs in particular are listed as a hazardous substance in the WFD, and their use has been banned in Europe since 2003. However, the use of an alternative, hexabromocyclododecane (HBCD), has consequently increased, especially in polystyrene foams and as thermal insulation in the building industry [56, 57, 59].

Several AT studies [56, 57, 60, 61] have analysed PBDE and HBCD concentrations in fish (barbel, bleak and others) upstream and downstream of industrialised areas in the Ebro. These studies found high values of PBDE and HBCD in fish and sediment downstream of industrialised towns (e.g. Monzón, a heavily industrialised town draining into the river Cinca, and the Barbastro industrial site on the

Vero river, a tributary of the Cinca) but low values upstream, implying pollutant sources from local industries. PBDE and HBCD contamination was highly correlated with fish length and weight, suggesting first that the contaminants were acquired from the same source, and second that they are bio-available and bioaccumulate [56, 57]. A further AT study on deca-BDE congeners produced from textile plants in an industrial park on the Vero River reported the highest levels of BDE-209 ever found in fish in the field [60]. Adverse effects of PBDEs on aquatic wildlife have not yet been widely studied; however, new AT studies suggest that PBDEs cause oxidative stress in the liver, neurotoxicity in the brain and histopathological effects in the liver and kidney of barbel fish [49]. As contaminants bioaccumulate throughout the aquatic food web, it is clear that this type of pollution poses a potential future threat both to aquatic wildlife and organisms higher in the food chain. PBDEs have already been detected in human biological tissues and fluids, especially breast milk, potentially through dietary intake [62].

High contaminant levels of PBDEs found in the Ebro sediments in relation to other European studies [63–65] are attributed to the small dilution factor of low flow rivers (e.g. the Vero river) [60]. The low flows also make water downstream of industrialised areas poorer in quality due to effluent discharges [49]. As the volume of river water represents an important dilution factor, there are direct implications of climate change for this type of pollution. If, as climate models project, air temperatures in the Ebro river basin increase and overall precipitation decreases in the future, less water will be available and as a consequence, less dilution of compounds will be possible although to date this has not been tested through modelling experiments. In this case, either alternative dilution methods need to be found or treatment efficiency needs to be improved for the ever-increasing diversity of pollutants discharged into the river system.

5.3 *Heavy Metals and Trace Elements*

In another AT study, Terrado et al. [15] characterised pollution patterns in different parts of the Ebro catchment. In the upper part of the Ebro, pollution was found to be mainly in the form of heavy metals (Zn, Cu, Cr, Pb, Cd and Hg), polycyclic aromatic hydrocarbons (PAHs), hexachlorocyclohexanes (HCHs) and trichlorobenzenes (TCBs). Etrophic conditions were also found. Pollution was found to source mainly from industry and urbanisation. The central Ebro was characterised by nutrient pollution such as the accumulation of Ca, Na, Mg and K, which highlighted the importance of salinisation effects from intensive irrigation and soils with high salt content. In the lower Ebro, organic [DDTs, hexachlorobenzene (HCB) and hexachlorobutadiene (HCBu)] and heavy metal (Hg, Cd, Zn and As) contamination was found to derive mainly from industrial and agricultural activities.

Normal agricultural practices generally cause an enrichment of heavy metals in soil, particularly Zn, Cu and Cd, due to the application of manure or its derivatives, compost or sludge and inorganic fertilisers and other human activities such as combustion of fossil fuels and waste incineration [66]. Such activities may increase the content of heavy metals in the soil of the Ebro to levels considered hazardous [66]. The total soil concentration of inorganic pollutants is not indicative of total toxic hazard, however, and it is more useful to assess the mobility and bioavailability of elements [67]. The mobility of inorganic pollutants (such as As, Zn, Pb and Cd) depends on physical, chemical and biological processes. Climatic changes such as more intense storm events, droughts or temperature increases can modify soil microbial activity which affects soil acidity, redox conditions and nutrient availability. This in turn affects the speciation, mobility, bioavailability and toxicity of heavy metals and trace elements in above-ground biomass, plant litter and soil [68, 69]. Increasing soil temperature and the reducing conditions in the soil, as a consequence of newly intensively irrigated soils or increased occurrence and duration of high water content in soils, for example, result in a significant increase in inorganic pollutant solubilisation [67, 70]. Soil-warming also increases the concentration of certain toxic elements (e.g. As, Cr and Zn) in the leaves of the main Mediterranean shrubs, due to either increased plant growth or extra photosynthetic capacity [69]. Drought also results in unequal changes to plant tissue between different plants [69], suggesting that under a drier climate, alternative species may become dominant. It is evident, therefore, that climate change may have serious effects on the toxicity of heavy metals and trace elements in polluted soils as pollutants already present at high levels in soils may become mobile and more bio-available.

5.4 Pesticides

The use of pesticides in agriculture is common in Spain due to its climate and soil conditions. For example, more than 30 different pesticides are used annually in vineyards in the Ebro to combat weeds, insects and fungi. Among these, atrazine, simazine, terbuthylazine, metolachlor and metalaxly are commonly applied in Rioja cultivation areas during April to October [50]. The Ebro Delta is also an important rice growing area and the main pesticides used for weed control in paddy fields include propanil, bentazone, MCPA and molinate, with fenitrothion and malathion used occasionally. Seafood farmers in the delta have complained about a loss of production in periods of rice cultivation (May to August) raising concerns about the toxicity of water in the area [71].

In several AT studies, pesticide levels in the Ebro were found to be high. Hildebrandt et al. [50] found a homogeneous contamination pattern from atrazine (and also from simazine from May 2000) in intensive Rioja cultivation areas throughout the Ebro. Nearer to the delta, Barata et al. [72] found high levels of bentazone, methyl-4-chlorophenoxyacetic acid, propanil, molinate and fenitrothion in water, while Kuster et al. [71] found low concentration levels of atrazine and simazine at the delta, but high levels of other pesticides used in rice cultivation. Importantly, Hildebrandt et al. [50] found that levels of pesticides in groundwater were higher than for surface waters, suggesting that aquifers are especially vulnerable to contamination. Keeping groundwater free from pollution is vital for surface water ecosystems [73].

Within the FOOTPRINT[2] (Functional TOOls for Pesticide RIsk assessmenNt and managemenT) project, Nolan et al. [74] conducted an extensive modelling study involving multiple soil types, timing of applied pesticides and variable precipitation patterns to identify key climatic factors influencing pesticide loss in soils by leaching. They found that although climatic factors influencing pesticide loss are specific to the soil–pesticide combination, pesticide loss increases with increasing clay content, increasing precipitation of variable duration, decreasing temperature (as indicated by climate variables and season of pesticide application) and increasing pesticide persistence. The predominant variable for pesticide mobility at Zaragoza was temperature which was found to be negatively correlated with pesticide loss. The timing of extreme weather events in relation to pesticide application was found to be very important, especially for drainage scenarios (as opposed to leaching) reflecting the rapid transport of pesticides to drains via macropores in soils of high clay content. It is obvious from this study that pesticide loss by leaching and drainage scenarios is highly influenced by climate, particularly storm events, which may mobilise pesticides downstream and could therefore be affected by changes in climate. On the other hand, more frequent or longer droughts may keep pesticides in the soil and higher temperatures may even enhance their degradation provided enough water is present to allow microbial activity. However, the likelihood of these changes and level of risk posed by the projected climate change highlighted in this chapter would require detailed modelling studies.

5.5 *Wastewater Treatment Plants*

Pharmaceuticals are contaminants which are continuously introduced to sewage waters through excreta, disposal of unused or expired drugs or directly from pharmaceutical discharges [75]. To measure the occurrence of 28 pharmaceuticals, including analgesics, anti-inflammatories, antibiotics, anti-histamines and β-blockers, Gros et al. [76, 82] monitored water quality upstream and the wastewater downstream of seven wastewater treatment plants (WWTPs) servicing the main cities in the Ebro Basin (Fig. 1, Table 2).

The removal rate of pharmaceuticals by WWTPs depends on factors such as reactor configuration, redox conditions, temperature, hydraulic retention time (HRT) and solid retention time (SRT). SRT is the average retention time of sludge

---

[2]FOOTPRINT (http://www.eu-footprint.org) is an EC 6th Framework Programme project aiming to develop computer tools to evaluate and reduce the risk of pesticides impacting on water resources in the EU: SSPI-CT-2005-022704.

Table 2 Wastewater treatment plants (WWTPs), the receiving waters where their effluents are discharged and the hydraulic retention time (HRT)

WWTP	Receiving river water	HRT (h)
WWTP1	Vallas	32
WWTP2	Ebro	18
WWTP3	Iregua	8
WWTP4	Arga	25
WWTP5	Ebro	10
WWTP6	Segre	6–10
WWTP7	Ebro	33

flocs before they are removed as excess sludge or lost through effluents and is the time microorganisms have to grow.

Gros et al. [76] found that HRT was one of the key parameters influencing the efficiency of WWTPs in removing pharmaceuticals; the higher the HRT, the greater the percentage of removal efficiency (%RE). For example, in Fig. 10, WWTP3 showed very poor elimination of the majority of compounds due to its lower HRT, while WWTP 1, 4 and 7 were most efficient due to the higher HRT. These results suggest that break down and retention mechanisms of pharmaceuticals in WWTPs need urgent attention to mitigate environmental loads.

WWTPs release compounds other than pharmaceuticals which may end up in the water or sediment of rivers downstream of WWTPs. For example, alkylphenols, such as nonlyphenol (NP), are non-ionic surfactants used in a variety of industrial and domestic applications which are associated with WWTPs and accumulate particularly in sediments [77, 78]. Part of the adsorbed NP, however, can desorb from the sediments into the water phase and as a consequence can become bioavailable again, causing a toxicological risk to aquatic ecosystems [79]. Preliminary AT studies suggest that this risk may be decreased by biodegradation but that this can only occur under aerobic conditions and not in anaerobic sediment. Flooding, brought about by an extreme storm event, for example, may therefore increase the toxicological risk as NP will dissolve in water but degradation will not occur due to limited oxygen concentrations [79].

Conventional WWTPs are, therefore, unable to remove wide ranges of pharmaceuticals and other compounds. For pharmaceuticals, although acute toxicity of aquatic organisms or chronic effects are unlikely with the present concentrations due to dilution effects, a wide range of pharmaceuticals are detected in the Ebro, and the overall toxicity of mixed pharmaceuticals may be high. Further studies are therefore required to assess the interactions of different compounds and the consequential health effects. In a similar manner to other pollutants, pharmaceuticals have a clear sensitivity to climate change through dilution effects, and the projected future decrease in annual precipitation could cause certain compound concentrations (e.g. anti-inflammatory diclofenac and β-blocker pranolol) to reach levels which may cause chronic effects [76].

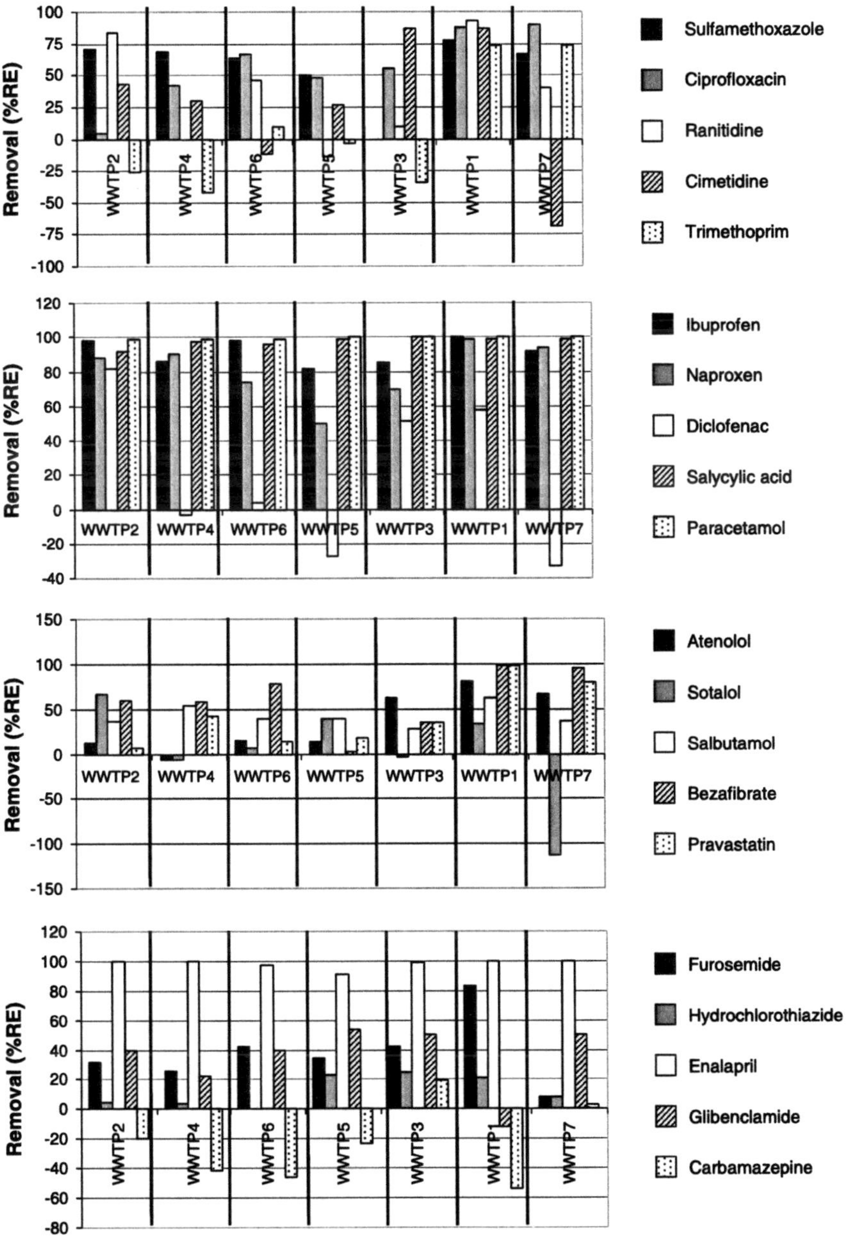

Fig. 10 Influence of hydraulic retention time (HRT) in pharmaceutical removal. The most ubiquitous pharmaceuticals detected are shown

6 Conclusions

This chapter provides an overview of recent and current research being carried out within the Ebro basin. In particular, it includes a summary and review of recent AT studies involving climate, water resources and environmental pollutants. In addition, it presents future climate scenarios for the Ebro and Gállego catchments derived from a multi-model ensemble and downscaled using a statistical downscaling technique. These were used for the generation of new hydrological simulations for the period 2071–2100. Also a qualitative assessment was made of how some environmental pollutants present in the Ebro basin may be affected by future changes in climate. Further detailed studies are required to assess the impact of climate change on pollutant fate and transport processes, however.

It is important to bring together these different studies and disciplines, which are often considered in isolation to each other, and take a more holistic approach to highlight their interdependencies and to provide an overview of the wide-ranging environmental issues affecting the Ebro. Such an interdisciplinary approach can help guide scientists, policy makers and others and provides an overview of the basin for management purposes. The sustainable management of a river at a basin scale and integrated basin management strategies are being encouraged by several directives such as the WFD, the Spanish AGUA plan and several others such as the Marine Strategy Framework Directive (2008/56/EC) and Directive 2007/60/EC concerning the assessment and management of flood risks applying to inland waters and coastal waters of the EU. It is important, therefore, to consider the basin in a holistic manner and, as this chapter has highlighted, to consider the effects of climate change in the long-term management of the basin.

The climate studies discussed above have shown that RCMs can simulate the current precipitation and temperature of the Ebro reasonably well, giving confidence that they can be used to model future climate scenarios. RCM results vary spatially, however, and differ most from each other, in their ability to simulate precipitation and temperature processes at altitude (being more consistent at lower elevations). These studies, based on an ensemble of different RCMs, suggest that the Ebro will become considerably hotter and drier in the future, especially in summer, with increases in long-term droughts. Although slightly wetter winters are projected, the average annual precipitation for the Ebro is projected to decrease between −14 and −62% for the A2 climate scenario for the 2071–2100 period, while in the Gállego average annual precipitation is projected to decrease by −18% to −22% between the upper and lower Gállego, respectively. Temperatures in the Gállego are also expected to increase by a daily average of about +4.8°C per year. These projected future climate scenarios have significant implications for water resource management strategies in the Ebro.

Hydrological studies have shown that as slightly more precipitation is expected in winter, reservoirs such as La Sotonera will not have sufficient capacity to store water in winter to be able to release sufficient water in summer to cope with the projected increased demands for irrigation due to the projected drier summers.

Modernisation of irrigation methods will therefore be required to conserve water resources (with unclear effects on salinisation), and extra dams may need to be built. With the simulated future water scarcity, water for hydroelectric use may also be threatened.

Persistent historical pollutants such as DDT and PBDEs continue to cycle in the ecosystem. High concentrations of these toxic pollutants are found in both sediments and fish in the Ebro, especially downstream of industrial areas. Dams accumulate pollutants as they form a natural barrier impeding both sediment transport and mixing of fish populations; so there is now a need to dredge sediments although this can release contaminants into the overlying water with significant health implications. DDT and PBDEs in particular are bio-available and can bioaccumulate, posing potential threats to aquatic wildlife and higher organisms.

The projected future climate changes will exacerbate problems of poor water quality in periods of low flow, and decreased water availability will lead to pollutant concentration due to a lack of dilution. Storm events can mobilise pollutants in water and sediments, while changing climate patterns and drought will also influence degradation, turnover, sorption and transport behaviour of pollutant contamination with unknown effects. Levels of pesticides in aquifers were found to be higher than for surface waters, implying that aquifers are especially vulnerable to contamination. The timing of extreme weather events in relation to pesticide application is also important as extreme events encourage rapid transport of pesticides via macropores especially in soils with high clay content [74]. In addition to pollution by POPs, heavy metals, trace elements and pesticides, WWTPs also contribute to pollution in the Ebro due to their inefficient removal of pharmaceuticals and other compounds. The HRT of WWTPs, a major factor concerning the removal rate of pharmaceuticals, may need to be increased to help decrease this type of pollution. Flooding events brought about by extreme storm events, for example, may also exacerbate the toxicological risk of compounds such as nonlyphenol, commonly found in sediments downstream of WWTPs, as it dissolves in water, thereby becoming bio-available, however, may not degrade due to anaerobic conditions.

Climate driven changes in the Ebro Basin catchment could, therefore, lead to significant environmental problems related to groundwater and surface-water quality and quantity, sediment quality and accretion (e.g. in the delta), pollution accumulation, its bioavailability and related health aspects, along with potential socio-economic issues such as population migration. As the Ebro contains one of the largest irrigated areas in Spain, water restrictions are likely to affect the agricultural sector mostly by limiting the range of crops capable of being grown there and forcing expensive modernisation of irrigation methods. Drought may increase plant stress and promote soil cracking and hence more rapid loss of fertilisers/pesticides under storm events. Soil organic matter levels may decrease in response to increasing temperature, and increased winter precipitation may stimulate plant disease requiring greater use of pesticides; higher temperatures, however, may also lead to faster pesticide degradation (see [4] and references therein). Water restrictions are also likely to affect the industry and energy sectors.

Building new dams or increasing existing dam water-storage capacities would allow more water storage; however, as dams block sediment transport, they could affect the Ebro delta if they further impede sediment accretion. On the other hand, they may act as pollutant traps associated with sediments. If so, removal and disposal of these environmental threats need to be considered.

The studies conducted in the Ebro and Gállego within the AquaTerra project are good examples of international collaboration in providing improved understanding of the investigation of the changing dynamics of hydrological systems and pollutant dynamics under a changing climate. Furthermore, the work conducted by the AquaTerra team has demonstrated that the integration of the climatological and hydrological communities is sufficiently well developed to able to produce valuable projections of future change for planners and managers and has laid an ideal foundation for the further development of such approaches in the analysis of changes in pollutants. Climate change communities and those engaged in the monitoring of water pollution should build upon the knowledge gained through projects such as AquaTerra and FOOTPRINT, so that through integration it becomes possible to understand the impacts of climate change on pollutants. The use of future climate scenarios may help ensure that water quality stays within the required legal limits and also ensure appropriate mitigation responses to current and potential threats to water resources.

Acknowledgements This work was supported by the European Union FP6 Integrated Project AquaTerra (Project no. 505428) under the thematic priority sustainable development, global change and ecosystems. Dr Hayley Fowler was supported by NERC Postdoctoral Fellowship award (2006–2009) NE/D009588/1. Meteorological data for the Ebro and Gállego catchments were supplied by a website supported by the Ministerio de Medio Ambiente, Madrid, and the Confederación Hidrográfica del Ebro (http://oph.chebro.es/). RCM data were obtained from the PRUDENCE project archive (http://prudence.dmi.dk/) which was supported by EU contract EVk2-CT2001-00132.

References

1. Van Der Zaag P, Gupta J, Darvis LP (2009) HESS opinions: urgent water challenges are not sufficiently researched. Hydrol Earth Syst Sci 13:905–912
2. WISE (2008a) Water notes on the implementation of the Water Framework Directive. Water Note 1 – Joining forces for Europe's shared waters: coordination in international river basin districts. http://ec.europa.eu/environment/water/water-framework/pdf/water_note1_joining_forces.pdf
3. Lucier A, Palmer M, Mooney H, Nadelhoffer K, Ojima D, Chavez F (2006) Ecosystems and climate change: research priorities for the US climate change science program. Recommendations from the scientific community. Report on an Ecosystems Workshop, prepared for the Ecosystems Interagency Working Group. Special Series No SS-92-06, University of Maryland Centre for Environmental Science, Chesapeake Biological Laboratory, Solomons, MD, 50 pp. http://www.usgcrp.gov/usgcrp/Library/ecosystems/eco-workshop-report-jun06.pdf
4. Bloomfield JP, Williams RJ, Goody DC, Cape JN, Guha P (2006) Impacts of climate change on the fate and behaviour of pesticides in surface and groundwater – a UK perspective. Sci Total Environ 369:163–177

5. CHE (2008a) Confederación Hidrográfica del Ebro. http://oph.chebro.es/DatosBasicosCHE.html. Accessed Oct 2008
6. CHE (2008b) Confederación Hidrográfica del Ebro. http://oph.chebro.es/Contenido-Climatologico.htm. Accessed Oct 2008
7. Ries JB, Hirt U (2008) Permanence of soil surface crusts on abandoned farmland in the Central Ebro Basin/Spain. Catena 72:282–296
8. Alcázar J, Palau A, Vega-García C (2008) A neural net model for environmental flow estimation at the river Ebro River Basin, Spain. J Hydrol 349:44–55
9. Tàbara JD, Ilhan A (2008) Culture as trigger for sustainability transition in the water domain: the case of the Spanish water policy and the Ebro river basin. Reg Environ Change 8:59–71. doi:10.1007/s10113-007-0043-3
10. Batalla RJ, Gomez CM, Kondolf GM (2004) Reservoir-induced hydrological changes in the Ebro River basin (NE Spain). J Hydrol 290:117–136
11. CHE (2000) Confederación Hidrográfica del Ebro: Dirección General de Obras Hidráulicas y Calidad de las Aguas. Asistencia Técnica: Q5017001 H. 2000-Ph-24-I. Los aprovechamientos en la cuenca del Ebro: Afección en el régimen hidrológico fluvial (in Spanish). ftp://oph.chebro.es:2121/BulkDATA/DOCUMENTACION/regimenhidro.pdf
12. CHE (2008c) Confederación Hidrográfica del Ebro: Plan hidrológico del rio Gállego – Documentación previa / not definitive version. http://oph.chebro.es/DOCUMENTACION/DirectivaMarco/Gallego/Gallego.htm
13. Herivaux C., Loubier S., Bouzit M., Strosser P., Chapman A., Bardos P., Ansink E., Ruijs A., Maring L., Cerrits L., Joziasse J (2005) Generic conceptual representation of river basin, and methodological guidelines to construct such a representation with a participatory approach. Deliverable No.: Integrator 1.3 of the AquaTerra project 505428 (GOCE). www.aquaterra.eugris.info
14. Subirats J, Font N, Costejá M (2002) Case study 1: Matarraña River Basin. Report for the European Water Regimes and the Notation of a Sustainable Status (EUWARENESS) project. Universidad Autonoma de Barcelona, Spain, 68 pp
15. Terrado M, Barceló D, Tauler R (2006) Identification and distribution of contamination sources in the Ebro river basin by chemometrics modelling coupled to geographical information systems. Talanta 70:691–704
16. Day JW Jr, Maltby E, Ibáñez C (2006) River basin management and delta sustainability: a commentary on the Ebro Delta and the Spanish National Hydrological Plan (Editorial). Ecol Eng 26:85–99
17. Négrel P, Roy S, Petelet-Giraud E, Millot R, Brenot A (2007) Long-term-fluxes of dissolved and suspended matter in the Ebro River Basin (Spain). J Hydrol 342:249–260
18. Sánchez-Archilla A, Jiménez JA, Valdemoro HI, Gracia V (2008) Implications of climatic change on Spanish Mediterranean low-lying coasts: The Ebro Delta case. J Coast Res 24:306–316
19. Le Treut H, Somerville R, Cubasch U, Ding Y, Mauritzen C, Mokssit A, Peterson T, Prather M (2007) Historical overview of climate change. In: Solomon S, Qin D, Manning M, Chen Z, Marquis M, Averyt KB, Tignor M, Miller HL (eds) Climate change 2007: the physical science basis. Contribution of Working Group I to the Fourth Assessment Report of the Intergovernmental Panel on Climate Change. Cambridge University Press, Cambridge
20. Randall DA, Wood RA, Bony S, Colman R, Fichefet T, Fyfe J, Kattsov V, Pitman A, Shukla J, Srinivasan J, Stouffer RJ, Sumi A, Taylor KE (2007) Climate models and their evaluation. In: Solomon S, Qin D, Manning M, Chen Z, Marquis M, Averyt KB, Tignor M, Miller HL (eds) Climate change 2007: the physical science basis. Contribution of Working Group I to the Fourth Assessment Report of the Intergovernmental Panel on Climate Change. Cambridge University Press, Cambridge
21. Fowler HJ, Blenkinsop S, Tebaldi C (2007) Linking climate change modelling to impacts studies: recent advances in downscaling techniques for hydrological modelling. Int J Climatol 27:1547–1578. doi:10.1002/joc.1556

22. Haylock MR, Cawley GC, Harpham C, Wilby RL, Goodess CM (2006) Downscaling heavy precipitation over the United Kingdom: a comparison of dynamical and statistical methods and their future scenarios. Int J Climatol 26:1397–1415
23. Khan MS, Coulibaly P, Dibike Y (2006) Uncertainty analysis of statistical downscaling methods. J Hydrol 319:357–382
24. Collins M (2007) Ensembles and probabilities: a new era in the prediction of climate change. Phil Trans R Soc A 365:1957–1970
25. Barth J, Grathwohl P, Fowler HJ, Bellin A, Gerzabek MH, Lair GJ, Barceló D, Petrovic M, Navarro A, Négrel Ph, Petelet-Giraud E, Darmendrail D, Rijnaarts H, Langenhoff A, de Weert J, Slob A, van der Zaan BM, Gerritse J, Frank E, Gutierrez A, Kretzschmar R, Gocht T, Steidle D, Garrido F, Jones KC, Meijer S, Moeckel C, Marsman A, Klaver G, Vogel T, Bürger C, Kolditz O, Broers HP, Baran N, Joziasse J, von Tümpling W, van Gaans P, Merly C, Chapman A, Brouyère S, Batlle AJ, Orban Ph, Tas N, Smidt H (2009) Mobility, turnover and storage of pollutants in soils, sediments and waters: achievements and results of the EU project Aqua-Terra. A review. Agron Sustain Dev 29:161–173
26. Christensen JH, Carter TR, Rummukainen CM, Amanatidis G (2007) Evaluating the performance and utility of regional climate models: the PRUDENCE project. Climate Change 81:1–6. doi:10.1007/s10584-006-9211-6
27. Jacob D, Bärring L, Christensen OB, Christensen JH, de Castro M, Déqué M, Giorgi F, Hagemann S, Hirschi M, Jones R, Kjellström E, Lenderink G, Rockel B, Sánchez E, Schär C, Seneviratne SI, Somot S, van Ulden A, van den Hurk B (2007) An inter-comparison of regional climate models for Europe: model performance in present-day climate. Clim Change 81(Suppl 1):31–52
28. Déqué M, Rowell DP, Lüthi D, Giorgi F, Christensen JH, Rockel B, Jacob D, Kjellström E, de Castro M, van den Hurk B (2007) An intercomparison of regional climate simulations for Europe: assessing uncertainties in model projections. Clim Change 81:53–70
29. Pope VD, Gallani ML, Rowntree PR, Stratton RA (2000) The impact of new physical parameterizations in the Hadley Centre climate model: HadAM3. Climate Dyn 16:123–146
30. Roeckner E, Arpe K, Bengtsson L, Christoph M, Claussen M, Dümenil L, Esch M, Giorgetta M, Schlese U, Schulzweida U (1996) The atmospheric general circulation model ECHAM-4: model description and simulation of present-day climate, Report No.218. Max-Planck Institute for Meteorology, Hamburg, 90 pp
31. Nakićenović N, Alcamo J, Davis G, DeVries B, Fenham J, Gaffin S, Gregory K, Gruebler A, Jung TY, Kram T, Lebre LaRovere E, Michaelis L, Mori S, Morita T, Pepper W, Pitcher H, Price L, Riahi K, Roehrl A, Rogner H-H, Sankovski A, Schlesinger M, Shukla P, Smith S, Swart R, VanRooijen S, Victor N, Dadi Z (2000) Special report on emissions scenarios: a special report of Working Group III of the Intergovernmental Panel on Climate Change. Cambridge University Press, Cambridge, p 600
32. Tebaldi C, Knutti R (2007) The use of the multi-model ensemble in probabilistic climate projections. Phil Trans R Soc A 365:2053–2075
33. New M, Dessai S, Lopez A, Wilby RL (2007) Challenges in using probabilistic climate change information for impacts and adaptation decision-making: an example from the water sector. Philos Trans R Soc 365:2117–2131
34. Mitchell TD, Carter TR, Jones PD, Hulme M, New M (2004) A comprehensive set of high-resolution grids of monthly climate for Europe and the globe: the observed record (1901-2000) and 16 scenarios. Tyndall Working Paper 55. Tyndall Centre, UEA, Norwich
35. Bürger CM, Kolditz O, Fowler HJ, Blenkinsop S (2007) Future climate scenarios and rainfall-runoff modelling in the Upper Gállego catchment (Spain). Environ Pollut 148:842–854
36. Fowler HJ, Ekström M, Blenkinsop S, Smith AP (2007) Estimating change in extreme European precipitation using a multimodel ensemble. J Geophys Res 112:D18104. doi:10.1029/2007JD008619
37. Bolle H-J (2003) Climate, climate variability, and impacts in the Mediterranean area: an overview. In: Bolle H-J (ed) Mediterranean climate. Springer, Berlin, p 372

38. Blenkinsop S, Fowler HJ (2007) Changes in European drought characteristics projected by the PRUDENCE regional climate models. Int J Climatol 27:1595–1610. doi:10.1002/joc.1538
39. Blenkinsop S, Fowler HJ (2007) Changes in drought frequency, severity and duration for the British Isles projected by the PRUDENCE regional climate models. J Hydrol 342:50–71. doi:10.1016/j.jhydrol.2007.05.003
40. Fowler HJ, Kilsby CG (2007) Using regional climate model data to simulate historical and future river flows in northwest England. Climate Change 80:337–367
41. de Elía R, Caya D, Côté H, Frigon A, Biner S, Giguère M, Paquin D, Harvey R, Plummer D (2008) Evaluation of uncertainties in the CRCM-simulated North American climate. Climate Dyn 30:113–132
42. Esteban-Parra MJ, Pozo-Vazquez D, Rodrigo FS, Castro-Diez Y (2003) Temperature and precipitation variability and trends in Northern Spain in the context of the Iberian Peninsula climate. In: Bolle H-J (ed) Mediterranean climate. Springer, Berlin, p372
43. Majone B, Alberici D, Bertagnoli A, Bellin A (2006) Structural uncertainty assessment of hydrological models: the semidistributed model GEOTRANSF. Eos Trans AGU 87(52), Fall Meet Suppl, Abstract H44D-04. http://www.agu.org/meetings/fm06/waisfm06.html
44. Majone B, Bertagnoli A, Bellin A, Botter G, Settin T, Marani M, Rinaldo A (2005) Structure of the modeling system with detailed description of the models. AquaTerra Deliverable C3.2.
45. Graveline N, Ansink E, Majone B, Grandmougin B, Van Duinen R, Petelet-Giraud E (2009). Integrating economics with quantitative and qualitative hydrology for simulating future scenarios on water resources and their salinity in the Ebro basin. AquaTerra Deliverable I3.11
46. Isidoro D, Quilez D, Aragues R (2004) Water balance and irrigation performance analysis: La Violada irrigation district (Spain) as a case study. Agric Water Manage 64(2):123–142
47. Lacorte S, Raldúa D, Martínez E, Navarro A, Diez S, Bayona JM, Barceló D (2006) Environ Pollut 140:471–482
48. Ocampo-Duque W, Sierra J, Ferre-Huguet N, Schuhmacher M, Domingo JL (2008) Estimating the environmental impact of micro-pollutants in the low Ebro River (Spain): an approach based on screening toxicity with Vibrio fischeri. Chemosphere 72:715–721
49. Raldúa D, Padrós F, Solé M, Eljarrat E, Barceló D, Riva MC, Barata C (2008) First evidence of polybrominated diphenyl ether (flame retardants) effects in feral barbel from the Ebro River basin (NE, Spain). Chemosphere 73:56–64
50. Hildebrandt A, Guillamón M, Lacorte S, Tauler R, Barceló D (2008) Impact of pesticides used in agriculture and vineyards to surface and groundwater quality (North Spain). Water Res 42:3315–3326
51. Ormad MP, Ratia JS, Rodriguez L, Ovelleiro JL (2008) Levels and distribution of DDT in the Cinca River (Spain). Water Environ Res 80:464–471
52. de la Cal A, Eljarrat E, Raldúa D, Durán C, Barceló D (2008) Spatial variation of DDT and its metabolites in fish and sediment from Cinca River, a tributary of Ebro River (Spain). Chemosphere 70:1182–1189
53. Rasenberg M, van de Plassche EJ (2002) Information dossier on DDT used for the production of Dicofol. Final Report. Ministerie van VROM. Royal Haskoning, Haskoning Nederland BV Environment. Available from: http://www.unece.org/env/lrtap/TaskForce/popsxg/2004/Dossier_DDT%20for%20dicofol%20production.pdf
54. WHO (1979) DDT and its derivatives: Environmental Health Criteria 9. WHO, Geneva
55. Schiedek D, Sundelin B, Readman JW, Macdonald RW (2007) Interactions between climate change and contaminants. Mar Pollut Bull 54:1845–1856
56. Eljarrat E, de la Cal A, Raldúa D, Duran C, Barceló D (2004) Occurrence and bioavailability of polybrominated diphenyl ethers and hexabromocyclododecane in sediment and fish from the Cinca River, a tributary of the Ebro River (Spain). Environ Sci Technol 38:2603–2608
57. Eljarrat E, de la Cal A, Raldúa D, Duran C, Barceló D (2005) Brominated flame retardants in *Alburnus alburnus* from Cinca River Basin (Spain). Environ Pollut 133:501–508
58. de Wit C (2002) An overview of brominated flame retardants in the environment. Chemosphere 46:583–624

59. Guerra P, Eljarrat E, Barceló D (2008) Enantiomeric specific determination of hexabromocyclododecane by liquid chromatography-quadrupole linear ion trap mass spectrometry in sediment samples. J Chromatogr A 1203:81–87
60. Eljarrat E, Labandeira A, Marsh G, Raldúa D, Barceló D (2007) Decabrominated diphenyl ether in river fish and sediment samples collected downstream an industrial park. Chemosphere 69:1278–1286
61. van Beusekom OC, Eljarrat E, Barceló D, Koelmans AA (2006) Dynamic modeling of food-chain accumulation of brominated flame retardants in fish from the Ebro river basin, Spain. Environ Toxicol Chem 25:2553–2560
62. Domingo J (2004) Human exposure to polybrominated diphenyl ethers through the diet. J Chromatogr A 1054(1–2):321–326
63. Allchin CR, Law RJ, Morris S (1999) Polybrominated diphenylethers in sediments and biota downstream of potential sources in the UK. Environ Pollut 105:197–207
64. de Boer J, Wester PG, van der Horst A, Leonards PEG (2003) Polybrominated diphenyl ethers in influents, suspended particulate matter, sediments, sewage treatment plant and effluents and biota from the Netherlands. Environ Pollut 122:63–74
65. Sellström U, Kierkegaard A, de Wit C, Jansson B (1998) Polybrominated diphenyl ethers and hexabromocyclododecane in sediment and fish from a Swedish river. Environ Toxicol Chem 17:1065–1072
66. Rodríguez JA, Nanos N, Grau JM, Gil L, López-Arias M (2008) Multiscale analysis of heavy metal contents in Spanish agricultural topsoils. Chemosphere 70:1085–1096
67. Joubert AVP, Lucas L, Garrido F, Joulian C, Jauzein M (2007) Effect of temperature, gas phase composition, pH and microbial activity on As, Zn, Pb and Cd mobility in selected soils in the Ebro and Meuse Basins in the context of global change. Environ Pollut 148:749–758
68. Deneux-Mustin S, Roussel-Debet S, Mustin C, Henner P, Munier-Lamy C, Colle C, Berthelin J, Granier-Laplace J, Leyval C (2003) Mobilite et transfert racinaire des elements en traces: influence des micro-organismes du spl. TEC & DOC, London
69. Sardans J, Peñuelas J, Estiarte M (2008) Warming and drought change trace element bioaccumulation patterns in a Mediterranean shrubland. Chemosphere 70:874–885
70. Bruand A (2005) Toward conditions favourable to mobility of trace elements in soils. C R Geosci 337(6):549–550
71. Kuster M, de Alda MJL, Barata C, Raldúa D, Barceló D (2008) Analysis of 17 polar to semi-polar pesticides in the Ebro river delta during the main growing season of rice by automated on-line solid-phase extraction-liquid chromatography-tandem mass spectrometry. Talanta 75:390–401
72. Barata J, Tauler R, Teixido E, Rieradevall M, Prat N, Riva MC, Soares AMVM, Barata C (2008) Combined use of biomarkers and in situ bioassays in Daphnia magna to monitor environmental hazards of pesticides in the field. Aquat Toxicol 87:310–320
73. WISE (2008b) Water notes on the implementation of the Water Framework Directive. Water Note 3 – Groundwater at risk: managing the water under us. http://ec.europa.eu/environment/water/water-framework/pdf/water_note3_groundwateratrisk.pdf
74. Nolan BT, Dubus IG, Surdyk N, Fowler HJ, Burton A, Hollis JM, Reichenberger S, Jarvis NJ (2008) Identification of key climatic factors regulating the transport of pesticides in leaching and to tile drains. Pest Manage Sci. doi:10.1002/ps.1587
75. Petrovic M, Hernando MD, Diaz-Cruz MS, Barceló D (2005) Liquid chromatography-tandem mass spectrometry for the analysis of pharmaceutical residues in environmental samples: a review. J Chromatogr A 1067:1–14
76. Gros M, Petrovic M, Barceló D (2007) Wastewater treatment plants as a pathway for aquatic contamination by pharmaceuticals in the Ebro river basin (Northwest Spain). Environ Toxicol Chem 26:1553–1562
77. Navarro A, Endo S, Gocht T, Barth J, Lacorte S, Barceló D, Grathwohl P (2008) Sorption of alkylphenols on Ebro River sediments: comparing isotherms with field observations in river water sediments. Environ Pollut. doi:10.1016/j.envpol.2008.08.007

78. Petrovic M, Lacorte S, Viana P, Barceló D (2002) Pressurized liquid extraction followed by liquid chromatography-mass spectrometry for the determination of alkylphenolic compounds in river sediment. J Chromatogr A 959:15–23
79. de Weert (2009) Availability and biodegradation of nonlyphenol in river sediment. In proceedings of AquaTerra final conference: processes-data-models-future scenarios. Tübingen, Germany, 3, p 22
80. Zambrano-Bigiarini M, Majone B, Bellin A, Isabella Bovolo C, Blenkinsop S, Fowler HJ (2010) Hydrological impacts of climate change on the Ebro River basin. Hdb Env Chem DOI 10.1007/698_2010_85
81. Eljarrat E, Raldúa D, Barceló D (2010) Origin, occurrence, and behavior of brominated flame retardants in the Ebro River basin. Hdb Env Chem DOI 10.1007/698_2010_70
82. Gros M, Petrovic M, Ginebreda A, Barceló D (2010) Sources, occurrence, and environmental risk assessment of pharmaceuticals in the Ebro river basin. Hdb Env Chem, DOI 10.1007/698_2010_72

Chemometric Analysis and Mapping of Environmental Pollution Sources in the Ebro River Basin

Marta Terrado, Damià Barceló, and Romà Tauler

Abstract Principal component analysis (PCA) and multivariate curve resolution alternating least squares (MCR-ALS) are shown to be powerful chemometric methods for the analysis of environmental monitoring data sets. They allow for the investigation, resolution, identification, and description of pollution patterns distributed over a particular geographical area, time, and environmental compartment. Historical data of the Ebro River basin available from the *Confederación Hidrográfica del Ebro* (CHE), which is the organization in charge of the basin, covering different years since 1992 for different environmental compartments were investigated to assess the contamination sources affecting the river basin. For ongoing contamination, data sets obtained from a 3-year extensive monitoring study (from 2004 to 2006) were analyzed. Agricultural practices were identified as the main source of surface and groundwater diffuse pollution, while sediments and soils appeared mostly polluted by a contamination pattern mainly loaded by polycyclic aromatic hydrocarbons (PAHs) of possible pyrolitic origin. Another pollution pattern related to past and ongoing industrial activities was detected to be principally stored in the sediment compartment. Moreover, a pollution source resulting from rice-related agricultural practices was identified in the Ebro River delta, with its highest levels detected in May, coinciding with pesticides application during the earlier stages of the rice cycle.

M. Terrado (✉) and D. Barceló
Department of Environmental Chemistry, Institute of Environmental Assessment and Water Research (IDAEA), Spanish Council for Scientific Research (CSIC), Jordi Girona 18-26, 08034 Barcelona, Catalunya, Spain
Catalan Institute for Water Research (ICRA), Parc Científic i Tecnològic de la Universitat de Girona, Edifici H2O, Emili Grahit 101, 17003 Girona, Catalunya
e-mail: mtcqam@idaea.csic.es

R. Tauler
Department of Environmental Chemistry, Institute of Environmental Assessment and Water Research (IDAEA), Spanish Council for Scientific Research (CSIC), Jordi Girona 18-26, 08034 Barcelona, Catalunya, Spain

D. Barceló and M. Petrovic (eds.), *The Ebro River Basin*,
Hdb Env Chem (2011) 13: 331–372, DOI 10.1007/698_2010_74,

Keywords Chemometrics, Contamination sources, Ebro River, Multivariate curve resolution, Principal component analysis

Contents

Abbreviations

AP	Alkylphenol
BOD	Biological oxygen demand
CHE	Ebro River hydrographical confederation
COD	Chemical oxygen demand
DDE	Dichlorodiphenyldichloroethylene
DDT	Dichlorodiphenyltrichloroethane
DL	Detection limit
EC	European Community
EU	European Union
GIS	Geographical information system
HCB	Hexachlorobenzene
HCBu	Hexachlorobutadiene
HCH	Hexachlorocyclohexane
ICA	Integrated quality network
K_{ow}	Octanol–water partition coefficient
MCR-ALS	Multivariate curve resolution alternating least squares
OC	Organochlorinated compound

PAH	Polycyclic aromatic hydrocarbon
PARAFAC	Parallel factor analysis
PCA	Principal component analysis
PMF	Positive matrix factorization
RCSP	Control network for dangerous substances
TCB	Trichlorobenzene
UV	Ultraviolet

1 Introduction

Diffuse and point pollution in the Ebro River basin (NE Spain) caused by pressure factors such as agriculture, industry, and human sewage, is an issue of great concern, since together with changes in climatic conditions and land use practices, it can have large scale adverse impacts over the quality of the basin. The Ebro River has a drainage area of approximately 85,000 km^2, discharging into the Mediterranean Sea, and comprising territories of nine autonomous communities, from Cantabria to Catalunya (see Fig. 1). Population is distributed in a very heterogeneous way all over the basin, with the major cities being Pamplona, Lleida, Logroño, Vitoria, and Zaragoza, which concentrate around 45% of the total population of the territory. It is the most important irrigated land in Spain, crops varying with relief and climate (see Fig. 1). Northernmost regions of Cantabria, País Vasco, Navarra, and La Rioja are dominated by woodland and pasture (in La Rioja vineyards are also important). Cereal crops dominate the central regions, in particular Aragón,

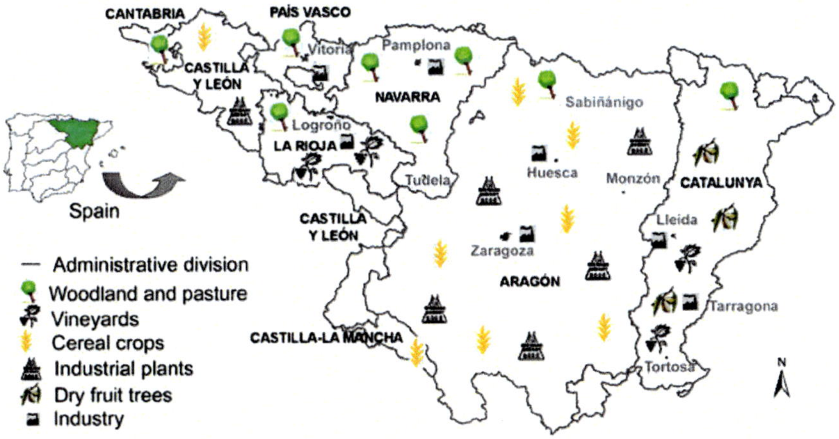

Fig. 1 Location of the Ebro River basin and division of the basin in autonomous communities with identification of the main urban centers and general agricultural and industrial activities

Castilla y León, and Castilla la Mancha. This part is also the most significant for industrial plants such as biomass crops and oilseed rape. In the southernmost regions of Castilla La Mancha and Catalunya, dry fruit trees and vineyards increase in significance, while the Ebro River delta supports a well-developed rice farming activity. Diffuse pollution originated by pesticides application in the basin has been widely studied [1–3]. A historical pollution from chemical plants manufacturing solvents and chlorinated pesticides in the southern part of the river basin is also well known [4]. Automobile, textile, food, and wood industry as well as mining activities are important in the northern part.

Environmental monitoring programs have been performed at different sampling sites and environmental compartments of the Ebro River basin during the last few years to assess the basin environmental quality status. From 1980, the *Confederación Hidrográfica del Ebro* (CHE) [5], which is the organization in charge of the management of the basin, has been performing an assessment of the environmental quality, first, through the *Red Integrada de Calidad de Aguas* (ICA network, which does not exist at present), and, from 1992, using the network *Red de Control de Sustancias Peligrosas* (RCSP), created to control the concentration of some specific priority compounds in the Ebro River basin according to the European directive 76/464/CEE. Nowadays, this network is constituted by 18 sampling sites, which are spread all over the river basin territory, and which are mainly focused in those areas considered potentially polluted by the release of dangerous substances into the environment.

During the last few years (from 2004 to 2006), with the intention of performing a more homogeneous sampling allowing comparison among different locations in different sampling campaigns, a new study has been developed in the framework of the European Project AquaTerra [6] (sixth EU Framework Program). Both, the CHE and the AquaTerra Project, have generated a large volume of concentration measurements of physical and chemical contaminants spread into the Ebro River basin, which are normally stored in large data tables or data matrices.

In order to derive useful environmental information from the data generated by the environmental monitoring programs, the application of modern chemometric methods based on new multivariate factor analysis [7] is very useful. The basic assumption of these methods when they are applied to environmental data tables is that each value of a measured variable in a particular sample is due to the sum of contributions from individual independent sources of different origin. Each one of these sources is characterized by a particular chemical composition profile and distributed among samples in a different way. As a result of the application of chemometric methods, the main point and diffuse sources of contamination in the environment and their origin may be identified and their distribution profiles among samples (geographical, temporal, among environmental compartments) characterized. The distribution of contamination sources and, their subsequent impact over the territory can be assessed by the use of geographical information systems (GISs, [8]). A GIS is a set of tools for the acquisition, storage, recovery, transformation, and visualization of spatial real-world data for specific purposes.

2 Chemometric Methods for the Analysis of Environmental Data

Environmental data tables or matrices obtained from environmental monitoring programs contain parameters of different nature (physical, chemical, biological, toxicological, etc.), and these data are characterized by a high variability coming from very different sources, normally unknown. The variability existing in environmental data sets can be associated to three different sources or origins:

(a) Natural sources, without important influences of contamination, varying geographically and temporally in a stochastic manner as a consequence of natural phenomena.
(b) Anthropogenic sources, originated by human activities, which interfere significantly with natural variability.
(c) Variability sources generated by experimental and measurement errors associated to the different stages that comprise the analytical process: sampling, pretreatment, and measurement of the sample and, finally, by the process of data analysis. The contribution of the experimental error to data variability will be higher when the determined concentrations are lower as well as when the magnitude of the sample diminishes.

The main objective of environmental research is obtaining objective information about the complex and various processes occurring in the environment. For this reason, the development of chemometric methods taking place during these last few decades becomes an important tool for the resolution of this type of problems. Nevertheless, environmental chemistry is a relatively recent discipline. Its general purpose consists in the application of statistical and mathematical tools to find an optimal way to solve environmental chemical problems, and to extract the maximum possible amount of information from experimental data. Therefore, it is a tool that allows performing a more rigorous research of the different processes occurring in the environment.

2.1 Structure of Environmental Data Tables

Data obtained from environmental monitoring programs can be classified, according to their complexity, in data ordered in one direction (one-way data), two directions (two-way data), three directions (three-way data), and in multiple directions (multiway data) [9, 10]. Scalar numerical data (one variable measured in one sample) would correspond to data ordered in zero direction (zero-way), while vector data (for instance, different variables measured in one sample or one variable measured in different samples) are ordered in one direction. When different variables are measured in different samples, obtained data can be ordered in two directions, that is, in a data table or data matrix. Finally, the compilation of different data matrices or data ordered in three directions constitutes what is called a data cube.

A data matrix is the structure most commonly found in environmental monitoring studies. In these data tables or matrices, the different analyzed samples are placed in the rows of the data matrix, and the measured variables (chemical compound concentrations, physicochemical parameters, etc.) are placed in the columns of the data matrix. The statistical techniques necessary for the multivariate processing of these data are grouped in a table or matrix, or use tools, formulations, and notations of the lineal algebra.

Depending on the data structure, different types of models are possible to be applied for data analysis. Thus, when data are ordered in one direction, linear univariant models can be applied (see (1)), and nonlinear models as well (see (2)). For data ordered in two directions, bilinear models can be applied (see (3)) or nonbilinear models. Finally, for data ordered in three directions, trilinear models can be applied (see (4)) or, failing that, nontrilinear models.

$$y_i = b_0 + x_i. \tag{1}$$

$$y_i = f_{\text{no lineal}}(x_i). \tag{2}$$

$$x_{ij} = \sum_{n=1}^{N} g_{in} f_{nj} + e_{ij}. \tag{3}$$

$$x_{ijk} = \sum_{n=1}^{N} g_{in} f_{jn} z_{kn} + e_{ijk}. \tag{4}$$

2.2 Preliminary Data Treatment

Before applying chemometric analysis to data, environmental data tables need a previous arrangement, which is often a very laborious part of the whole data analysis. Data disposal entails different processes such as data arrangement, homogenization, and data transformation which, in each case, depend on the format in which data have been initially obtained, their volume, etc.

Following are among the most common problems to be solved:

(a) Presence of missing values. These are unknown values or blank values, for which the considered compounds have not been analyzed. In case the number of missing values is low and they are randomly distributed within the data table, they can be estimated taking into account the variations observed in variables (matrix columns) and in samples (matrix rows). Once the estimation has been made, it is important to check that values are meaningful from an environmental point of view (for instance, checking that negative values have not been obtained for variables that can only be positive), or that their magnitude is not extremely different from what should be expected according to the observed variation in the measures. For this reason, if the value to be predicted is an outlier, its estimation using these procedures would not be accepted unless it corresponds to a systematic variation previously detected for the rest of the data.

(b) Values below the detection limit (<DL). These values are replaced by half the considered detection limit (DL/2). This has been identified as a good estimation, better than the simple substitution by zero or DL, besides implying little computation effort [11]. Other proposals, such as substitution of these values by the ones directly obtained from instrumental measures (noncensored data) are considered even better. However, sometimes the values of the signals experimentally measured by the analytical instruments are not available because of the nondetection of the signal, or for the reason that even when the signal has been detected, it has been censored by the analyst for low statistical reliability.

2.3 *Data Pretreatment*

To make the extraction of useful environmental information from experimental data easier, the application of data preprocessing or pretreatment techniques is often necessary. The selection of the most suitable pretreatment technique will depend on various factors inherent to every data set, such as the way data have been obtained, the kind of environmental information the analyst is looking for, or the *a priori* knowledge about the nature of data. There is no unique pretreatment method to be generally applied to the analysis of all environmental data matrices. The choice of the most suitable one will depend on the type of phenomena or environmental problem under study. Some pretreatments typically applied are mean-centering, scaling, autoscaling, and logarithmic transformation:

Mean-centering consists in extracting the mean value of the variable to each one of the values of the original variable. In this way, each variable in the new data matrix (centered matrix) presents a mean equal to zero.

$$z_{ij} = x_{ij} - \overline{X}_j \quad i = 1, ..., I \quad j = 1, ..., J. \tag{5}$$

$$\bar{X}_j = \frac{\sum_{i=1}^{I} x_{ij}}{I}, \tag{6}$$

where z_{ij} is the new centered value, x_{ij} the original value of the variable, and $\bar{x}_j$ is the mean of the x_i values of variable j. Mean-centering makes a translation of the origin of coordinates from zero to the mean value of the data. In this way, variations from the mean are easier to visualize, since that part of the information from variables which do not change and which are considered constant in the data matrix is eliminated. Nevertheless, mean-centering makes the calculation of the apportionment or quantitative proportional assignment of the different contamination sources in samples more difficult because of the presence of negative values. Mean-centering is encouraged in the case of some chemometric methods such as principal component analysis (PCA [12]) to investigate variance sources, but it is not recommended for multivariate resolution methods [13], positive matrix factorisation (PMF [14]), or PARAFAC (parallel factor analysis [15]), when applying nonnegativity constraints.

Scaling consists in dividing each value of a variable by its standard deviation.

$$z_{ij} = \frac{x_{ij}}{s_j} i = 1, ..., I \quad j = 1, ..., J, \tag{7}$$

where z_{ij} is the new scaled value, x_{ij} the original value, and s_j the standard deviation of the variable:

$$s_j = \sqrt{\frac{\sum_{i=1}^{I} (x_{ij} - \bar{x}_j)^2}{I - 1}}. \tag{8}$$

$\bar{x}_j$ corresponds to the mean of variable j values. Scaling is a data pretreatment widely used for the analysis of environmental data sets. The obtained scaled variables present a more homogeneous distribution being, therefore, more easy to compare. Moreover, the standard deviation of variables, once they have been scaled, is equal to 1, and they have the same relative importance. However, special attention has to be placed on those variables with very low or constant values in nearly all their measurements since, in these cases, scaling can increase their relative importance in an artificial way (because they are divided by a very low standard deviation making the scaled values very large), leading to erroneous results and conclusions. In these situations, it is better to directly eliminate these variables or, in case their inclusion in the analysis is regarded as necessary to keep the data structure (for instance in multiset and multiway data analysis), it is better to give them a very low weight (for instance, dividing all the values of the variable by a high value, such as 20, instead of dividing them by the variable's standard deviation). Scaling is a data pretreatment especially useful when data of different variables are expressed in different units and magnitudes. In these cases, variables with higher values dominate the observed variance of data in an undesirable way and, for this reason, variables with the lower values are masked. When this happens, it is preferable to scale data to decrease the exaggerated effect of variables presenting higher magnitudes. On the other hand, environmental data have associated uncertainties which normally have not been determined. Care with these uncertainties has to be taken when scaling is applied since, like in the case of very low values, these errors can be amplified in an important way.

Autoscaling consists in the simultaneous application of mean-centering and scaling. The distribution of the values of variables obtained after applying autoscaling is similar to the case of scaling but, at the same time, variables experience a translation of their origin because of mean-centering. Autoscaling is the preferred data pretreatment in the application of methods such as PCA, when the main interest focuses in the description and investigation of data variance. However, like in the case of mean-centering, autoscale generates negative values which will prevent the application of procedures applying nonnegativity constraints.

Logarithmic transformation acts increasing the weight of variables having the lowest values with respect to variables presenting the highest ones. Like scaling, it is a useful procedure in the case of variables having different magnitudes, scales, or measurement units. This procedure is often recommended in the literature for skewed data sets, such as data generated in environmental studies, in which an important part of the values of variables are low, and a small number of high measurements exist. By applying logarithm, a more symmetrical data distribution is obtained. However, the logarithmic transformation can originate unwanted effects, since the internal data structure can be affected, and the possible nonlinearity as well as the number of components necessary to explain data variance can be increased. Once logarithmic transformation is applied to the data, the additional application of the other data pretreatments described in this section (mean-centering, scaling, or autoscaling) is also possible, obtaining a different distribution of the values of variables.

2.4 Multivariate Data Analysis Methods

Two different chemometric methods for the analysis of environmental monitoring data sets are presented for the investigation of pollution patterns distributed over particular geographical areas, times, and environmental compartments.

2.4.1 Principal Component Analysis

PCA [12, 16] is a multivariate statistics method frequently applied for the analysis of data tables obtained from environmental monitoring studies. It starts from the hypothesis that in the group of original data, there is a set of reduced factors or dominant components (sources of variation) which influence the observed data variance in an important way, and that these factors or components cannot be directly measured (they are hidden factors), since no specific sensors exist for them or, in other words, they cannot be experimentally observed.

Generally, a significant amount of the information contained in the experimentally measured variables is redundant (correlated with other variables) or nonrelevant. PCA allows the transformation of this set of variables into a new set of noncorrelated variables, which is more easily interpretable, and underlines the

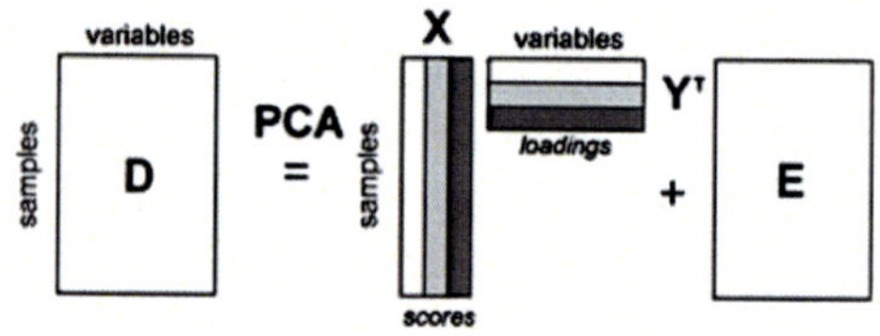

Fig. 2 PCA decomposition of D data matrix following a bilinear model for a number of components $N = 3$

most relevant underlying information contained in data. PCA is a chemometric method which assumes a bilinear model to explain the observed data variance using a small number of components. Using the linear combination of the original variables, a new group of noncorrelated orthogonal variables is deduced (loadings), which makes possible the calculation of new axes (orthogonal) of data representation (scores), in which the data variance is explained in a decreasing order.

PCA attempts to summarize the relevant information contained in the analyzed data matrix, distinguishing it from noise or error. Mathematically, the original (experimental) data matrix **D** is decomposed, following a bilinear model, into the product of two orthogonal data matrices, **X** and $\mathbf{Y}^{\mathrm{T}}$ (see Fig. 2)

$$D = XY^{\mathrm{T}} + E = D^{*} + E, \tag{9}$$

where **D** is the matrix of experimental data, **X** the matrix of scores (map of samples), $\mathbf{Y}^{\mathrm{T}}$ the matrix of loadings (map of variables), and **E** the error matrix (9). The product of the scores and loadings matrices will reproduce the original data matrix for a determined number N of components (number of columns in **X** and of rows in $\mathbf{Y}^{\mathrm{T}}$). Because of the fact that components in PCA are obtained in a decreasing order of explained data variance, the more relevant information in the data set is always concentrated in the first generated principal components. It is for this reason that the latter components can be omitted without a significant loss of variance or relevant information, since they explain only a small part of the variance related to noise and experimental error. Since the solution to (9) is ambiguous, the matrix decomposition in this equation has to be performed under some constraints. In the case of PCA, matrix decomposition is performed under orthogonal constraints, loadings normalization, and maximum explained variance for the successive extracted components. Under these constraints, PCA provides unique solutions. However, these solutions are abstract linear combination of the true experimental variance sources and, although they are very useful for data exploration and summary, in many cases they can be difficult to interpret in environmental terms.

In general, the number of components N is selected at the point where the addition of a new component does not give relevant additional information within the context of the studied problem or, in other words, when this component explains experimental noise only. Those components explaining proportions of small variance are not investigated, and they are assumed to be mainly related to small background contributions or to noise and experimental error. The selected number of principal components will be related to the independent number of sources or patterns of variation (contamination) which are present in the analyzed data.

According to the magnitude of the retained variance and the contribution that the original variables make to each component, the environmental meaning of the identified components can be deduced, and the approximate level of error contained in the experimental data can also be determined. In this context, the displaying of scores (matrix **X**) and loadings (matrix $\mathbf{Y}^{\mathrm{T}}$) obtained from PCA decomposition of the original data matrix **D** are extremely useful.

The value of loadings indicates the magnitude of the contribution of each original variable to each component or factor (new axes of representation). When environmental data are considered, loadings can indicate the chemical composition of the identified contamination sources. Variables with high loading values for the same component covary. If they have the same sign, they covary positively, whereas they covary negatively when signs are opposed (inverse covariation). Sample representations in the new space defined by the new components or factors are called scores. Their representation allows sample grouping according to their similarity as well as detection of those samples presenting extreme values. From an environmental point of view, scores contain information about the distribution of contamination sources in samples (geographical and temporal distribution, and distribution among environmental compartments). The error matrix **E** contains the part of data variation which is not explained by the PCA model and that is consequently a result of the experimental error. In general, a correct interpretation of the different sample grouping obtained in the score graphs needs to be performed together with the corresponding loading graphs.

2.4.2 Multivariate Curve Resolution Alternating Least Squares

Multivariate curve resolution methods (MCR [17]) describe a family of chemometric procedures used to identify and solve the contributions existing in a data set. These procedures have been traditionally applied for the resolution of multiple chemical components in mixtures investigated by spectroscopic analysis techniques [18].

Multivariate curve resolution alternating least squares (MCR-ALS) (as well as PCA) decomposes a data matrix assuming a bilinear model based on (9) (see PCA section). MCR-ALS uses softer constraints than PCA, making loading and score profiles more easily interpretable and reasonable from an environmental point of view. Constraints more frequently applied to environmental data during the MCR-ALS bilinear matrix decomposition are nonnegativity, normalization of loadings, and if possible, trilinearity [19]. This last constraint is restricted to the case of augmented data matrices, when the same samples and variables are measured in different sampling campaigns, which implies that the data set can be arranged in a regular data cube or three-way data structure [7]. The main drawback of MCR-ALS and other MCR methods is that the solutions provided by these methods, although they are more easily interpretable and closer to the true underlying sources of data variance than in the case of PCA, are not unique and the effect of constraints must be carefully examined.

One of the best ways to improve MCR-ALS solutions is its extension to the simultaneous analysis of multiple data sets obtained from different experiments, situations, or conditions. This extension improves the reliability of the solutions decreasing and even eliminating the previously mentioned ambiguities completely. This extension to the simultaneous analysis of multiple data sets is possible through the augmentation or concatenation of the individual data matrices corresponding to individual experiments. A data matrix can be augmented in the row direction (row-wise), in the column direction (column-wise), or simultaneously in the row and column directions (row and column-wise) to generate a new multimatrix structure, where matrices with the same number of rows, columns, or both at the same time, are concatenated in the appropriate direction [20]. The first advantage of the matrix augmentation is that those resolution characteristics which are better for one or some of the analyzed matrices will always have a positive effect over the resolution of other more complex data matrices. The extension of the bilinear model to a column-wise augmented data matrix (in this example, constituted by the concatenation of four individual data matrices) is given in (10):

$$D_{\text{aug}} = \begin{pmatrix} D_1 \\ D_2 \\ D_3 \\ D_4 \end{pmatrix} = \begin{pmatrix} X_1 \\ X_2 \\ X_3 \\ X_4 \end{pmatrix} Y^{\mathrm{T}} + \begin{pmatrix} E_1 \\ E_2 \\ E_3 \\ E_4 \end{pmatrix} = X_{\text{aug}} Y^{\mathrm{T}} + E_{\text{aug}}. \tag{10}$$

In this equation, whereas the same loading matrix ($\mathbf{Y}^{\mathrm{T}}$ matrix) is common for the different individual data matrices $\mathbf{D}_k$, $k = 1, 2, 3, 4$, four different score matrices $\mathbf{X}_k$, $k = 1, 2, 3, 4$ are considered to explain the variation in $\mathbf{D}_{\text{aug}}$. Since these four $\mathbf{D}_k$ matrices have equal sizes (same number of rows or samples and of columns or variables) they can also be arranged in a three-way data cube, with the four data matrices in the different slabs of this cube. However, in the frame of the MCR-ALS method and of the general bilinear model in (10), it is preferable to consider them to be arranged in the column-wise augmented data matrix $\mathbf{D}_{\text{aug}}$.

In the case of data following a trilinear model structure [7], the profiles in the score matrices $\mathbf{X}_k$ in $\mathbf{X}_{\text{aug}}$ of (10) will present the same shape in the score profiles (or, in other words, the relative distribution of contamination patterns among samples in the different data matrices is the same) in the different individual data matrices $\mathbf{D}_k$ (obtained in the different sampling campaigns). In this case, the data set fulfills the trilinear model [7, 15, 21], which can be described by (11) (in this equation, super-index t in D_k^t and X_k^t matrices indicates that they conform to the trilinear model):

$$D_k^t = X^t Z_k Y^{\mathrm{T}} + E_k = X_k^t Y^{\mathrm{T}} + E_k, \tag{11}$$

where $\mathbf{Z}_k$ is a diagonal matrix containing, in its diagonal, the relative amounts of each component score profiles, which are uniquely defined by $\mathbf{X}^{\mathbf{t}}$. Therefore, individual X_k^t matrices are simply obtained by $X_k^t = X^t Z_k$, $k = 1, 2, 3, 4$.

According to the bilinear model described in (10), the loading matrix $\mathbf{Y}^T$ was forced to be the same (invariant) for all the different $\mathbf{D}_k$ matrices, $k = 1, 2, 3, 4$ in $\mathbf{D}_{aug}$. This means that the composition profiles for the different pollution patterns are assumed to be the same during the different sampling campaigns. In contrast, the score matrix $\mathbf{X}_{aug}$, has individual score matrices $\mathbf{X}_k$ which are different (not invariant), allowing therefore the identified pollution patterns to have a different distribution during the different sampling campaigns. However, when matrices $\mathbf{D}_k$ and $\mathbf{X}_k$ have all the same number of rows (the same samples are analyzed in different campaigns), each column of the $\mathbf{X}_{aug}$ matrix can be properly arranged to give a new matrix with a number of rows equal to the number of sampling sites and with a number of columns equal to the number of analyzed conditions (sampling campaigns, see Fig. 3). Rearranging and averaging [20], the columns of the score matrix $\mathbf{X}_k$ gives the new common scores and relative scale profiles ($\mathbf{X}^t$ and $\mathbf{Z}_k$, respectively in (11)) for the considered component. These two profiles describe the variation captured by ALS in the two modes (sampling sites and campaigns) for that particular component. The Kronecker product [21] of these two new vectors gives a new augmented long score vector, which substitutes the corresponding column of the $\mathbf{X}_{aug}$ score matrix during the ALS optimization (see Fig. 3). This procedure forces the shape of the score vectors ($\mathbf{X}^t$ in (11)) to be the same for the different $\mathbf{X}_k$ matrices (different campaigns). Pollution patterns are therefore forced to have the same samples (spatial) distribution during the different analysis campaigns.

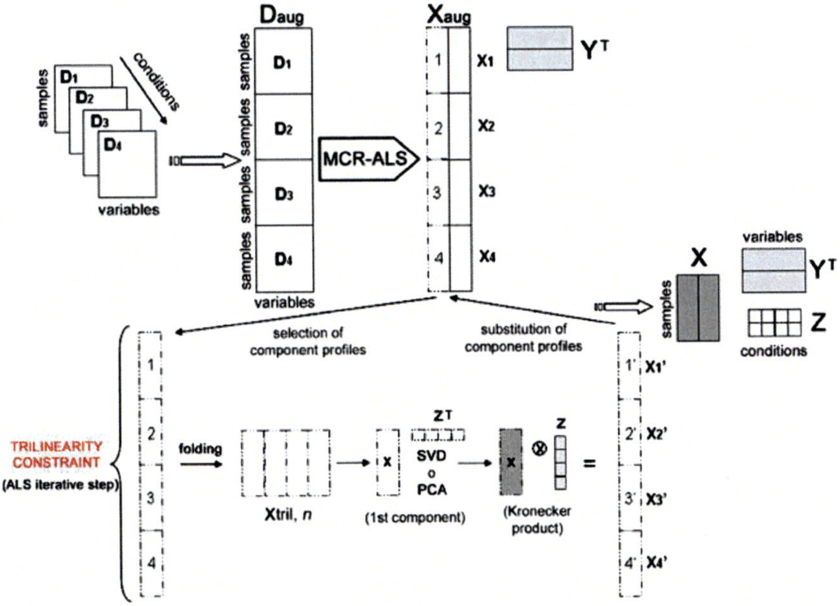

Fig. 3 Data matrix arrangement, augmentation, and resolution using MCR-ALS and application of the trilinearity constraint

Besides, intensity variation of the component is captured in the third mode loadings ($\mathbf{Z}_k$ in (11)), showing the scale differences of this component during the four sampling campaigns. When this procedure is applied to all the resolved components, it is equivalent to performing a trilinear decomposition like in the PARAFAC [15] method. The main advantage of using this procedure in MCR-ALS compared to PARAFAC is that it can be easily adapted to data sets of different complexity and structure, trilinear or nontrilinear, providing optimal least squares solutions. Since the trilinearity constraint can be applied optionally to one or two or all the components, the method used in MCR-ALS covers the intermediate situations existing between a pure bilinear model and a completely trilinear model.

2.5 *Geostatistical Methods*

Environmental data are found to be typically distributed in space and time. It is for this reason that knowing the value of a specific attribute such as the concentration of a contaminant, lacks interest when its location or the moment of its measurement is unknown and not taken into account during the data analysis procedure. In this sense, geostatistics [22, 23] offers a set of statistical tools which allow incorporating spatial and temporal tendencies of different measurements during data processing. Therefore, the spatial continuity of any natural phenomenon can be described. Apart from allowing interpolation of the value of a particular attribute in those locations where it has not been measured, geostatistical techniques allow modeling of the uncertainties generated during the process of prediction of this value.

Geostatistics is on the basis of the concept of stochastic functions, in which a particular set of measured values are considered spatially dependent random variables. A random variable is a variable which can adopt a series of different values according to a probability distribution.

In random functions, each realization of the function can be conceived as the sum of a structured component and an erratic or stochastic component. The structured component is the one assuring that the observed values have a systematic variation or, in other words, that if we are, for instance, in an area where different measurements above a normal value have been obtained, there exist a high probability for additional measurements to be high as well. On the other hand, the random component is the one making difficult the exact prediction of these hypothetical measurements, since unpredictable fluctuations exist.

There are various algorithms aiming to predict the value of specific attributes in specific nonmeasured locations, such as the so-called kriging method. Kriging assumes the spatial variation of a specific variable to be statistically homogeneous through a particular surface. It is a geostatistical approach which quantifies the spatial correlation structure among the different locations in the study area according to their separation distance.

3 Chemometric Analysis of Historical Contamination in the Ebro River Basin

A chemometric analysis of historical data available from the CHE, which is the organization in charge of the management of the Ebro River basin, has been performed using PCA. Data covering different years since 1992 for water, sediments (SEs), and fish have been investigated.

3.1 Data Description and Arrangement

Data were divided into six different groups according to the physical compartment (water, SEs, and biota) and the analyzed variables (metals, organic compounds, and physicochemical parameters). Analyzed parameters and sampling sites were not the same for all compartments and years for reasons of data availability (see Fig. 4 for sample location and Table 1 for sample identification). SE and water sample data covering years from 1996 to 2003 were selected, since before 1996 the data set was too much incomplete. Time interval for biota was reduced, covering only years from 2000 to 2003 for metals and from 1999 to 2002 for organic compounds.

3.1.1 Metals

Concentrations of metals in SEs were arranged in a data matrix of size 9 × 8 (rows × columns), rows corresponding to sampling sites and columns to the

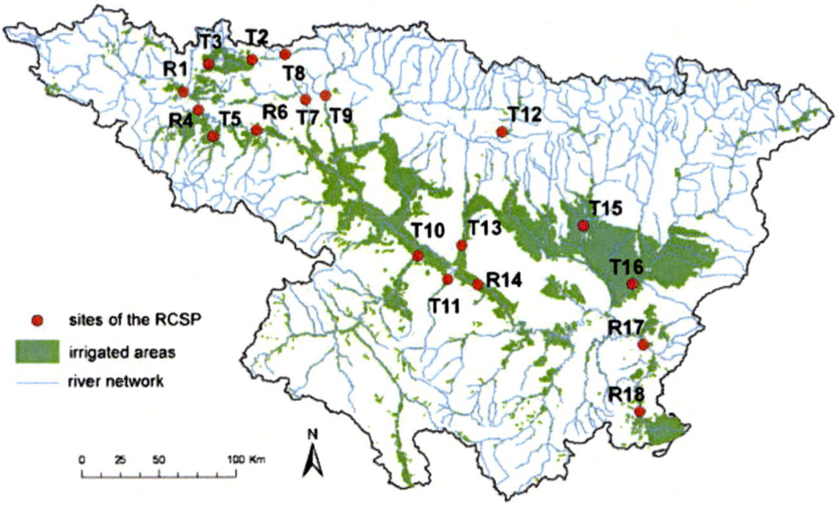

Fig. 4 Location of sampling sites of the RCSP in the Ebro River basin

Table 1 Identification of sampling sites of the RCSP in the Ebro River basin

Station code[a]	Location description
R1 (SP7)	Ebro River in Miranda de Ebro (Burgos)
T2 (SP18)	Zadorra River in Salvatierra (Álava)
T3 (SP8)	Zadorra River in Vitoria Trespuentes (Álava)
R4 (SP11)	Ebro River in Conchas de Haro (La Rioja)
T5 (SP17)	Najerilla River in Nájera (downstream) (La Rioja)
R6 (SP12)	Ebro River in Logroño (downstream) (La Rioja)
T7 (SP13)	Ega River in Arinzano (Navarra)
T8 (SP10)	Araquil River in Alsasua (Navarra)
T9 (SP6)	Arga River in Puente la Reina (Navarra)
T10 (SP16)	Jalón River in Grisén (Zaragoza)
T11 (SP15)	Huerva River in Fuente de la Junquera (Zaragoza)
T12 (SP1)	Gállego River in Jabarrella (Huesca)
T13 (SP14)	Gállego River in Villanueva de Gállego (Zaragoza)
R14 (SP2)	Ebro River in Presa de Pina (Zaragoza)
T15 (SP5)	Cinca River in Monzón (downstream) (Huesca)
T16 (SP4)	Segre River in Torres de Segre (Lleida)
R17 (SP3)	Ebro River in Ascó (Tarragona)
R18 (SP9)	Ebro River in Tortosa (Tarragona)

[a]SP – RCSP nomenclature; R and T – nomenclature used in the AquaTerra Project. "R" stands for sampling sites located in the principal flow while "T" stands for sampling sites located in any of their tributaries

concentrations of the analyzed metals in milligram/kilogram of SE. Analyzed metals were As, Cd, Cr, Cu, Hg, Ni, Pb, and Zn. Samples were arranged in different data matrices according to sampling years from 1996 to 2003, giving seven individual data matrices of metals in SE (1999 excluded).

Concentrations of metals in fish were arranged in a data matrix of 9 × 8, rows corresponding to sampling sites and columns to the concentrations of the analyzed metals and one physical parameter. Analyzed variables were Cd, Cr, Cu, Hg, Pb, Se, Zn, and fish length. Concentrations of analyzed metals were given in milligram/kilogram of biota, and fish length was in centimeters. Data were arranged according to the years of sampling from 2000 to 2003, giving four individual data matrices of metals in fish.

Concentrations of metals in surface water (SW) were arranged in a data matrix of 10 × 13, rows corresponding to the sampling sites (not coinciding in all the years) and columns to the concentrations of analyzed metals in milligram/liter of water. Analyzed metals were As, Ca, Cd, Cr, Cu, Fe, K, Mg, Mn, Na, Ni, Pb, and Zn. Samples were arranged according to sampling years from 1996 to 2003, giving eight individual data matrices of metals in SW.

3.1.2 Organic Compounds

Concentrations of organic compounds in SEs were arranged in a data matrix of size 9 × 12, rows corresponding to sampling sites and columns to the concentrations of the analyzed organic compounds in microgram/kilogram of SE. Analyzed organic compounds were HCHs, DDTs, and HCB (pesticides); HCBu and TCBs (volatile organo-halogenous compounds); and naphthalene, fluoranthene, benzo(a)pyrene, benzo(b)fluoranthene, benzo(g,h,i)perylene, benzo(k)fluoranthene, and indeno (1,2,3-cd)pyrene (polycyclic aromatic hydrocarbons, PAHs). Samples were arranged according to sampling years from 1996 to 2003, giving seven individual data matrices of organic compounds in SE (1999 excluded).

Concentrations of organic compounds in fish were arranged in a data matrix of size 9 × 12, rows corresponding to sampling sites and columns to the concentrations of the analyzed organic compounds and one physical parameter. Analyzed variables were HCB, HCHs, *o,p*-DDD, *o,p*-DDE, *o,p*-DDT, *p,p*-DDD, *p,p*-DDE, *p,p*-DDT, and DDTs (pesticides); TCBs and HCBu (volatile organo-halogenous compounds); and length. Concentrations of analyzed organic compounds were given in microgram/kilogram of biota and fish length was in centimeters. Samples were arranged according to sampling years from 1999 to 2002, giving four individual data matrices of organic compounds in fish.

3.1.3 Physico-Chemical Parameters

Values of physico-chemical parameters in water were arranged in a data matrix of size 10 × 17, rows corresponding to sampling sites and columns to concentrations of the analyzed parameters, each one with its own specific units. Analyzed parameters were alkalinity (mg/L of $CaCO_3$), chlorides (mg/L of Cl^-), cyanides (mg/L of CN^-), total coliforms (MPN/100 ml), conductivity at 20°C (μS/cm), biological oxygen demand (BOD_5, in mg/L of O_2), chemical oxygen demand (COD_{UV}, in mg/L of O_2), fluorides (mg/L of F), suspended matter (mg/L), NH_4^+ (mg/L), NO_3^- (mg/L), O_2 (mg/L), pH, PO_4^{3-} (mg/L), $SO_4^=$ (mg/L), and water and air temperature (°C). Samples were arranged according to sampling years from 1996 to 2003, giving eight individual data matrices of physico-chemical parameters in water.

Each data matrix corresponded to 1-year sampling campaign and contained a specific number of samples (rows) with various variables measured in each sample (columns). The selected data pretreatment method was autoscaling, which is a combination of data mean-centering and scaling. To perform PCA, matrices of each analyzed variable in each physical compartment were concatenated one on the top of the other, giving a new augmented data matrix. Obtained augmented data matrices for subsequent analysis were MS (metals in SE), MF (metals in fish), MW (metals in water), OS (organic compounds in SE), OF (organic compounds in fish), and PW (physico-chemical parameters in water).

3.2 *Results and Discussion*

PCA results are summarized in Fig. 5, which shows the loading plots characterizing the main contamination patterns in every analyzed data set and their explained

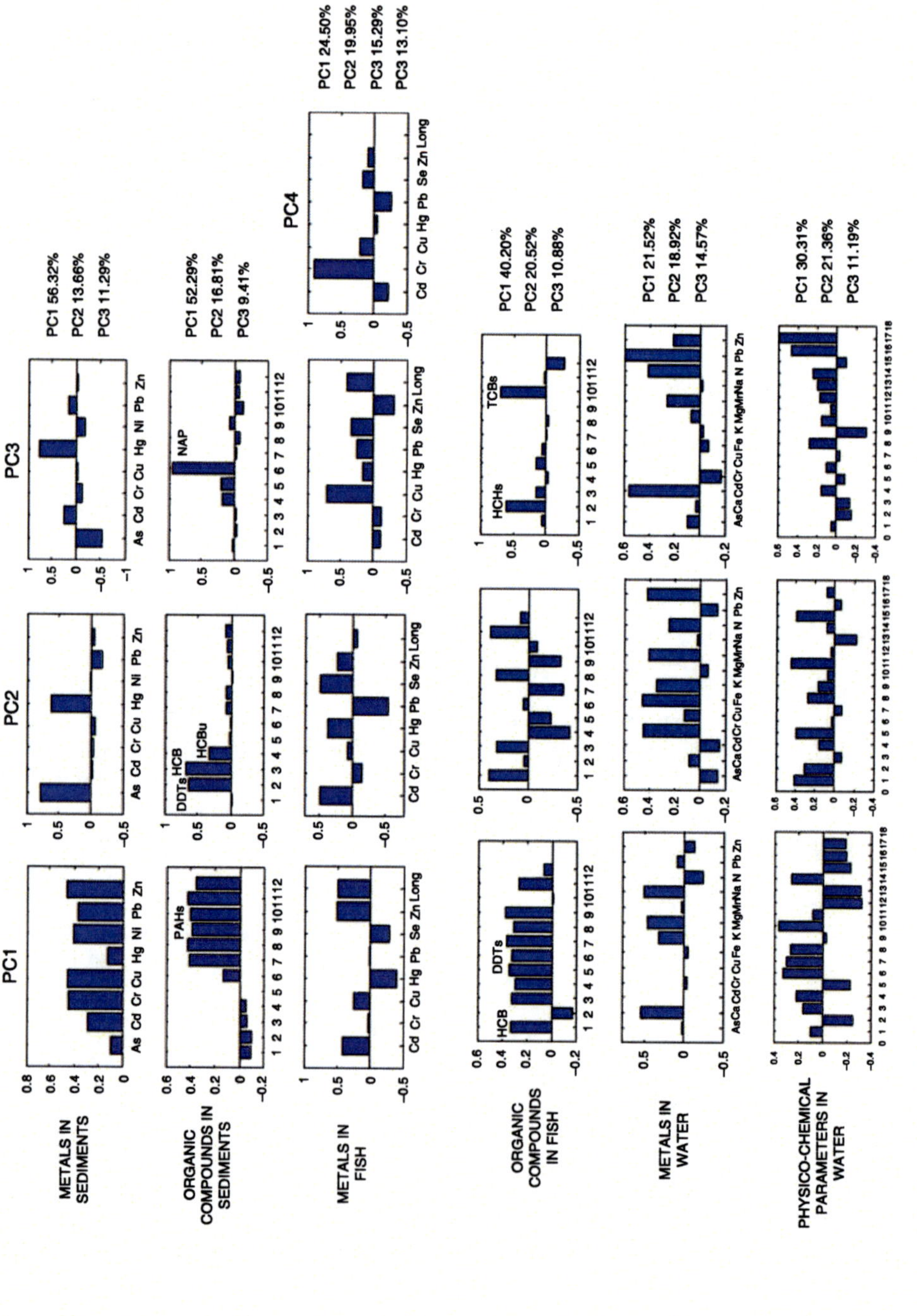

Fig 5. (Continued)

variance. High positive and negative loadings indicate that these variables have an important contribution to the considered principal component. Whenever possible, interpretation of principal components is done in terms of contamination sources operating over the Ebro River basin.

3.2.1 Metals

In the SE compartment, PC1 (explaining 56.3% of data variance) describes a general contamination pattern of metals which is mostly characterized by the simultaneous presence of Zn, Cu, Cr, Ni, and Pb. The contribution of this pattern is especially relevant at sampling site T3 in the Zadorra River in Vitoria Trespuentes (Álava, País Vasco). It is an irrigated land close to the city of Vitoria, where an important industrial activity is located. Industry could possibly be the cause for this heavy metal contribution [24], but care has to be taken in its interpretation, since in some cases these metals can also be attributed a natural origin. PC2 (13.7% of data variance) is loaded by a high contribution of As and Hg, the two metals covarying positively in this pattern. R17 in the Ebro River in Ascó (Tarragona, Catalunya) was the sampling site presenting a higher contribution of this pattern. It corresponds to an irrigated land, where contaminants from agricultural activities are possibly released into the river which, in addition, receives a very strong industrial pressure. While As is a metal frequently found as a component in different types of fertilizers, Hg can be generally attributed to an industrial origin. PC3 (11.3% of data variance) is also loaded by As and Hg but, in this case, the two metals covariate negatively (that is, in samples where a high content of one element is found, the level of the other element is lower). Thus, site R18 in the Ebro River in Tortosa (Tarragona, Catalunya) maintains the high content of Hg detected in the sampling site immediately upstream (R17), although As concentration in this site is lower.

Assessment of the geographical distribution of pollution in the river using fish samples is more difficult than in the case of SEs, probably because of their mobility, which in some cases can lead to confusing results. PC1 identified in fish samples defines a contrast between two inversely correlated contamination profiles. Positive

Fig. 5 Main contamination sources identified by PCA for sediments, fish, and suface water in the Ebro River basin, and explained variances for each principal component. Variable identification. Organic compounds in sediments: 1, summatory of hexachlorocyclohexanes (HCHs); 2, summatory of DDTs (DDTs); 3, hexachlorobenzene (HCB); 4, hexachlorobutadiene (HCBu); 5, summatory of trichlorobenzenes (TCBs); 6, naphthalene; 7, fluoranthene; 8, benzo(a)pyrene; 9, benzo(b) fluoranthene; 10, benzo(g,h,i)perylene; 11, benzo(k)fluoranthene; 12, indene(1,2,3-cd)pyrene. Organic compounds in fish: 1, hexachlorobenzene (HCB); 2, summatory of hexachlorocyclohexanes (HCHs); 3, *o,p*-DDD; 4, *o,p*-DDE; 5, *o,p*-DDT; 6, *p,p*-DDD; 7, *p,p*-DDE; 8, *p,p*-DDT; 9, summatory of DDTs (DDTs); 10, summatory of trichlorobenzenes (TCBs); 11, hexachlorobutadiene (HCBu); 12, fish length. Physico-chemical parameters in water: 1, alkalinity; 2, chlorides; 3, cyanides; 4, total coliforms; 5, conductivity at 20°C; 6, biological oxygen demand; 7, chemical oxygen demand; 8, fluorides; 9, suspended matter; 10, total ammonium; 11, nitrates; 12, dissolved oxygen; 13, phosphates; 14, sulfates; 15, water temperature; 16, air temperature loadings are mostly described by Zn and Cd contribution whereas negative loadings are characterized by Hg and Se. Together, these two tendencies describe the 24.5% of the total data variance. Both Cd and Zn can have industrial and agricultural origins. High contributions of positive PC1 loadings are detected at R18 sampling site which corresponds to the Ebro River in Tortosa (for years 2001 and 2003), T9 in the Arga River in Puente la Reina (for 2003), and R17 in the Ebro River in Ascó (for 2001). Hg covariates with Se at sampling sites T12, corresponding to Gállego River in Jabarrella (Huesca), T5 in the Cinca River in Monzón downstream (Huesca), and T13 in the Zadorra River in Vitoria-Trespuentes (Álava). While it is possible that Se comes from natural sources, Hg appears in the environment by anthropogenic means only. PC2 (describing a 19.9% of data variance) also describes a contrast between two contamination profiles. Positive loadings are characterized by Cd, Se, and Hg contribution while negative loadings are mainly loaded by Pb. Positive loadings in PC2 present their major influence in the Ebro River mouth, at locations R17 and R18 (where Hg concentrations are higher than the average concentration of this metal detected in fish and SEs). Pb was mainly detected in R14 in the Ebro River in Presa de Pina (Aragón) and T9 in the Arga River in Puente la Reina (Navarra). PC3 (15.3% of data variance) is described by the inversely correlated contribution of Cu and Zn. Cu presence is important at R1 in the Ebro River in Miranda de Ebro (Burgos) for years 2002 and 2003, and at T15 in the Cinca River in Monzón (Huesca) for 2003. R1 is located in an urban area while T15 is close to an industrial emplacement, where industry can contribute to the contamination described by this PC. Zn contamination is important at R14 for year 2001 and at R17 and R18 during years 2000 and 2001. PC4 (13.1% of data variance) is mostly defined by Cr contribution. R14 in Zaragoza is the sampling site where a highest contribution of this contamination pattern has been found.

For water, a first contamination pattern of metals is defined by PC1 (21.5% of data variance) with high loadings for Ca, Na, Mg, and K. This cannot be regarded as a source of contamination since these metals present no toxic effect and only reflect the salinity content of water. The central area of the river basin, grouping sampling sites located around Zaragoza (T10, T13 and T14), presents the richest content in nutrients and salinization because of the nature of parental rocks and the intensively developed irrigation agriculture. PC2 (18.9% of data variance) is characterized by the presence of Cr, Fe, Mn, and Zn. PC3 (14.6% of data variance) is loaded by Cd, Pb, and Ni. The most polluted sampling sites by PC2 and PC3 contamination patterns are T3 in the Zadorra River in Vitoria Trespuentes and R17 and R18 at the lowest course of the river.

3.2.2 Organic Compounds

In the SE compartment, PC1 (explaining 52.3% of data variance) describes a contamination pattern of PAHs, except for the naphthalene compound. This compound is the most volatile within this group, and presents a slightly different chemical behavior. The pattern of PAHs is detected at high levels in the upper part of the Ebro River, mainly at sampling site T9 in the Arga River in Puente la Reina, followed by R1 and T3, which correspond to the Ebro River in Miranda de Ebro (Burgos) and the Zadorra River in Villodas (Álava), respectively. These sites are located very close to the cities of Vitoria and Pamplona, which have an important industrial activity. PC2 (16.8% of data variance) describes a contamination pattern mostly loaded by pesticides: DDTs and HCB (and also with some contribution of HCBu). It mostly affects the lowest course of the Ebro River, with higher inputs at locations R17 and R18. The Ebro delta is one of the most polluted wetlands by organochlorinated compounds (OCs) in Spain. Although the proportion of industry in this area is low compared to that in other areas, it is the zone where the highest amounts of rice farming specific pesticides are applied (molinate, bentazone, fenitrothion, etc.). Moreover, other industrial products have been continuously released directly into the river flow with very little or no depuration processes for years. PC3 (9.4% of data variance) is a rather specific contamination pattern for naphthalene, whose origin can be related to industrial activities. T3 and R1 are the sampling sites presenting the highest contribution of this contamination pattern.

The group of PAHs has not been included to assess the fish compartment since these contaminants were not available for some sampling campaigns, and too many values were found below the detection limit. PC1 identified in fish (40.2% of data variance) describes the same contamination pattern as PC2 in SE, loaded by DDTs, HCB, and HCBu, with its highest contribution presented at T15 and R17 sampling sites. PC2 (20.5% of data variance) defines a contrast between two inversely correlated groups of compounds. Compounds with positive loadings are HCB, HCBu, *o,p*-DDD, and *p,p*-DDT, whereas other DDTs (*o,p*-DDE, *p,p*-DDE and *o,p*-DDT) have negative loadings. Therefore, this principal component shows a distinction between the behavior and distribution of different DDTs. Whereas R17 and R18 are affected by the compounds with positive loadings, at T15, compounds with negative loadings are the ones which are more important. PC3 (10.9% of data variance) is characterized by a high contribution of HCHs and TCBs.

3.2.3 Physico-Chemical Parameters

Three principal contamination patterns have been derived for physico-chemical parameters in water, explaining as much as 62.9% of the total data variance. PC1 (30.3% of data variance) describes two different contributions that covariate inversely. The first one is described by variables with positive loadings: $NH4^+$, BOD_5, COD_{UV}, PO_4^{3-}, and fluorides. These characteristics correspond to waters with high contents of organic matter and, therefore with low oxygen concentrations and subsequent eutrophication problems. Sampling sites more affected by this positive contamination profile are located at the upper course of the river, and correspond to T3 in the Zadorra River in Vitoria Trespuentes (País Vasco) and R4 in the Ebro River in las Conchas de Haro (la Rioja). Both are located after the river has flowed through urban areas receiving important effluents from industrial complexes and urban sewage. The inverse PC1 profile is defined by the negative loadings and has a high contribution of the variable pH, oxygen, chlorides, $SO_4^{=}$, and conductivity. PC2 (21.4% of data variance) is defined by high NO_3^-, alkalinity, conductivity, and $SO_4^{=}$ values. Nitrates can come from animal wastes and agriculture, but it is important to remark that in this study their concentrations in water have not exceeded the EC drinking water limit of 50 ppm in any sample. PC2 mostly affects the central part of the Ebro River, at Zaragoza surroundings. Salinization is especially evident in this area due to the high salt content of soils and underlying geology. PC3 (11.2% of data variance) describes a contamination pattern with positive and negative loadings. Positive loadings show a contribution described by high temperatures in water and air, and also by a relatively high fluoride contribution. It is important at sampling sites close to the mouth of the river: R17 in the Ebro River in Ascó, R1, Ebro River in Tortosa, as well as T16 in the Segre River in Torres de Segre, and R14 in the Ebro River in Presa de Pina. The area at the lower course of the river is, in fact, a region whose climate is characterized by Mediterranean temperatures. Anthropogenic activities also have important effects on water temperature in the lower course. Besides, the nuclear power station located in Ascó, at 4 km downstream from Flix, is responsible for a 3°C increase in river annual temperature [25]. Also important is the effect of reservoirs present at the lower part of the river basin: Mequinensa, Riba-Roja and Flix. Negative loadings in PC3 define a pattern characterized by the presence of suspended matter, which is mainly found in T13 in the Gállego River in Villanueva de Gállego (Aragón) and sites R1 and R4, at the upper course of the river. Suspended matter reduces the amount of light passing through water, which can affect photosynthesis and subsequently reduce the dissolved oxygen in water.

4 Chemometric Analysis of Ongoing Contamination in the Ebro River Basin

A chemometric analysis of data generated in the environmental monitoring sampling campaigns carried out within the Framework of the European Integrated Project AquaTerra (sixth EU RTD Framework Programme) has been performed. The AquaTerra project aimed to provide the scientific basis for an improved river basin management through a better understanding of the river–SE–soil–groundwater (GW) system as a whole [6].

An integrated analysis of four different environmental compartments – SW, GW, SE, and soil – is performed applying MCR-ALS. The concentration of four different families of compounds – OCs, PAHs, pesticides, and alkylphenols (APs) – were analyzed during six different sampling campaigns (from year 2004 to 2006) at various locations distributed within the entire Ebro River basin. SW and GW were sampled twice a year, in spring and fall, whereas SE and soil were only sampled in fall.

Because of their particular physico-chemical properties, every compound was not detected in all compartments. Thus, polar pesticides were usually detected in aquatic compartments, while hydrocarbons and OCs were mainly detected in SE and soil. Consequently, the use of different variables (concentrations of chemical compounds) in the different compartments will make it difficult to achieve a complete view of the pollutant dynamics. Only APs were the group of compounds detected in all the analyzed environmental compartments.

4.1 Data Description and Arrangement

Concentration units were given in microgram/liter of water and in microgram/kilogram of SE/soil. The sampling network for SW and SE coincided geographically, and the same for GW and soil (see Fig. 6 for sample location and Table 2 for sample identification).

Six different individual data matrices (one per sampling campaign: sw_1, sw_2, … , sw_6) were obtained for SW, with rows (samples) and columns (variables) coinciding in every data matrix. These individual data matrices were column-wise appended, one on top of another, keeping the same number of columns and originating a new augmented SW data matrix containing 138 samples in total (23 samples analyzed in each of the 6 sampling campaigns) (see Fig. 7). This data arrangement can be written in a concise way using MATLAB notation programming language as [sw_1, sw_2, … , sw_6], where $\mathbf{sw}_k$ ($k = 1, 2, \ldots , 6$) corresponds to the different SW individual matrices from each sampling campaign, and the semicolon ";" notation is used to indicate that the different data matrices are column-wise concatenated, and that they are supposed to share the same vector space. Likewise, SW matrix was divided in two new submatrices SW_1 and SW_2, which were obtained from

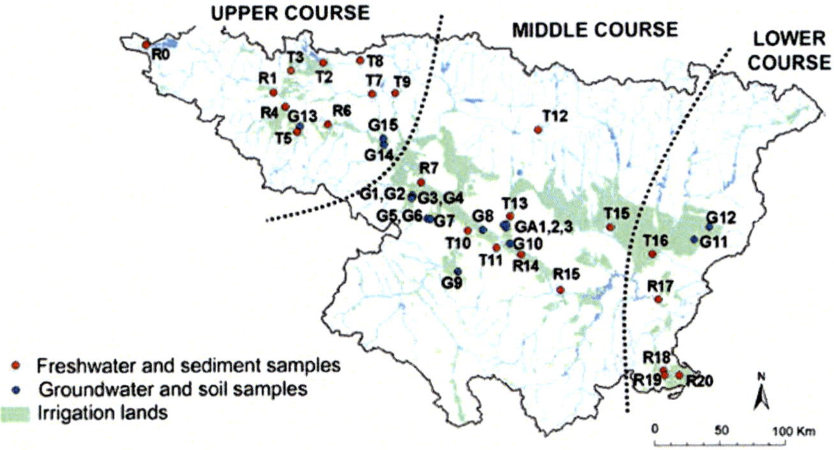

Fig. 6 Spatial location of sampling sites from the AquaTerra Project surveillance monitoring over the Ebro River basin

Table 2 Sampling sites identification

Freshwater/sediment	Groundwater/soil
R0: Ebro in Reinosa (Cantabria)	G1: Cascante (Navarra)
R1: Ebro in Miranda de Ebro (Burgos)	G2: Cascante (Navarra)
T2: Zadorra in Audinaka (Álava)	G3: Monteagudo (Navarra)
T3: Zadorra in Villodas (Álava)	G4: Monteagudo (Navarra)
R4: Ebro in Haro (La Rioja)	G5: Maleján (Zaragoza)
T5: Najerilla in Nájera (La Rioja)	G6: Maleján (Zaragoza)
R6: Ebro in Logroño (La Rioja)	G7: Ainzón (Zaragoza)
T7: Ega in Estella (Navarra)	G8: Sobradiel (Zaragoza)
R7: Ebro in Tudela (Navarra)	G9: Alfamén (Zaragoza)
T8: Araquil in Alsasua (Navarra)	G10: Movera (Zaragoza)
T9: Arga in Puente la Reina (Navarra)	G11: Mollerussa (Lleida)
T10: Jalón in Grisén (Zaragoza)	G12: Tornabous (Lleida)
T11: Huerva in Zaragoza (Zaragoza)	G13: Uruñuela (La Rioja)
T12: Gállego in Caldearenas (Huesca)	G14: San Adrián (Navarra)
T13: Gállego in San Mateo de Gállego (Zaragoza)	G15: Andosilla (Navarra)
R14: Ebro in Presa da Pina (Zaragoza)	GA1: Villanueva de Gállego (Zaragoza)
R15: Ebro in Sástago (Zaragoza)	GA2: Villanueva de Gállego (Zaragoza)
T15: Cinca in Alcolea de Cinca (Huesca)	GA3: Villanueva de Gállego (Zaragoza)
T16: Segre in Torres de Segre (Lleida)	
R17: Ebro in Flix (Tarragona)	
R18: Ebro in Tortosa (Tarragona)	
R19: Ebro in Amposta (Tarragona)	
R20: Ebro in Delta del Ebro (Tarragona)	

individual matrix concatenation (see Fig. 7). SW_1 data matrix contained the data matrices sampled in summer (23 samples analyzed in 3 sampling campaigns: sw_1, sw_3, and sw_5), while SW_2 contained those sampled in fall (23 samples analyzed in 3 sampling campaigns: sw_2, sw_4, and sw_6). Samples were numbered from north to south, "R" indicating "Ebro River" and "T" indicating "tributary." Fourteen variables were measured in every sample: organophosphate compounds (diazinon, dimethoate, ethion, and tributylphosphate), triazines (atrazine, desethylatrazine, simazine, terbutryn, and terbuthylazine), an anilide (propanil), chloroacetanilides (alachlor and metolachlor), and APs (octylphenol and nonylphenol).

Six different individual data matrices were obtained for GW (gw_1, gw_2,..., gw_6) but, in this case, rows (samples) were not common for all data matrices. A new GW data matrix was obtained after individual matrix concatenation containing 92 samples in total (see Fig. 7). In this case, the number of samples for the different sampling campaigns was not coincident (10, 16, 17, 15, 17, and 17 locations were sampled, from first to sixth campaigns respectively). Seven variables (all of them detected in SW as well) were measured in every GW sample: an organophosphate compound (tributylphosphate), triazines (atrazine, desethylatrazine, simazine, and terbuthylazine), and APs (octylphenol and nonylphenol).

Three different individual data matrices (one per sampling campaign: se_1, se_2, and se_3) were obtained for SE, with rows and columns coinciding in every data matrix. A new SE data matrix was obtained after individual matrix concatenation

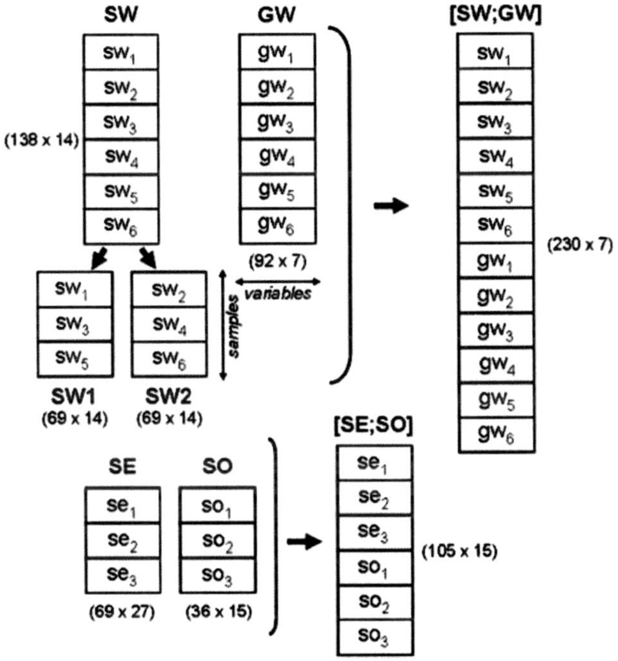

Fig. 7 Arrangement of the different analyzed column-wise concatenated data matrices. In capitals, the augmented data matrices: SW surface water, GW groundwater, SE sediment, SO soil, [SW; GW] surface and groundwater, and [SE;SO] sediment and soil. In lower case letter, the individual data matrices corresponding to each sampling campaign (i.e., sw_1, data matrix corresponding to the first sampling campaign for SW)

contained 69 samples in total (23 samples analyzed in 3 sampling campaigns, coinciding with SW campaigns that sampled in fall) (see Fig. 7). Twenty-seven variables were measured in every sample: PAHs (naphthalene, acenaphtylene, acenaphtene, fluorene, phenanthrene, anthracene, fluoranthene, pyrene, benzo(a) anthracene, and benzo(g,h,i)perylene), APs (octylphenol and nonylphenol), an organophosphate compound (tributylphosphate), and OCs (α-HCH, hexachlorobenzene, 2,4-DDE, 4,4-DDE, 2,4-DDD, 4,4-DDD, 2,4-DDT, and 4,4-DDT).

Finally, three additional individual data matrices were obtained for soil (so_1, so_2, and so_3), in this case with the same number of samples (rows) for each of them. A new soil data matrix (SO) was obtained after individual matrix concatenation containing 36 samples in total (12 samples analyzed in 3 sampling campaigns) (see Fig. 7). Fifteen variables (all of them detected in SE as well) were measured in every sample: PAHs (acenaphtylene, phenanthrene, anthracene, fluoranthene, pyrene, benzo(a)anthracene, chrysene, benzo(b)fluoranthene, benzo(a)pyrene, indeno (1,2,3-cd)pyrene, dibenzo(a,h)anthracene, and benzo(g,h,i)perylene), an organophosphate compound (tributylphosphate), and an OC (4,4′-DDE).

Matrices SW and GW were then column-wise appended giving the new matrix [**SW;GW**] and the same for matrices SE and SO, giving [**SE;SO**] matrix (see

Fig. 7). Only variables coinciding in both compartments (SW and GW on one side, and SE and SO on the other side) are included in the matrix concatenation. Noncoincident variables are not considered in this new arrangement of compartmental matrices. Matrix **[SW;GW]** contained 230 samples in total (138 SW samples and 92 GW samples) and seven variables were measured in every sample (coinciding with the ones measured in GW). Matrix **[SE;SO]** contained 105 samples in total (69 SE samples and 36 soil samples), and fifteen variables were measured in every sample (coinciding with the ones measured in soils).

4.2 Results and Discussion

This section is divided in four parts describing the results obtained by the application of MCR-ALS to the analysis of the different environmental data matrices corresponding to various compartments: Sect. 4.2.1, for SW (SW_1 and SW_2 augmented matrices); Sect. 4.2.2, for surface and GW (**[SW;GW]** augmented matrix); Sect. 4.2.3, for SE (SE augmented matrix); and Sect. 4.2.4, for SE and soil (**[SE;SO]** augmented matrix).

4.2.1 Surface Water (SW1 and SW2 Data Matrices)

MCR-ALS analysis of SW data matrices was performed separately: on one hand, the augmented matrix SW_1 (69 samples × 14 variables), containing data monitored during summer campaigns and, on the other hand, the augmented matrix SW_2 (69 samples × 14 variables), corresponding to fall campaigns. For these two augmented data matrices, five different patterns of contamination were identified (some of them coinciding in both seasons) explaining around 75% of the total data variance in both cases. Obtained composition profiles (loadings) are displayed in Fig. 8. Variables (14 organic compounds) are identified with a number in the x axis. Variables with higher loadings are directly labeled in the plot for clarity. In the y axis, the relative contribution of every scaled variable to the identified contamination pattern is given. Even though the study was made for the whole river basin in both seasons, a division of the area into three subregions (the upper course or river source, the middle course, and the lower course or river mouth) was suggested from the obtained results. Thus, in order to simplify the environmental interpretation of data, the discussion of results is presented in Fig. 8 according to this division of the area of study. As MCR loadings do overlap (they are not orthogonal like in PCA), the sum of the variances explained by each separate component exceeds the variance explained by the whole model. An indication of the relative variance explained by each component is also given in Fig. 8. Contamination patterns have been arbitrarily ordered, without taking into account the amount of explained variance.

In the upper course, a contamination pattern associated to agricultural practices, mostly described by the contribution of terbutryn and terbuthylazine, was identified

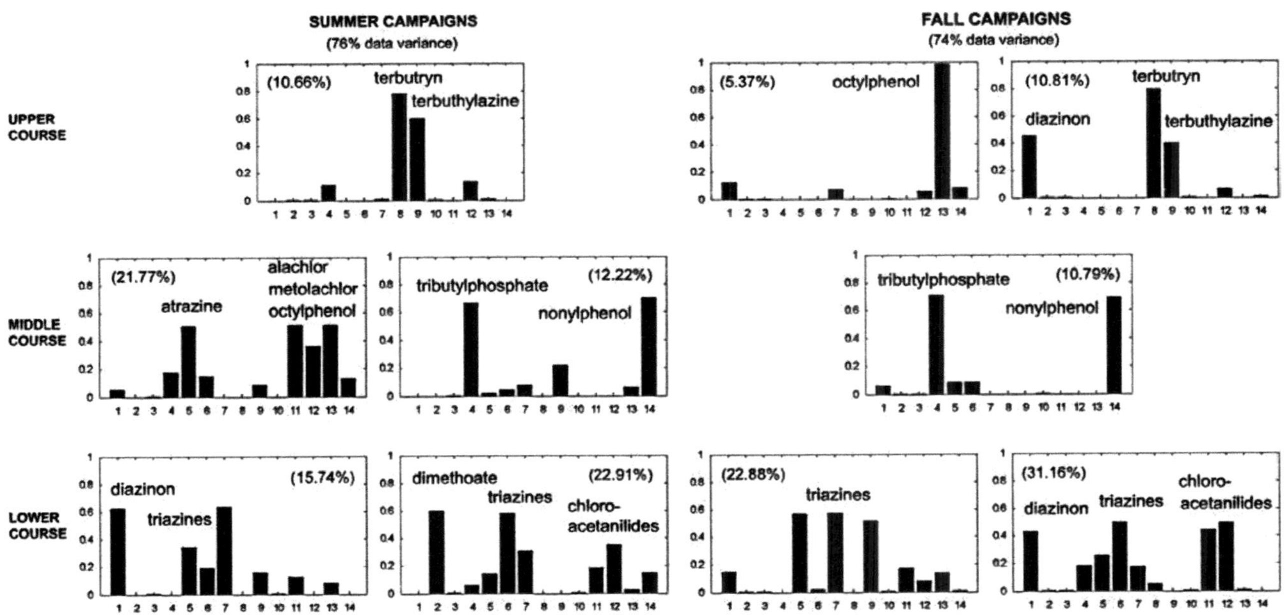

Fig. 8 Composition of the identified patterns of contamination (loadings) in SW of the Ebro River in summer and fall campaigns from 2004 to 2006. Variables identification: 1, diazinon; 2, dimethoate; 3, ethion; 4, tributylphosphate; 5, atrazine; 6, desethylatrazine; 7, simazine; 8, terbutryn; 9, terbuthylazine; 10, propanil; 11, alachlor; 12, metolachlor; 13, octylphenol; 14, nonylphenol

for both, summer and fall campaigns (see upper course patterns in Fig. 8). This contamination pattern was also identified in fall, presenting as well, a little contribution of diazinon. A second industrial and/or urban contamination pattern, mostly loaded by octylphenol, was detected in this course from the analysis of fall campaigns.

In the middle course, a contamination pattern described by the contribution of atrazine, alachlor, metolachlor, and octylphenol (this last variable presenting a lower loading than in the pattern identified in the upper course, mostly loaded only by nonylphenol), was obtained from the analysis of summer campaigns (see middle course patterns in Fig. 8). This component was associated to agricultural practices taking place in the central region. In addition, another contamination pattern loaded by tributylphosphate and nonylphenol, which are compounds often related to industry and urbanization, was identified in both, summer and fall campaigns.

In the lower course, four agricultural contamination patterns were resolved, two of them in summer and two in fall. They were all characterized by the presence of triazines in their chemical composition (see lower course patterns in Fig. 8). Variables diazinon and the group of chloroacetanilides also contributed to define the agricultural contamination pattern resolved in the lower course in both seasons. However, while in summer campaigns diazinon appeared in one of the identified patterns and chloroacetanilides in the other, both compounds appeared together in the same contamination pattern in fall campaigns. Neither industrial nor urban contamination was detected in SW samples analyzed in the lower course of the Ebro River basin.

As it has been already studied [1, 26], triazines resulted to be the most intensively applied pesticides over the Ebro River basin, and also the most ubiquitously found. They are mainly used for fruit bearings, vineyards, and olive trees. Normally, the period of pesticides application in the Ebro basin corresponds to the months of May to September [1]. A contamination pattern loaded by triazines together with alachlor and metolachlor was identified in summer in the central region of the basin, in agreement with their application over cereals (wheat) and industrial crops (sun-flower). However, the highest contribution of the triazines pattern was detected for sample T16, around Lleida (a well-known agricultural area), and this high level was maintained until reaching the Ebro delta. Contamination patterns loaded by terbutryn and terbuthylazine were resolved in the upper part of the basin. These compounds are frequently applied over vineyards [1]. As their period of application can vary depending on the climate, they were mostly detected in summer or in fall, depending on the year of sampling and the different subregions of the basin. In the upper course, a contamination pattern related to industrial and urban activities was resolved as well, which was mostly loaded by APs. These compounds are released in the environment by direct urban or industrial input or via sewage treatment plants effluents [27]. The same type of contamination due to industry and urbanization was also resolved in some locations in the middle course of the river, mainly around Zaragoza, a highly populated city with a considerable industrial activity.

4.2.2 Surface and Groundwater Simultaneous Analysis ([SW;GW] Data Matrix)

Two main patterns of contamination were resolved in the MCR-ALS analysis of the [**SW**;**GW**] data matrix (230 samples × 7 variables). Composition profiles (loadings) of the resolved components are shown in Fig. 9 (plots on the left). Variables are identified in the x axis. In the y axis, the relative contribution of every scaled variable to the identified contamination pattern is given. Temporal and spatial sample distribution profiles of contamination patterns (scores) are displayed in Fig. 9 (plots on the right). Samples are represented in the x axis for the two compartments, GW and SW, successively ordered from the first to the sixth campaign and, within each campaign, from North-West to South-East. First, third and fifth campaigns were sampled in summer while second, fourth and sixth were sampled in fall. In the y axis, the contribution of every resolved contamination pattern (described in the loadings) to each of the analyzed samples is displayed. A contamination pattern attributed to agricultural practices was clearly distinguished from another associated to an urban and/or industrial origin. The agricultural contamination pattern was mainly described by the contribution of triazines (atrazine, desethylatrazine, simazine, and terbuthylazine). This contamination was generally found at higher levels in GW than in SW (see the upper score plot in Fig. 9). Triazines have been classified as lixiviable pesticides according to the GW Ubiquity Score (GUS) index [28], with a medium mobility and persistence which make them able to pollute both surface and GW if there is rain after their application on crops or if irrigation networks exist in the area. Thus, soil becomes a potential source for GW pollution. As displayed in the upper score plots in Fig. 9, the agricultural contamination of triazines presented different tendencies in their contribution to samples depending on the environmental compartment being analyzed. While a rather constant temporal behavior was detected in GW during all the analyzed sampling campaigns, the tendency for SW was different. Contamination by triazines in SW was always higher in summer than in fall campaigns. This evidenced that triazines were applied over crops in the Ebro River basin during the months of May and June (assuming a break through time of a few weeks or months). Moreover, SW presents a more sensitive response to pesticides application than GW. Levels of triazines detected in SW were higher during summer campaigns (after they were applied), in contrast to their detection in fall campaigns, when levels appeared to be lower. The highest contribution of triazines contamination in SW was located downstream from the central part of the river basin. Besides, as SW flows faster than GW (which travels only a few centimeters per year), contaminants are more rapidly transported. In GW, levels of triazines contamination were quite constant during the whole sampling period, not indicating a punctual but an accumulative contamination effect. A contamination pattern coming from industrial and urban activities was resolved, which was mostly loaded by tributylphosphate, octylphenol, and nonylphenol compounds (see the lower loading plot in Fig. 9). These substances are normally linked to industrial and household applications, being, most of the times, directly discharged into the river. As their

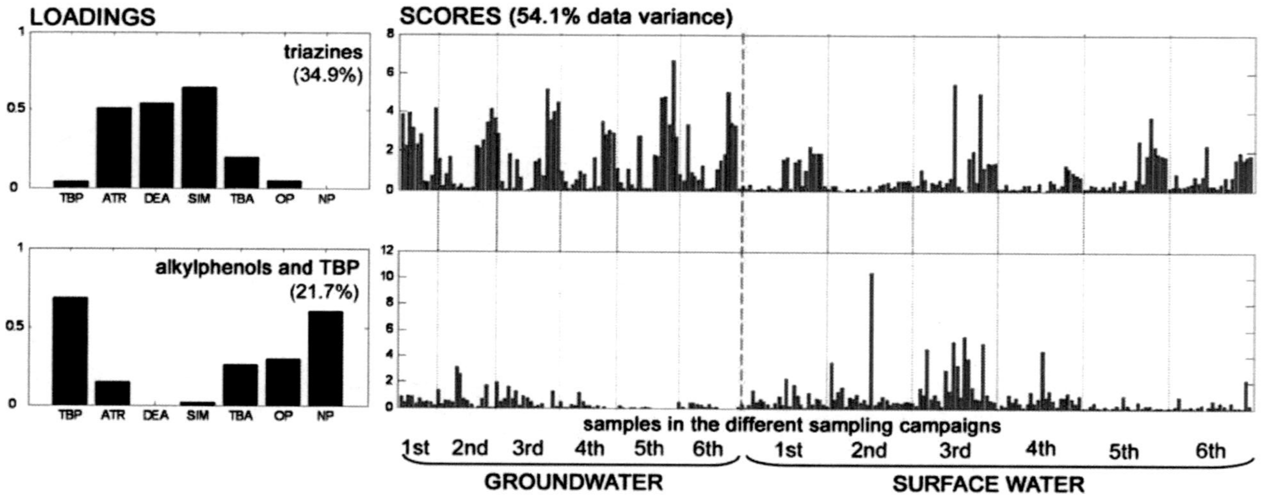

Fig. 9 Composition of the identified patterns of contamination (loadings) in surface and groundwater of the Ebro River basin and patterns contribution to the analyzed samples (scores) from year 2004 to 2006. Samples ordered for both compartments from first to sixth sampling campaigns and, for each campaign, from NW to SE. Campaigns 1, 3, and 5 sampled in summer and 2, 4, and 6 sampled in fall

octanol–water partition coefficient (K_{ow}, indicating the ratio of the concentration of a chemical in octanol and in water at equilibrium) is high, they present a low capacity to be dissolved in water and therefore, they are difficult to lixiviate. In contrast to agricultural contamination, the contamination originated by urban and industrial practices was found at slightly higher levels in SW than in GW (see the lower score plot in Fig. 9). Its levels were lower in year 2006 for both of the analyzed compartments. No seasonal tendency was identified for this pollution pattern, neither in surface nor in GW.

4.2.3 Sediment (SE Matrix)

Three main patterns of contamination were resolved by MCR-ALS analysis of the augmented SE data matrix (69 samples × 27 variables). Figure 10 shows the geographical representation of results over the map of the Ebro River basin. In this figure, the chemical composition (loadings) of the contamination patterns is given in bars, and the importance of the contribution of each pattern on every sample (scores) is represented by dots presenting a different color and size according to their relative importance. One of the resolved patterns of contamination was loaded by the contribution of most of the PAHs analyzed in the SE compartment. As no trilinearity constraint was applied over this component during the MCR-ALS analysis, the spatial distribution of PAHs contamination resulted to be different for the three analyzed campaigns. It is for this reason that three different plots are given in Fig. 10 to display PAHs spatial distribution (from 2004 to 2006). A contamination pattern loaded by the group of OCs was resolved as well, which presented an important contribution of hexachlorobenzene, and a last resolved pattern of contamination was characterized by a high contribution of APs and, in a less important amount, by those PAHs having a higher number of rings. For the display of these two last contamination patterns, only one plot was necessary (see Fig. 10), since the trilinearity constraint was applied over them during the MCR-ALS analysis. This constraint forces the components to adopt the same distribution profile (the same shape of the scores plot) during the three different analyzed campaigns. The selection of the components over which to apply the trilinearity constraint was performed allowing for the obtaining of a good model fit which, at the same time, allowed a good environmental data interpretation.

PAHs are widespread environmental contaminants resulting from combustion, discharge of fossil fuels, and automobile exhausts [29]. As they are hydrophobic substances, they are strongly adsorbed to the organic fraction of SEs and soils. A different spatial distribution of PAHs was obtained for each of the three analyzed years. However, the upper course of the Ebro River was the most affected area by this contamination during the whole period of study. In Fig. 10, larger dots represent higher contributions of this PAHs contamination pattern than smaller dots. In the year 2004 (upper map on the left of Fig. 10), samples R0 (the closest to the river source) and T8 (an industrial place located in Navarra) were the most affected sites by PAHs contamination. As a result of its location, R0 was not

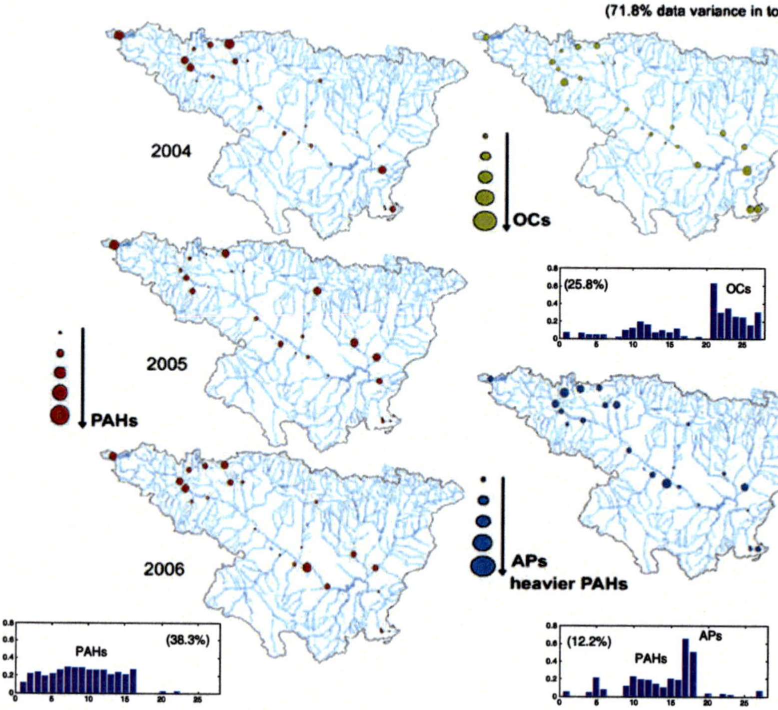

Fig. 10 Composition and spatial distribution of the main patterns of contamination identified in sediment of the Ebro River basin from year 2004 to 2006. Different temporal distribution of the PAHs pattern of contamination over the territory and constant distribution in time of the APs and heavier PAHs as well as the OCs pattern. *Big circles* representing higher levels of pattern contribution than small circles. Variables identification: 1, naphthalene; 2, acenaphtylene; 3, acenapthene; 4, fluorene; 5, phenanthrene; 6, anthracene; 7, fluoranthene; 8, pyrene; 9, benzo(a) anthracene; 10, chrysene; 11, benzo(b)fluoranthene; 12, benzo(k)fluoranthene; 13, benzo(a)pyrene; 14, indeno(1,2,3-cd)pyrene; 15, dibenzo(a,h)anthracene; 16, benzo(g,h,i)perylene; 17, octylphenol; 18, nonylphenol; 19, tributylphosphate; 20, α-HCH; 21, HCB; 22, 2,4-DDE; 23, 4,4-DDE; 24, 2,4-DDD; 25, 4,4-DDD; 26, 2,4-DDT; 27, 4,4-DDT

expected to be a polluted site but a rather clean site. Apart from these locations, other sites in the middle course (close to the cities of Zaragoza and Lleida), were also identified as specially affected by this type of contamination in years 2005 and 2006 (central and lower maps on the left of Fig. 10). As ratios obtained from the contribution of different PAHs can help to recognize the origin of contamination [29], the ratios phenanthrene/anthracene (Phe/Ant) and fluoranthene/pyrene (Flu/Pyr) in the identified pattern of PAHs contamination were calculated with the aim to deduce their possible source in the Ebro River basin. In this case, the loadings obtained in the MCR-ALS resolution of the PAHs pattern of contamination were used instead of raw concentrations to calculate the ratio. Since data were scaled previously to perform the analysis, loadings were then unscaled to eliminate the effect of the scaling transformation on variable proportions (applying the inverse operation used in scaling, that is, multiplying every scaled variable concentration by the standard deviation of the values of this variable without scaling; see more details about this operation in Terrado et al. [30]). According to Benlahcen et al. [29], ratios of Phe/Ant under 10 and ratios of Flu/Pyr over 1 are indicators of pyrolitic origin (combustion processes) in contrast to petrogenic origin. Results obtained in this way in the present study suggest that PAHs contamination identified in the SE compartment was mostly of pyrolitic origin.

The distribution of the OCs contamination pattern is presented in the upper map on the right of Fig. 10. These substances were used as pesticides in the past, but their application is banned at present. OCs were specially detected in the lower course of the Ebro River basin. Sample R17 in Flix, Tarragona, was the one presenting a higher contribution of this contamination pattern, since this area has a historically well-known contamination problem caused by OCs produced as subproducts of the local industry. Samples R19 and R20, located downstream of R17 and very close to the Ebro River delta, presented a high OCs contamination as well.

The APs contamination pattern, which is related to industry and urbanization, was widely distributed over the whole river basin (lower map on the right of Fig. 10). The highest contribution of this contamination was mainly detected close to the big cities such as Zaragoza, Lleida, and Vitoria, among others, where the main sources for this type of contamination (urbanization, industry, and wastewater treatment plants) coexist.

4.2.4 Sediment and Soil Simultaneous Analysis ([SE;SO] Data Matrix)

Three main patterns of contamination were resolved by MCR-ALS analysis of [SE;SO] data matrix (105 samples $\times$ 15 variables). Composition profiles (loadings) of the resolved components are shown in Fig. 11 (plots on the left). Variables are identified with a number in the x axis. In the y axis, the relative contribution of every scaled variable to the identified contamination pattern is given. Temporal and spatial sample distribution profiles of the contamination patterns (scores) are represented in Fig. 11 (plots on the right). In the x axis, samples are identified for the two compartments, SE and SO, successively ordered from first to third campaign and, within each campaign, form North-West to South-East. The y axis displays the contribution of every resolved contamination pattern to samples.

Two of the resolved contamination patterns were mostly loaded by PAHs. One of them was loaded by PAHs of 3–5 rings (the majority of analyzed PAHs) while the other had contribution of only the heaviest ones (5–6 rings), presenting hardly any contribution of PAHs with less than five rings. The third resolved contamination pattern in SEs and soils was mostly loaded by tributylphosphate (TBP), 4,4-DDE (substances which did not appear in the previous identified patterns), and some PAHs with 4–5 rings.

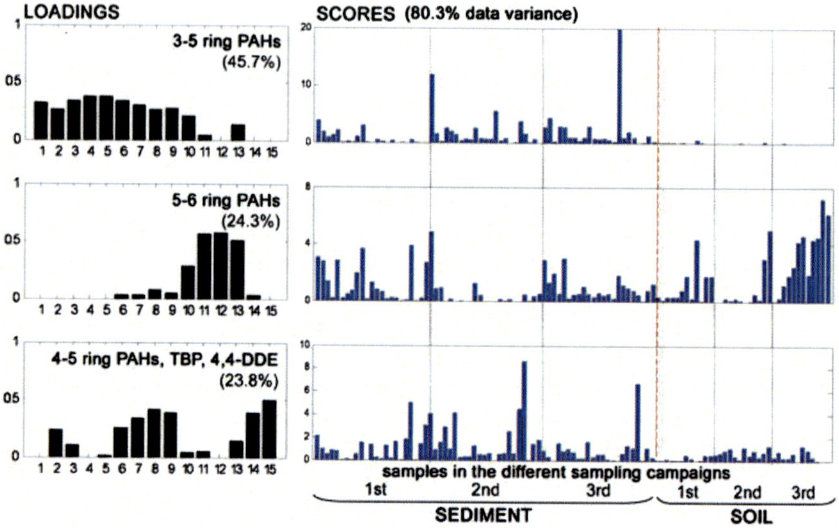

Fig. 11 Composition of the identified patterns of contamination (loadings) in sediment and soil of the Ebro River basin and patterns contribution to the analyzed samples (scores) in fall from year 2004 to 2006. Samples ordered for both compartments from first to third sampling campaigns and, for each campaign, from NW to SE. Variables identification: 1, acenaphtylene; 2, phenanthrene; 3, anthracene; 4, fluoranthene; 5, pyrene; 6, benzo(a)anthracene; 7, chrysene; 8, benzo(b)fluoranthene; 9, benzo(k)fluoranthene; 10, benzo(a)pyrene; 11, indeno(1,2,3-cd)pyrene; 12, dibenzo (a,h)anthracene; 13, benzo(g,h,i)perylene; 14, tributylphosphate; 15, 4,4-DDE

The pattern of contamination described by lighter PAHs presented a higher contribution over SE samples than over soil samples (see score plots in Fig. 11). Even though from observing Fig. 11, this contamination pattern seemed not to present any contribution in soil, it was only an effect caused by the simultaneous scaling of both compartments. The contribution of PAHs contamination was much more reduced in soil samples (but also detected). Ratios of different PAHs suggested, as well, a pyrolitic instead of a petrogenic origin for this contamination pattern. As it was already found in the SE compartment, samples R14 (close to Zaragoza) and R0 (close to the river source) presented an important contribution of the PAHs contamination pattern. No clear temporal tendency was observed for the lighter PAHs in SEs (see the upper plot on the right of Fig. 11). Soils presented higher levels of the contamination pattern defined by heavier PAHs (especially in the third sampling campaign). Samples GA1 and GA2 (close to Zaragoza), and G14 (close to Navarra, in the upper course) were the most affected ones. From previous studies [31], a higher contribution of heavier PAHs indicates a local pollution in contrast to the pollution originated by atmospheric transport, occurring mostly for the lightest PAHs. Soil samples in this study were taken from agricultural soils, characterized by intense vineyard and corn productions. As discussed in Hildebrandt et al. [32], burning of plants is a common practice in Spain to eliminate plant residues and it is also used as fertilizer, being a potential source of PAHs in agricultural soils. Nevertheless, the omission of this practice results in soil impoverishment. TBP and DDE contribution (especially important in the third identified pattern of contamination) was mainly detected in the lower course of the river, at higher levels in SEs than in soils. DDE is the principal metabolite of DDT, which was used in Spain during the 80s as insecticide, presenting a large persistence in the environment. Again, as in the SE compartment, R17 was identified as the sample with the highest levels of this type of organic contamination.

5 Chemometric Analysis of Pesticides Pollution in the Ebro River Delta During the Rice-Growing Season

The Ebro delta (Catalunya, NE Spain) is a geographical area of 20,600 ha. mainly used for rice cultivation but which also contains other crops such as orchard and fruit trees (at the sides of the river since they need sweet water to subsist). It contains a network of irrigation and drainage channels devoted to rice farming. Through the irrigation channels, water arrives to the fields and it is collected afterwards for the drainage channels. The whole hydrological cycle of the delta is man controlled by means of upstream barrages and water canalization. Two main channels (see Fig. 12), one on each side of the river, bring the water from Xerta, 25 km upstream, to the rice cultivation system of the delta. This area receives large amounts of pesticides, especially during the main growing season of rice (from May to August).

The multivariate method MCR-ALS has been used to analyse data in order to identify the main sources of organic pollution affecting the Ebro River delta. Subsequently, an interpolation procedure has been also applied to obtain distribution maps from the punctual resolved data (corresponding to the score values obtained from MCR-ALS).

5.1 Data Description and Arrangement

Four samplings were carried out monthly from May to August 2005, to include the period in which more pesticides were used in the region of the Ebro River delta. The concentration of 17 different pesticides was determined in every sample (see Table 3). Variables cyanazine and malathion were removed from the original data matrix before further analysis since very few samples contained concentrations above the detection limit. Only those sampling sites coinciding in the four samplings were selected for analysis. In this way, data tables had identical dimensions, variables, and sampling sites coinciding every month, and leading to a three-way data structure [7]. Experimental details regarding analytical determinations and

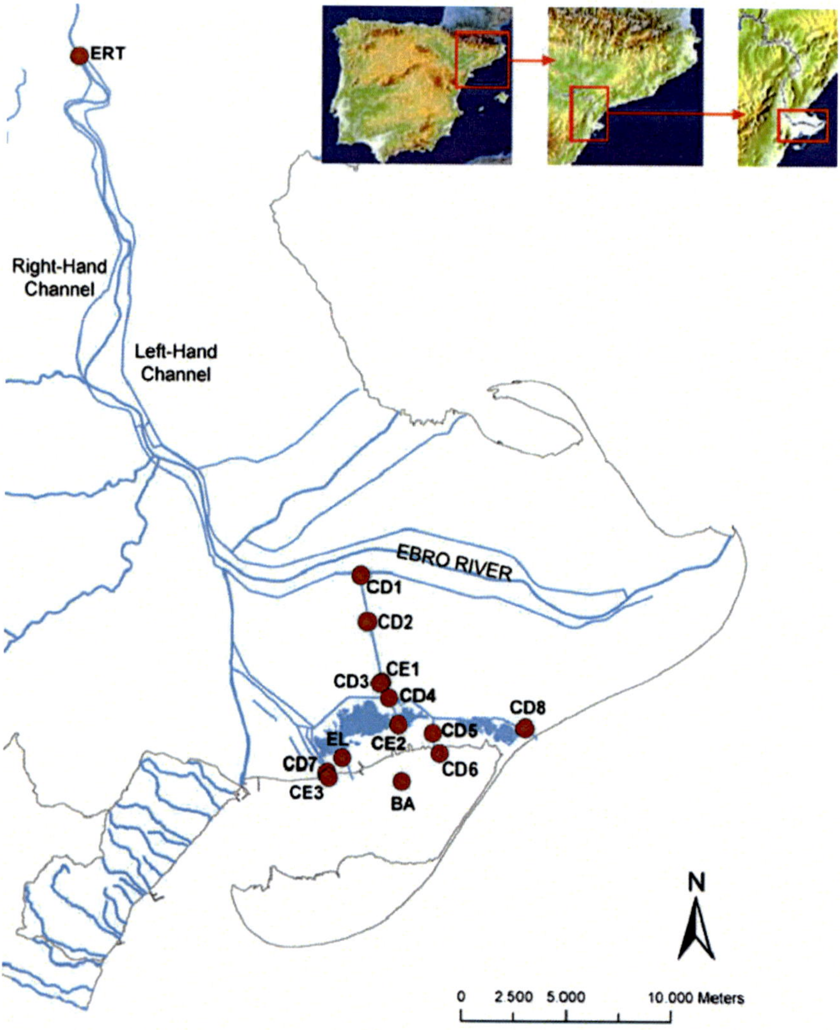

Fig. 12 Location and identification of sampling sites in the irrigation and drainage network of the Ebro delta. Sampling sites identification: ERT, Ebro river at Tivenys; CD1, Drainage channel 1; CD2, Drainage channel 2; CD3, Drainage channel 3; CD4, Drainage channel 4; CD5, Sèquia de l'Ala-Before Gates; CD6, Sèquia de l'Ala-After Gates; CD7, Sèquia de Campredó; CD8, Sèquia de Baladres; CE1, Irrigation channel 1; CE2, Irrigation channel 2; CE3, Irrigation channel 3; EL, Llac de l'Encanyissada; BA, Badia d'Alfacs

sampling techniques can be found in Kuster et al. (2008) [33]. The 11 sampling sites were located in the right hemi-delta of the Ebro River, the largest river in Spain. They were distributed at different locations within a network of irrigation and drainage channels devoted to carrying and collecting water for rice farming.

Table 3 Identification of pesticides analyzed in the study

Pesticides		Group
1.	MCPA	Acidic herbicides
2.	Mecoprop	
3.	2,4-D	
4.	Bentazone	
5.	Simazine	Triazine
6.	Isoproturon	Phenylureas
7.	Chlortoluron	
8.	Atrazine	Triazine
9.	Diuron	Phenylurea
10.	Propanil	Anilide
11.	Molinate	Thiocarbamate
12.	Alachlor	Chloroacetanilides
13.	Metolachlor	
14.	Fenitrothion	Organophosphates
15.	Diazinon	

Water used for field irrigation is then transported to littoral lagoons, like in the case of the Badia d'Alfacs, where important mussel farms are located (see Fig. 12).

Data collected during the four samplings were arranged in four experimental data tables or matrices ($\mathbf{D_1}$, $\mathbf{D_2}$, $\mathbf{D_3}$, and $\mathbf{D_4}$, one per month) of 11 rows (sampling sites) and 15 columns (concentrations of the different pesticides). The four data matrices were combined, keeping their columns (variables) common, and giving an augmented data matrix $\mathbf{D_{aug}}$ of dimension 44 × 15.

6 Results and Discussion

Three components were identified from the analysis of the scaled $\mathbf{D_{aug}}$ (44 × 15). The selection of two and four components was discarded since the total explained data variance was much lower with two components, and no additional environmentally relevant information was added with four components [30].

The three components extracted from MCR-ALS application to $\mathbf{D_{aug}}$ are displayed in Fig. 13. The first two components included a trilinearity constraint while the third component did not. Trilinearity was imposed on the first two because a repeated pattern in the scores (matrix **X**) was observed. In other words, the same spatial source distribution was identified during the four analyzed months. The third component did not show this repeated pattern of spatial distribution. Only a slight decrease of explained variance was observed when the trilinearity constraint was applied (R^2 turned from 77.0 to 75.1%). However, the interpretation of the results was preferred in this case because it was significantly simplified for the trilinear components which are described by a single spatial distribution profile during the 4 months of the study. On the left side of Fig. 13, loading profiles (matrix $\mathbf{Y^T}$) are shown. Variables are displayed on the *x* axis while the *y* axis indicates their

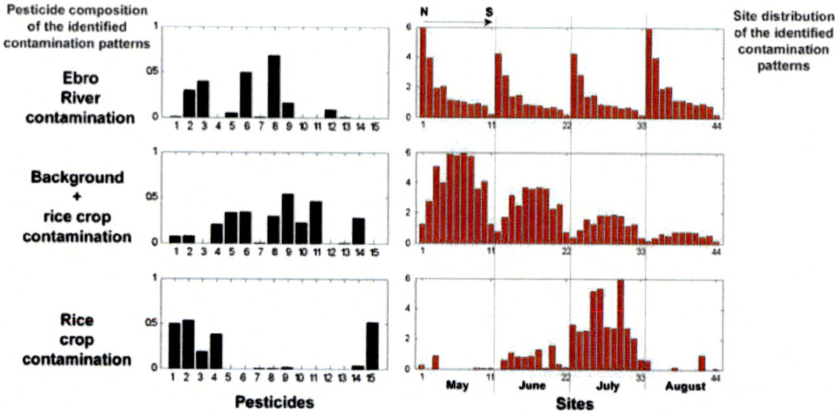

Fig. 13 Contamination patterns identified (MCR-ALS resolved loading profiles) in the Ebro River delta from May to August 2005. On the left: loadings in the second mode (normalized for variables to unit norm) describing the composition of the contamination patterns. Variable identification in Table 1. On the right: loadings in the first and third modes (mixed) describing spatial and temporal distribution of the contamination patterns. Sampling sites ordered from North to South, for the four analyzed months displayed consecutively (May: 1–11; June: 12–22; July: 23–33; August: 34–44)

contribution to the identified component. The three resolved components were identified as follows: (a) a first component describing a contamination pattern coming from the Ebro River, with larger contributions from pesticides not typically applied over the rice crop (mainly mecoprop, 2,4-D, simazine, isoproturon, atrazine, diuron, and alachlor); (b) a second component describing a contamination pattern with a larger contribution of pesticides typically applied over the rice crop, mixed with other pesticides with a more generalized use in agriculture; and (c) a third component describing a contamination pattern with contribution of diazinon and the group of acidic herbicides (MCPA, mecoprop, 2,4-D, and bentazone), some of which are specifically applied over rice fields.

On the right side of Fig. 13, the score profiles (matrix **X**) resolved by MCR-ALS are shown. Samples in the score plots are displayed on the x axis ordered from North to South for the four consecutively analyzed months. The y axis indicates the contribution of the identified component (on the left) to each one of the samples. Since trilinearity was applied to the first and second components, the same pattern of spatial distribution is presented in the score profiles of these two components over the 4 months, only changing their relative scale. These differences in their scale were small for the first component, which is interpreted to have a rather constant behavior over the investigated period of time (as well as over space). On the contrary, the second resolved component presented a clear decreasing scale trend from May to August. Finally, since the trilinearity constraint was not applied to the third component, its spatial distribution appeared to be significantly different over time. The highest levels of this third contamination pattern were attained in July, followed by June, whereas in May and August, the contamination was scarce.

The spatial distribution of the resolved components (matrix **X**) was integrated in the geographical information system ArcGIS (v.9.0, 2005, ESRI, Redlands, CA, USA) and displayed by means of geostatistical techniques [23] in Fig. 14. Note that the spatial distribution displayed by the geographical information system in this work refers to the distribution (scores) of the main underlying multivariate (multicomponent) patterns and sources of contamination resolved by MCR-ALS, which are not directly measurable and which are also of high environmental relevance.

The resolved pattern of contamination coming from the Ebro River main flow, decreased significantly as water flowed down through the delta channels (from North to South), indicating that the origin of the contamination was located outside the deltaic area. Since this pattern was loaded by pesticides with a generalized use, it was not associated with any specific agricultural rice crop practice. Another identified pattern of pesticide contamination was attributed to background pollution together with pollution originating from agricultural rice crop practices. It presented a low contribution close to the Ebro River, and increased significantly as water flowed through the rice fields. The highest levels of this contamination pattern were detected in May (see graph of scores in Fig. 13), which coincides with the application of pesticides during the earlier stages of the rice cycle. The use of these pesticides probably ceased by the end of May or, if their use continued past this point in time, they were likely applied at lower levels. The third identified contamination pattern, which was attributed to more specific agricultural practices over the delta rice fields, had its largest contribution to samples during the month of July (month displayed in Fig. 14 for this third pattern). The highest levels were detected in the drainage channels CD4 and CD7. The effect of water repumping was more evident during this month, when the amount of water in the irrigation channels decreased and water from the drainage channels (already polluted by pesticides) was used again to irrigate the rice fields.

7 Conclusions

The potential of the chemometric methods PCA and MCR-ALS for the analysis of databases obtained from environmental monitoring studies has been shown.

Environmental quality assessment of the Ebro River basin from 1996 to 2003 has been performed looking for different contamination sources and trends in data. Three different geographical areas showing different tendencies have been roughly delimited according to the distribution of contamination patterns identified by PCA. A correlation between land uses and characteristics of the contaminants affecting river water, SE, and biota has been attempted. Area 1, in the upper course of the river (sampling sites R1, T3, R4, and T9) is characterized by high inputs of contamination sources from industry and urbanization: these sources are described by heavy metals (Zn, Cu, Cr, Pb, Cd, and Hg, which sometimes can also have a natural origin), PAHs, HCHs, TCBs, and eutrophic conditions (high content of NH_4^+, MES, organic matter, PO_4^{3-}, etc.). Area 2 covers the middle course of the

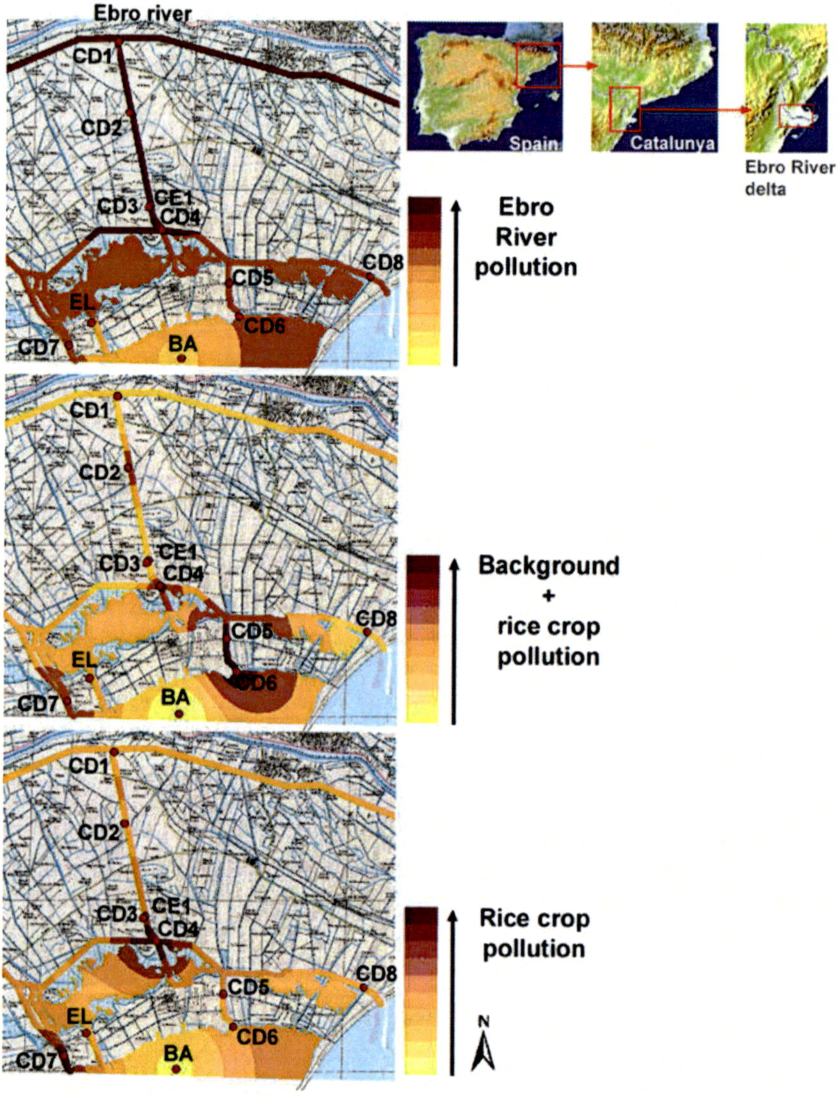

Fig. 14 Spatial distribution of the identified contamination patterns (MCR-ALS resolved score profiles) in the Ebro River delta from May to August 2005 (distribution of the third pattern was only displayed for the month of July, since the trilinearity constraint was not applied to this profile). Identification of sampling sites: CD1 to CD4, drainage channels; CD5, Sèquia de l'Ala-before Gates; CD6, Sèquia de l'Ala-after Gates; CD7, Sèquia de Campredó; CD8, Sèquia de Baladres; CE1, irrigation channel; EL, Llac de l'Encanyissada; BA, Badia d'Alfacs. Darker areas describe larger contribution of the identified contamination patterns while lighter areas describe lower contribution

Ebro River (sampling sites T10, T11, T13, and R14). The high accumulation of nutrients in water, like Ca, Na, Mg, and K as well as large values of alkalinity, conductivity, NO_3^-, $SO_4^=$, etc., show that this region is affected by an important salinization effect, identifying lower inputs from contamination sources derived in Area 1. Soils and underlying geology in the middle course present a high salt content which, together with intensive irrigation techniques from agricultural activities, has worsened the problem. Area 3 is located in the Ebro lower course, from T15 in Monzón to the mouth of the river (sampling sites T15, T16, R17, and R18). In this area, organic and heavy metal contamination sources come mainly from industrial activities although agriculture (rice farming) is also responsible for a high release of pesticides into the environment. Contamination sources of metals such as Hg, Cd, Zn, and As, as well as an organic source of DDTs, HCB, and HCBu, have a high contribution in this lower part of the river.

Agricultural practices have been identified as the main responsible source of ongoing surface and GW pollution in the Ebro River basin. A main contamination source of triazines was detected for the whole period of study, from 2004 to 2006, since they are intensively used and distributed all over the territory of the basin. Moreover, the burning of weed and plant residues, which is a common practice in the area, constitutes a second different contamination source loaded by PAHs, mostly affecting SE and soil compartments. In general terms, SE and GW reflect the accumulated historical contamination, providing a good fingerprint of past activities in the area, while SW and soil reflect the contamination from more recent practices. Contamination by APs coming from industry and urbanization or via sewage treatment plants was also an important source of pollution affecting mainly SE and aquatic compartments but not presenting a significant impact over soil. On the other hand, a specific contamination related to agricultural practices applied over the rice crop was identified in the Ebro River delta. Contamination sources identified in the Ebro delta presented a different composition from agricultural sources identified in the rest of the basin.

Acknowledgements This work was supported by the European Union FP6 Integrated Project AQUATERRA (GOCE 505428) and the Spanish Ministry of Education and Science (CTQ2006-15052).

References

1. Claver A, Ormad P, Rodríguez L, Ovelleiro JL (2006) Chemosphere 64:1437
2. Mañosa S, Mateo R, Guitart R (2001) Environ Monit Assess 71:187
3. Terrado M, Kuster M, Raldua D, Lopez de Alda M, Barcelo D, Tauler R (2007) Anal Bioanal Chem 387(4):1479
4. Lacorte S, Raldúa D, Martínez E, Navarro A, Díez S, Bayona J, Barceló D (2006) Environ Pollut 40:471
5. CHE, www.chebro.es

6. Barth JAC, Grathwohl P, Fowler HJ, Bellin A, Gerzabek MH, Lair GJ, Barceló D, Petrovic M, Navarro A, Négrel P, Petelet-Giraud E, Darmendrail D, Rijnaarts H, Langenhoff A, de Weert J, Slob A, Frank E, Gutierrez A, Kretzschmar R, Gocht T, Steidle D, Garrido F, Jones KC, Meijer S, Moeckel C, Marsman A, Klaver G, Vogel T, Bürger C, Kolditz O, Broers HP, Baran N, Joziasse J, von Tümpling W, Van Gaans P, Merly C, Chapman P, Brouyere S, Batlle Aguilar J, Orban P, Tas N, Smidt H (2009) Agron Sustain Dev 29:161
7. Smilde A, Bro R, Geladi P (2004) Multi-way analysis with applications in the chemical sciences. Wiley, Chichester
8. Burrough PA, McDonnell RA (1998) Principles of geographical information systems. Oxford University Press, Oxford
9. Sánchez E, Kowalski BR (1988) J Chemom 2:247
10. Sánchez E, Kowalski BR (1988) J Chemom 2:265
11. Farnham IM, Singh AK, Stetzenbach KJ, Johannesson KH (2002) Chemom Intell Lab Syst 60:265
12. Jolliffe IT (2002) Principal component analysis. Springer, New York
13. Tauler R (1995) Chemom Intell Lab Syst 30:133
14. Paatero P, Tapper U (1994) Environmetrics 5:111
15. Bro R (1997) Chemom Intell Lab Syst 38:149
16. Wold S, Esbensen K, Geladi P (1987) Chemom Intell Lab Syst 2:37
17. Tauler R, Smilde A, Kowalski BR (1995) J Chemom 9:31
18. de Juan A, Tauler R (2003) Anal Chim Acta 500:210
19. Tauler R, Marqués I, Casassas E (1998) J Chemom 12:55
20. Tauler R, Maeder M, De Juan A (2009) Multiset data analysis: extended multivariate curve resolution. In: Brown S, Tauler R, Walczak R (eds) Comprehensive chemometrics, vol 2. Elsevier, Oxford, pp 473–505
21. Sánchez E, Kowalski BR (1990) J Chemom 4:29
22. Cressie N (1993) Statistics for spatial data. Wiley, New Jersey
23. Goovaerts P (1997) Geostatistics for natural resources evaluation. Oxford University Press, New York
24. Moore J, Ramamoorthy S (1984) Heavy metals in natural waters. Applied monitoring and impact assessment. Springer, New York
25. Prats J, Val R, Armengol J, Dolz J (2007) Limnetica 26(2):293
26. Hildebrandt A, Lacorte S, Barceló D (2007) Anal Bioanal Chem 387:1459
27. Céspedes R, Lacorte S, Ginebreda A, Barceló D (2008) Environ Pollut 153:384
28. Gustafson DI (1989) Environ Toxicol Chem 8:339
29. Benlahcen KT, Chaoui A, Budzinski H, Bellocq J, Garrigues P (1997) Mar Pollut Bull 34:298
30. Terrado M, Barceló D, Tauler R (2009) Environ Sci Technol 43:5321
31. Fernández P, Grimalt JO (2003) Chimia 57:514
32. Hildebrandt A, Lacorte S, Barceló D (2009) Arch Environ Contam Toxicol 57:247
33. Kuster M, de Alda MJ López, Raldúa D, Barata C, Barceló D (2008) Talanta 75:390

Experiences and Lessons Learned on the Implementation of the Water Framework Directive in Selected European River Basins

M.S. Díaz-Cruz, R.M. Darbra, A. Ginebreda, P.D. Hansen, S. Sabater, L. Galbiati, E. Capri, C. Vale, and D. Barceló

Abstract River basins worldwide are under pressure from economic activities. In Europe, the two main factors hindering the achievement of good chemical and ecological status of European river basins are pollution, mainly coming from agriculture, and hydromorphology (e.g. for navigation, hydroelectricity and flood control). The economic activities affect the chemical and ecological status of rivers, lakes and groundwater and deplete available soil, sediments and water resources. The wide range of these activities and the eco-hydrological complexity of many river basins, both in terms of the functioning of the soil–sediment–water system and

M.S. Díaz-Cruz (✉) and A. Ginebreda
Institute of Environmental Assessment and Water Research-Spanish Council of Scientific Research (IDAEA-CSIC), Jordi Girona 18-26, 08034 Barcelona, Spain
e-mail: sdcqam@cid.csic.es

D. Barceló
Institute of Environmental Assessment and Water Research-Spanish Council of Scientific Research (IDAEA-CSIC), Jordi Girona 18-26, E-08034 Barcelona, Spain
Catalan Institute for Water Research (ICRA), Emili Grahit, 101, Edifici H2O Parc Científic i Tecnològic de la Universitat de Girona, E-17003 Girona, Spain

R.M. Darbra
Polytechnic University of Catalonia, Av. Diagonal, 647, 08034 Barcelona, Spain

P.D. Hansen
Department of Ecotoxicology, Technische Universitaet Berlin, Keplerstrasse 4-6, D-10589 Berlin, Germany

S. Sabater
Catalan Institute for Water Research (ICRA), Emili Grahit, 101, Edifici H2O Parc Científic i Tecnològic de la Universitat de Girona, E-17003 Girona, Spain

L. Galbiati
Agència Catalana de l'Aigua (ACA), Provença 204-208, 08036 Barcelona, Spain

E. Capri
Università Cattolica del Sacro Cuore, Via Emilia Parmense, 84, 29100 Piacenza, Italy

C. Vale
Instituto de Investigaçao das Pescas e do Mar (IPIMAR), Avenida de Brasilia, 1400 Lisbon, Portugal

D. Barceló and M. Petrovic (eds.), *The Ebro River Basin*,
Hdb Env Chem (2011) 13: 373–424, DOI 10.1007/698_2010_80,

of the links between water quantity, quality and economic activities, make the integrated management of river basins extremely complex. Approaches to this management across Europe have been dramatically impacted by the introduction of the European Water Framework Directive (WFD). The WFD promotes the integrated management of water resources based on the natural geographical and hydrological unit of the river basin rather than administrative or political boundaries. In this chapter, experiences and recommendations on the implementation of the WFD in a number of representative river basins across Europe are described and regarded as a support on the implementation of the WFD in the Ebro river basin.

Keywords Ecosystem services, European river basins, Risk assessment, Risk management, River basin management plans, Water framework directive

Contents

1 Introduction

1.1 General Aspects of Water Management and Policy Under the WFD

The increasing demand by citizens and environmental organisations for cleaner rivers and lakes, groundwater and coastal beaches has been evident for considerable time. It has recently been reconfirmed by a representative opinion poll Euro barometer in all 25 EU countries.

This demand by citizens is one of the main reasons why the European Commission has made water protection one of the priorities of its work.

European Water Policy has undergone a thorough restructuring process, which has ultimately culminated with the approval by the European Parliament of the so-called Water Framework Directive (WFD) (Directive 2000/60/EC) [1].

The WFD introduces a new water quality concept by assuming ecological effects as a basis of control rather than focussing on agents with a potential for adverse effects. Therefore, the assessment of water quality must be defined directly in terms of "functioning and structure of ecological systems" rather than be only based on chemical contamination. In this frame, a water body is an environmental good to be protected and not a resource to be exploited, and the biological–ecological quality assumes a prevailing role. According to WFD, the water quality assessment is a site-specific procedure to characterise the environment, according to the definitions of ecological and chemical status of water bodies that are proposed. The objective is the "good status" that depends on the overall combination of potential "stressors" capable to affect environmental quality [and not the concentration of a chemical, even if for management reasons environmental quality standards (EQS) for priority chemicals are set].

This allows an integrated management of the environment, surface water and groundwater but also sediments and soil. These two last compartments are sink for many pollutants as well as an important source of water pollution.

To achieve these goals, the WFD proposes the water management by river basin – the natural geographical and hydrological unit – instead of according to administrative or political boundaries.

It is now necessary to assess the whole river–groundwater–soil–sediment system "with the objective to achieve a good ecological status" and "good chemical status" for all waters at large scale and to develop a strategy through a River Basin Management Plan (RBMP) [2].

The development of a new environmental quality objective with the aim to preserve the biologic community, both in its functional and structural features, leads to the need for more refined and sensitive approaches for assessing environmental risk. The main issues and milestones established in the WFD are indicated below (Fig. 1).

According to the proposed calendar of the WFD, after the preliminary characterisation done on previous years, the programme of measures (POM) was ready in 2009 and included in the subsequent RBMP.

1.2 Soil and River Basin Management

Soil protection in the EU is significantly lagging behind current implementation of water management.

Before 1970, the soil was generally considered as an environment with an almost infinite self-purifying capacity. Political attention to soil problems had to await

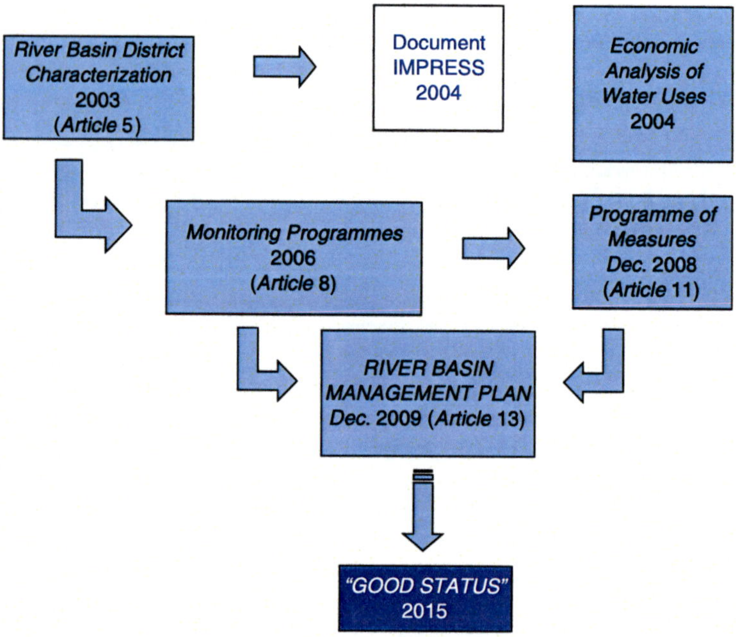

Fig. 1 WFD implementation: major issues and milestones

major incidents with contaminated sites in the early 1980s of the previous century, like Love Canal, New York State; Times Beach, Missouri; and Lekkerkerk, the Netherlands.

The Council of Europe highlighted the need for soil protection in an international perspective already in 1972. Proposals for international action suggested the following principles:

- Recognition that soil is a common heritage and non-renewable resource.
- Integration of soil protection into other environmental policies.
- Rational use of soil and careful management of soil.
- Respect for multi functionality through harmonisation of surface land use.
- The reversibility rule (impact on soil quality by man should be reversible).

It is noteworthy that most threats to soil have strong relation with the water management. Soil protection may be needed to keep surface water, ground waters and sediment in an ecologically satisfactory state.

The main functions of soil [3] within a river basin are its sink and source functions, which can be described by filtering, buffering and transformation activities between the atmosphere and the underground, thus protecting sediments, groundwater and surface water against contamination.

Contaminated land policies need to result in sustainable solutions, which will restore the usability and economic value of the land. These solutions can be characterised by the following three elements.

1.2.1 Suitability for Use

This is achieved by reducing human health risks and ecological risks as far as necessary to permit the safe (re)use of the land. It is focussed on quality requirements of the land for uses and functions.

1.2.2 Protection of the Environment

For example, preventing further spread of pollution by surface water and groundwater. Environmental protection of soils as a resource may also lead to policies favouring redevelopment of brown fields over green fields.

1.2.3 Long-Term Care

Sustainable solutions minimise the burden of aftercare. Endless pump and treat solutions or containment walls that require control and maintenance forever may be less desirable in view of the amount of aftercare required [4].

From an environmental point of view, soil and water interact too much to be managed separately. Successful implementation of the WFD has to involve land management, and sound management of soil and water resources calls for harmonisation of spatial planning with the environment. In this way land uses can be optimised with respect to soil quality and hydrological situation.

2 EU Research Projects in Relation to WFD Implementation: State of the Art

2.1 Bringing Science into Legislation and Management

River basin management is a matter of public concern, usually under the responsibility of public authorities. In such a context, the "rules of the game" are those stated in the current legislation. Therefore, legislation/regulations provide the basis for management practices. In turn, legislation is constructed from different inputs (scientific, social, economical, etc.).

In the environmental field, it can be generally established the following chain:

Scientific Results $\rightarrow$ Regulations $\rightarrow$ Public Management

Science is always "one step beyond" legislation and for that reason scientific results are thus a starting point. This constitutes a serious weakness that needs attention: How scientific results can be efficiently transferred into legislation?

How can we have an acceptable legislation in terms of consistency with the "state of the (scientific) art"? How can we minimise the gap between science and legislation? These are not simple and straightforward questions to answer.

Of course, the new oncoming legislation realises the relevance of the problem, and, for instance, the WFD imposes itself (see Article 20) to be reviewed and adapted to scientific and technical progress every 6 years. At the same time, current scientific research projects under the VI Programme strongly encourage the participation of end-users and stakeholders (see Sect. 2.2). Therefore the need for an approach from both sides (science and regulators) is recognised. However, it is still not efficient enough and new ways of bringing science into regulations and (subsequently) to management need to be developed.

2.2 *Overview of EU Relevant Projects*

The WFD requires a scientifically sound diagnosis and forecasting of environmental pollution on freshwater and marine ecosystems, and exploitation for the development of management options, preventive policies and remedial activities. As a consequence, several EU-funded projects, based on river basin-related subjects, are working to develop tools to assist European water managers to comply with the WFD. In this section, we provide a brief overview of some of the EU-funded projects that are considered most relevant regarding the WFD, namely SOCOPSE, MODELKEY, AQUATERRA, GABARDINE, ECORIVER, SWIFT and NOMIRACLE.

SOCOPSE (http:// www.socopse.se/) was an EU-funded project (2006–2009) within the sixth framework programme (FP6-2005-global-4, Topic II 3.1), with the overall objective to support the implementation process for the WFD by providing guidelines and a decision support system (DSS) for the management of priority substances (PS).

To fulfil this overall objective, the following activities were carried out:

- To conduct a material flow analysis for selected PS.
- To evaluate available and emerging measures and management options for PS.
- To develop a decision support tool for identification and selection of relevant measures on European, national and regional levels.
- To evaluate different potential measures by applying the decision support tools in case studies.
- To facilitate the development of collective action plans (i.e. RBMPs) involving all stakeholders (industries, authorities, citizens and NGOs).
- To disseminate results to stakeholders and to strongly interact with industrial organisations, research networks, authorities and NGOs.

This project focused on a set of PS (11 of the 33 identified by the WFD as PS) representing different uses and emission sources. These include industrial chemicals, pesticides and unintentionally released compounds. Most of the selected PS belong to the group Priority Hazardous Substances [tributyltin, di(2-ethylhexyl) phthalate, polybrominated diphenyl ethers, nonylphenol, atrazine, isoproturon, polycyclic aromatic hydrocarbons (PAH) including anthracene].

Another important objective of SOCOPSE was to maintain contacts with on-going research projects already involving SOCOPSE partners and networks related to the implementation of WFD and related topics. The project was performed in close cooperation with the industrial sector, the different authorities and other stakeholders.

MODELKEY (http:// www.modelkey.org/) was an EU-funded project (2005–2010) within the sixth framework programme (FP6-2005-global-4, Topic II 3.1) that aimed to deliver "models for assessing and forecasting the impact of environmental key pollutants on freshwater, and marine ecosystems and biodiversity."

The main objectives of MODELKEY were:

- To identify the key toxicants impacting marine and freshwater ecosystems on a site and basin scale as a crucial basis for scientifically sound risk assessment and decision support on risk management, remedial action strategies and preventive policies for the mitigation of harmful effects.
- To provide a better understanding of toxic impacts on aquatic ecosystems, cause–effect relationships between changes in biodiversity and the impact of environmental pollution as causative factor as well as the underlying processes. This included the assessment of sub-lethal effects in vitro and in vivo as early warning strategies and of their strength to predict potential hazards to the ecosystem.
- To assess, forecast, and mitigate the risks of key toxicants on fresh water and marine ecosystems and their biodiversity at a river basin and adjacent marine environment scale focusing on DSSs for the selection of the most efficient management options to prevent effects on biodiversity and to prioritise contamination sources and contaminated sites.

It comprises a multidisciplinary approach to develop interlinked tools for a better understanding between insufficient ecological status and the impact of environmental pollutants as causative factor. Furthermore, MODELKEY aimed at the assessment and forecasting of the risks of key pollutants on fresh water and marine ecosystems at a river basin scale. New modelling tools including generic exposure assessment models, mechanistic models of toxic effects in simplified food chains, integrated diagnostic effect models based on community patterns, predictive component effect models applying artificial neural networks and GIS-based analysis of integrated risk indexes have been developed and linked to a user-friendly DSS for the assessment and prioritisation of risks, contamination sources and contaminated sites (Fig. 2).

In order to establish a link among the exposure of freshwater and marine ecosystems to environmental toxicants and observable effects on these ecosystems, there are basically two approaches: (1) the deterministic approach that focuses on an understanding of functions, processes and mechanisms and (2) the stochastic approach that focuses on an identification of relationships by statistical means.

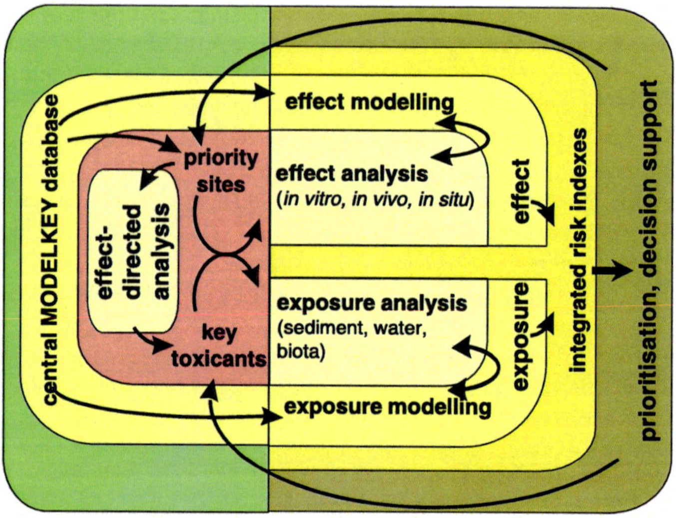

Fig. 2 Diagram of the MODELKEY approach [5]

For both approaches, MODELKEY developed innovative and integrative models and experimental tools.

AQUATERRA (http://www.eu-aquaterra.de/) was another EU-funded project (2004–2009) within the sixth framework programme (FP6-2005-global-4, Topic II 3.1). AQUATERRA aimed to provide the scientific basis for an improved river basin management through a better understanding of the river–sediment–soil–groundwater system as a whole, by integrating both natural and socio-economics aspects at different temporal and spatial scales. The AQUATERRA models integrated the key biogeochemical, climatic and hydrological processes over relevant scales in time and space. With this integrated modelling system, AQUATERRA provided the basis for improved river basin management, enhanced soil and groundwater monitoring programmes and the early identification and forecasting of impacts on water quantity and quality during this century. The project involved practitioners and end-users to elaborate operational tools for the different stakeholders (policy makers, river basin managers, regional and urban land planners, etc.). Field study areas were in the river basins of the Ebro, the Meuse, the Elbe and the Danube, as well as in the small 3-km^2 French agricultural catchments of the Brévilles spring (Fig. 3) [6–9]. In support to EU policy matters, AQUATERRA took into account socio-economic issues and their legal implementations.

Within the frame of the EU-funded project GABARDINE (2005–2008; http://www.geoservice.gr/), the identification of alternative sources of water and the feasibility, both environmental and economic, of their utilisation were explored. Alternative water sources to be artificially recharged comprised: surface water runoff, treated effluent and imported water.

Furthermore, brackish water bodies that are present in many aquifers could be used after desalination. Different test sites were selected for GABARDINE project,

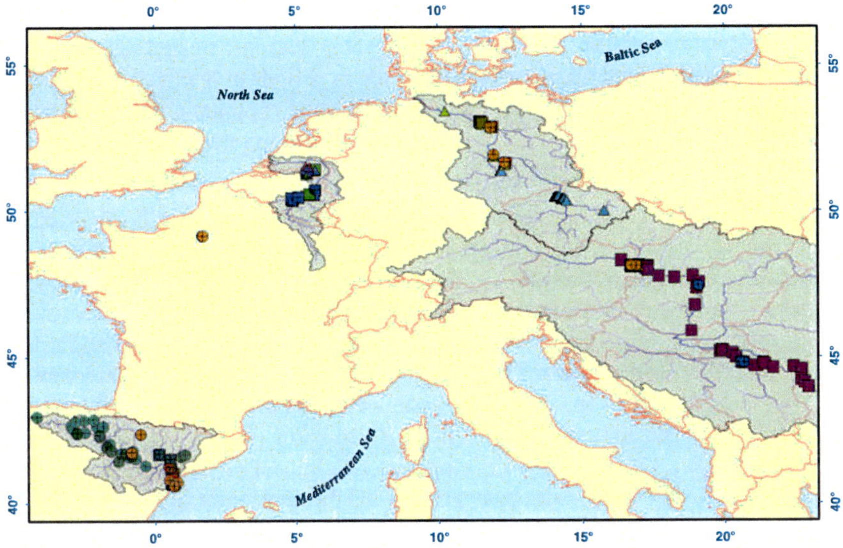

Fig. 3 Overview of the fieldwork carried out in the four AquaTerra River Basins of the Ebro, the Danube, the Elbe and the Meuse as well as the small French catchments of the Brévilles Spring. *BOKU* Universität für Bodenkultur Wien, *TNO* Netherlands Organisation for Applied Scientific Research, *TUM* Technische Universität München, *ETHZ* Eidgenössische Technische Hochschule Zürich, *UHP* Université Henri Poincaré Nancy, *VITO* Flemish Institute for Technological Research, *VUV* Masaryk Water Research Institute Prague, *UFZ* Umweltforschungszentrum Halle-Leipzig, *VU* Vrije Universiteit Amsterdam, *TUHH* Technische Universität Hamburg-Harburg, *CSIC* Consejo Superior de Investigaciones Cientificas, *ULAN* Lancaster University, *TUB* Eberhard Karls Universität Tübingen, *WUR* Wageningen University, *BGC* AquaTerra Unterprojekt BIOEGEOCHEM. Map created by David Kuntz

each representing a different aspect of the problem. In Portugal, the Campina de Faro aquifer (Algarve region) was selected for a case study to achieve groundwater quality improvement by injecting surface water.

The major objectives of GABARDINE project were as follows:

- To explore the viability of supplementing existing water resources in semi-arid areas with alternative sources of water that could be exploited in the context of an integrated water resources management approach.
- To investigate the feasibility of using aquifers as the primal facility for the large scale storage of these alternative water sources and investigate techniques for their artificial recharge and injection of the produced alternative water, including monitoring of water quality and purification by natural attenuation and filtration processes.
- To evaluate and quantify the potential impact of degrading factors, such as climate change, changes in the quality of water and salt water, on the global quality and usability of the resource, by developing tools for risk mapping, modelling and monitoring, and to propose measures for preventing or

minimising and mitigating their impact. The alternative water sources are surface water surpluses generated during rainy seasons, treated effluent, surpluses of desalinated water that are expected in periods of low water demand or high water availability (from natural resources) and exploitation of saline water bodies that could be used for adequate agricultural practices or used as raw material for low-cost desalination [10].

The ecotoxicological assessment of wastewaters as contribution to river basin management was addressed, within the framework of the EU-funded ECORIVER demonstration project (2002–2005). The objective of ECORIVER was to demonstrate the importance of using toxicological tests for waste management and to set up an appropriate technology for the control and planning of industrial and urban waste water network systems. The actions carried out included the assessment of the ecotoxicity of urban and industrial waste water, the evaluation of the impact of the wastewater channelled to the Trancão River basin (from 17 industries from different industrial sectors), the development of a model to predict the potential benefits obtained by wastewater management and the development of a proposal for the establishment of ecotoxicity criteria to be included in national legislation [11].

SWIFT-WFD project (2004–2007; www.swift-wfd.com) aimed to support the WFD by the production of quality control tools for validation purposes of screening methods, an inventory of existing screening methods currently used or under development for water monitoring, the comparison of screening test (chemical and biological) methods through laboratory-based (tank experiments) and/or field interlaboratory studies based on a selection of reference aquatic ecosystems at European scale, and with classical laboratory-based analyses to validate their results and demonstrate their equivalence (in terms of statistical comparison procedure) for parameters regulated by the WFD.

NOMIRACLE (2004–2009, http://viso.jrc.it/nomiracle/) provided support to the development and improvement of a coherent series of methodologies underpinned by mechanistic understanding, while integrating the risk analysis approaches of environmental and human health. The project delivered understanding of and tools for sound risk assessment, developing a research framework for the description and interpretation of combined stressor effects that leads to the identification of biomarkers and other indicators of cumulative impacts.

3 Pressures and Impacts on Selected EU River Basins

3.1 Pressure/Impact Assessment

The description of a water body and its catchment area underpins the pressures and impacts analysis. Useful information includes climate, geology, and land and soil use. In addition to a general description of the water body, it is essential to identify the driving forces that may be exerting pressures on the water body. Driving forces are sectors of activities that may produce a series of pressures. Information describing driving forces and pressures will be required for both surface and groundwater. For instance, agricultural activity may exert a pressure on both. In the case of surface waters, the WFD requires only significant pressures to be identified, i.e. those contributing to an impact that may result in the failing of an objective. For groundwater, however, it requires a general analysis of the pressures associated with the risk of failing to meet objectives. Thus, although the processes are described separately and differently for surface and ground waters, a similar approach for the identification of pressures can be adopted. This requires an understanding of the nature of the impact that may result from a pressure and appropriate methods to monitor or assess the relationship between impact and pressure.

It is also essential to consider that a specific pressure will not always cause a particular impact. Scale, both temporal and spatial, is one of the issues that determines the impact of a pressure. Other characteristics of the catchment area of the water body may also have an influence and the particular characteristics relates to the nature of the pressure.

3.1.1 Pressures/Impacts Characterisation

The assessment of whether a pressure on a water body is significant must be based on knowledge of the pressures within the catchment area, together with some form of conceptual understanding of water flow, chemical transfer and biological functioning of the water body within the catchment system. An alternative is that the conceptual understanding is embodied in a set of simple rules that indicate directly if a pressure is significant. One approach is to compare the magnitude of the pressure with a criterion, or threshold, relevant to the water body type. Such an approach cannot be valid using one set of thresholds across Europe since it fails to recognise the particular characteristics of the water body and its vulnerability to the pressure.

Assessing the impacts on a water body requires some quantitative information to describe the state of the water body itself and the pressures acting on it. The type of analysis is dependent on the data that are available. However, it has to be considered that many pressures do not cause a clear-cut impact, but substantially change the probability of adverse conditions. Thus, the impact assessment requires an estimate of which change in the probability of occurrence of favourable circumstances represents a threat to the ecosystem. Monitoring data may indicate that there are no current impacts. This information reveals that none of the pressure identified in the initial screening process is significant, or that the time lag required for a pressure to give rise to an impact has not yet passed. The latter is likely to be of particular importance when assessing groundwater bodies in which pollutants travel very slowly.

3.1.2 Modelling Approaches

Modelling approaches allow impact to be estimated and should be considered complementary to monitored data from the water body. Simple and reliable modelling approaches are available for all the water body types considered in the WFD. These models can represent various aspects of the flow regime, hydromorphology and hydrochemistry of the water body, either separately or within an integrated framework. Relevant considerations when selecting a type of model are the data availability and the time and funds available. In general, the more complex the model, the greater the data requirements and the greater the time and costs needed to complete it.

In situations with no observed data, one possible means to evaluate status is to use a similar analogous site for which data are available, and to assume that the assessment made from the observed data can be applied validly to both sites. The site for which data are available must have good status according to the WFD. For instance, bodies subject to similar pressures and with similar characteristics could be grouped. It is noteworthy that proximity cannot be taken on its own as an indication of similarity since the features of the catchments, such as climate, topography and ecology, can change abruptly.

3.1.3 Groundwater Bodies

Regarding groundwater bodies, the concept "potential impact" is introduced by the WFD to describe the effects that a pressure is likely to have on them again on the context of failing the directive's objectives. For ground waters impacts are a consequence of both the magnitude of the pressure (pollution, abstraction, artificial recharge) and the susceptibility of the groundwater to that pressure.

3.1.4 Tools in the Pressures/Impacts Analysis

At present there is no single tool capable of performing a complete pressure and impacts analysis for all types of water body, and it is unlikely that such a tool will eventually exist. The results from more than one tool may have to be integrated to undertake a complete pressure and impact analysis of a water body. To date, there are only available tools for a limited number of pressure types, mostly dealing with organic and nutrient pollution loads. Many developments are required for hydrological pressures, for instance. Quantifying the pressure would be done using monitoring data. However such data do not exist in many circumstances. Hence, the existing tools use alternative information to quantify the pressure.

A large number of tools for modelling impacts in rivers have been developed and calibrated; however, most of them simulate physical–chemical mechanisms and do not help assess the new issues introduced by the WFD, as, for instance, the impact of changes in hydrological regime or morphology.

Groundwater vulnerability maps, based on a regional assessment using an index-based system, can be used as a screening tool to rapidly assess the relative scale of impacts arising from pressures. They may be useful for assessing whether groundwater bodies are at risk from pollution sources at initial characterisation.

In summary, a successful pressures and impact assessment will be a study in which there is a proper understanding of the objectives, a good description of the water body and its catchment area (including monitoring data), and a knowledge of how the catchment system functions.

3.2 *Case Studies*

3.2.1 Mediterranean River Basins: Llobregat, Ter and Po Rivers

Llobregat (156 km length, 700 Hm^3/year mean flow) and Ter (208 km length, 845 Hm^3/year mean flow) rivers are two examples of Mediterranean river basins (Fig. 4). The reason why both rivers are considered as only one system relies on the fact that both are interconnected and supply drinking water to Barcelona

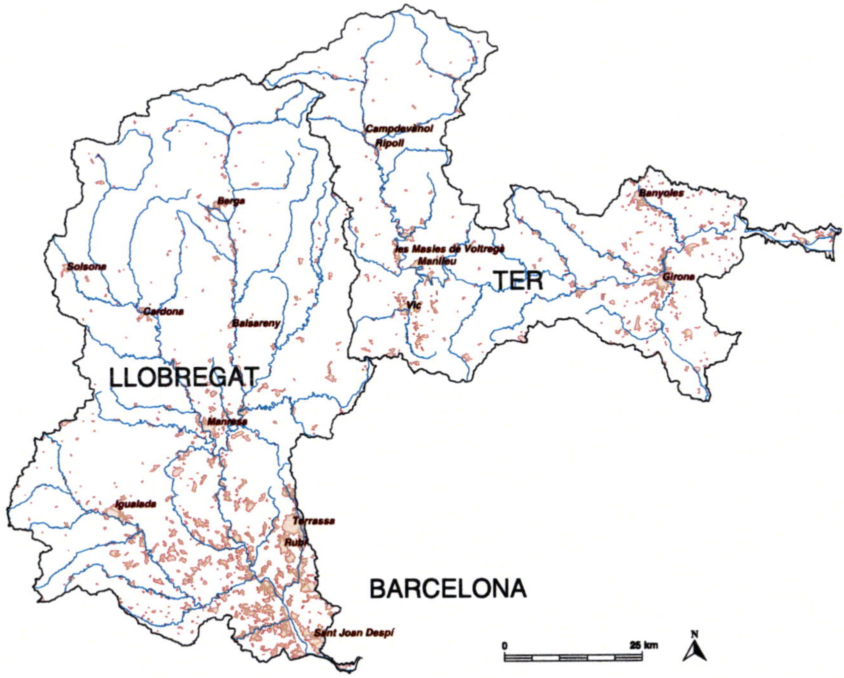

Fig. 4 Llobregat and Ter river basins overview map [12]

metropolitan area (Catalonia, Spain), which represents 5.500.000 inhabitants. Although both rivers share many characteristics, they exhibit some differences: while Llobregat is subjected to industrial, mining and urban pressures, Ter is more impacted by agricultural practices. Their hydrological regimes are those of typical Mediterranean river basins, i.e. they are subjected to sequential seasonal events showing high flow regime variation, causing sudden floods and severe droughts related to climate changes.

Main pressures on these two basins are as follows:

- Extreme events: floods and droughts
- Water demand management
- Drinking water supply (Barcelona metropolitan area, Costa Brava, etc.)
- Industry
- Agriculture (low river and delta parts)
- Urban and industrial pollution (Barcelona area)
- Groundwater over extraction
- Nutrient pollution
- Hydromorphological alterations (hydropower plants, destruction of riparian quality, biological discontinuity, ecological minimum flows, etc.)
- Marine intrusion (groundwater salinisation)
- Ecosystem and biodiversity conservation: presence of invasive species
- Unregulated mining waste disposal (salt mining in the Llobregat basin)

The expected evolution in the water supply points out the increment of water demand as consequence of both the rise of population and the scarcity of water. The latter is related to the increasing occurrence of droughts periods, as a consequence of the climate change (Fig. 5) and/or changes in land use [12].

In order to manage the associated impacts, future plans foreseen to look for different sources of water, focusing on non-conventional resources, such as desalinisation and reuse of seawater and wastewater, respectively, together with a higher use of ground water. Therefore, the need for connecting surface and groundwater was identified as an important issue in water management. In addition, artificial recharge, hydraulic barrier to prevent saline intrusion from sea water and

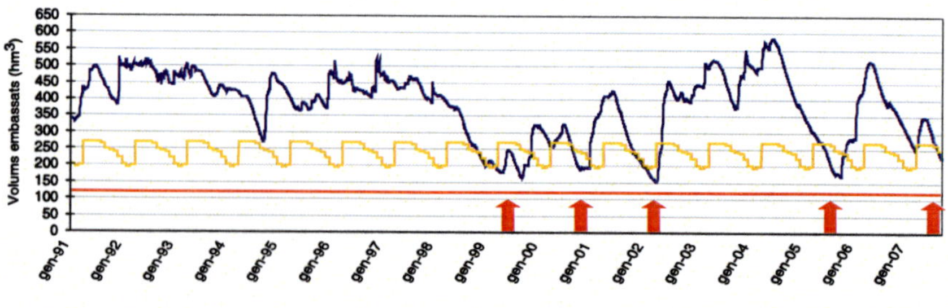

Fig. 5 Evolution of water reserves on the reservoirs of Llobregat–Ter system. Increasing occurrence of drought periods [12]

optimal sustainable pumping rates pose a great potential to improve the actual situation.

With the aim of recovering the good status of water bodies, management programmes and measures have to be developed and its economic cost evaluated. In any case, a reference scenario and a set of goals have to be defined. As an example, the Water Catalan Agency (ACA) has used an application of the pressure-impact analysis tool (physical–chemical model coupled with a multi-criteria routine) to evaluate the cost effect (on the impacts of the pressures) and efficiency of single measures in reaching the WFD objectives [13].

Moreover, the application of this tool permitted to evaluate the effect and efficiency of combinations of measures. Using multi-criteria analysis during the decision-making process allows taking into consideration several criteria such as: technical efficiency of the measures, cost of their application, political criteria and socio-economic issues.

This model implements Qual2k to simulate river flow and the behaviour of selected water pollutants. Qual2k is a well-known and well-referenced model and is used by the US EPA since the end of the 1970s. It simulates the physical and chemical reactions of pollutants coming from a sort of sources (point and diffuse).

With this approach, the ACA aims to define and optimise the treatment typologies for each of the 1,300 wastewater treatment plants, which will be operative in the period 2011–2015 to achieve the WFD goals. Moreover, this methodology allows the optimisation of both investment and management cost of the wastewater treatment plants along Catalonia (Spain).

There exist other models to, for instance, assess aquifers status and future scenarios. As an example, the chloride concentrations (salinity) in the Llobregat aquifers have been investigated [14]. The model applied in this approach was proven to be useful for the following:

- Integrating broad range of data
- Assessing scenarios
- Designing (optimally) corrective measures

Taking in consideration the features of this model, it could also be useful to assess other water pollutants in the future.

In order to have a broader picture of Mediterranean river basins, Po River (Italy), one of the largest in the south of Europe (652 km length), was considered [15].

One of the most important impacts identified in Po river basin was the loss of water quality as consequence of the extended use of agrochemicals. These compounds are of high concern since they are responsible for adverse effects on human health and the environment. The risk associated with each substance is evaluated during its authorisation process. However, it is not possible to assess the cumulative effects of all the agrochemicals currently in use. Within this scenario, it is mandatory to perform constant measurement of such substances.

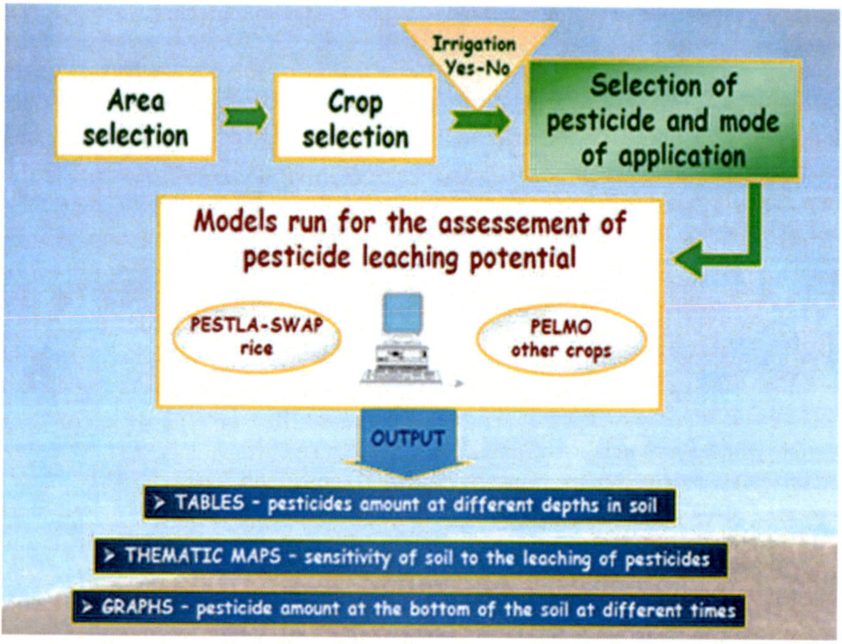

Fig. 6 SuSAP decision support system [15]

Some Italian regions started different actions to identify the driving forces of the processes involved in agrochemicals movement. Furthermore, tools to simulate the behaviour of such substances at different scales were developed, aiming at defining pesticide use permissions and restrictions at a regional level, planning monitoring programmes and optimising the study budget by focussing on sampling in the areas where higher agrochemicals concentrations were likely to be found. SuSAP (Fig. 6) is a Decision Support System (DSS) working at different spatial scales, which allows the users and decision makers at regional and local scales to map the soil vulnerability to agrochemicals leaching. This is done at a farm level in order to identify the most sustainable crop protection strategies. This system integrates existing data (soil type, climate, crop type and agrochemicals used) and leaching models in GIS software. This assessment identifies sensitive areas to set permission/restrictions leading to a correct monitoring plan by considering both, the scientific point of view (spending better) and the economic point of view (saving money).

3.2.2 Dommel/Meuse River Basin

The Dommel is a small tributary (1,350 km^2) of the Meuse River. It originates in Belgium, crosses the border in the south of the Netherlands (Kempen region) and confluences with the Meuse near the city of Den Bosch (Fig. 7). The annual water transport is 420 million cubic metre rainwater and 90 million cubic metre of industrial and public wastewater. The river drains an intensive agricultural and heavily populated area (593,000 inhabitants). In the Belgium part of the Kempen, just before crossing the border (Neerpelt), it also drains an area where several zinc factories are located. Also in the Dutch part of the Kempen a zinc factory is located. Especially in this trans-boundary region, the Dommel is valuated as an important nature reserve.

The emphasis of the WFD on good ecological and chemical status might lead to overlooking the impacts on human beings, as they are very dependent on the water for drinking, recreation (swimming) and their urban environment. Hence, a main question for the water board is: what kind of risks are posed from the water system of the "Dommel" to the inhabitants of the area or people who recreate in or along the river?

Other possible sources of risk for humans are as follows:

- Heavy metals. For instance, in 2005, the total load of zinc in the "Dommel" was 21,580 kg, whereas the load of cadmium was about 525 kg, coming from emission sources upstream in Belgium.
- Pathogens, pharmaceuticals, industrial chemicals, etc. from the effluent of the wastewater treatment.
- Toxins from blue algae and botulism.
- *Escherichia coli* bacteria from sewage spills.

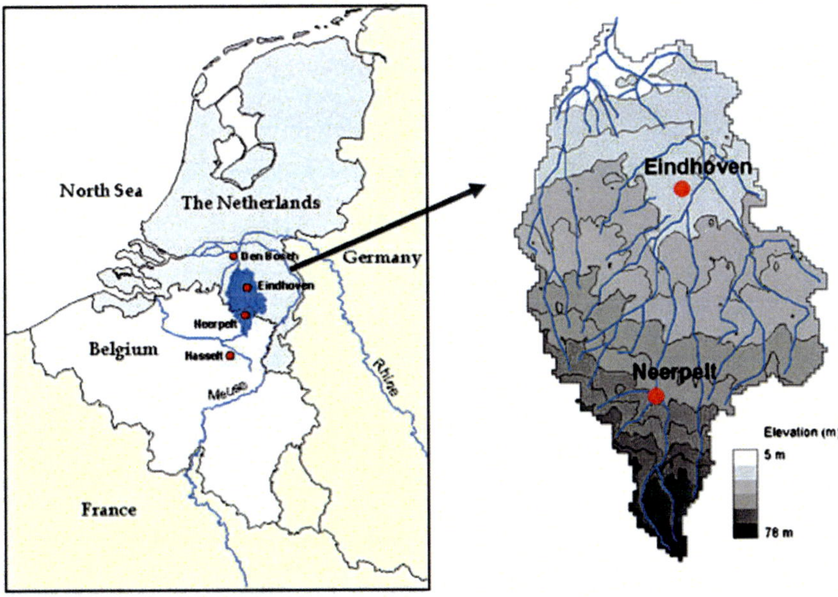

Fig. 7 Dommel river basin overview map [16]

3.2.3 Sava/Danube River Basin

Sava is the third longest tributary and the largest by discharge tributary of Danube. The length of the river from the main source in Slovenia to the river mouth by Belgrade is more than 860 km. The basin covers most of Slovenia, Croatia, Bosnia and Herzegovina and northern Serbia. The size of the basin is more than 96,000 km^2 and the average discharge is 1,600 m^3/s (Fig. 8).

Sava is nowadays navigable for large vessel up to Slavonski brod (377 km) and for small vessels up to Sisak (583 km)

The Sava river basin is of great significance because of its outstanding biological and landscape diversity. It hosts a large complex of alluvial wetlands and large lowland forest complexes. The Sava is a unique example of a river where some of the floodplains are still intact, supporting both flood alleviation and biodiversity.

Four Ramsar sites, namely Cerkniško Jezero (SLO), Lonjsko Polje (CRO), Bardača (B&H) and Obedska Bara (SER), and numerous important bird and plant areas, protected areas at the national level, and Natura 2000 sites have been designated in the Sava river basin.

The Ramsar site and the important bird area Lonjsko Polje Nature Park (510 km^2) represent mainly palustrine-riverine wetlands located within the floodplains of the Middle Sava river basin (Central Posavina, Croatia). It represents the largest maintained inundation area of all the Danube River catchment and, at the same time, the key facility of the flood control system of the entire Sava river basin (including Bosnia and Herzegovina, and Serbia and Montenegro).

The most important drivers impacting Lonjsko Polje Nature Park are navigation and land transport, urbanisation, towns (Zagreb, Velika Gorica, Sisak), industries such as refinery Sisak, fertiliser factory Kutina and upstream hydropower plants (existing and planned). Thus, the main pressures are drainage, pollution and urbanisation. The input of industrial and municipal wastewater, especially from the big cities, Zagreb and Ljubljana, is of concern, since there only are few wastewater treatment plants in the countries at the Sava, which means that less than 5% of the wastewater is treated. Moreover, pollution from former industrial and old landfill sites might be washed to the Sava River in case of floods.

Another important pressure is the overexploitation of gravel (taken out for construction of houses, streets, etc.) carried out since the end of the 1980s (e.g. in Hungary, Danube). Gravel extraction, in Bosnia and Herzegovina, will be forbidden for a certain period to avoid overexploitation. It will be allowed only for cleaning the path for navigation.

It was a common view that issues related to sediment balance are of major importance for river basin management, in general, and for the Danube, in particular.

An example for sediment balance is found near Bratislava (Slovakia), where the gravel transport is interrupted by the dams and the reservoir of the Gabcikovo hydropower station. Thus, the river bed downstream is lowering and wetlands along the Danube are impacted.

Projects to improve navigation likely would cause major impacts on hydrological and morphological aspects. People in this region lived with the water for a long

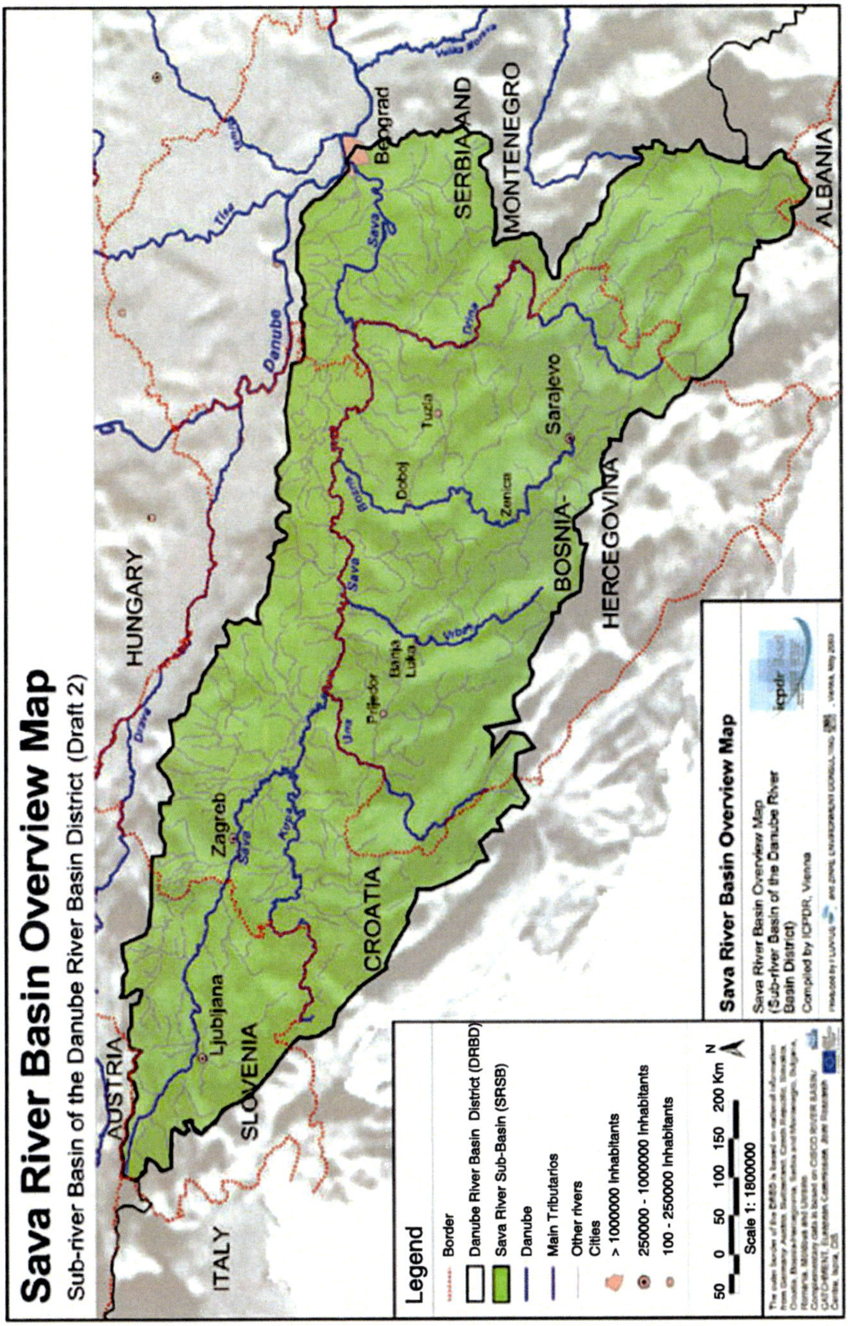

Fig. 8 Sava river basin overview map [17]

time, despite the increasing industrial use and major changes of the system concerning navigation. And thus, the key question at stake is how to sustain living with water in the region of Lonsko Polje. If there are no plans to cope with these changes, the direct impact will be a decline in population.

4 From Monitoring Data to Water Body Status Diagnosis

4.1 Monitoring Programmes Under the WFD

Changes in climatic conditions, land use practices and soil and sediment pollution have large scale adverse impacts on water quantity and quality. Based on the understanding that water is an inherited good that has to be protected and used in a sustainable way, the EU WFD demands for good status in European waters till 2015. More in detail, the objective for surface waters (epicontinental, coastal and transitional) is the achievement of the good ecological and chemical status, while groundwater bodies must achieve good qualitative and quantitative status. For each status, appropriate indexes, elements and metrics are defined depending on the typology of the water bodies (WB), which are subsequently combined in order to establish the final qualification of the WB, under the "one out-all out" principle (Table 1).

The introduction of this new kind of assessment has supposed new monitoring challenges for European water managers, as aquatic ecosystems undergo several pressures, both natural and anthropogenic. Elevated levels of numerous chemicals are frequently detected in European surface waters, and, hence, chemical stress may be one of the driving forces for an insufficient ecological status and reduced biodiversity of freshwater and marine ecosystems due to environmental pollutants. With the WFD, the concept of chemical status gained a new perspective, as the impact of contaminants in biological elements is considered of utmost relevance. To date, no generally accepted approach exists to establish this link between exposure of freshwater and marine ecosystems to environmental toxicants and observable effects on these ecosystems [5].

According to the WFD, in order to establish a coherent and comprehensive overview of the water status, the design of monitoring programmes is mandatory. They must have been operative from the beginning of 2007, and will be extended during a 6-year period (2007–2012). The monitoring programmes, as important part of the RBMPs, were of key relevance for the definition of the programmes of measure, established by the end of 2009 (Figs. 1 and 9).

Monitoring programmes ultimately provide the following:

- Data from the reference sites to provide class boundary conditions to be established for all groups of water body types.
- The data to enable the classification of all individual water bodies.
- The means of monitoring progress with the implementation of the RBMPs and associated programmes of measures and the basis for their subsequent revision.
- Early warning of new problems.

Table 1 Elements and indexes used for the qualification of the status of the different classes of water bodies [18]

	R	L	Rs	Es	C	B	G	Elements
Ecological status								
Biological index	x	x	x	x	x	x		Macro-invertebrates
	x	x	x	x	x	x		Marine phanerogams
	x	x			x	x		Macro algae
		x			x			Fishes
								Phytoplankton
								Phyitobenthos
Hydromorphological index	x	x	x	x	x			Continuity
	x	x		x				Hydrology
	x			x				Morphology – river forest
								Modification of the coastline
Physicochemical index	x	x	x	x	x	x	x	Thermal conditions
	x	x	x	x	x	x	x	Transparency
	x	x	x	x	x	x	x	Oxygenation conditions
	x	x	x	x	x	x		Salinity
	x	x	x	x	x	x		Nutrient conditions
		x	x		x	x		Acidification status
Chemical status								
Chemical index	x	x	x	x	x	x	x	Priority substances
	x	x	x	x	x	x		Discharged substances
Quantitative status								
Quantitative index			x				x	Piezometric level

R rivers, *L* lakes and wetlands, *Rs* reservoirs, *Es* estuaries (*transitional*), *C* coastal, *B* bays (*transitional*), *G* groundwater

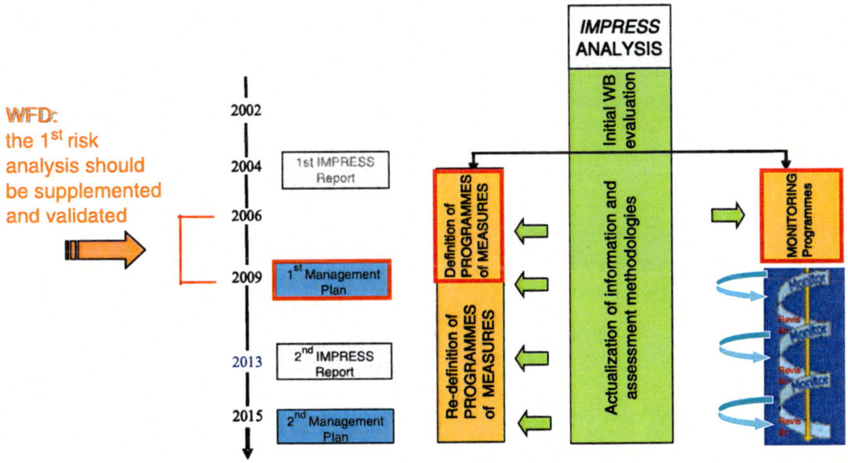

Fig. 9 Role of monitoring programmes in the WFD implementation [19]. IMPRESS Guidance 2004

Monitoring programmes are classified as surveillance, operational and investigative, depending on the purpose seek. Surveillance monitoring is carried out in order to have a general insight on the quality status of the WB, whereas operational is designed for the control of WB under risk of failing the WFD's objectives due to the impact of existing pressures/impacts, or in order to test the effects of the applied measures. Finally, investigative monitoring is applied on specific cases (i.e. critical situations related with unknown causes, natural catastrophes, accidents, etc.). Furthermore, in protected areas the foregoing monitoring criteria will be complemented with those specific requirements established by the particular legislation which has given rise to the protection.

Within the WFD, monitoring plans must be established for a comprehensive coverage of transitional and coastal water bodies [20]. The monitoring activities to be carried out constitute a serious additional workload on the technical and scientific communities, and due to logistic and/or financial constraints it may be necessary to prioritise different monitoring activities according to the management issues available. The two main concerns in the design of a monitoring programme are: (1) water sampling stations within each river basin district should be sufficient in number, and (2) observations should be frequent enough to provide an assessment of the overall water status.

The core of the monitoring programmes is based on the combination of the following:

- The network of control points belonging to each typology of WBs
- The elements to be controlled (elements, analytical profiles, protection requirements, etc.)
- The pre-established frequencies of work for every element

The problem to be faced by the monitoring programmes can be formulated in a logistic perspective as how to allocate the available resources in order to give response to the different control requirements in the most efficient way.

$$\text{Elements/Analytical Requirements} \times \text{Points} \times \text{Frequencies} \rightarrow \text{Work Schedule}$$

4.2 Ecological Aspects

The objective of the WFD is to implement an optimal *ecological integrity* in European water bodies. Overall, this is a very ambitious objective that needs to be fulfilled by 2015, and in the meantime the development of methodologies have evidenced both the potentials and limitations of the so-called ecological status approximation.

The *ecological status* is a utilitarian concept that is largely based on the combination of the hydromorphological, chemical and biological estimates. This is a rather novel approach combining multiple indicators, well beyond of precedent estimations, which were mostly based on a limited set of chemical indicators. However, the concept needs to be reshaped and adapted to the complexity of many situations. This has been evidenced throughout the deployment of the WFD, and in particular in the assessment of effects of multiple stressors (e.g. nutrients, water level fluctuations, organic pollutants). There is an increasing number of non-autogenic toxicants that enter the aquatic ecosystems. In most cases, knowledge on toxicant effects on biological communities is scarce. On the other hand, other stressors interfere with toxicants. In particular, nutrient effects on the organisms might interact with the toxicants. Furthermore, water flow variation associated with global change (flow variation, rising demand of water resources) do affect the general functioning of ecosystems.

The estimation of the ecological status of inland waters is largely based on the composition and abundance of the main biological communities inhabiting them. The WFD considers the relevance of phytoplankton, phytobenthos, macroinvertebrate and fish. Zooplankton has not been considered, and this being a rather relevant omission in slowly moving systems. Nevertheless, all these structural (abundance and composition *functions*) elements are general descriptors of the configuration of the biological communities, which allow to determine a general description of their resistance and adaptation to the effects of disturbances. This view (though informative because it provides the response of the main components of the biological communities in river systems) is incomplete because it does not include the *functions* that biological communities perform in the environment.

The functions that the communities perform result from the interactions between them and with the environmental variables. Among the several functions that the biological communities are responsible for, we may quote some, which are relevant to the ecological state of the system. Among them, the use of inorganic materials, sequestration of toxicants, oxygen production and consumption, mineralisation of organic matter, etc. can be considered. Accordingly, some of these functions are recognised as *Services of the ecosystem* [21, 22] (Table 2), and as such their value is fully acknowledged.

A debate on whether structure or function is the most relevant approach in the monitoring of river ecosystems or in the definition of the ecological state should be undertaken. The two aspects of the ecosystem are completely intertwined and as such one is feeding the outcome of the other and vice versa (Fig. 10). In practical terms, both approaches need to be considered to obtain a real image of the response

Table 2 Functions and services of river ecosystems. Classification according to the Millennium Ecosystem Assessment

Provisioning values	Regulating and supporting values	Cultural values
Water resources	Gas regulation	Aesthetic and spiritual
Food production	Climate regulation	Educational
Energy production	Disturbance regulation	
	Nutrient recycling	
	Material processing	

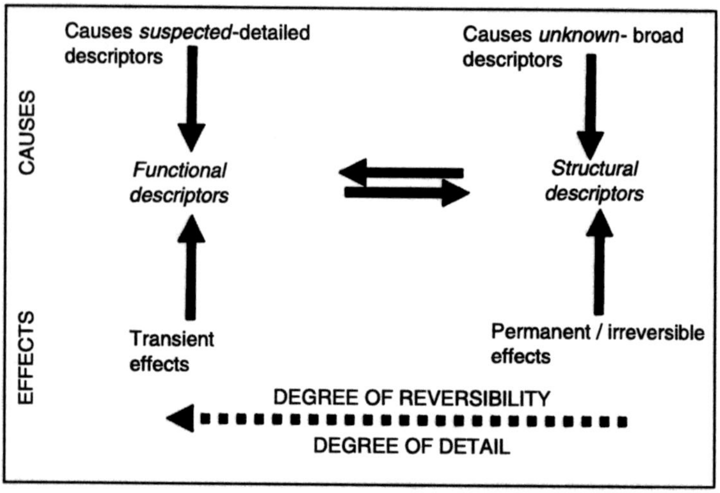

Fig. 10 Linkage of structural and functional descriptors in the use of biological communities to monitor stress in river systems

of the ecosystem to stressors. This assertion will be exemplified with the use of biofilms as monitors of stressors in river systems.

4.2.1 Using Structure and Function to Monitor Complex Situations: The Biofilm as an Example

Biofilms are a complex mixture of bacteria, algae and other organisms (Fig. 11). In well-illuminated environments, micro-algae (phytobenthos) make up the largest fraction of the biofilm biomass, which plays a vital role as a primary producer. However, in non-lit environments, heterotrophs (bacteria, protozoa) account for the greatest proportion within the biofilm. The composition and abundance of phytobenthos (periphyton or autotrophic biofilm) have a recognised role by the WFD.

Biofilms are the first to interact with dissolved substances such as nutrients, organic matter and toxicants. Furthermore, they are sensitive to hydrological variations (droughts and floods, hydraulic connectivity loss, etc.). As a consequence of their sensitivity, biofilms can be used to detect the early effects that they might cause on the ecosystem [23]. The biofilms integrate the influences of environmental conditions from early stages to over extended periods of time, mainly because of the small size and rapid growth of the organisms, which integrate them. Because of their extended distribution in the river, they are also representative at the spatial scales of the basin and account for both the local and diffuse influences. Disturbances produce transient or irreversible effects, but the responses on biofilms occur fast, justifying the use of biofilms as good indicators of the ecological state of the system.

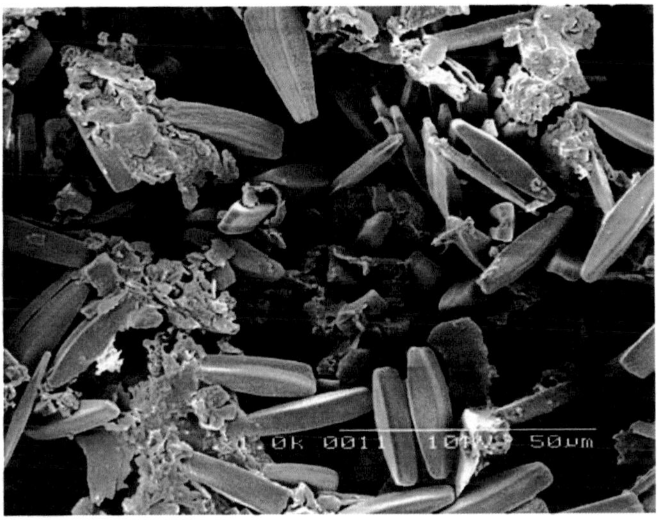

Fig. 11 Image of an autotrophic biofilm obtained by means of a scanning electronic microscope

Structure

The use of biofilms as indicators of the ecological state of the river systems requires approaching as much as possible on the responses of the natural community in its natural environment. This approach is much opposed to the standard laboratory tests, and as such has both advantages and disadvantages. Among the latter, standardisation is a pending matter. However, the use of natural communities has an added value since it includes not single species, but the complex combination that organisms perform in nature. In a community-based approach, the response of all species is included, the outcome being the emerging response of a significant part of the ecosystem.

Communities react to physical disturbances, nutrients or toxicants by changing their community composition, usually favouring the most tolerant taxa. This effect might be detected at the large group level and sometimes at the species level. Among the latter, some toxicants may even alter the morphology of the taxa composing the biofilm community [24]. However, most of the effects do not raise acute responses. The use of indices has gained increasing popularity because of their easiness to use and the summarising response that they offer. In particular, those applied to diatoms are increasingly used and widely implemented in national surveillance networks (e.g. France, Germany, United Kingdom, Spain, the Netherlands). Diatom indices are elaborated as the sum of the particular sensitivities and autoecological preferences of the taxa of the diatom community. Diatom indices have been used to relate diatom community structure to pollution, mostly to summarise the effects of eutrophication or other disturbances. However, the ability of diatom indices to indicate the separate effects under multiple stressor situations (e.g. nutrients, water flow, toxicants) is limited, since the taxa react to the overall effect of these disturbances. Furthermore, the indices need to be refined or regionally adapted to reflect the ecoregional attributes, this work being under progress.

Even more general than the indices of community structure, chlorophyll *a*, growth rate (differences in algal cell densities) or standing crop (biomass, expressed as dry weight – DW, or ash-free dry weight – AFDW) have been commonly used as descriptors of biofilm biomass and may express the long-lasting effect of nutrients, toxicants and other stressors on these communities. Again, their ability to reflect separate effects is limited.

Structural variations may be also produced at the microscopical scale and are able to produce significant improvements in our understanding of stressor effects. Observation of the biofilm architecture and characterisation of the different fractions (i.e. algae, bacteria, mucopolysaccharides) may be useful to identify particular effects of toxicants to selective components of the biofilm. The use of confocal laser scanning microscopy remains promising [25].

The determination of effects on the natural communities, while realistic, introduces a large source of variation that needs to be counteracted by progress on standardisation. The use of *artificial substrata* (glass or tiles depending on the type of effect to be analysed), which may be left for colonisation in rivers for a given period of time, is a step to be considered for standardisation. Artificial substrata provide a reduction in the natural variability of natural substrata and have other advantages such as reduction in the time of sampling and the ability to obtain quantitative samples. However, their implementation requires a thorough testing in a variety of situations.

There is still a long way to go in the harmonisation of indices, applicability of new techniques and standardisation. These are essential steps in order to advance in the detection of the stressor effects by means of structural descriptors. However, even if these questions might be solved, the ability of structural descriptors to detect effects is limited. Many stressors occur in low concentrations, in acute episodes, or have side effects that are not reflected in the structure (composition, abundance) of the biofilm. In these cases where chronic effects would not occur, or are hidden, a finer scale of detection is required.

Function

Metabolic and physiological approaches in the detection of cause–effect relationships may be useful in the early assessment of stressors on biofilms. Physiological responses may be much faster and might reflect transient effects on the biofilms, while those that are more persistent may be reflected in the structural components (Fig. 10).

General descriptors may be related to the metabolism responses in the biofilm. Biofilm algae have several mechanisms to counterbalance the damage caused by the toxicants. Environmental stress produces oxidative damage in the cells, which can be tracked down by means of the analysis of many enzymes (superoxide dismutase, catalase, peroxidase, etc.) that function as effective quenchers of reactive oxygen species (ROS).

The photosynthetic activity of biofilms is one of the most commonly used approaches to detect effects on biofilms. The methods based on photosynthetic activity are obviously adequate in the case of stressors that affect the performance of photosynthesis, either directly or indirectly.

Measurement of exoenzymatic activities is potentially useful in detecting the effects of toxicants on heterotrophic biofilm communities. Sensitivity and direct relationship with organic matter use and, therefore, microbial growth make extracellular enzyme activities a relevant tool to assess the toxicity of specific compounds. Use of novel approaches that combine enzymatic and microscopic tools (e.g. ELF-phosphatase) may be extremely useful to detect anomalies at the subcellular scale.

In spite of all of this variety of approaches, covering a wide array of metabolism pathways, limitations also exist. Differences in the vulnerability of biofilms have been found to depend on the age, community composition and succession status of the community. In dense biofilms the transfer of contaminants may be limited, resulting in decreased bioavailable concentrations of nutrients or toxicants for the algae. Biofilms show an inverse relationship between metal toxicity and biomass accrual [26], and a similar relationship has been established with nutrients. Therefore, the colonisation time or *biofilm thickness* are relevant factors to be included in the procedure uses.

Although the effect of nutrients and "classical" toxicants (e.g. heavy metals, herbicides) is well known, that of the so-called non-PS in biofilms is still largely unknown. Furthermore, the combination of effects, which operate at the basin scale, requires complex approaches. Therefore, the response of biofilms to these situations should trigger the development of new applications and higher standardisation in the use of biofilms. The standardisation of methods and procedures is a challenge for the future research on the use of natural and laboratory biofilms.

Structure and function need to be jointly considered in the assessment of effects of stressors on river systems. It has been shown that the two sets of parameters offer complementary information since they cover different time scales and responses. This being shown in the case of biofilms is not a unique characteristic of them, but it might be applied to all other biological communities (e.g. macroinvertebrates, fish). These differ from the biofilm in its higher size and life span, and therefore in their integrative capacity to reflect effects in one part of the ecosystem. Higher traffic levels in addition to biofilms should be considered to study the whole ecosystem. In all of these biological compartments, the combined use of descriptors may amplify our ability to predict the effect of stressors on river basins.

4.3 Chemical Aspects

The official list of priority pollutants threatening the aquatic environment is presently based on that reported on Directive 2008/105/EC [27]. It was the output resulting from a previously done "risk assessment" study carried out by the Fraunhoffer Institute (COMMPS procedure) [28, 29], using monitoring data gathered throughout many European river basins.

The WFD, so far, has identified 33 priority hazardous pollutants (PHS), for which Environmental Quality Standards (EQS) have been set. To some extent, these EQS can be met through the establishment of emission control measures. These PHS may originate from several different sources and activities. The main sources of toxic substances to water bodies in Europe may be categorised as agriculture, sewage treatment plants, urban runoff, industry, contaminated lake/river sediment, soils and landfills. Input via atmospheric transport and deposition has also been identified as an important source both far from and close to source areas. Many of the PS are today banned in Europe, but due to their persistence they are still present in the environment [30].

A schematic drawing of the potential pathways of PS in industry (primary producers, secondary producers, applications) and release patterns are presented in Fig. 12.

Assessments of the quality status of water systems are required as part of the WFD. These assessments include data on the concentrations of contaminants in different compartments such as water and sediment. It is, however, perhaps more important is to assess the impact these contaminants may have on the ecosystem and to predict future impacts. The impact of contaminants is determined by the behaviour of chemicals in the system, determining exposure of biota and the

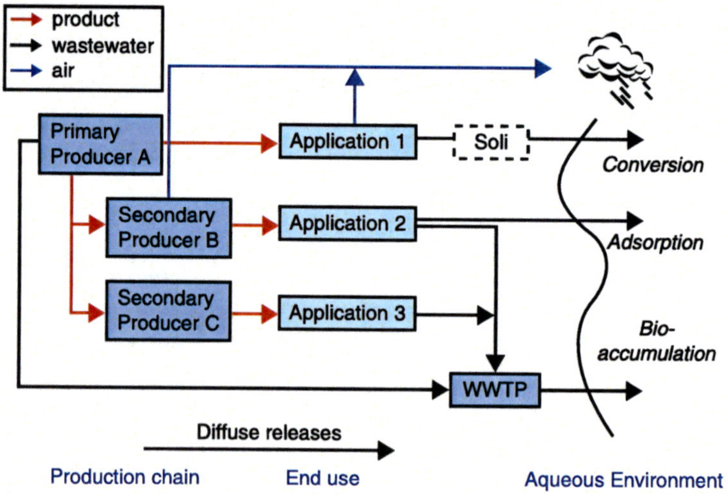

Fig. 12 Potential pathways of PS in industry [29]

response of organisms to this exposure. This means that processes relevant for chemical contaminant must be characterised and included in the assessment. This will enable an assessment of how the system is coping with chemical stress and how this will develop, in other words the resilience of the system. Important aspects of the behaviour and effects of chemical contaminant include partitioning processes such air/water, air/soil and sediment/water exchange, their bioavailability and their degradation. It is of course particularly important to study the long-term behaviour of POPs. For a complete picture (assessment), it is essential to not only know the current levels of chemical contaminants in the different compartments of a river basin, but the different processes that can affect or determine these levels should also be characterised. This will enable assessment and prediction of how the basin system will respond to changes in emissions.

Sediments are important compartments for many organic contaminants in the aquatic environment, in particular for hydrophobic POPs such as PAHs and PCBs. Sediments have been recognised as important sinks for these compounds but with the reduction in levels of them in water, the question arises of whether the older highly contaminated sediments will function in the future as secondary sources of the compounds or whether burial by recent, cleaner sediment will prevent exchange with the water phase. This will depend on the strength of turbulence/bioturbation and on anthropogenic influences such as dredging.

Sediments can also serve as potential exposure routes for aquatic food chains through the bioaccumulation of contaminants by benthic organisms. The potential of sediment contaminants to expose organisms in sediments and the water column is determined by their bioavailability. The bioavailability or bioaccessibility of non-polar organic contaminants is determined by how strongly they are bound to organic matter in soil and sediment [31, 32]. This fact should be taken into account in a realistic assessment of the environmental risks of these contaminants (Fig. 13).

Predictions of the impact of organic contaminants should also consider the role the biodegradation may play in removing chemicals from the system and thereby reducing exposure. Biodegradation is not only of importance in removing chemicals from waste WWTP effluents but also for removing chemicals with long residence times in the river basin system. Many of the WFD PS fall into this category, because of either their inherent persistence or their strong sorption to sediment, or both.

Although many of the WFD priority pollutants are regarded as being persistent, this is not an absolute property. Biodegradation has been demonstrated for many of these compounds, although it is often slow even under optimum conditions. Examples of such compounds are surfactants, phthalate esters, polycyclic aromatic compounds, pesticides and even organochlorine compounds such as PCBs. Biodegradation of surfactants has been studied extensively, mostly in connection to waste water treatment. In recent years the focus has shifted to their biodegradation in river systems. Many of the surfactants in current use, such as alkyl ethoxylates and linear alkylbenzene sulphonates, are readily biodegradable during wastewater treatment and in the environment [33, 34], although the high production levels of these compounds results in measurable levels being present in surface water and

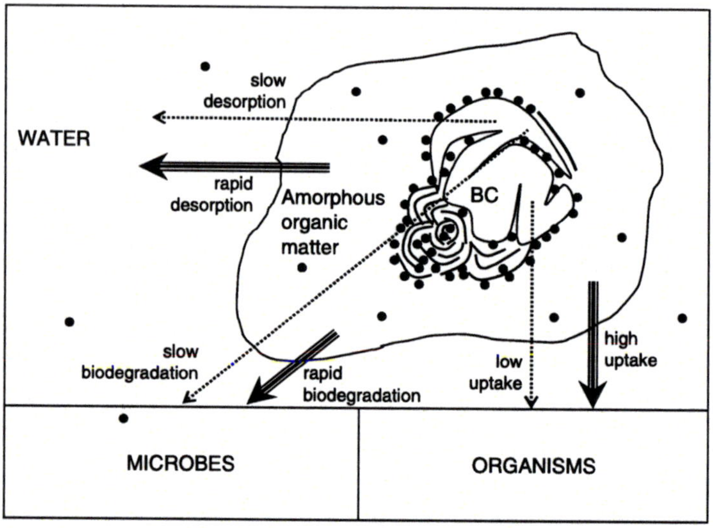

Fig. 13 Representation of the sorption of organic compounds to two forms of sediment organic matter, amorphous organic matter and black carbon (BC), and its effect on bioavailability

sediment. In some cases, such as the alkyl phenol ethoxylates, there exists concern about the effects of their degradation products (nonyl and octylphenol), which have been identified as estrogenic compounds. The cationic surfactants are in general poorly studied.

PAHs are a major group of pollutants of direct fossil fuel origin. The PAHs are a particularly important group of priority pollutants in sediments and soil. The biodegradation of these compounds has been studied extensively [35, 36]. The lighter can be degraded by a variety of microorganisms, but the heavier PAHs only seem to undergo partial transformation. Since PAHs accumulate in sediments, more information is required on their degradation potential under the anaerobic conditions common in sediments. This topic has only recently been started to be addressed [37]. The influence of sorption on the biodegradation of PAHs in sediments is a topic requiring further clarification.

Pesticides are another major group of pollutants but in this case consisting of many chemical classes. As far as the WFD priority pollutants are concerned, the most important groups of pesticides are the organophosphorous and the organochlorine pesticides. In general, the organophosphorous pesticides do not have a very high persistence in aquatic systems [38], but they may be of concern because of their relatively high toxicity. Atrazine and the other triazine pesticides have also been studied extensively and are known to be biodegradable [39]. Nevertheless, the compounds are still present in river systems many years after being banned [40] and must therefore be included in current assessments.

Chlorinated organic compounds are notoriously persistent and are therefore prominent on the list of priority pollutants. However, this is not always the case, and many of the less chlorinated analogues in particular can be degraded by microorganisms, although this is often slow and incomplete [41, 42]. In contrast, highly chlorinated compounds can be dehalogenated under anaerobic conditions. This is in fact an exothermic reaction and there are anaerobic microorganisms able to use the energy released from the dechlorination of some compounds as a form of respiration (sometimes referred to as chlororespiration) [43]. This is potentially significant since many organochlorines are highly hydrophobic and therefore accumulate in anaerobic sediment [44]. Examples of chemicals for which anaerobic dechlorination has been demonstrated include chlorobenzenes, PCBs and even PCDDs. Whether similar reactions occur for compounds containing other halogens as substituents remains to be seen.

Finally, if the search is limited to the list of regulated priority pollutants, there is a high risk of missing something. In fact, taking into account that several thousands of chemicals are currently and commonly used, restricting our control to 33 can be clearly insufficient and need to be considered with some caution.

New pollutants are now under concern, constituting the so-called emerging pollutants (i.e. human and veterinary drugs, personal care products, etc.). These compounds sometimes are polar and not persistent, but their continuous input into the environment allows qualifying them as "pseudo-persistent."

Therefore, the list of PS should be kept "alive," thus remaining open to continuous updating. Quick procedures for inclusion of new pollutants on the official lists are necessary.

4.3.1 Sampling Issues

Sampling is a key point on any analytical process, especially when applied to real cases. Many possibilities remain open: spot samples, integrated samples, use of continuous passive sampling "devices," frequency of sampling, sampling site selection and result handling are some of the pending questions requiring further scientific advice, in order to be applied on practical cases. Therefore, monitoring of water bodies required by the WFD should incorporate the state of the art in this field.

The assessment of the spatial variability and the time scales of processes relevant to water quality status is necessary for all types of monitoring programmes (surveillance, operational and investigative as explained in Sect. 4.1). The adequate measurements are harder to achieve in transitional waters because of the natural variability of water properties often difficult to evaluate the magnitude and extension of anthropogenic effects. It is well known that the determination of instantaneous concentrations of physical–chemical parameters, metals or organic contaminants in water is unlikely to reflect appropriately the range of values existing in river and transitional waters; therefore, advanced observation or sampling devices for chemical monitoring of water quality have to be developed and further implemented in the programmes. As an example, for a short-time observation, a new device has been developed at highly resolution at IPIMAR [2], (Lisbon,

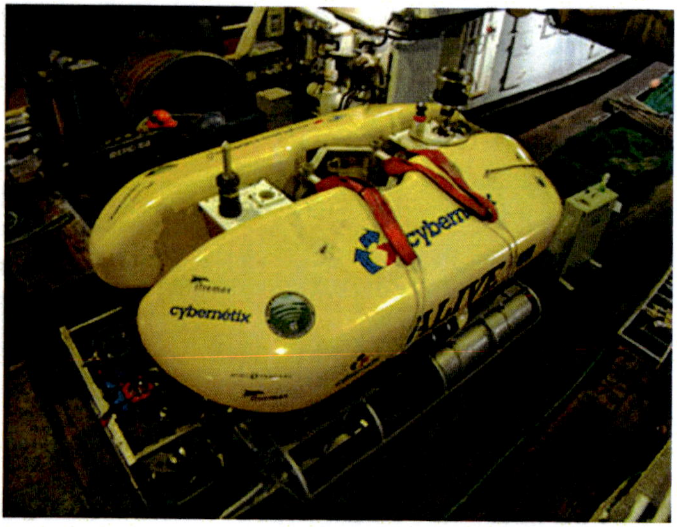

Fig. 14 Autonomous unit vehicle, in situ sampling device

Portugal) under the project SEABEE, recently. Data on salinity, temperature, nutrients, turbidity and chlorophyll *a* measured in situ by coupling probes and a portable auto-analyser to an autonomous unit vehicle that crosses transitional and coastal waters were obtained (Fig. 14). The coupling of those devices allows registering alterations in water quality at selected time scales, and to screen conditions existing at problematic areas. Results showed day–night fluctuations of chemical properties in eutrophic waters, changes across water fronts and alterations at confined rocky areas [45].

The use of sensors and biosensors on a continuous real-time measurements (online monitoring systems), at increasingly more and more affordable costs, introduces new possibilities on monitoring that cannot be underestimated. In turn, those new technologies generate new problems, mostly related to the amount of results available and their interpretation.

Although under the common practice the use of grab sampling in the last years has been usually recommended, DG research has funded several projects (STAMPS, SWIFT) to promote the use of passive sampling. The introduction of passive sampling has been one of the most important developments over the last few years in environmental sampling to the analysis of aquatic phase chemicals. This technique has a longer history in air sampling where it enables the determination of low concentrations of gas phase semi-volatile components. Several types of passive sampling devices are available for organic and inorganic contaminants [46], and some of these are being applied routinely to water quality monitoring.

For the monitoring of water-borne contaminants, the main advantages of passive sampling are convenience and the fact that sampling is time integrated and is

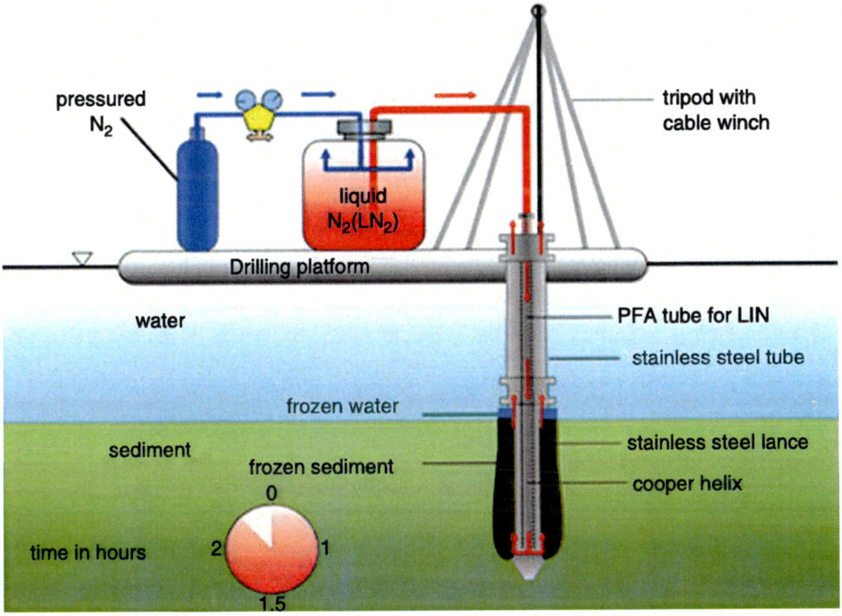

Fig. 15 Sampling device for sediments

therefore less influenced by short-term variations than the more traditional grab sampling approach. Furthermore, the concentrations being monitored are those of dissolved phase chemicals and are therefore the biologically relevant ones. Passive sampling may not completely replace conventional techniques, but gives important additional information. There is therefore place for both in the WFD monitoring toolbox.

Passive samplers [47] are very useful to determine compounds when their concentrations fluctuate over short-time scales. There are two classes of passive samplers; equilibrium samplers that are useful where concentrations of pollutants do not fluctuate markedly, and kinetic samplers that provide time weighted averaged concentrations even where levels of pollutants can change markedly over time. These sampling devices allow to measure concentrations of freely dissolved analytes, and not total concentrations, and have the potential to mimic uptake by organisms, thus providing more biologically relevant information. Pollutant fractions that are bound to particulate matter or to dissolved organic carbon are not available for accumulation by the devices.

New sampling methods are also required for the monitoring of solid samples. The Environmental Specimen Bank (ESB) of Germany has developed devices for sampling suspended particulate matter and sediments. In both purposes, sampling and freezing are carried out in the field. For example, Fig. 15 shows the device used in sediment monitoring [48].

5 Risk Assessment and Risk Management: Bridging the Gap Between Chemistry and Ecology

For risk assessment and risk assessment tools, new recommendations are described in the Technical Guidance Document of EU – Edition 2, in the new EU Chemicals Legislation REACH and in the status report for toxicological methods of the European Centre for the Validation of Alternative Methods (ECVAM). In the EU, risk assessment of chemicals are driven by the requirement of directive 93/67/EEC on risk assessment for new notified substances and Commission Regulation No. 1488/94 on risk assessment for existing substances (TGD). The EU decided in 2001 to develop a new chemical policy strategy called REACH (Registration, Evaluation and Authorisation of Chemicals) to complete this task by 2012 [49, 50]. The so-called bioassays and toxicity tests are complementary tools to the instrumental analytics to protect the ecosystem (risk of disintegration of functional and structural effects) and human health.

Table 3 describes the main parts of an environmental risk assessment (ERA) that are based on the two major elements: characterisation of exposure and characterisation of effects [27, 51]. ERA uses a combination of exposure and effects data as a basis for assessing the likelihood and severity of adverse effects (risks) and feeds this into the decision-making process for managing risks. The process of assessing risk ranges from the simple calculation of hazard ratios to complex utilisation of probabilistic methods based on models and/or measured data sets. Setting of thresholds such as EQS and quality norms (QN) [27] relies primarily on

Table 3 Comprehensive strategy for the risk assessment and management for the sustainable development for aquatic life and human health protection in surface water bodies

Dose-biomonitoring (exposure) – exposure assessment			Effects-biomonitoring (response)	
↓			↓	
Data of exposure using chemical analysis →	Fate of the compounds (binding and transport) by characterisation of the matrix → time →	Bioavailability (intake and distribution) → time →	Short, long-term effects – primary molecular damage, biochemical functional effects using biomarkers → time →	Population effects (environment and human) – fitness, diversity, reproduction, stress factors, mortality
↓	↓	↓	↓	↓
[EQS-QN] criteria for aquatic life and human health protection to characterise reference conditions and to prove ecological status class boundaries in surface water				
Exposure data acquisition	Exposure pathways	Exposure scenarios	Risk analysis	Risk estimate
↓	↓	↓	↓	↓
Risk assessment and risk management				

ecotoxicity and effects data. QN are defined as being concentrations of substances in water or biota which, when reached or exceeded, can be scientifically proven to cause adverse effects.

5.1 Harmonisation Protocols

There is a broad variety of chemical analysis and bioassay methods available to generate ecotoxicity and effects data [49]. Bioassays are complementary to chemical analysis and able to detect the complex toxic potential of bioavailable multiple unknown contaminants at low concentrations. For an integrated assessment, it is necessary to use a complementary combination of several test methods. A tiered testing is suggested in a hierarchical approach covering the cellular, species, population and community level with a wide range of sensitivity. The evaluation of those hierarchical systems is quite complex, and the results are difficult to understand and to compare along international water ways. To solve this problem, many so-called harmonisation studies are done. A much better strategy is to use the advantage of the national or even better international standards and harmonisation protocols, e.g. after International Organisation for Standardisation (ISO) or CEN (European Organisation for Standardisation).

For standardised instrumental analytical methods, i.e. biomarkers, biosensors and bioassays, there are well-established standard protocols on the national level, e.g. under Association Francaise de Normalisation (AFNOR), British Standard Institute (BSI), DIN (German Organisation for Standardisation), etc., and all those standards are formed by ISO-Working Groups and by validation studies into ISO – and CEN – Standards. Normal accredited and well-qualified laboratories should be able to perform the monitoring.

The protocols of analytical methods and bioassays include the sampling and preparation steps of the test matrix before the test procedures. The sampling should be conducted in accordance with ISO 5667-16. There are already available harmonised protocols according to Hansen et al. [49]. The statistics of the ecotoxicity data should be conducted in accordance with ISO/CD 20281.

The ISO protocol for the biochemical response EROD (ISO 23893-2/AWI) as a recent example of a bioanalytical (biomarker) [49, 50] method standardised under ISO for fish needs harmonisation with the other test systems and between the laboratories (users) before implementation. Use of biomarkers (biochemical responses) in multi-arrays for environmental monitoring according to Hansen et al. [50] is complementary to chemical analysis since they can alert for the presence of ecotoxic compounds. Bringing into the WFD, the effect-related approaches concerning bioassays and biomarkers are only relevant in the context of the QN of environmental relevant substances and the "good chemical status." But it is rather difficult to transfer the monitored biochemical responses or biomarkers into an operational effect-related standard. They serve as the basis for environmental protection against hazardous substances. In relation to hazardous substances, QN must take into consideration the following goals for protection.

Aquatic communities: Effects on aquatic organisms as a result of both short and long-term exposure and accumulation in aquatic organisms.

By definition a bioconcentration in the organisms are relevant, if the bioconcentration factor (BCF) is higher than 100.

The communities include in particular bacteria, lower aquatic plants (algae), higher aquatic plants, organisms fish feed on (e.g. water flea, amphipods etc.) and fish. They participate in the self purification of waters (reduction of residual pollution from effluent discharges like industrial drainage) and maintain the natural biological equilibrium.

Commercial and sport fishing: The management goal in relation to this use is the maintenance and/or restoration of the diversity of the communities of importance to fishery, such as fish-food organisms and site-specific fish populations including the conditions for their natural reproduction.

Moreover, water quality is to be maintained or restored in such a way that human beings do not suffer any health damage as a result of the pollutant content of the fish they consume. The ecotoxicological data required in order to protect aquatic organisms and fish populations are the so-called No Observed Effect Concentration (NOEC) data.

A compensation, uncertainly, or safety factor has to be introduced (Table 4) to take into account the uncertainty associated with extrapolating results to the real environment.

Recognised test methods include methods developed by internationally recognised agencies such as DIN, ISO, CEN or OECD. As a general principle, the lowest test result for the most sensitive species has to be used as the starting point for the risk assessment and for the derivation of the water and biota quality norms. The toxicity data used for the risk assessment have to be examined critically with respect to validity and relevance.

In Fig. 16, the "single substance approach" concerning exposure and effects is depicted. There are comparable approaches and strategies for mixtures and effluents (lowest identified dilution step). There are further diagnostic instruments needed, which allow to identify the possible causes of deteriorated water quality and biota. By these instruments, it will be possible to gather further information about the effectiveness of possible measures. Examples for further diagnostic tools

Table 4 Safety factors and available ecotoxicological data [1]

Available data	Safety factor
At minimum one acute assay at one trophic level (algae, daphnia or fish)	1,000
One long-term, chronic toxicity assay (NOEC): with fish or daphnia	100
Two long-term, chronic toxicity assays (NOEC) at two trophic levels: algae and/or daphnia and/or fish	50
Three long-term, chronic toxicity assay at three species (NOEC): algae, daphnia and fish (three trophic levels)	10

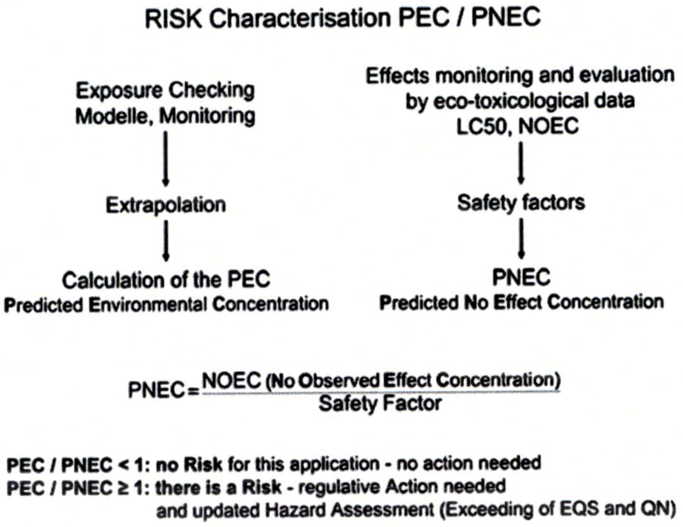

Fig. 16 Risk characterisation by exposure and effects (Table 4)

could be a battery of biomarkers, biosensensors [49] and biofilms. For sediment monitoring of effect related biochemical responses (biomarkers) and the relevant endpoints are demonstrated at the molecular level by Hansen et al. [49]. But there are even more principles associated with the different scales of the biochemical responses relating to an acting ecosystem.

The problem of calculating the PNEC (Fig. 16) is very often the limitation of acute toxicity data for aquatic organisms. Nowadays, effect assessment and risk characterisation can be assisted by analytical techniques (Table 5). New methods for and strategies of bioeffect monitoring are known to yield valid ecotoxicological information [1, 27, 51], QN [27] and diagnostic effects such as genotoxicity, neurotoxicity, immunotoxicity and cell toxicity in suborganismic systems [49] (Table 5). Toxicogenomics and proteomics assist in documenting the mode of action of chemicals and in classifying those for risk management on a case-by-case basis [52] and exposure scenarios [49]. At present, toxicogenomics and proteomics are only rarely applied to the environmental sciences. The major difficulty for ecotoxicologists is the lack of genome information for the most commonly used test organisms.

The link between the ecological/ecotoxicological risk assessment and the risk management frameworks is demonstrated. The ecological risk assessment consists of seven interactive elements (Fig. 17). The quantitative and descriptive science used to conduct ERA (Table 5) does not answer, in a direct way, the question of what should be done to manage the risk. Science determines adversity, but the public determines acceptability (Fig. 18). But "acceptable risk" is a highly subjective and relative term. It is time and space-specific and depends upon definitions of quality of life and robustness of the environment.

Table 5 Selected biochemical responses and diagnostic tools for assessment of the environmental biota health at the molecular level [50]

Method	Characteristic of health
Measurement of blue and green fluorescence of NADH and FAD in living tissues	Metabolic state of mitochondria, cells or tissues respiration and glycolysis
Quantitative fluorescent cytochemistry	DNA, RNA, proteins and lipids content
Using permeable fluorogenic substrates of enzymes, specific inhibitors, and kinetic analysis	Enzyme activity in living cells in situ: (a) Non-specific esterase (b) Detoxifying enzymes (c) Marker enzymes
Using special fluorescent anionic markers	Alterations of permeability of plasma membranes, epithelial layers and histo-haematic barriers
Using specific fluorescent transport substrates, inhibitors and kinetic analysis	State of carrier-mediated transport system for xenobiotics elimination
Using fluorescent xenobiotics or fluorescent analogue of xenobiotics	Xenobiotics distribution, extra and intracellular accumulation and storage
Using special fluorescent xenobiotics or fluorescent analogues of xenobiotics	State and function of xenobiotic-binding proteins
Vital tests with acridine orange or neutral red	State of lysosomes and cell viability
Metachromatic fluorescence of intercalated or bound acridine orange, 590/530 nm microfluorometry	Functional rate of nuclear chromatin, DNA denaturation
Complete cyto and histopathological examination	Early pathological alterations and signs of environmental pathology
Electron microscopy	Cell structures and organoids
Cytogenetic examinations	Detection of environmental genotoxicity and clastogenicity
Mass spectrometry (MALDI/TOF/MS; ESI-TOF-MS/MS)	Identification and detection of membrane proteins, epitope-binding areas of proteins

5.2 *Ecosystem Services*

Many questions in the ecological domain are unanswerable according to Sutherland et al. [54]. But it is important to ask these questions again and again in the context of the greater synergism of practice, research and policies. Many environmental issues require competent answers and a prompt policy and/or management response. Regarding applied aspects in ecology, many ecologists are not willing to understand that while ecology and ecological principles are necessary they alone are insufficient when dealing with the relationship between human societies and nature. To understand this relationship, it is necessary to include cultural, social and economic sides. The largest threat (as regards the future of mankind as well as the future of what remains of natural environments) will in the long run prove to be human growth and population growth. The question after all this increasing threats by man is how can "ecology serve" and guarantee the quality of life in the future. When discussing ecosystem services [54] we have to put costs on these services that typically have been regarded as free.

A comprehensive strategy is necessary to protect the ecology and elucidate the benefits of protected habitats, wetlands, landscape, etc. in comparison with

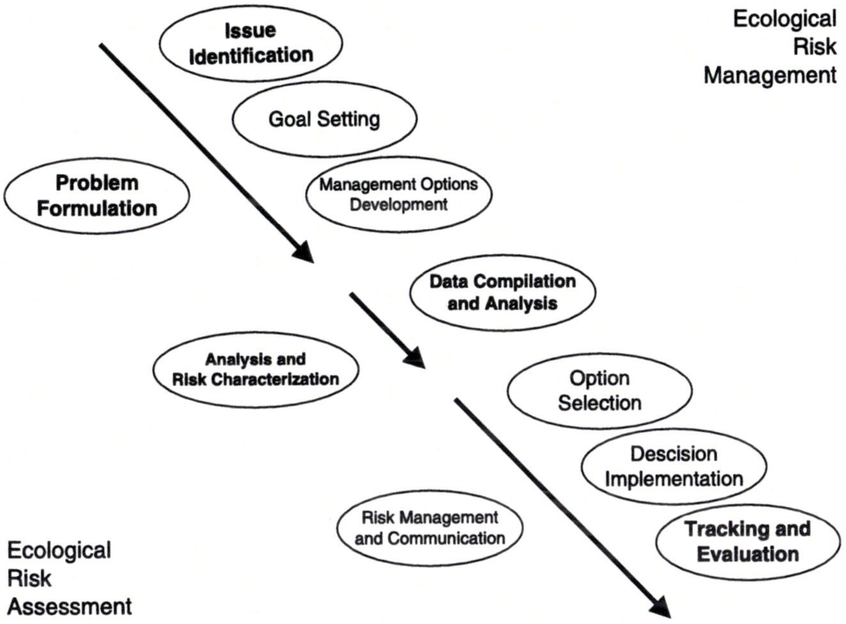

Fig. 17 Relationship of the ecological risk management framework to the ecological risk assessment framework (modified after Stahl et al. [53])

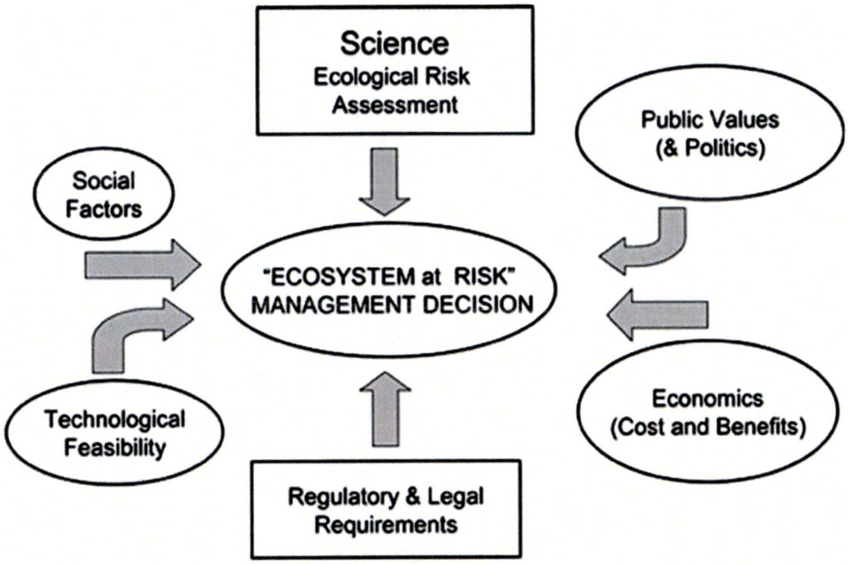

Fig. 18 Inputs to the "Ecosystem at risk" management decisions (modified after Stahl et al. [53])

non-protected areas. The first step will be to characterise reference conditions, for example, in urban heavily modified areas. The importance of the quality components defined by the potentials and limitations of the "good ecological status," "good chemical status" [1, 27, 51] and "quality of life" has to be demonstrated by, "on site effects monitoring" tools to promote an environmentally sensitive and sustainable use of the resources in urban areas. The study contributes to the resource management, evaluation of the interaction of ecosystems and urbanisation as well as quantification of risk.

The environmental risk assessment in urban systems has two major elements: characterisation of effects and characterisation of exposure. In a close integration of ecosystem-related aspects in the planning process of urban projects and in the urban design criteria, the key research questions that define the sustainable quality criteria are mainly regarding possibilities of reducing water consumption without decreasing the comfort of living combined with an integration of environmental impact assessment (EIA, EMEA [52]) and an ecological risk analysis. Therefore, special regard is paid to natural ecological conditions and socio-economic factors.

Aiming at a sustainable use of water and urban resources under extreme temperature and precipitation situations, a specific water quality has to be developed so that the water management guarantees a water quality for drinking water and irrigation in agriculture, aquaculture and protecting ecosystem services to human health. The main contribution is the quantification of risk. The indicators will be surface water, irrigation water, runoff water and waste water management. Finally, guidelines for environmental management (standard operation procedures, manuals for monitoring and risk assessment and communication) should be established through a DSS (Fig. 19). The operations in this field will disseminate solutions for municipalities and authorities and their decision makers.

In considering the impact of either natural stress or manmade stress, we always encounter detoxification, disease defence, regulation and adaptation processes. This situation makes the assessment approach by biomarkers rather complicated. On the other hand, symptoms analysis including functional (behaviour, activity and metabolism) and structural changes in organism (cellular, tissue and organs), the biomarkers do have a significant ecological assessment potential. For landscape planning and environmental management, it is necessary to get significant data from biochemical responses for relevant and sustainable actions.

The biosensors together with the effect-related parameters or biochemical responses for environmental monitoring are very complex, but they will give a clear picture of the health status of the investigated system.

The effect-related parameters or biochemical responses (biomarkers) are very complex, but they will give a clear picture of the health status of the investigated system. The "Ecosystem health" is defined as being synonymous with "environmental integrity," from which it follows that the scope of Ecosystem Health research encompasses all the tools and approaches which are efficacious in increasing the cognitive, curative, and preventive knowledge, which has as its goal the preservation of environmental integrity. Ecosystem health research thus directs its attention to the prediction of reversible and irreversible damages, which human or

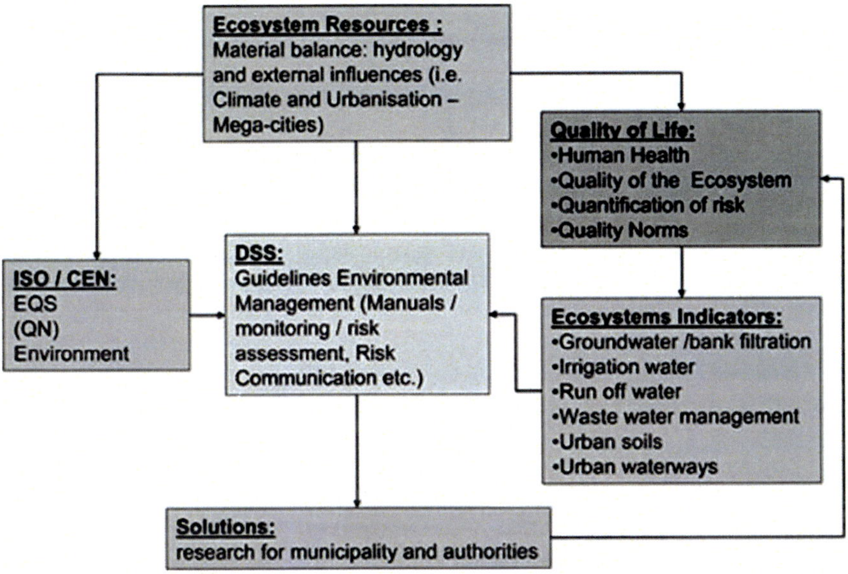

Fig. 19 Guidelines for environmental management through a decision support system for ecosystem services and sustainability

other activities could potentially inflict on the environment. For the assessment of ecosystem health, very promising biomarker approaches are centred on quantifying biochemical effects in organisms and populations. But finally to quantify the "ecosystem services," the potentials and limitations, this will be demonstrated with practical examples to find convincing answers concerning the benefits of protected ecosystems and related questions [54]:

- What are the benefits of protected habitats in terms of water resources relative to a non-protected area?
- What is the role of biodiversity in maintaining ecosystem functions?
- How does soil biodiversity influence and response to above-ground biodiversity?
- What is the role of marine biota and benthic–pelagic coupling in the carbon cycling and primary production?
- How can we measure natural capital (renewable and non-renewable resources)?

6 Analysis of First Generation RBMPs

6.1 Identification of Key Issues

A key issue is defined as a threat or cause that hampers the achievement of the "good status" as aimed for by the WFD. Usually such key issues are identified on the basis of monitoring data, ecological and chemical, databases of emissions, and water system modelling.

The three main key issues that are found to be common in several river basins across Europe relate to:

- *Hydromorphology*: The overregulation of rivers for power generation, irrigation, flood protection and urban supply has resulted in severe morphological alteration of rivers causing the disruption of their ecological functioning. Despite hydromorphology is taken into account within the WFD as part of the ecological status, it has to be pointed out that there is no specific EC policy (guidance or guidelines) on how – and how much – to undo hydromorphological alterations.
- *Chemical pollution*: From a qualitative point of view, an important common issue is the excess of chemicals (fertilisers and pesticides) associated with agrarian and livestock farming diffuse sources and the point-source discharge of (treated and untreated) sewage water. Industrial point-source pollution is not addressed as a relevant issue, but it is assumed that industries operate following the legal rules and only the incomplete elimination of industrial chemicals in urban wastewater treatment plants is perceived as a problem.
- *Climate change*: Quantitative problems arise from an overestimation of the existing water resources, which presumably will be reduced because higher plant evapotranspiration associated with climate changes. The overexploitation of groundwater bodies is identified as a relevant issue in vast areas. Moreover, in the Mediterranean region the IPPC predictions foresee a reduction of overall rainfall; therefore, besides water scarcity, which will doubtlessly be a key issue, other issues would be an increase in extreme hydrological events, severe droughts and floods. The latest may lead to the remobilisation of historic contamination associated with sediments that lay at the bottom of rivers or on the flood plains. This remobilisation may impact the chemical and ecological status of river basins, but a lot of research questions still remain here.

Other unsolved key issues relate to:

- *Eutrophication*: Problems with eutrophication of river basins do not rely on lack of research, but on the effectiveness of the measures that are difficult to quantify. Furthermore, due to the diffuse nature of this key issue, it is also difficult to select appropriate measures.
- *Sediments*: Quality and quantity of sediments are an issue of concern, both in relation to the stored sediments in dams, and also concerning the possible remobilisation of sediment bound, historic contaminants as consequence of floods events.
- *Impact on biodiversity*: Water level regulation and shortage of fresh water, for instance, due to an increase in brackish water may affect biodiversity. Furthermore, the introduction of invasive species may also result in a decrease of biodiversity as consequence of competitive subsistence processes.

As general issues of concern, also other aspects can be pointed out:

- Lacking analysis and monitoring data of processes in the natural system: in some circumstances there are no, or not enough, monitoring data collated, for instance, due to a long-term duration of a number of processes.
- Lacking methods for increasing the acceptance of selected measurements and measures: these methods are needed to convince politicians and support their decision making. The methods are also expected to be helpful in rising public awareness and understanding and hence to gain public support for the measures.
- Lacking international and local cooperation: an insufficient cooperation and coordination between different implementation scales. This leads to discrepancies in prioritisation of measures and their effects, depending on the scale considered. This lack of common objectives occurred with harmonisation of measurement protocols and with selection and implementation of measures between countries and institutions during the implementation of first generation RBMPs.
- Difficulties in finding a good balance for the interests of all stakeholders, such as water *versus* land use interests. This is particularly the case where measures meet public opposition (a measure perceived useful by the water manager might be perceived negative by the public. For example, in the Netherlands, for nature restoration, water managers in general want to raise groundwater tables. However, "wet enough for nature" is often perceived by farmers as "too wet for farming"). The interaction between water management (river basin) and land management is not fully elaborated up to now. In both domains, actions are taken that do not take all important inter-compartment effects into account (e.g. intensified biomass production *versus* erosion/water retention in land management or restoration of flood plains/wetlands in river basin management).

6.2 Selection of Measures

In general, the first generation of RBMPs only seems to contain measures of which the river basin managers have the long-time practical experience that they will work. Hence, it seems that no innovative, not yet fully proven, measures are implemented yet.

Measures are considered as actions and/or interventions for protection of a sustainable functioning of the river basins. Coherence in national programmes of measures could be improved by the generation of new ones, among other aspects, instead of just compiling local measures. Furthermore, measures often make sense on a certain scale level. Scale issues relate to either space or time. Scale issues at the local, national and international level can be related to spatial planning. For time scale issues, there is a need for foresight in addition to monitoring, which could result in river basin characterisation. Subsequently, there is a need for system indicators which help to monitor the changes of the system in time.

Water managers apply cost-benefit analyses of the potential measures in the selection process. Difficulty arises since the costs of measures are more or less clear, in contrary to the uncertain effects/benefits they can achieve, especially dealing with ecological or hydromorphological measures. Hence, management of uncertainties requires a compromise between precision and "roughness" in order to mitigate actual and emerging problems. For this reason, deeper knowledge of cost-effectiveness of measures is necessary, for instance, through dynamic, multi-scale and multi-dimensional evaluation tools.

The decision on the selection of measures has to be made by politicians/managers. Scientists can provide transparency for the decision makers by clarifying the current and potential future status of the system, the possible measures and their ecological, economical and social consequences. Scientists should learn the language and the interests of the public, because often information handed over from the scientists/ water managers to the public is perceived not reliable and/or relevant for them, moreover taking into account that public participation is mandatory according to the WFD. It is clear that public participation is important for the implementation of measures and for acceptance, because measures can have an impact on specific interests of the public (see example described in the previous section).

For the three key issues described in the previous section, potential measures to be taken are presented below.

Related to the hydromorphological key issue, one of the main uncertainties is the feasibility (technical but mainly economic) of the restoration measures. As already mentioned, the lack of policy (guidelines, guidance) on how to restore hydromorphology is identified. Guidelines on the percentage of the basin to restore are desirable but also recommendations how to deal with, for instance existing dams in the basin and potentially the need to build new ones.

Related to chemical pollution – referring to all kind of contamination (mineral and organic) – there is a clear distinction between point-source pollution and diffuse pollution. It appears that it is easier to take measures for point-source pollution, for instance, the improvement of the wastewater treatment plants, even if the treatments for specific compounds (pesticides, emerging compounds, etc.) still need further research. Measures for diffuse pollution can be more complex because some of them require real political decisions, for instance to interfere on agricultural practices to reduce inorganic and organic fertilisers.

As regard to climate change, a question appears about the possibility to take real measures or if only possible measures can be envisaged at that moment. Concerning floods, measures are based on the results of the different existing climate scenarios to know the amount of water to be retained during extreme events to prevent catastrophic flooding: for example, one proposed measure in the Netherlands is to temporarily retain water in restored areas (riparian areas, wetlands, flood plains, etc.). One identified gap is the possible negative impact of this water storage through remobilisation of historical contamination present in floodplain soils/sediments due to redox changes.

There are some other issues related to the selection of measures. For instance, concerning drought, one measure is the prioritisation of water use in case of severe drought. In Spain, where severe drought can be expected, possible measures are water reuse and artificial recharge of aquifers by injection of reclaimed wastewater.

However, so far there is a lack of knowledge concerning the potential geochemical and ecological impact of the artificial recharge, in both unsaturated zone and the aquifer itself. Another measure can be the relocation of certain crops in higher altitude because of a temperature increase.

6.3 Next Generation RBMP: New Elements to be Included

For the next generation RBMPs some points to be improved or integrated can be identified. These relate to the following.

6.3.1 Integrative Modelling

Better connection and closer interrelation between technical, economical and social aspects of the RBMPs. Besides integration, water managers would like to have more flexibility in the prioritisation of these aspects. With regard to the natural system itself, there is a need to focus more on groundwater and its connection (integration) with the rest of the aquatic system, i.e. sediment/soil/surface water/groundwater and the land system.

6.3.2 Economic Analysis

Economic analysis/valuation of specific functions or services provided by the natural system (i.e. the ecosystem) and of the (societal) costs and benefits of the measures proposed to restore these services, contributes to the transparency of the decisions. Concerning the system, a better definition of water services is needed in order to evaluate recovery costs of water uses and environmental costs. The economic analysis should also include a valuation of water and ecosystem services. For instance, the delivery of clean water by the ecosystem (e.g. through its filtering capacity or capacity to degrade contaminants) is one of these services.

6.3.3 Monitoring

The next generation RBMPs should start with the idea of the whole river basin and should continue with the information that is already available from the first generation RBMPs and information that is mandatory by law. Furthermore, additional information could be generated by monitoring or modelling the system. A common, i.e. basin wide, data base and information system should be built and used to store the acquired monitoring data. As a consequence, this proposed common information system requires harmonisation of measurement, protocols and guidelines (integrated tools) for monitoring. The design of better diagnostic assessment tools for possible measures could be helpful to save money in the monitoring programmes. Special attention could be paid to enable technologies such as multi-stressor analysis and bioassays.

6.3.4 Internationalisation

Trans-boundary cooperation/adjustment of plans was perceived to be largely lacking in the preparation of the first generation of RBMPs and for the next generation it is a challenge to improve it. There is a clear need of harmonisation of legislation at EU level. Administrative coordination on several disciplines should be enhanced, for instance, with regard to the harmonisation of data and integration of the RBMPs with the Floods and Marine Directives. At this level, international river basin commissions could play an important role.

The water managers could be supported by providing a platform for international exchange of their experiences, amongst others through the availability of a data base of existing platforms/organisations involved in RBMPs (e.g. IWRM-net, WFD-CIS, WISE-RTD, etc.).

6.3.5 Climate Change

Climate change scenarios will probably be a major topic to be fully considered in the next planning cycle. At the present stage, the debate on the adequacy of the hydrological series used as a planning reference has turned around the use of long (60 years) and short (25 years) series. Precipitation shortness combined with higher plant evapotranspiration in the near future is introduced in the discussion by scientists and NGOs as an argument for a precautionary approach to be applied to water use in the decision-making process.

6.3.6 System Understanding

Next plans should more reflect the understanding of the interactions and dependencies of the aquatic and land systems with their individual drivers and pressures (e.g. increased irrigation and change in land use). By achieving this, they will be prepared to better prioritise measures to be taken, assess their effectiveness and to take into account the impact of potential future scenarios within the river basins.

6.3.7 Public Participation

Intensification and real effective public participation are expected in the development of the next generation(s) of RBMPs.

6.4 Knowledge Gaps and Remarks

A number of scientific and management knowledge questions are still open:

- Risk based river basin management has to take into account the interaction between aquatic systems and land systems. This system interaction, covering natural and anthropogenic drivers, is not fully established. Changes in land use due to new incentives (e.g. industrial biomass use) highlight the close dependency between both sectors of management.
- Climate change will impact on water quality and quantity and hydromorphology and may thus remobilise historic pollutants from river bottom sediments and/or floodplain soils. However, the extent of this impact is still largely unclear and there is also a lack of feasible measures to mitigate the (potential) impacts.
- Population growth and unpredictable climate changes will pose high demands on water resources in the future. Even at present, surface water is certainly not enough to cope with the water requirement for agricultural, industrial, recreational and drinking purposes. In this context, the usage of groundwater has become essential; therefore, their quality and quantity have to be carefully managed. Groundwater artificial recharge can guarantee a sustainable level of groundwater, whereas strict quality control of waters intended for recharge will minimise contamination of both the groundwater and aquifer area. However, to date, it is still unclear what the exact geo/chemical effects will be of artificial recharge on aquifers and unsaturated zone.
- Emerging pollutants (including degradation products). Sources, mobility, toxicity, monitoring, measures. In the past, research has focused on priority pollutants, such as POPs, pesticides and toxic metals. Only recently the attention of the scientific community has started to shift to emerging contaminants. Therefore, a major challenge will be to identify the chemicals which potentially will become dangerous in the future. It has to be cleared if it is sufficient to look just for persistent, high flux, toxic, endocrine active compounds. For most of the occurring emerging contaminants, risk assessment and ecotoxicological data are not available, and therefore it is difficult to predict what health effects they may have on humans, terrestrial and aquatic organisms and ecosystems.
- Integrated assessment of policy to clarify social, economic and biophysical drivers and pressures. In the area of environmental and resource management and in policies aiming at sustainable development, multi-criteria decision analysis seems to be an appropriate policy tool since it allows taking into account a wide range of assessment criteria. The management of a policy process also involves many layers and kinds of decisions, and requires the construction of a dialogue process among many social actors, individual and collective, formal and informal, local and non local. An outcome of this discussion is that the political and social framework must find a place in evaluation exercises.
- In many circumstances, it is very difficult to predict the potential effect of measures, especially on biology/ecology (hydromorphology, redesign of water bodies, pollutants), mainly as consequence of the time scale of the processes.

- The application of new technology/methodology in measurements and measures, for instance, the application of ecological engineering and bioremediation, could allow an improvement in the reliability and in the volume of data collected. The assessment of potential synergy of measures and their application might also be an interesting approach to expand the effects of measures as well as to reduce their cost of implementation. In addition, the development and assessment of diagnostic tools for multi-pressure situations will simplify the design and application of measures.
- A reliable economic analysis and evaluation of measures, especially for innovative measures with only a short-term practical experience, is required, since the cost of the water is an issue of major importance. With regard to the system, a better definition of water services is needed in order to evaluate recovery costs of water uses and environmental costs.
- Most efforts should be done to establish a direct connection between groundwater and the rest of the aquatic system (surface water, sediment, soil) in order to ensure an integrated management of the river basins.
- There is still a lack of knowledge on ecosystem services integrated concepts. The ecology of recovery (for instance, how long does recovery of a measure or due to a measure take? When does a system recover?) is also a concept still not well understood. For instance, for instance, the most important factor for the recovery of a population may be difficult to identify since there can be multiple causes involved. Studies discriminating these are scarce, especially comparisons of the significance of the species traits and of its interactions with other species for its ability to recover.
- It is critical under specific circumstances to find a consensus between environmental protection and technological advance.
- It is necessary to integrate different languages and multidisciplinary scientific knowledge. Different sciences and different scale use different languages; we should not narrow it down, but on the contrary, we should enrich our languages.
- Knowledge communication from experts to politicians and public in general. Often information handed over from the water managers to the public is found not that reliable and/or relevant for the public. This is due to several reasons such as the use of too complicated language, the lack of transparency of the decision process and the lack of a clear view on the costs.

Although it is clear that EU river basins differ a lot (in economy, ecology, culture and social issues), they also share a lot of common problems. Such differences also occur within one trans-boundary river basin and even within a same country. Thus, putting a lot of emphasis on communication between all parties that have a stake in (are affected by) the implementation of the RBMPs and specifically the implementation of the POM. Harmonisation – at the river basin scale – of data collection and processing and of the selection, implementation and monitoring of the effectiveness of measures will certainly improve the communication and improve also the understanding of the functioning of the soil–sediment–water systems. Moreover, problems with communication between scientist–managers–public may be solved by integrating different "languages" (public in general, scientists, and politicians) and knowledge from the different scientific disciplines involved, and also intensifying public participation. The application of new innovative technologies and methodologies as well as the synergy of measures will probably help in the optimisation of economical efforts to be taken to achieve and maintain a good quantitative and qualitative status of aquatic systems. Climate change will result in an increase in frequency and severity of draughts and floods events. Climate change is one issue that will gain relevance in next RBMPs associated with pollution as consequence of the concentration of pollutants due to water shortage, inefficient depuration of pollutants in water treatment plants and remobilisation of pollutants as consequence of floods. Main knowledge gaps were identified concerning emerging pollutants (sources, effects, measuring the impact at local scale, determination of metabolites and breakdown products), cost-effectiveness of measures (economic analysis, application of models, new possibilities for measures and potential synergy among them, increasing certainty on the effects of measures) and ecosystem services (ecology of recovery, ecological engineering, effect of measures on ecology).

Furthermore, it seems that the first generation RBMPs reflect a rather "low" level of ambition and will include only measures, that river managers have been applying for a long time with practical experience and they know that are effective. Not much innovative approaches are included. The key objective of the WFD (to achieve a good status in all European waters by a set date) is extremely ambitious. This calls for a POM that matches this ambition. Thus, the question arises whether we will ever be able to reach that objective.

References

1. Directive 2000/60/EC of the European Parliament of the Council, of 23 October 2000, establishing a framework for Community action in the field of Water Policy. Off J Eur Com. L327/1. 22/12/2000
2. Vale C, Rosa-Santos F, Walter C et al (2007) Advanced observation devices for chemical monitoring of water quality. In: 1st Riskbase Thematic Workshop, Monitoring and Assessment of River Pollutants: a step towards the implementation of risk based management plans. Oral presentation, Lisbon, May 17–18
3. Blum WEH (2007) Role of soils in river basin management. In 1st Riskbase Thematic Workshop, Monitoring and Assessment of River Pollutants: a step towards the implementation of risk based management plans. Oral presentation, Lisbon, May 17–18
4. Vegter Joop J (2007) From soil contamination to soil protection and management of soil and water resources. In 1st Riskbase Thematic Workshop, Monitoring and Assessment of River Pollutants: a step towards the implementation of risk based management plans. Oral presentation, Lisbon, May 17–18
5. Von der Ohe P, Brack W (2007) MODELKEY – models for assessing and forecasting the impact of environmental key pollutants. In: Proceedings of the RISKBASE 1st Thematic Workshop, Lisbon, pp 36–40

6. Barth JC, Steidle D, Kuntz D et al (2007) Deposition, persistence and turnover of pollutants: first results from the EU project AquaTerra for selected river basins and aquifers. Sci Total Environ 376:40–50
7. Eljarrat E, Barcelo D (2006) Quantitative analysis of polychlorinated n-alkanes in environmental samples. Trends Anal Chem 25:421–434
8. Ratola N, Lacorte S, Alves A et al (2006) Analysis of polycyclic aromatic hydrocarbons in pine needles by gas chromatography–mass spectrometry. Comparison of different extraction and clean-up procedures. J Chromatogr A 1114:198–204
9. Meyer M, Hermand JP, Asch M et al (2006) An analytic multiple frequency adjoint-based inversion algorithm for parabolic-type approximations in ocean acoustics. Inverse Probl Sci Eng 14:245–265
10. Lobo-Ferreira JP (2007). Gabardine project – groundwater artificial recharge based on alternative sources of water: advanced integrated technologies and management. Objectives and Portuguese test site achievements. In: Proceedings of the RISKBASE 1st Thematic Workshop, Lisbon, pp 54–59
11. Cunha MA, Brito S, Moura I et al (2007) Ecotoxicological assessment of wastewaters in the framework of ECORIVER project: contribution to River Basin Management. In: Proceedings of the RISKBASE 1st Thematic Workshop, Lisbon, pp 87–91
12. Ginebreda A (2008) Ter and Llobregat river basins. The water supply of the Barcelona Area. In: 2nd Riskbase WP 1b Thematic Workshop, Bridging Science, Water Management and Practical Experiences in European River Basins Oral presentation, Budapest, May 15–17
13. Galbiati L (2008) Main pressures threatening the achievement of the WFD's objectives. Optimization of the PoM to reach a good physical-chemical state. In: 2nd Riskbase WP 1b Thematic Workshop, Bridging Science, Water Management and Practical Experiences in European River Basins Oral presentation, Budapest, May 15–17
14. Carrera J (2008) Baix Llobregat aquifers: understanding, modeling, management implications. In: 2nd Riskbase WP 1b Thematic Workshop, Bridging Science, Water Management and Practical Experiences in European River Basins Oral presentation, Budapest, May 15–17
15. Capri E (2008) The implementation of the Water Framework Directive in Italy: state of the art on plant protection products (PPPs) monitoring and model estimation. In: 2nd Riskbase WP 1b Thematic Workshop, Bridging Science, Water Management and Practical Experiences in European River Basins Oral presentation, Budapest, May 15–17
16. Kolkmann A (2008) Waterboard De Dommel. In: 2nd Riskbase WP 1b Thematic Workshop, Bridging Science, Water Management and Practical Experiences in European River Basins Oral presentation, Budapest, May 15–17
17. Ibrahimpasic A (2008) A brief introduction to the Sava River basin. In: 2nd Riskbase WP 1b Thematic Workshop, Bridging Science, Water Management and Practical Experiences in European River Basins Oral presentation, Budapest, May 15–17
18. Ginebreda A (2007) WFD Monitoring Programmes of the Hydrological River Basins of Catalonia, Spain. In: Proceedings of the RISKBASE 1st Thematic Workshop, Lisbon, pp 74–78
19. Rocha F (2007) Monitoring programme related to the implementation of the water framework directive. In: Proceedings of the RISKBASE 1st Thematic Workshop, Lisbon, pp 50–53
20. Ferreira J, Vale C, Soares V et al (2007) Monitoring of coastal and transional waters under the E.U. Water Framework Directive. Environ Monit Assess 135:195–216
21. Constanza R, d'Arge R, de Groot R et al (1997) The value of world's ecosystem services and natural capital. Nature 387:253–259
22. Millenium Ecosystem Assessment (2005) Ecosystema and Human Well-being: Syntesis. Island Press, Washington, DC, ISBN 1-59726-040-1
23. Sabater S, Guasch H, Ricart M et al (2007) Anal Bioanal Chem 387:1425–1432
24. Guasch H, Paulsson M, Sabater S (2002) J Phycol 38:241–247
25. Lawrence JR, Swerhone GDW, Wassenaar LI et al (2005) Canadian J Microbiol 51:655–661
26. Admiraal W, Blanck H et al (1999) Water Res 33:1989-1995

27. Directive 2008/105/EC of the European Parliament and of the Council of 16 December 2008 on environmental quality standards in the field of water policy, amending and subsequently repealing Council Directives 82/176/EEC, 83/513/EEC, 84/156/EEC, 84/491/EEC, 86/280/EEC and amending Directive 2000/60/EC of the European Parliament and of the Council. Off J Eur Com. L348/84 24/12/2008
28. Anonymous (1999) (COMMPS study): Study on the prioritization of substances dangerous to the aquatic environment. Office for Official Publications of the European Communities (ISBN 92-828-7981-X). Luxembourg, 1999
29. Revised proposal for a list of priority substances in the context of the water framework directive (COMMPS Procedure). Final report. Declaration ref. 98/788/3040/DEB/E1. Fraunhoffer-Institut Umweltchemie und Ökotoxicologie. Schmallenberg, Germany, June 1999
30. Munthe J, Brorström-Lundén E, Jöborn A et al (2007) A. Source Control of Priority substances in Europe (SOCOPSE). In: Proceedings of the RISKBASE 1st Thematic Workshop, Lisbon, pp 32–35
31. Semple KT, Doick KJ, Jones KC et al (2004) Defining bioavailability and bioaccessibility of contaminated soil and sediment is complicated. Environ Sci Technol 38:228A–231A
32. Parsons J, Belzunce MJ, Segarra G et al (2007) Characterisation of contaminants in sediments – effects of bioavailability on impact. In: Barcelo D, Petrovic M (eds) Sustainable management of sediment resources. Sediment quality and impact assessment of pollutants. Elsevier, Amsterdam
33. Knepper TP, Barceló D, de Voogt P (2003) Analysis and fate of surfactants in the aquatic environment. Elsevier, Amsterdam
34. Scott MJ, Jones M (2000) The biodegradation of surfactants in the environment. Biochim Biophys Acta 1508:235–251
35. Sutherland JB, Rafii F, Khan AA et al (1995) Mechanisms of polycyclic aromatic hydrocarbon degradation. In: Young LY, Cerniglia CE (eds) Microbial transformation and degradation of toxic organic chemicals. Wiley, New York
36. Kanaly RA, Harayama S (2000) Biodegradation of high-molecular-weight polycyclic aromatic hydrocarbons by bacteria. J Bacteriol 182:2059–2067
37. Meckenstock RU, Safinowski M, Griebler C (2004) Anaerobic degradation of polycyclic aromatic hydrocarbons. FEMS Microbiol Ecol 49:27–36
38. Pehkonen SO, Zhang Q (2002) The degradation of organophosphorus pesticides in natural waters: a critical review. Crit Rev Environ Sci Technol 131:17–72
39. Wackett LP, Sadowsky MJ, Martinez B et al (2002) Biodegradation of atrazine and related s-triazine compounds: from enzymes to field studies. Appl Microbiol Biotechnol 58:39–45
40. Capri E, Calliera M (2007) Pesticide monitoring in Italy and impact on environmental legislation/policy implementation. In: Proceedings of the RISKBASE 1st Thematic Workshop, Lisbon, pp 60–63
41. Neilson AH (1996) An environmental perspective on the biodegradation of organochlorine xenobiotics. Int Biodet Biodeg 37:3–21
42. Bhatt P, Kumar MS, Mudliar S (2007) Biodegradation of chlorinated compounds – a review. Crit Rev Environ Sci Technol 37:165–198
43. Dolfing J (2003) Thermodynamic considerations for dehalogenation. In: Häggblom MM, Bossert ID (eds) Dehalogenation: microbial processes and environmental applications. Kluwer Academic Publishers, Boston
44. Gil O, Vale C (2001) Evidence for polychlorinated byphenyls dechlorination in the sediments of Sado estuary, Portugal. Marine Poll Bull 42:453–461
45. Pereira P, de Pablo H, Vale C et al (2009) Metal and nutrient dynamics in a eutrophic coastal lagoon (Óbidos, Portugal): the importance of observations at different time scales. Environ Monit Assess 149:150–161
46. Vrana B, Mills GA, Allan IJ et al (2005) Passive sampling techniques for monitoring pollutants in water. Trends Anal Chem 24:845–868

47. Greenwoord R (2007). Passive sampling devices for advanced monitoring of pollutants. In: Proceedings of the RISKBASE 1st Thematic Workshop, Lisbon, pp 101–103
48. Ricking M (2007) Sampling methods of SPM and sediments within the environmental specimen bank (ESB) of Germany. In: Proceedings of the RISKBASE 1st Thematic Workshop, Lisbon, pp 98–100
49. Hansen PD, Blasco J, de Valls A et al (2007) Biological analysis (bioassays, biomarkers, biosensors). In: Barcelo D, Petrovic M (eds) Sustainable management of sediment resources. Sediment quality and impact assessment of pollutants. Elsevier Publishers, Amsterdam
50. Hansen P-D (2003) Biomarkers. In: Markert BA, Breure AM, Zechmeister H (eds) Bioindicators & biomonitors. Principles, concepts and applications. Elsevier, Amsterdam
51. European Commission, Common Implementation Strategy fort the Water Framework Directive (2000/60/EC), Guidance Document No.7: Monitoring under the Water Frame Work Directive, Produced by Working Group 2.7 – Monitoring, 2007
52. EMEA, European Agency for the Evaluation of Medical Products. Guideline on environmental impact assessment (EIAS) for veterinary medical products-phase I. CVMP/VICH/592/98-Final London 2000
53. Stahl RG Jr (2001) Introduction and background to the development of a framework for ecological risk management. In: Ralph G, Stahl I et al (eds) Risk management: ecological risk-based decision-making. SETAC Technical Publications Series, Pensacola, FL
54. Sutherland WJ (2006) The identification of 100 ecological questions of high policy relevance in the UK. J Appl Ecol 43:617–627

Index

D. Barceló and M. Petrovic (eds.), *The Ebro River Basin*,
Hdb Env Chem (2011) 13: 425–431, DOI 10.1007/698_2010,